Nature-Based Solutions and Water Security

Nature-Based Solutions and Water Security

An Action Agenda for the 21st Century

Edited by

Jan Cassin
Forest Trends, Washington, DC, United States

John H. Matthews
Alliance for Global Water Adaptation, Corvallis, OR, United States

Elena Lopez Gunn
ICATALIST S.L., Madrid, Spain

ELSEVIER

Elsevier
Radarweg 29, PO Box 211, 1000 AE Amsterdam, Netherlands
The Boulevard, Langford Lane, Kidlington, Oxford OX5 1GB, United Kingdom
50 Hampshire Street, 5th Floor, Cambridge, MA 02139, United States

Notices
Knowledge and best practice in this field are constantly changing. As new research and experience broaden our understanding, changes in research methods, professional practices, or medical treatment may become necessary.

Practitioners and researchers must always rely on their own experience and knowledge in evaluating and using any information, methods, compounds, or experiments described herein. In using such information or methods they should be mindful of their own safety and the safety of others, including parties for whom they have a professional responsibility.

To the fullest extent of the law, neither the Publisher nor the authors, contributors, or editors, assume any liability for any injury and/or damage to persons or property as a matter of products liability, negligence or otherwise, or from any use or operation of any methods, products, instructions, or ideas contained in the material herein.

Library of Congress Cataloging-in-Publication Data
A catalog record for this book is available from the Library of Congress

British Library Cataloguing-in-Publication Data
A catalogue record for this book is available from the British Library

ISBN: 978-0-12-819871-1

For information on all Elsevier publications
visit our website at https://www.elsevier.com/books-and-journals

Publisher: Candice Janco
Acquisitions Editor: Louisa Munro
Editorial Project Manager: Pat Gonzalez
Production Project Manager: Sruthi Satheesh
Cover Designer: Christian Bilbow

Typeset by SPi Global, India

Working together to grow libraries in developing countries
www.elsevier.com • www.bookaid.org

Dedication

To Tom Hille, for your support.

To Austin J. Matthews and Jackson Fox, for your future.

Contents

Section 1: Introduction to nature-based solutions and water security

Chapter 1: Setting the scene: Nature-based solutions and water security 3

Jan Cassin and John H. Matthews

Chapter 2: History and development of nature-based solutions: Concepts and practice 19

Jan Cassin

Section 2: State of science and knowledge

Section 3: *State of policy and governance*

Chapter 6: Why governments embrace nature-based solutions:
The policy rationale...*109*

Kathleen Dominique, Nathanial Matthews, Lisa Danielson, and John H. Matthews

Chapter 10: Protecting source waters in Latin America 215

Marta Echavarria, Jan Cassin, and Jose Bento da Rocha

Chapter 11: Mobilizing funding for nature-based solutions: Peru's drinking water tariff 241

Cheyenne Coxon, Gena Gammie, and Jan Cassin

Contents

Chapter 12: Urban blue spaces, health, and well-being..........................263

Paula Vandergert, Pauline Georgiou, Lisa Peachey, and Sam Jelliman

Chapter 13: Learning from indigenous and local knowledge: The deep history of nature-based solutions..........................283

Jan Cassin and Boris F. Ochoa-Tocachi

Jan Cassin, John H. Matthews, Elena Lopez-Gunn, Leah L. Bremer, Cheyenne Coxon,
Kathleen Dominique, Marta Echavarria, Gena Gammie, Roxane Marchal, Alex Mauroner,
Naabia Ofosu-Amaah, Denielle M. Perry, Eleanor Sterling, Ingrid Timboe, Paula Vandergert,
Kari Vigerstol, and Sophie Trémolet

Contributors

Robin Abell Conservation International, Arlington, VA, United States
Monica A. Altamirano DELTARES, Delft, Netherlands
Nureen F. Anisha Oregon State University, Corvallis, OR, United States
Brooke Atwell The Nature Conservancy, Arlington, VA, United States
Eugenio Barrios Mexican National Water Commission (CONAGUA), Mexico City, Mexico
Michael Becker Critical Ecosystem Partnership Fund, Arlington, VA, United States
Jose Bento da Rocha Secretary of State of Environment and Sustainable Development, Goias, Brazil
Kate Brauman University of Minnesota, Minneapolis, MN, United States
Leah L. Bremer University of Hawai'i at Mānoa, Honolulu, HI, United States
Wouter Buytaert Imperial College London, London, United Kingdom
Jan Cassin Forest Trends, Washington, DC, United States
Nora van Cauwenbergh IHE Delft, Delft, Netherlands
Cheyenne Coxon Forest Trends, Washington, DC, United States
Lisa Danielson Government of Canada, Ottawa, ON, Canada
Ernesto Dela Cruz Asian Development Bank, Manila, Philippines
Kathleen Dominique OECD, Paris, France
Mia Ebeltoft Climate Risk Advisory, Oslo, Norway
Marta Echavarria Ecodecision, Quito, Ecuador
Jehanne Fabre Danone, Paris, France
Paul Fleming Microsoft, Redmond, WA, United States
Gena Gammie Forest Trends, Washington, DC, United States
Pauline Georgiou Sustainability Research Institute, University of East London, London, United Kingdom
Nina Graveline INRAE, Montpellier, France
Sujith Sourab Guntoju Alliance for Global Water Adaptation, Corvallis, OR, United States
Sam Jelliman Sustainability Research Institute, University of East London, London, United Kingdom
Bonnie Keeler University of Minnesota, Minneapolis, MN, United States
Peter van der Keur Geological Survey of Denmark and Greenland, Copenhagen, Denmark
Elena Lopez-Gunn ICATALIST S.L., Madrid, Spain
Roxane Marchal Caisse Centrale de Réassurance, Paris, France

Michael Matosich   The Nature Conservancy, Arlington, VA, United States

John H. Matthews   Alliance for Global Water Adaptation, Corvallis, OR, United States

Nathanial Matthews   Global Resilience Partnership, Stockholm, Sweden

Alex Mauroner   Alliance for Global Water Adaptation, Corvallis, OR, United States

Beatriz Mayor   ICATALIST S.L., Madrid, Spain

Pedro Zorrilla Miras   ICATALIST S.L., Madrid, Spain

David Moncoulon   Caisse Centrale de Réassurance, Paris, France

Jason Morrison   Pacific Institute, Oakland, CA; UN Global Compact CEO Water Mandate, New York, NY, United States

Raul Muñoz   Inter-American Development Bank, Washington, DC, United States

Florentina Nanu   Business Development Group, Bucharest, Romania

Boris F. Ochoa-Tocachi   Imperial College London, London, United Kingdom; Regional Initiative for Hydrological Monitoring of Andean Ecosystems (iMHEA), Lima, Peru; ATUK Consultoría Estratégica, Cuenca, Ecuador

Naabia Ofosu-Amaah   The Nature Conservancy, Arlington, VA, United States

Pua'ala Pascua   Center for Biodiversity and Conservation, American Museum of Natural History, New York, NY, United States

Lisa Peachey   Sustainability Research Institute, University of East London, London, United Kingdom

Denielle M. Perry   School of Earth and Sustainability, Northern Arizona University, Flagstaff, AZ, United States

Kathryn Pharr   International Water Policy Consultant, London, United Kingdom

Eleanor Sterling   Center for Biodiversity and Conservation, American Museum of Natural History, New York, NY, United States

Ingrid Timboe   Alliance for Global Water Adaptation, Corvallis, OR, United States

Sophie Trémolet   The Nature Conservancy, Arlington, VA, United States

Paula Vandergert   Sustainability Research Institute, University of East London, London, United Kingdom

Tara Varghese   PepsiCo, Purchase, NY, United States

Kari Vigerstol   The Nature Conservancy, Arlington, VA, United States

Adrian Vogl   Natural Capital Project, Stanford University, Stanford, CA, United States

Rebecca Walker   University of Minnesota, Minneapolis, MN, United States

Josh Weinberg   Stockholm International Water Institute, Stockholm, Sweden

Authors' biographies

Ms. Robin Abell is Conservation International's freshwater lead. She joined Conservation International from The Nature Conservancy's Global Water program, where she led an ambitious project to demonstrate and quantify the benefits of source water protection. Before that, Robin spent 17 years with World Wildlife Fund's Conservation Science Program, where she led a team that produced foundational publications on freshwater conservation. She holds a BA from Yale University and an MS in aquatic ecology from the University of Michigan's School of Natural Resources and Environment.

Dr. Monica Altamirano specializes in systems thinking and public-private partnerships (PPP) at Deltares. With 16 years of experience in designing and managing programs in water, sanitation, transportation, environment, and capacity-building in Europe, Asia, Latin America, and Africa, she is a specialist in analyzing the institutional framework for infrastructure finance. Monica has advised governments around the world (Netherlands, Bangladesh, Peru, Philippines, Mexico, Indonesia, Spain, and Finland) on how to catalyze private sector investments in infrastructure, water security, urban resilience, and climate adaptation. Monica holds a master of science in systems engineering, policy analysis, and management, and a PhD in economics of infrastructure (innovative contracts in the road sector) from Delft University of Technology (TU Delft).

Ms. Brooke Atwell is the conservation strategy manager, water security for The Nature Conservancy. In this position she brings innovation and financing to the water sector to meet the needs of people without neglecting the needs of nature. Brooke has a strong background in collaborative strategic planning, program management, developing NBS for water security projects, economic and financial analyses for NBS, strategic communications, and operational and relationship management in cross-cultural, multidisciplinary settings. She has a bachelor of science degree in environmental science from the University of Virginia.

Ms. Nureen Faiza Anisha is an environmental researcher specializing in freshwater systems and their overlap with climate change. A water resources engineer by training, she has worked on several projects on climate change mitigation and adaptation in the water sector in Bangladesh and as a research fellow with the Alliance for Global Water Adaptation (AGWA). Nureen is currently a doctoral candidate in environmental sciences at the Oregon

State University where her research is focused on understanding coupled human and natural systems' dynamics in the wetlands of Bangladesh.

Mr. J. Eugenio Barrios O. has 30 years of experience in water resource management, with a solid experience on environmental flows and water allocation policies. He is a former deputy director general of the Mexico National Water Commission. He had several positions at WWF Mexico and Latin America; the most relevant as a director of the Freshwater Program where he led the National Environmental Water Reserves Program, for the National Water Commission and the Inter-American Development Bank. He has a professional engineering degree from the University of Mexico (UNAM), and a master of science in environmental engineering from the University of Illinois Urbana-Champaign United States.

Mr. Michael Becker is the executive director of Nature Invest and the leader of the Regional Implementation Team—CEPF Cerrado for the International Institute of Environmental Education. Michael is responsible for the regional implementation of the Critical Ecosystem Partnership Fund (CEPF) in the Cerrado Biodiversity Hotspot, including mainstreaming biodiversity protection into the public and private sector business practices in the region. Michael previously served as the World Wildlife Fund's Conservation Director in Brazil. Michael has a master of science in environmental engineering from the Brandenburgische Technische Universitiät Cottbus-Senftenberg and a master of business administration from the Fundação Getulio Vargas.

Mr. José Bento da Rocha is currently the undersecretary of environmental licensing and water resources for the State of Goiás, Brazil, and formerly worked for the Water, Power and Basic Sanitation Regulatory Agency for the Federal District (ADASA) in Brazil. He has a master's degree from the Universidade Estadual de Goiás.

Dr. Kate A. Brauman is a 2020–21 Water and Climate Resilience Fellow at the US Department of Defense through a AAAS Science Technology Policy Fellowship, where she works on water resilience at military installations. Kate remains affiliated with the University of Minnesota's Institute on the Environment, where she was the lead scientist for the Global Water Initiative and a member of the Natural Capital Project science team. Kate was a coordinating lead author for the Global Assessment of the Intergovernmental Platform on Biodiversity and Ecosystem Services (IPBES). She received her doctorate from Stanford University and her undergraduate degree from Columbia University.

Dr. Leah Bremer is an environmental science and policy specialist with the University of Hawai'i Economic Research Organization (UHERO) and the Water Resources Research Center (WRRC) at the University of Hawai'i at Mānoa. She is also cooperating faculty with the Department of Geography and Environment, the Department of Natural Resources and Environmental Management, and the Biocultural Initiative of the Pacific at the University Hawai'i at Mānoa, and she is a research affiliate with the Gund Institute of the Environment

(University of Vermont) and Fundación Cordillera Tropical, an NGO in Ecuador. She holds a BA in psychology, an MS in conservation biology, and a PhD in geography.

Dr. Wouter Buytaert is a professor of hydrology and water resources at the Department of Civil and Environmental Engineering, Imperial College London. His research is situated at the interface between hydrological process understanding, water resources management, and sustainable development. He has a particular interest in the impact of land use change and climate change on the terrestrial water cycle, and the use of participatory processes to support decision-making on water resources and water-related risks in data-scarce regions. He is involved in several research projects on nature-based solutions, with a specific focus on their application to promote water security in the tropical Andes. Wouter graduated with a master of engineering/master of science in environmental engineering from the University of Leuven, Belgium, in 2000 and obtained a PhD from the same university in 2004.

Dr. Jan Cassin is director of Forest Trends' Water Initiative, where she leads the organization's work on scaling nature-based solutions for sustainable water management and climate adaptation. An ecologist with expertise in terrestrial and freshwater systems, her career has emphasized applied research and practice in a wide range of water resource management issues, including: integrated watershed management; wetland, stream, and floodplain restoration; environmental flows; natural infrastructure design and assessment; ecosystem services quantification; and innovative financial mechanisms, such as wetland banking and investments in watershed services. She has worked as an environmental and water resources consultant, as a research assistant at MIT's Center for Policy Alternatives, as a biologist for The Nature Conservancy, and on tree planting and firefighter crews for the US Forest Service. Jan received her BA from the University of Colorado and her doctorate from the University of Michigan.

Ms. Cheyenne Coxon is a senior communications associate at Forest Trends, working across all Forest Trends initiatives on project management, publications, outreach strategy, and storytelling. Cheyenne is a biologist by training and, before joining Forest Trends, worked as a laboratory researcher and animal care assistant at the UC Davis Veterinary Medical Teaching Hospital, One Health Institute, and Smithsonian Conservation Biology Institute. More recently, she completed a Science Communication Fellowship at the National Center for Ecological Analysis and Synthesis (NCEAS) and a land use and water quality risk assessment for Denver Water. Cheyenne has a bachelor of science in animal science from UC Davis and a master of environmental science and management from the Bren School of Environmental Science & Management (UC Santa Barbara).

Ms. Mia Ebeltoft is managing director at Climate Risk Advisory and deputy director with Finans Norge, Norway's Financial & Insurance Services Association. Mia is an attorney with a demonstrated history of working in the financial services industry, focusing on climate change/climate risk, private-public projects, disaster risk reduction and disaster risk management, and insurance systems. Mia has an international legal education and experience having studied law at University of Oslo, University of North Dakota and University of Aix-en-Province/France.

Ms. Marta Echavarria, is the founder of EcoDecisión, a social firm committed to conserving natural landscapes and changing the way the formal economy values nature. Marta has worked in the design, negotiation, and development of financial and institutional mechanisms for environmental management and watershed conservation in various countries. Marta was instrumental in the development of the first water fund, FONAG, in Quito, Ecuador. She has received international recognition as a social entrepreneur and is involved in several advisory committees and NGO boards. Marta has a BA in environment studies and received the first MA in development studies from Brown University.

Ms. Lisa Danielson is a senior policy advisor on environment and natural resources with the Government of Canada. She has a background working with international organizations, government, and academic institutions on topics including climate change adaptation, energy development, and Indigenous policy. Prior to her current position, Lisa spent over 3 years at the Organization for Economic Cooperation and Development (OECD) in France leading several work streams on topics such as coastal adaptation and nature-based solutions. She holds a master of public policy from Simon Fraser University and a BA in political science from the University of British Colombia.

Dr. Ernesto Ocampo Dela Cruz is an environmental consultant for Asian Development Bank Projects on transport. He is also a consultant for the development of a practitioner's guide for ADB on nature-based solutions. Previously he was associate professor and chairman of the Department of Chemical Engineering, University of the Philippines, a visiting professor of Rutgers University, New Jersey, and visiting scientist of Osaka University, Japan. He has a PhD in chemical engineering from the University of the Philippines.

Ms. Kathleen Dominique is a senior policy analyst at the OECD and leads the OECD's program of work on financing water, including managing the Roundtable on Financing Water—a dedicated platform to engage policy makers and the finance and investment community. Since joining the OECD in 2010, Kathleen has led work on water resources allocation, groundwater, water management in the context of climate change adaptation, biodiversity policy reform, and country-level reviews. She also coordinates the OECD's Working Party on Biodiversity, Water & Ecosystems. Kathleen has a master of public policy from Science Po (France) and a bachelor's in business administration from the University of Notre Dame (United States).

Ms. Jehanne Fabre is the head of global water stewardship at Danone. Jehanne's role is centered on developing Danone's water policy, emphasizing water stewardship, circular economy, and regenerative agriculture. As Danone ecosystem fund coordinator for Asia, she managed a 12 million Euros projects portfolio in China, India, and Indonesia. Jehanne leads high-level consultations, multistakeholders' coalitions, fundraising efforts, and designs new business models. She holds a master of public policy analysis and a master of international cultural exchange strategies from Sciences Po Lyon.

Mr. Paul Fleming is the water program manager lead in Microsoft's Corporate Environmental Sustainability group. In that capacity, Paul is responsible for developing and advancing the company's water commitments through its operations, stewardship strategies, products, platforms, and policies. Paul has been an active leader on water and climate issues, serving on federal advisory committees such as the Federal Advisory Committee for the Sustained National Climate Assessment and chairing groups such as the Water Utility Climate Alliance. Prior to his current role, Paul developed and directed the Seattle Public Utilities' (SPU) Climate Resiliency Group where he was responsible for directing SPU's climate research initiatives, assessing climate risks, mainstreaming adaptation and mitigation strategies and establishing collaborative partnerships. Paul has a degree in economics from Duke University and a MBA from the University of Washington.

Ms. Gena Gammie is associate director of Forest Trends' Water Initiative and deputy chief of party, Natural Infrastructure for Water Security (Peru). Gena works collaboratively with civil society, government, and professional associations to systematically address barriers to realizing effective watershed investments and leads Forest Trends' work on analyses and tools for evaluating the cost-effectiveness of green infrastructure, with the aim of promoting performance-based investments in green infrastructure by mainstream water resources managers. Gena holds an MSc with distinction in environmental policy and regulation from the London School of Economics, where her dissertation focused on multistakeholder platforms for cooperative water resources management.

Dr. Pauline Georgiou is a social anthropologist with a PhD from Goldsmiths, University of London. She is interested in human/place relations and has focused on cultural perspectives of nature, borders, heritage, and tourism. She works at UEL's Sustainability Research Institute looking at the implementation of nature-based solutions through various projects and is cofounder of EM|Path, a social enterprise focusing on innovative research methodologies for creative engagement.

Dr. Nina Graveline is an agricultural and water economist with the French National Institute for Agriculture (INRAE) and the French Geological Survey. She provides research and policy advice on environmental and water management and adaptation to climate change in the agriculture sector, including the integration of uncertainty in water management and planning decisions (robust decision-making), the concepts and determinants of adaptation and resilience to climate change or to natural hazards, and the insurance value of nature-based solutions. Nina has a PhD in agricultural economics from AgroParisTech, a master of environmental and resource economics from the Institut National Agronomique de Paris Grignon, and a master of environmental economics from AgroParisTech.

Mr. Sujith Sourab Guntoju is an Indian urban and environmental planner who graduated from the School of Planning and Architecture, New Delhi. His research interests are water resource management and planning, water-food-energy nexus, urban and disaster resilience, and geo-informatics in environmental planning. He is a young professional, researcher, consultant, and youth environmental activist. He strongly advocates for the rights of youth and mainstreaming youth into environmental decision-making internationally through organizations such as UNMGCY, YOUNGO, World Youth Parliament for Water, PCCB Network, Mock COP26, and more. He is also the founder of an online networking hub for planning professionals across Asia called Asian City Planners Network.

Mr. Sam Jelliman is a research assistant at the Sustainability Research Institute at the University of East London. He works in a range of areas, including decarbonization, nature-based solutions, air pollution, sustainable transport, and behavior change. Sam has experience working on environmental project in the United Kingdom, New Zealand, and Mongolia. Sam studied environmental science at the University of East Anglia.

Dr. Bonnie Keeler is an assistant professor at the University of Minnesota's Humphrey School of Public Affairs, where she is affiliated with the Center for Science, Technology, and Environmental Policy, the Institute on the Environment, and the Natural Capital Project. Dr. Keeler works at the intersection of sustainability science and environmental economics, with particular expertise in water management and policy. At the Humphrey School, Keeler's team partners with state and federal agencies, community-based organizations, and other stakeholders to codevelop solutions to environmental and social challenges. Current projects include mainstreaming equity in water resource management, mitigating the effects of green gentrification in cities, and understanding the value of water-related ecosystem services.

Dr. Elena Lopez-Gunn is the founder and director of ICATALIST, a Spanish start-up specialized in climate change adaptation and sustainability. Elena has been a Cheney Fellow at the University of Leeds in the United Kingdom, and she has worked at the Water Observatory of the Botin Foundation, a think-tank undertaking targeted research to provide new evidence on key policy questions both in Spain and globally. She was scientific coordinator of the H2020 NAIAD project on the Natural Insurance Value: assessment and demonstration (www.naiad2020.eu). Her current interests are focused on how to have a deeper understanding on the role nature can play in risk management and generate cobenefits at different scales. Elena has a PhD from King's College (London) and a master of philosophy in development and the environment from the University of Cambridge.

Ms. Roxane Marchal is a catastrophic risk analysis with the CCR Groupe (Caisse Centrale de Réassurance). She works on analyses to estimate and model risk, such as damage functions, damage curves, vulnerability and mapping using insurance data, as well as risk management and preventive measures, such as nature-based solutions and natural hazards assessment. Roxane was a research fellow on the EU Horizon 2020 NAIAD project on the insurance value of nature. She has a master in European studies from the University of Geneva.

Mr. Michael Matosich is a corporate strategy associate, global water security at The Nature Conservancy. His professional pursuits have centered around environmental sustainability—from coordinating US environmental, science, technology and health policy for 22 countries in Central America and the Caribbean to advising some of the world's largest multinational corporations on global water risk. At The Nature Conservancy, Michael provides research and analysis in support of corporate water stewardship, nature-based solutions for watersheds, and water sustainability goals. Prior to joining TNC, Michael was a graduate consultant with SwissRe, researching the economic impacts of climate change on agriculture-dependent communities. Michael holds a bachelor of political science and international relations from UC Davis, and a master of international affairs and international economics from Johns Hopkins University.

Dr. John H. Matthews is the executive director and cofounder of the Alliance for Global Water Adaptation (AGWA), and he is one of the world's leading experts on climate adaptation, water resilience, and nature-based solutions. His work blends technical and policy knowledge for climate adaptation and water management for practical implementation. John has worked on 5 continents and more than 30 countries and published in many policy, scientific, and technical journals and books, including *Science*, *Nature Climate Change*, and the *PLoS* family. John's work primarily targets decision-making frameworks for adapting water infrastructure and ecosystems to climate impacts, including integrating natural infrastructure and nature-based solutions with gray infrastructure. Under his leadership, the AGWA network directly advises many national governments, corporations, the European Union, and key financial institutions on water-related climate risks, while AGWA has also become the official voice of the water community in the UNFCCC. John is a Senior Water Fellow at Colorado State University, and a courtesy faculty member of the Water Resources Graduate Program at Oregon State University. Previously, John started and directed global freshwater climate adaptation programs for WWF and Conservation International. He has PhD in ecology from the University of Texas.

Dr. Nathanial (Nate) Matthews is chief executive officer at the Global Resilience Partnership, a global initiative of over 60 partners that aims to help millions of people in the Sahel, the Horn of Africa, and South and Southeast Asia better adapt to shocks and chronic stresses and invest in a more resilient future. Nate is a multidisciplinary scientist who has written 2 books and over 60 publications across resilience, risk, water, energy, agriculture, and natural resource management. He previously worked at the CGIAR, UNEP, the International WaterCentre, NGOs, and the private sector. Nate is a Senior Visiting Fellow at King's College London and a Fellow of the Royal Geographic Society. He has a PhD in geography from King's College London.

Mr. Alex Mauroner is the network director at the Alliance for Global Water Adaptation (AGWA), an international NGO working to develop and mainstream technical approaches and enabling policies around climate-resilient water management. At AGWA, Alex helps coordinate the network of 2000+ members as well as a number of projects around technical approaches to climate risk and various capacity-building initiatives. He is also a cohost, producer, and editor of AGWA's ClimateReady Podcast. Prior to his current role, he served as a research associate for AGWA from 2014 to 2019. Alex holds a BA in biology from Westminster College and a professional science master's, environmental science from Oregon State University, where he concentrated on natural resources management.

Dr. Beatriz Mayor is senior researcher at ICATALIST in areas related to climate change adaptation and social and economic innovation for sustainability. She also contributes to the development and tracking of new project proposals and opportunities. Beatriz is an environmental scientist and holds a PhD on the water-energy-food nexus. She has recently focused on nature-based solutions for climate adaptation within the frame of the H2020 NAIAD and GeoEra TACTIC projects. She previously worked at the International Institute for Applied Systems Analysis (IIASA) based in Vienna, where she developed a multidimensional assessment of water-energy-land (WEL) nexus relevant technologies (desalination, irrigation) covering cost, technology diffusion and scaling trends, and WEL nexus trade-offs in the frame of the IS-WEL project.

Dr. David Moncoulon is head of modeling and R&D at Caisse Centrale de Réassurance (CCR), France's public reinsurer. An agronomy engineer with a doctorate in hydrology, David joined CCR in 2007 as a member of the Technical Studies State Guaranty team. At CCR David Moncoulon is primarily in charge of the development of mathematical models for natural perils covered by CCR's natural disaster scheme. He is also in charge of managing scientific partnerships with the research teams working in these areas, for example, *Météo France*, the National Research Institute of Science and Technology for Environment and Agriculture (*IRSTEA*), and the French Geological Survey (*BRGM*).

Since 1993, **Mr. Jason Morrison** has been with the Pacific Institute, a nonprofit, nonpartisan sustainability policy research center based in Oakland, California. In October 2016, Jason began his role serving as president of the organization. Prior to that, he directed the Institute's Corporate Sustainability Program, where he studied the policy implications of private sector sustainability initiatives, with a focus on freshwater-related business risks and on sustainable water management in the business community. Under a Memorandum of Understanding between the Pacific Institute and the UN Global Compact, Jason serves as the head of the CEO Water Mandate and has been supporting the initiative since 2008 with applied research, event organization, and other services. He is a cofounder of the Alliance for

Water Stewardship (AWS), a global initiative that has developed and is now implementing a freshwater certification program to advance responsible water practices by water providers and large-scale users. Jason holds a master in energy and environmental studies from Boston University and a BA in philosophy from the University of California at San Diego.

Dr. Raúl Muñoz Castillo is an environmental engineer and currently serves as climate change adaptation specialist and senior water specialist at the Inter-American Development Bank (IDB). He holds a PhD from the University of Maryland, and his research is focused on the water-energy-food (WEF) nexus of biofuel production in Brazil. Raúl has more than 15 years of experience working on environmental and water resources management, with broad international experience working in the private sector, public sector, multilateral organizations, and academia, with a strong focus in water infrastructure and water policy. He leads the Water Funds Program, the Bank's flagship project on green infrastructure and nature-based solutions, the Water Security Agenda, and the Bank's Initiative for Transboundary and International Waters. He has published, as lead author and as coauthor, numerous publications on water, sustainability, and environmental issues.

Ms. Florentina Nanu is the managing partner at the Business Development Group. She has a background in economics with a specialization in international relations. Florentina was key expert in the Horizon 2020 NAIAD project focused on promotion of the ecosystems services for mitigation of extreme water-related events, ecological (re)construction, and sustainable development of local communities. Her main contributions to the project were to support demonstration of nature-based solutions in local conditions and to bridge toward the insurance sector for identification of win-win solutions for risk reduction and increased resilience to climate challenges. Previously, she was involved in the elaboration of the Romanian National Strategy for Territorial Development 2014–20 and of the Climate Change Adaptation component of Romania's Climate Change Strategy 2011–20 as senior expert conducting on-site research, consultations, trainings, and elaborating public policy documents.

Dr. Boris Ochoa-Tocachi is a hydrologist and civil engineer with both academic and professional experience in Andean hydrology, hydrological monitoring and modeling, and water resources management. He is a research associate at Imperial College London, United Kingdom, and CEO of ATUK Strategic Consultancy, Ecuador. Boris has served as an international consultant for the Regional Initiative for Hydrological Monitoring of Andean Ecosystems (iMHEA) and for the Community Tourism Network of Austral Ecuador (Pakariñan), and as a researcher at Universidad de Cuenca, Ecuador. In the Consortium for the Sustainable Development of the Andean Ecoregion (CONDESAN), he worked in projects such as MOUNTAIN-EVO: Adaptive Governance of Mountain Ecosystem Services for Poverty Alleviation Enabled by Environmental Virtual Observatories, PRAA: Project for Adaptation to the Accelerated Retreat of Glaciers in the Tropical Andes, and CGIAR's

CPWF: Challenge Program on Water and Food—Andes. Boris has a PhD in civil engineering research (2019) and an MSc with distinction in hydrology and water resources management (2014) from Imperial College London. He also holds a civil engineering degree with distinction (2011) from Universidad de Cuenca, Ecuador.

Ms. Naabia Ofosu-Amaah is senior corporate engagement advisor, global water at The Nature Conservancy (TNC). Naabia focuses on developing and implementing the strategy for engaging companies to further TNC's global water goals. This includes research on priority companies, support of corporate engagement in TNC's regional programs, and representing TNC in water coalitions for industry. Naabia also serves as the relationship manager for a few of TNC's relationships with key Fortune 500 companies. Naabia joined TNC in January 2017 after 7 years at the Global Environment & Technology Foundation (GETF), where she managed public-private partnerships in water, including The Coca-Cola Company's $65M Replenish Africa Initiative (RAIN) and programs with the US Agency for International Development, Diageo, plc and WaterHealth International. Ms. Ofosu-Amaah holds a master of philosophy in environment, society and development from the University of Cambridge and a BA in environmental science and public policy with honors from Harvard University.

Ms. Pua'ala Pascua is a visiting scientist at the Center for Biodiversity and Conservation of the American Museum of Natural History, specializing in locally and culturally centered approaches to natural resource management. Pua'ala is of Native Hawaiian descent and currently resides in Hawai'i. She holds a master's degree in natural resource and environmental management from the University of Hawai'i at Mānoa and bachelor's degrees in Hawaiian studies and marine science from the University of Hawai'i at Hilo.

Ms. Lisa Peachey graduated from the landscape architecture master's program at UEL in 2020 and is employed as a graduate landscape architect at Vogt, London. Her thesis on the role that the design of natural spaces can have in supporting mental health—and in particular grief—was nominated for the Landscape Institute Student Thesis award in 2020 and funded by Landscape Institute Student Travel award in 2019 and a Bloedel Reserve creative residency. She was employed in 2018 by the Sustainability Research Institute at UEL to support their work on the Healthy New Towns Project, mapping blue spaces at Barking Riverside and engaging in community research.

Dr. Denielle Perry obtained her PhD in geography at the University of Oregon, United States. She is a water resource geographer and director of the Free-flowing Rivers Lab in the School of Earth and Sustainability at Northern Arizona University. Her research largely focuses on policies that influence both the development and protection of riverine resources. She is particularly interested in how conservation policies can be used as climate adaptation policy as well as for the re-Indigenization of water governance. She conducts fieldwork across the Americas, Europe, and Asia, and serves as cochair of the international Durable River Protections Coalition. She has published/produced scientific papers, book chapters, films, datasets, and policy resolutions on river conservation, impacts of development on riverine ecosystems, and water resource governance broadly.

Ms. Kathryn Pharr is an international water consultant. Now based in the United Kingdom, she has consulted with WaterAid UK on their water and climate work, Arup on their City Water Resilience Approach, and the Alliance for Global Water Adaptation (AGWA) on international water resources and climate change policy. After earning a master of science in analytical chemistry, she worked with the US Department of State on international policy for science and technology innovation and international water issues for over 6 years. She has researched science policy and water resources at the University of Oxford and coordinated their Oxford Water Network. In 2019, she founded the Community of Women in Water to support women professionals around the world working on water challenges.

Dr. Eleanor J. Sterling is Jaffe chief conservation scientist at the American Museum of Natural History's Center for Biodiversity and Conservation. Building on her interdisciplinary training and over 30 years of field experience in Africa, Asia, Latin America, and Oceania, her work focuses on systems thinking approaches; the intersection between biodiversity, culture, and languages; the factors influencing ecological and social resilience; equity, inclusion, diversity, and justice issues; and the development of indicators of well-being in biocultural landscapes. She is an adjunct professor at Columbia University's Department of Ecology, Evolution, and Environmental Biology, and deputy vice-chair for IUCN's World Commission on Protected Areas Core Capacity Development group where she coleads working groups on Indigenous Peoples and Local Communities and on Capacity Development Evaluation. She received her BA degree from Yale College, and MPhil and a dual degree PhD in anthropology and forestry and environmental studies from Yale University.

Ms. Ingrid Timboe is the policy director at the Alliance for Global Water Adaptation (AGWA). Based in the United States, she coordinates AGWA's ongoing efforts to mainstream water into national and global climate policy. She also cohosts the *ClimateReady* podcast. Ingrid holds a master of science in water resources management from Oregon State University and a bachelor of arts in international affairs from Lewis and Clark College. Prior to graduate school, Ingrid spent 6 years at the World Wildlife Fund in Washington, DC, working with the Global Policy and Conservation Science teams.

Ms. Sophie Trémolet is water security Europe director for The Nature Conservancy, where she leads the development and implementation of a program to accelerate investment in nature-based solutions for water security in Europe and globally. She is in charge of building strategic relationships with Europe-based public and private funders, utilities, or regulators so as to support a gradual shift toward greater incorporation of nature into water security investments. She holds a masters in economics and business from Sciences-Po (Paris) and a masters in international affairs, with a specialization in economic and political development from Columbia University (New York).

Dr. Nora van Cauwenbergh is an expert in water and sustainable development with over a decade of experience in academia and the private sector. She is passionate about contributing to a green and inclusive society through codesign of innovative natural resources management considering different value systems. In the water sector, she has worked on participatory design and use of software tools and protocols for stakeholder negotiation in water management. In the private sector, she has performed water/energy audits at different scales and codesigned and implemented green buildings. Her recent work in Latin America and EU focuses on integration of nature-based solutions and green infrastructure in basin and city planning to reduce flood and drought risk. She has academic positions in VUB and IHE Delft, where she teaches various water management topics and is business owner and lead business development green solutions in the private sector. Nora has a PhD in environmental sciences and land planning.

Dr. Peter van der Keur is senior scientist at the Geological Survey of Denmark and Greenland (GEUS), Department of Hydrology. He is hydrologist by training with an educational background in hydrological modeling and soil physics. He currently is lead for the EU financed GeoERA—TACTIC project on climate change impact on groundwater and adaptation strategies. Recently he has been involved in projects on hydrologic impact of climate change adaptation and disaster risk reduction, including the effect of nature-based solutions in the EU financed NAIAD project. Previously, he worked on topics on capacity-building in environmental sciences, disaster risk reduction and water governance, adaptive water resources management, including field experience from Europe (incl. Greenland), South Africa, and Malaysia.

Ms. Tara Varghese is the senior manager for Global Water Stewardship at PepsiCo, where she focuses on corporate sustainability, water stewardship, regenerative agriculture, and climate change adaptation. Prior to joining PepsiCo, Tara was the director of Water and Development Programs at the Global Environment & Technology Foundation, a natural resources management consultant at The Nature Conservancy and a researcher at the Ladakh Ecological Development Group. She has a master of environmental management from Yale University and a bachelor of arts from Case Western Reserve University.

Dr. Paula Vandergert is a Senior Research Fellow in the Sustainability Research Institute, University of East London. Paula is currently part of the Horizon 2020-funded program Connecting Nature, which is working with 11 cities across Europe on scaling up nature-based solutions implementation. Other recent work includes an action research project on Blue Spaces in Thames Ward, Barking, as part of the NHS England Healthy New Towns program. Prior to joining UEL, Paula was senior sustainable design advisor at CABE, the UK government advisor on architecture, urban design, and public space, where she managed the development of their Sustainable Cities program. Her PhD from the London School of Economics and Political Science examined sustainable forest management from a property rights perspective. Paula also spent over 10 years as director of Forests Monitor, where she worked with civil society organizations worldwide on forestry issues.

Ms. Kari Vigerstol is the director of water security science and innovation for The Nature Conservancy's Global Water program, strategically strengthening and advancing the science behind nature-based solutions for water security, source water protection, and water scarcity. Over the last 20 years, Kari has advanced sustainable water management in dozens of critical basins around the world through application of technical skills and multistakeholder partnership building. Kari is a licensed professional engineer and holds a BS and an MS in civil and environmental engineering from Rice University and the University of Washington, with a focus in water resource planning and management.

Dr. Adrian Vogl leads the Securing Freshwater program at the Natural Capital Project, and has over 10 years of experience working across disciplines in collaborative scientific teams, engaging with policymakers to advance ecosystem services approaches in the United States, Latin America, South Asia, and Kenya. Her focus is on how land and forest management impact water resources, particularly in the face of changing and uncertain climate conditions. Her work is problem-driven, with an emphasis on developing standardized approaches and tailored tools (e.g., RIOS, ROOT) to assess and map ecosystem services and their values and connect these benefits to communities. Adrian received her PhD in aquatic resources from Texas State University-San Marcos, and her BA from the University of Arizona in cultural anthropology.

Ms. Rebecca Walker is a PhD student at the University of Minnesota's Humphrey School of Public Affairs studying urban environmental planning. Her research considers the intersection of housing, equity, and the environment in American cities. This research uses a critical GIS framework to interrogate landscapes of urban environmental injustice, the historic housing and environmental policies that produced them, and how sustainability interventions today might challenge or deepen spatial inequalities. Her research questions and methods are guided by the voices and leadership of environmental justice activists in Minneapolis and beyond.

Mr. Josh Weinberg is program manager at the Water Resources Department at SIWI, where he currently coordinates SIWI's work under several EU programs, including Policies, Innovations and Networks for China Europe Water Cooperation (PIANO) and NAture Insurance value: Assessment and Demonstration (NAIAD). Josh previously led assignments under the EU-China Policy Dialogue Support Facility and the EU-China Environmental Sustainability Program. Josh has extensive experience working in China, Europe, and globally on broad range of issues within water and environmental governance. He is former managing editor and current editorial board member of *Stockholm Water Front Magazine*. Josh holds a master of science in natural resource management and governance from the Resilience Center at Stockholm University.

Dr. Pedro Zorrilla Miras is senior researcher and consultant at ICATALIST, leading projects in the search of sustainability from a socio-ecological approach. He is an environmental scientist, holds a PhD in ecology and environment, and has worked in the scientific and business spheres. Pedro is currently working in the H2020 project RESET developing spatial modeling, artificial intelligence, and environmental sensing technologies to better understand more sustainable ways of farming and urban development. He has recently finished a Marie Skłodowska-Curie grant from the European Commission to investigate adaptation to climate change of small farmers in Mozambique. He has previously worked among other projects in the H2020 project NAIAD on the insurance role of ecosystems for climate change adaptation, evaluating the consequences of deforestation on the well-being of the rural population in Mozambique (ACES project), on the valuation and mapping of ecosystem services (Spanish Millennium Ecosystem Assessment and Cañada Real Conquense, central Spain), and in the management of groundwater in central Spain (NEWATER project).

Foreword

As we write this foreword in spring 2021, society looks to be on the verge of the biggest global infrastructure investment in a generation. And it could not be more necessary. It is not just that our current infrastructure is outdated and in ill repair, but it is also that the planet's natural systems, which sustain our economies, communities, and health, are changing in unprecedented ways, adding additional layers of urgency and complexity to this long-deferred problem. In this new reality, the old approach to addressing infrastructure challenges will not suffice. Transformative investments are instead needed to speed the transition to a low-carbon economy, prevent new global pandemics, and hasten attainment of the Sustainable Development Goals—and it must all be done on a planet with depleted natural resources and growing climatic unpredictability. It is the right moment, then, to ask: How do we define infrastructure in the modern era? What do we want it to do? And for whom?

This book sets out an agenda for understanding nature's place in this conversation, for finally giving nature its due as essential infrastructure. The book intentionally focuses on water, which will, for nearly all of us, be the face of climate change—either through more frequent and severe storms, droughts, wildfires, and floods or through rising sea levels. Specifically, its authors highlight the power of "nature-based solutions," a category of water management techniques that harness or mimic natural processes to restore nature and increase resilience. A city might, for example, install new vegetated gardens to absorb stormwater during heavy rains. Or a water utility might invest in restoring forests surrounding its drinking water reservoir, with an eye to better water quality and less erosion into the reservoir.

Nature-based solutions are not new ideas. Indigenous peoples have employed similar technologies and passed down rich traditions rooted in nature-based water management for millennia. And nature-based solutions have long had the attention of forward-thinking planners, conservationists, water utility managers, and land users.

Unfortunately, these tools—which offer real and tangible solutions for today's challenges—have been largely overlooked by the mainstream. This book provides an important contribution to the conversation on how we can change that fact, bring nature-based solutions to the necessary global scale, and provide a more rigorous definition and framework for designing and evaluating "natural infrastructure" at this most critical time.

This book is practical, but it is also quietly profound. Western society has long considered nature as something separate from itself, a resource needing protection from the destructive impacts of humanity, at best. The agenda laid out here helps us flip that script to finally consider nature as our partner, our collaborator. As the authors so eloquently write, "nature-based solutions are a way to renegotiate our core values through the medium of water itself."

To that end, the nature-based strategies described here become a force that can bring down old partitions and fuel new and unusual coalitions between, for example, conservationists and engineers, upstream and downstream communities, a ministry that manages agriculture, and another ministry that oversees planning or finance. Uptake of these tools can inspire our societies to confront and begin to heal deep injustices, including a lack of respect for traditional knowledge of Indigenous peoples and rural communities, unequal access to green spaces and clean water, insecure land rights, gender-based inequities, and the absence of recognition or compensation for good stewardship of the land.

A nature-based solution, in this light, is not only the protected forest, replanted wetland, or new cover crop. It is also the act of addressing the institutions, policies, data, norms, and governance systems that foster and protect these activities. Planting trees is easy. Changing institutional culture among stakeholders or making sure communities have secure tenure on the lands they have protected for generations—that is the hard work. This book serves as a guide not only for the practical application of nature-based solutions but also for working in and changing these complex systems from the ground up.

Among many challenges, 2020 saw a global pandemic, the second-warmest temperatures on record, catastrophic forest fires and hurricanes, and a spike in global forest loss—all of which served as a chilling reminder that we cannot continue to test planetary boundaries. As we stand on the precipice of transformational global investment in our shared infrastructure, we have a rare opportunity to change course—a chance to mainstream nature-based solutions, especially those for water management, as a central part of a green and robust recovery. We have the tools, momentum, and (increasingly) public opinion on our side. Now is the time to act. Together with the authors, we are confident that this book can help guide our way forward to a greener, healthier, and more just world.

Michael Jenkins
CEO and President, Forest Trends
Jennifer Morris
CEO, The Nature Conservancy

Introduction to nature-based solutions and water security

CHAPTER 1

Setting the scene: Nature-based solutions and water security

Jan Cassin[a] and John H. Matthews[b]

[a]Forest Trends, Washington, DC, United States, [b]Alliance for Global Water Adaptation, Corvallis, OR, United States

But then, when the land was still pristine, today's mountains supported high hills, and what we call the Stony Plains were full of rich earth, and in the mountains, there was a good deal of timber, of which there are clear indications even now. Some of the mountains can sustain only bees these days, but it was not long ago that they were wooded and even now the roofs of some of our largest buildings have rafters cut from these areas and these rafters are still sound. There were also many tall, cultivated trees, and the land offered a vast amount of pasture for animals. What is more, the land enjoyed the annual rain from Zeus, not lost, as now, when it flows off of the bare earth into the sea. Rather, much of it was retained, since the earth took it in within itself, storing it up in the earth's retentive clay, releasing water from the high country into the hollows, and supplying all regions with generous amounts of springs and flowing rivers. That what we are now saying about the land is true is indicated by the holy sanctuaries, which are situated where this water used to spring up.

(Plato, 2008, Timaeus and Critias, 111a–d).

A global water crisis and water insecurity

The connection between water security and natural systems such as forests, wetlands, and rivers, has long been recognized. History is replete with examples of how ecosystem degradation and loss have impacted people and civilizations through water. Throughout recorded history, local water-related crises have followed deforestation, with erosion and accumulation of sediment in rivers resulting in increased flooding as well as the loss of soil fertility and declining agricultural production (Marsh, 1864; Montgomery, 2007). The world, however, now has a *global* water crisis and successfully addressing it will require a radical transformation of the way water is managed, by prioritizing working *with* the natural systems on which water security ultimately depends.

Nature-Based Solutions and Water Security. https://doi.org/10.1016/B978-0-12-819871-1.00003-8

The COVID-19 pandemic that began in late 2019 has also had an influence on how we envision water security. Certainly, many recommendations for reducing transmission such as handwashing assumed ample, clean, and accessible water. More broadly, the appearance of broader economic impacts and the shortage of basic household and essential medical supplies led to widespread discussion about vulnerabilities in supply chains, planning, and coordination, spanning local to international scales. As a result, the pandemic led many people to see transport, energy, utilities, manufacturing, and medical processes as systems that operate and interact in complex ways. Water is often the connecting element within and between these systems. For many groups, pandemic planning is now emerging as a new component of water security, while water security is also now seen as critical to responding to both the medical and economic impacts of major disease outbreaks. The linkages between water security, epidemics, and nature-based solutions (NBS) are more tenuous, but they are also appearing, such as the German Ministry for Economic Cooperation and Development's (BMZ) program "One Health,"[a] which launched in May 2020 to coordinate policy and action between human health, ecological resilience, and climate change, with water explicitly identified as the connection between all three domains. Similar initiatives are reportedly under discussion in multilateral and bilateral aid and in national and intergovernmental policy initiatives (Matthews and Dela Cruz, 2020).

For the past 10 years, water-related risks have led the World Economic Forum's global risk assessment in terms of both likelihood and severity of impact (World Economic Forum, 2019). Between 2009 and 2019, water supply crises and/or water hazards related to extreme weather have routinely been in the top five risks in terms of likelihood and severity of impact. In addition, both water and climate constitute risks that are overriding and increase the likelihood or impacts of other risks. Water and climate in particular are closely linked. People will feel the effects of climate change mostly through water. From more and longer or deeper droughts impacting food and energy production, loss of life and damages from more extreme storms and flooding, or the unpredictability of rainfall that can lead farmers to abandon land and spurs internal and international migration, fundamentally, climate is water (Nyugen and Hens, 2019; Jiménez Cisneros et al., 2014). Water provision is critical for economic development, underpinning agricultural production and fisheries, supporting biodiversity, energy generation, and water-based transportation, meaning that the water crisis is a systemic risk that affects all major global challenges (WWAP, 2016; World Economic Forum, 2011; World Economic Forum, 2019).

Water security and the contours of the water crisis

The water crisis comprises a constellation of global threats affecting supplies of available water, its quality, and the frequency and severity of water-related hazards, which all

---

[a] https://www.bmz.de/en/press/aktuelleMeldungen/2020/oktober/201027_pm_035_World-Health-Summit_-Minister-Mueller-announces-creation-of-a-new-One-Health-priority-area-at-the-BMZ/index.html

contribute to growing water insecurity for people and ecosystems. Water security as defined by the United Nations is the "capacity of a population to safeguard sustainable access to adequate quantities of acceptable quality water for sustaining livelihoods, human well-being, and socio-economic development, for ensuring protection against water-borne pollution and water-related disasters, and for preserving ecosystems in a climate of peace and political stability" (UN-Water, 2013). Water security is multidimensional and dynamic. Implicit in concepts around water security is the need to manage risk and uncertainty. An explicitly risk-based approach can capture these dynamic and multidimensional aspects, particularly in the context of uncertainty about shifting economic, political, social, and climate conditions. Indeed, some have even defined water security as an "acceptable level of water-related risks" (Grey and Sadoff, 2007; Grey et al., 2013) or "a tolerable level of water-related risk" to society regarding four categories of hazards: shortages or scarcity; inadequate quality; excess or flooding; and eroded or reduced resilience of freshwater systems (OECD, 2013a; Garrick and Hall, 2014).

Rather than being simply a function of the physical availability of water (too much, too little, poor quality), water security is an outcome of the dynamics of coupled social-ecological (human-water) systems. As such, human agency, expressed in aspects of water governance, social power, water infrastructure, and the political and social institutions influencing water management, is central to water security and the water crisis (Bakker and Morinville, 2013; Srinivasan et al., 2017; Staddon and Scott, 2018). This broader framing means that questions around equity, sustainability, and effectiveness over time need to be addressed. Whose security is important (which communities or sectors, people and ecosystems, present generation or future generations as well)? Who decides what levels of risk and uncertainty are acceptable? How will risks evolve with time? What management options (infrastructure, governance) are most likely to successfully address the multiple dimensions of water security across the greatest number of uncertain futures?

The contours of the water crisis

A growing human population and increasing resource demands are putting pressure on Earth's critical systems, including the hydrological system, and threatening to breach planetary boundaries that constitute a "safe operating space" for humanity (Rockström et al., 2009; Steffen et al., 2015). The water crisis is directly related to a majority of the planetary boundaries originally identified by Rockström et al. (2009): climate change, interference in phosphorus and nitrogen cycles, global freshwater use, land-system change, rate of biodiversity loss, and chemical pollution. The water crisis, like water security, is thus multifaceted and driven by many evolving, interconnected factors.

Despite this complexity, the dimensions of the crisis can be described by a small set of syndromes that capture the suite of driving factors: (1) increasing demand for water for domestic, agricultural, and industrial uses from a growing human population; (2) a history of unsustainable water management in many regions, strongly influenced by governance systems that often result in ineffective tradeoffs; (3) environmental impacts that have degraded the freshwater and terrestrial ecosystems that provide critical hydrological services; and (4) a lack of or an inadequate infrastructure (Srinivasan et al., 2012; Wada et al., 2016; WWAP, 2015; WWAP, 2018). Cutting across all of these syndromes is climate change and the ways in which climate change itself will alter the impact of other drivers on the water crisis, sometimes reducing but often amplifying nonclimate drivers (Jiménez Cisneros et al., 2014). For example, demand for domestic and agricultural water use from growing populations is outstripping renewable supplies in some regions, resulting in overexploitation of groundwater resources. On top of this, as temperatures increase under climate change, growing crops will need more water, evaporative loss from reservoirs will increase, and water demand from the energy sector will increase as demand for air-conditioning during heat waves skyrockets and water transport/pumping grows, further increasing the imbalance between demand and available supplies in many regions (Elliott et al., 2014).

Growing demand, threatened supplies, unsustainable water management

A growing human population and rapidly developing economies are putting increasing demands on available water supplies, while mismanagement and pollution continue to diminish the amount of water that is actually usable (Guppy and Anderson, 2017; Makarigakis and Jimenez-Cisneros, 2019). Global water demand is expected to increase by more than 50% by 2050 to meet increased needs for people, agriculture, and energy production (OECD, 2012). Given the current distribution of freshwater resources, by 2050 more than 40% of the world's population will live in river basins with severe water stress. Groundwater exploitation is already unsustainable in many regions, with 20% of the world's aquifers overexploited, about 33% in distress, and groundwater levels declining around major cities and farming regions (Gleeson et al., 2012; Richey et al., 2015). Almost half the water used for irrigation is provided by groundwater and at least 50% of the world's population relies on groundwater for drinking water supplies (WWAP, 2012). Groundwater withdrawals that exceed replenishment rates can eliminate the capacity of aquifers to serve as important buffers to offset surface water scarcity in times of drought (Burek et al., 2016). Exports of virtual water, or the water embedded in food or other exports, can increase local water stress if virtual water flows out of arid or semi-arid regions (Sun et al., 2016). Perverse incentives can favor the intensive use of fossil or slow-recharging groundwater resources, creating long-term or even permanent loss of water capital.

Pollution further affects water availability, with discharges from industry and untreated wastewater, and agricultural and urban runoff rendering some surface and groundwater sources unusable, damaging human and ecosystem health (Rosenzweig et al., 2014). Globally, at least 80% of wastewater is released directly into rivers, lakes, or the ocean without adequate treatment, resulting in major health and environmental risks (WWAP, 2017).

Sediment, nutrient, and chemical runoff from agriculture has major effects on water quality, contributing to algal blooms and dead zones in lakes and coastal waters. With current practices, global food production will need to increase by as much as 60% by 2050 to meet the needs of an estimated 9.8 billion people, meaning that water quality impairment from agricultural runoff is expected to increase as well (WWAP, 2015; Makarigakis and Jimenez-Cisneros, 2019).

These stresses on water availability in combination with inadequate water supply and treatment infrastructure disproportionately affect poorer and more marginalized populations. Already about one-third of the world's people (1.2 billion) do not have access to safe drinking water, and 4.5 billion lack access to the level of sanitation services targeted by the water and sanitation goal in the 2030 Sustainable Development Agenda (UN High Level Panel on Water). Women and girls bear the brunt of water collection in households without piped in water (WWAP, 2017). Of the people without improved sanitation, 70% live in rural areas, and a lack of services disproportionately affects women and girls who suffer repercussions in terms of both health and personal danger (Guppy and Anderson, 2017).

Impaired ecosystems, increasing hazards

Ecosystems that underpin human well-being are under increasing threat (IPBES, 2019). A number of recent reports document the extent of alteration to the environment: 75% of the planet's land surface has been significantly altered; about 50% of the extent of coral reefs has been lost in the past one hundred years; and more than 85% of the area of the world's wetlands has been lost (IPBES, 2019; Environment, 2019). Ecosystem loss and degradation impacts the capacity to continue to deliver critical water-related services; mitigating risks and ensuring clean, reliable flows by capturing, filtering, infiltrating, and storing water (MEA, 2005; Brauman et al., 2007; WWAP, 2018; Harrison et al., 2016; IPBES, 2019). Degraded watersheds providing drinking water for the world's largest cities impact the water supplies of about 700 million people and are estimated to cost USD 5.4 billion per year in water treatment costs (McDonald et al., 2016). The loss of these ecosystem services puts further stress on water supplies and increases hazards, exposing more people to water scarcity, poor or dangerous water quality, and flooding (WWAP, 2018).

Natural disasters have been occurring more and more frequently (Word Economic Forum, 2019) and are expected to continue to occur more frequently with climate change. The vast majority, 90%, of these disasters are water-related (UNISDR, 2016). Global economic losses from urban flood damages are USD 120 billion per year (UNHLP, 2015). Water-related losses (agriculture, health, livelihoods, property) could reduce global GDP by as much as 6% by 2050 in some regions, spurring sustained, negative growth (World Bank, 2016). Floods alone account for 43% of natural disasters. Loss of ecosystems increases vulnerability to hazards, while at the same time reducing the resilience or adaptive capacity of social-ecological systems. With more and more people living in cities and on floodplains, the loss of wetlands, deforestation, rising sea levels, and more severe storms, many more people will be vulnerable to flood disasters each year—an estimated 2 billion by 2050 (WWAP, 2012). Droughts are occurring more frequently and for longer periods with climate change, increasing physical water scarcity, damaging agricultural production, and contributing to increasingly severe wildfires (Spinoni et al., 2017; Williams et al., 2015).

Can old solutions succeed for emerging challenges?

Beginning on a large scale in the nineteenth century, water security in more developed countries and major cities has been secured through large, usually public, investments in built or "gray" infrastructure. The system of storage reservoirs, pipes, and pumps that distribute irrigation water to fields or potable water to homes, along with treatment plants to ensure water of acceptable quality, have resulted in huge gains in public health, agricultural production, and energy generation over the past two centuries (see Melosi, 2000 for United States examples; Juuti et al., 2007 for global examples). Gray infrastructure continues to be the dominant approach to water security in both developed and developing countries, as priorities are to repair or replace aging and/or build new, centralized gray infrastructure.

Despite gray infrastructure's undeniable current and future relevance (Muller et al., 2015), it has significant social and environmental impacts. Large dams for example, while providing storage to regulate supplies, have displaced communities, eliminated riverine and floodplain habitats and species, and resulted in the loss of ecosystem services such as fisheries and flood mitigation (Richter et al., 2010; Auerbach et al., 2014). Gray infrastructure is capital- and energy-intensive, lasting decades or centuries, and often inflexible in operations and design, planned to address a narrow set of water issues under well-defined and generally static conditions—which themselves are often based on historical hydro-climatic assumptions that can be easily violated even with modest amounts of climate change (Poff et al., 2016; Milly et al., 2008; Hallegate, 2009).

In effect, gray infrastructure is developed with a narrowly optimized vision of the future in mind that risks losing its economic benefits as operational parameters decline, making tradeoffs for equity and ecological resilience even more challenging over time. This inflexibility means that gray infrastructure is vulnerable to failure, especially under rapid climate and land-use changes. Large, expensive infrastructure projects can become "stranded assets" unable to function or deliver their designed services when conditions change (OECD, 2013b). In fact, in much of the developed world, the success of gray infrastructure in delivering clean water to people may mask the extent to which freshwater ecosystems have been impacted (Vorosmarty et al., 2015; Abell et al., 2019; Harrison et al., 2016). As Vorosmarty et al. (2015) argue, this "impair-then-repair" pattern means that an "expensive technological curtain" can hide the state of impairment of water resources; when people see clean water at the tap, the degree to which hydrological ecosystem services have been lost is obscured.

Rethinking the nature of infrastructure

To address threats to water security, conventional approaches to water infrastructure and water management need to be reconsidered in light of emerging trends. Indeed, some "new" challenges such as rapid anthropogenic climate change should prompt us to compare the scale of proposed solutions against the scale of the threats.

Even as our concerns around ecological issues, equity, and rapid social changes such as the rise of megacities and climate migration have increased, our solution set for governance and infrastructure has remained relatively narrow. We assume that most traditional resource management and economic optimization frameworks can be adjusted and modestly altered to accommodate what may be profound challenges to those approaches. The attention to ecosystem services, "co-benefits," environmental and social safeguards, adaptation and resilience, and new sources of finance and business models have not led to radically different outcomes or more than light adjustments in our use of natural resources, especially the eco-hydrological landscape. Have we settled for reform instead of revolution? Small-scale pilot "green" projects instead of working at scale by adjusting program-level guidelines and procedures? What might a more radical approach look like?

While many advocates for NBS focus on accumulating (and re-accumulating) evidence designed to persuade economists, engineers, and politicians that ecosystems have value and can perform sufficiently to supplement or replace traditional infrastructure roles, an alternative approach may be to reconceive the nature of infrastructure itself in light of a forward-looking, complex, and integrative vision of economic development. A deeper critique might ask if highly optimized approaches to designing long-lived infrastructure can even meet their intended goals over an operational lifetime lasting decades. Gray infrastructure is arguably the crystallization of our assumptions about the future, about governance, and about lasting economic needs made manifest in concrete, steel, and stone. Have we made tradeoffs between efficiency, need, impact, and function in ways that reflect the complexity of the world we live in now, much less the world of 50 years from now? If we are unable to determine what the future might look like with confidence, should we instead reconceive infrastructure in ways that are more open-ended, can grow or shrink, and can respond dynamically to dynamic economies and ecosystems? Will this century see more or less change relative to the twentieth century?

Given these questions, our minimization of the range of solutions we use to meet economic, social, and environmental challenges seems weak and limited. NBS are inherently flexible, and represent an additional set of options for politicians, communities, and planners that

---

Box 1.1   Nature-based solutions

Concepts and definitions around NBS have been evolving over the past 10–15 years. The International Union for the Conservation of Nature (IUCN) defines NBS as: "actions to protect, sustainably manage, and restore natural and modified ecosystems that address societal challenges effectively and adaptively, simultaneously providing human well-being and biodiversity benefits" (Cohen-Shacham et al., 2016). IUCN's NBS concept includes a broad range of solutions or approaches such as ecosystem protection and restoration, managed production landscapes, and urban green infrastructure. NBS are intended to address water security as well as challenges such as biodiversity conservation, climate adaptation, disaster risk reduction, and food security (Cohen-Shacham et al., 2019).

The UN World Water Assessment Programme considers that NBS are inspired and supported by nature and use, or mimic, natural processes to contribute to the improved management of water. NBS can involve conserving or rehabilitating natural ecosystems and/or the enhancement or creation of natural processes in modified or artificial ecosystems (WWAP, 2018).

The European Commission considers NBS to be living solutions inspired by, continuously supported by and using nature, designed to address various societal challenges in a resource efficient and adaptable manner and to provide simultaneously economic, social and environmental benefits (Maes and Jacobs, 2017).

NBS concepts are closely related to the ecosystem services approach, ecosystem-based management, ecosystem-based adaptation, natural infrastructure, and green infrastructure concepts.

"Nature-based solutions" is still somewhat of an open term and an umbrella concept. NBS comprise a gradient of systems or practices, from more to less "natural," for example, protected natural forest ecosystems, forest restoration, semi-natural managed forests, agroforestry systems, urban forests, parks, and green infrastructure.

---

we have largely neglected to date (Matthews et al., 2019; Smith et al., 2019), while our gray designs reflect an overconfidence in our ability to know the future. Perhaps the real sustainability gap is in our lack of imagination and the conservatism of our decision-making processes, which did not solve problems of the past century very effectively and often seem profoundly inadequate and incommensurate to the problems of this century.

This book focuses on one major family of solutions that have been consigned to a corner of our water security portfolio. Nature-based approaches (Box 1.1) that work with natural systems are seen as critical strategies for developing a successful and sustainable path to water security (Cohen-Shacham et al., 2016; WWAP, 2018). NBS can be more cost-effective alternatives to gray infrastructure, improve the effectiveness and operable life of gray infrastructure (e.g., reducing sedimentation that limits storage capacity in reservoirs), avoid social and environmental impacts, and deliver multiple additional benefits. Yet gray infrastructure will always be needed. The choice is not, either "green" or "gray" but ensuring that we are realistically considering the full range of options we have available, making honest and realistic tradeoffs. If we have an assumption as authors, it may be that a reconsideration of how we define water security as a long-term issue will often result in projects that blend green and gray for people and ecosystems (McCartney and Dalton, 2015; Palmer et al., 2015).

The emergence of nature-base solutions for societal challenges

Interest in NBS has grown in recent years because they can be cost-effective and sustainable approaches for addressing multiple pressing societal challenges. These challenges include water security (Abell et al., 2019; WWAP, 2018), urban water resilience (Arup and Shift, 2019), climate change (Jones et al., 2012; Griscom et al., 2017), disaster risk reduction (Gunnell et al., 2019; Kalantari et al., 2018; Sebesvari et al., 2019), biodiversity conservation (Harrison et al., 2016; Boelee et al., 2017), and sustainable development (Vörösmarty et al., 2018; Wood et al., 2018). Investments and on-the-ground implementation of these solutions have also grown steadily over the past 10 years. Global investments in watershed services by downstream communities reached USD 25 billion in 2015 and grew at a rate of about 12% between 2010 and 2015 (Bennett and Ruef, 2016). In Peru, water utilities have allocated more than USD 130 million for natural infrastructure investments to improve water security under climate change, including traditional water harvesting practices that work with nature (Ochoa-Tocachi et al., 2019).

In particular, interest in and implementation of NBS have expanded rapidly over the past 2–3 years. The diversity of organizations embracing these approaches, including businesses, conservation nonprofits, and multilateral development banks, is generating a growing body of experience and knowledge (see, for example, the Nature-Based Solutions Initiative, https://www.naturebasedsolutionsinitiative.org/). The next few years will be a critical time to leverage emerging knowledge and practice to realize the full potential of NBS. In 2020, countries will be updating their national commitments on climate change under the Paris Agreement. This opens opportunities, if designed and implemented appropriately, to jointly address water and climate through significantly increasing the role of natural climate solutions in both mitigation and adaptation (Seddon et al., 2019). The 15th COP for the Convention on Biodiversity will emphasize making the post-2020 framework more ambitious than the current strategic plan, with key opportunities to integrate biodiversity within related international frameworks on water security, climate, disaster risk reduction, and sustainable development. With 10 years remaining until the world should have achieved the ambitious 2030 Sustainable Development Goals (SDGs), 2021 will be an important time to take stock, evaluate progress, and refocus strategies to meet those goals. With water and nature in the center of most SDGs, nature-based approaches for water should be front and center in the SDGs.

At the same time there is a need to refine our understanding of the potential and limitations of NBS. There remain significant gaps in our knowledge of the effectiveness of individual

NBS to deliver the desired outcomes (Eggermont et al., 2015; Cohen-Shacham et al., 2019; Nesshover et al., 2017). The effectiveness of wetlands for flood mitigation, for example, depends on attributes of the individual wetland and its surrounding landscape (e.g., type of wetland, size, position in the landscape) as well as the magnitude of the flood event. NBS will not necessarily be protective in the case of large flood events, and it is important to be accurate about what these solutions can and cannot do. Concepts around NBS are still being refined, and there is a need to clarify which practices should be considered NBS and which should not and a need for clarifying principles to ensure that NBS, when implemented, are effective, equitable, and sustainable (Cohen-Shacham et al., 2019; Cassin and Locatelli, 2020).

A reference volume on nature-based solutions and water security

While there has been much progress on NBS, much work remains local, context-dependent, and fragmented, with limited avenues for connecting learning across organizations or regions. While NBS are cross-cutting solutions that address a range of societal challenges, we focus here on water as an enabler of solutions across multiple challenges. Insights and lessons from NBS for water can inform the broader NBS framework of analysis. Identifying generalizable approaches or "rules for success" are in embryonic stages, and dissemination of innovations from one place to another has been slow. While a diversity of experimentation and approaches will be needed to fully scale NBS, the lack of a shared strategic roadmap can lead to wasteful duplication of efforts or a tragic isolation, even ghettoization, that makes progress toward systemic solutions difficult to measure and hides gaps or bottlenecks to scaling effective approaches. This volume continues a series of dialogues among experts at recent World Water Week and World Water Forum meetings. The goal of those dialogues, and this volume, is to support decision-makers and practitioners in this growing field by: (1) synthesizing the accumulating, collective global experience with NBS to provide a practical reference on the current state of knowledge, policy, and practice; and (2) present a consensus reflection to guide the kind of strategic actions that are needed for implementation of NBS at scale.

This volume:

- provides a resource for decision-makers and practitioners wanting to engage with NBS;
- synthesizes in one place current evidence, science, policy, and practice as a readily accessible reference on NBS;
- provides critical case studies that show how NBS have been successfully implemented around the world, with specific lessons for confronting challenges in planning, designing, financing, implementing, and managing NBS; and
- presents a synthesis of lessons for what is next for mainstreaming and scaling NBS.

Rather than attempt to present a single vision for NBS, we have cultivated a diversity of current perspectives, approaches, and opinions.

Organization of this book

Section 1, Nature-based solutions and water security, Chapter 2: Following this introduction to NBS and water security, Chapter 2 presents an overview of NBS and describes the evolution of NBS concepts and emerging efforts to develop more clarity, standardization, and guidance principles (e.g., NBS that are effective, equitable, and sustainable) to guide analyses and research, evaluate projects, and communicate trends and progress toward scale for the diversity of practices that fall under NBS.

Section 2, State of NBS science and knowledge, Chapters 3–5. This section will present reflections or deep dives into the current state of science and knowledge informing NBS practice in three key areas: (1) water risks and water security; (2) NBS and climate change; and (3) NBS, sustainable development and equity. Chapter 3 reviews the current state of evidence around the role of NBS in regulating water quantity, water quality, and reducing water hazards (e.g., flooding and drought). While there are gaps in our knowledge about how specific NBS will perform, the chapter provides valuable guidance for moving ahead under uncertainty. Chapter 4 discusses the emerging focus on natural climate solutions (mitigation and adaptation) and presents an overview of framing concepts and the potential and limitations of NBS for climate adaptation. Chapter 5 covers the relationships between NBS, the SDGs, and equity. NBS designed to improve water security also have the potential to be a positive force for greater gender equality, social justice, economic opportunities, and resilience for vulnerable communities, and it can be a core strategy for achieving multiple SDGs. The chapter emphasizes the importance of explicitly addressing social justice and the three dimensions of equity—distributional, procedural, and recognitional—in designing and implementing NBS.

Section 3, State of Policy and Governance, Chapters 6–9. In Chapter 6, the authors address the policy rationale for governments to embrace NBS, discuss some distinct policy challenges limiting wider adoption of NBS, and discuss some of the levers of change for creating more enabling policies. Chapter 7 discusses how water-centric NBS projects have been applied at the national and transboundary level to meet targets for climate change adaptation and mitigation, disaster risk recovery, biodiversity conservation, and sustainable development. The authors then review the value that NBS bring to global policy frameworks—both as a solution to complex, societal issues and as a mechanism for addressing some of the coordination and implementation issues those frameworks face at local, national, and international levels. Chapter 8 explores how the concept of legibility, or how society and governments *know* and *represent* natural resources, has informed the development of national policies in the United States that protect free-flowing rivers and the many beneficial services they provide people. What makes nature legible to us influences what is valued and managed, with lessons for NBS policies from the development of wild and scenic river policies in the United States.

Section 4, State of practice: NBS projects, policies, and tools in action, Chapters 9–15. Case studies from a diversity of geographies and illustrating a range of applications related to water security are presented to show how NBS are being implemented to address water security (and co-benefits for climate, development, and health). Cases will describe why NBS were embraced, how the business case for NBS was made to decision-makers or water users, how the NBS solutions were designed, implemented, and financed, how key issues and bottlenecks were overcome, and what lessons were learned. Chapter 9 presents NBS in the context of source water protection in North America, while Chapter 10 discusses NBS and source water protection in Latin America. Chapter 11 presents the evolution of innovative policies and institutional changes in Peru that are leading to the mainstreaming of NBS at national and local levels in the drinking water sector, climate action plans, and sustainable development (green economy). Chapters 12 and 13 introduce two new areas of focus for NBS studies. Chapter 12 covers blue spaces in cities and their role in health and the revitalization of neighborhoods. Chapter 13 reviews the contribution of indigenous and local knowledge and reveals the deeper history of NBS for water security. Chapter 14 reviews the ways in which businesses are engaging with NBS as key strategies in corporate water stewardship, along with the challenges that businesses face in integrating NBS into their water stewardship practices. Chapter 15 discusses the state of practice in funding and financing NBS and the range of options for scaling NBS funding.

Section 5: Emerging innovations, moving to scale. This section presents two chapters on emerging innovations: roles the insurance sector can play in mainstreaming NBS and the "greening" of project development to normalize NBS within project planning and procurement processes. Chapter 16 reviews the five potential "hats" the insurance industry can wear in enabling NBS: as owners of data on risk; as providers of novel nature-based insurance products; as investors in NBS; as insurance and data innovators; and as partners in promoting prevention, hazard mitigation, and resilience. Chapter 17 examines the rationale for mainstreaming NBS within economic development pathways, identifying the processes and actions by which low- and middle-income countries, in particular, can more systematically bring NBS to scale as a complement to traditional or "gray" infrastructure. Finally, in Chapter 18 the contributors representing a network of NBS practitioners reflect on the potential and future direction of NBS for water security.

References

Abell R, Vigerstol K, Higgins J, Kang S, Karres N, Lehner B, Sridhar A, Chapin E. Freshwater biodiversity conservation through source water protection: quantifying the potential and addressing challenges. Aquat Conserv Mar Freshwat Ecosyst 2019;29:1022–38.

Arup, Shift TR. The city water resilience approach: city characterisation report, Cape Town; 2019.

Auerbach DA, Delsenroth DB, McShane RR, McCluney KE, Poff NL. Beyond the concrete: accounting for ecosystem services from free-flowing rivers. Ecosyst Serv 2014;10:1–5.

Bakker K, Morinville C. The governance dimensions of water security: a review. Phil Trans R Soc A 2013;371:20130116.

Bennett G, Ruef F. Alliances for green infrastructure: state of watershed investments 2016. Washington, DC: Forest Trends Ecosystem Marketplace; 2016.

Boelee E, et al. Overcoming water challenges through nature-based solutions. Water Policy 2017;19:820–36.

Brauman KA, Daily GC, Duarte TK, Mooney HA. The nature and value of ecosystem services: an overview highlighting hydrologic services. Annu Rev Env Resour 2007;32:67–98.

Burek P, Satoh Y, Fischer G, Kahil MT, Scherzer A, Tramberend S, Nava LF, Wada Y. Water futures and solution—fast track initiative final report. In: IIASA Working Paper. Luxemburg, Austria: IIASA; 2016.

Cassin J, Locatelli B. Guía para la Evaluación de Intervenciones en Infraestructura Natural para la Seguridad Hídrica: Escala de Efectividad, Equidad y Sostenibilidad. Forest Trends; 2020. Available at: https://www.forest-trends.org/publications/guia-ees/.

Cohen-Shacham E, Walters G, Janzen C, Maginnis S, (eds.). Nature-based solutions to address global societal challenges. Gland, Switzerland: IUCN; 2016. https://doi.org/10.2305/IUCN.CH.2016.13.en.

Cohen-Shacham E, Andrade A, Dalton J, Dudley N, Jones M, Kumar C, Maginnis S, Maynard S, Nelson CR, Renaud FG, Welling R, Walters G. Core principles for successfully implementing and upscaling nature-based solutions. Environ Sci Policy 2019;98:20–9.

Eggermont H, Balian E, Azevedo JMN, Buemer V, Brodin T, Claudet J, Fady B, Grube M, Keune H, Lamarque P, Reuter K, Smith M, van Ham C, Weisser WW, Le Roux X. Nature-based solutions: new influence for environmental management and research in Europe. GAIA 2015;24(4):243–8.

Elliott JD, Deryng C, Muller K, Frieler M, Konzmann D, Gerten M, Glotter M, Florke Y, Wada N, Best S. Constraints and potentials of future irrigation water availability on agricultural production under climate change. Proc Natl Acad Sci 2014;111:3239–44.

Environment UN. Global Environment outlook—GEO-6: Healthy planet. Nairobi: Healthy People; 2019.

Garrick D, Hall JW. Water security and society: risks, metrics, and pathways. Annu Rev Env Resour 2014;39:611–39.

Gleeson T, Wada Y, Bierkens MFP, van Beek LPH. Water balance of global aquifers revealed by groundwater footprint. Nature 2012;488:197–200. https://doi.org/10.1038/nature11295.

Grey D, Sadoff CW. Sink or swim? Water security for growth and development. Water Policy 2007;9:545–71.

Grey D, Garrick D, Blackmore D, Kelman J, Muller M, Sadoff C. Water security in one blue planet: twenty-first century policy challenges for science. Phil Trans R Soc A 2013;371:20120406.

Griscom BW, et al. Natural climate solutions. 114. Proc Natl Acad Sci; 2017. p. 11645–50.

Gunnell K, et al. Evaluating natural infrastructure for flood management within the watersheds of selected global cities. Sci Total Environ 2019;670:411–24.

Guppy L, Anderson K. Water crisis report. Hamilton, Canada: United Nations University Institute for Water, Environment, and health; 2017.

Hallegate S. Strategies to adapt to an uncertain climate change. Glob Environ Chang 2009;19:240–7.

Harrison IJ, Green PA, Farrell TA, Juffe-Bignoli D, Saenz L, Vorosmarty CJ. Protected areas and freshwater provisioning: a global assessment of freshwater provision, threats and management strategies to support human water security. Aquat Conserv Mar Freshwat Ecosyst 2016;26(Suppl. 1):103–20.

IPBES. IPBES global assessment on biodiversity and ecosystem services. Draft 2019;31:2019. May.

Jiménez Cisneros BE, Oki T, Arnell NW, Benito G, Cogley JG, Döll P, Jiang T, Mwakalila SS. Freshwater resources. In: Field CB, Barros VR, Dokken DJ, Mach KJ, Mastrandrea MD, Bilir TE, White LL, editors. Climate Change 2014: Impacts, Adaptation, and Vulnerability. Part A: Global and Sectoral Aspects. Contribution of Working Group II to the Fifth Assessment Report of the Intergovernmental Panel on Climate Change. Cambridge, United Kingdom and New York, NY, USA: Cambridge University Press; 2014. p. 229–69.

Jones HP, Hole DG, Zavaleta ES. Harnessing nature to help people adapt to climate change. Nat Clim Chang 2012;2:504–9.

Juuti PS, Katko TS, Vuorinen HS. Environmental history of water—global views on community water supply and sanitation. London, UK: IWA Publishing; 2007.

Kalantari Z, et al. Nature-based solutions for flood-drought risk mitigation in vulnerable urbanizing parts of East-Africa. Curr Opin Environ Health 2018;5:73–8.

Maes J, Jacobs S. Nature-based solutions for Europe's sustainable development. Conserv Lett 2017;10(1):121–4. https://doi.org/10.1111/conl.12216.

Makarigakis AK, Jimenez-Cisneros BE. UNESCO's contribution to face global water challenges. Water 2019;11:388–404.

Marsh GP. Man and nature: Physical geography as modified by human action. New York, New York: Charles Scribner; 1864.

Matthews JH, Dela Cruz E. A practitioner's guide to nature-based solutions. In: Asian Development Bank, technical assistance Consultant's report. Project number: 50159-001. October 2020; 2020.

Matthews JH, Matthews N, Simmons E, Vigerstol K. Wellspring: sources water resilience and climate adaptation. Arlington, VA: The Nature Conservancy; 2019. https://www.nature.org/content/dam/tnc/nature/en/documents/Wellspring_FULL_Report_2019.pdf.

McCartney M, Dalton J. Built or natural infrastructure: a false dichotomy. In: Thrive blog. CGIAR research program on water, land and ecosystems (WLE); 2015. website. wle.cgiar.org/thrive/2015/03/05/built-or-natural-infrastructure-false-dichotomy.

McDonald RI, Weber KF, Padowski J, Boucher T, Shemie D. Estimating watershed degradation over the last century and its impact on water-treatment costs for the world's large cities. Proc Natl Acad Sci 2016;113:9117–22.

MEA (Millenium Ecosystem Assessment). Ecosystems and human well-being. Washington, D.C.: Island Press; 2005.

Melosi MV. The sanitary city. The Johns Hopkins University Press; 2000.

Milly PCD, Betancourt J, Falkenmark M, Hirsch RM, Kundzewicz ZW, Lettenmaier DP, R. J. Stouff er. Stationarity is dead: whither water management? Science 2008;319(5863):573–4.

Montgomery DR. Dirt: the erosion of civilizations. Berkeley, Los Angeles, London: University of California Press; 2007.

Muller M, Biswas A, Martin-Hurtado R, Tortajada C. Built infrastructure is essential. Science 2015;349(6248):585–6.

Nesshover C, Assmuth T, Irvine KN, Rusch GM, Waylen KA, Delbaere B, Haase D, Jones-Walters L, Keune H, Kovacs S, Krauze K, Kulvik M, Rey F, van Dijk J, Vistad OI, Wilkinson ME, Wittmer H. The science, policy and practice of nature-based solutions: an interdisciplinary perspective. Sci Total Environ 2017;579:1215–27.

Nyugen AR, Hens L. Human ecology of climate change hazards in Vietnam: risks for nature and humans in lowland and upland areas. Cham, Switzerland: Springer Climate; 2019.

Ochoa-Tocachi BF, Bardales JD, Antiporta J, Perez K, Acosta L, Mao F, Zulkafli Z, Gil-Rios J, Angulo O, Grainger S, Gammie G, DeBievre B, Buytaert W. Potential contributions from pre-Inca infiltration infrastructure to Andean water security. Nat Sustain 2019;2:584–93.

OECD (Organisation for Economic Cooperation and Development). Environmental Outlook. Paris: OECD Publ; 2012.

OECD (Organisation for Economic Cooperation and Development). Water security for better lives. Paris: OECD Publ; 2013a.

OECD (Organisation for Economic Cooperation and Development). Water and climate change adaptation: policies to navigate uncharted waters. Organization for Economic Cooperation and Development Paris: OECD Publ; 2013b.

Palmer MA, Liu J, Matthews JH, Mumba M, D'Odorico P. Manage water in a green way. Science 2015;349(6248):584–5.

Plato. Timaeus and Critias. In: Penguin Classics. Penguin Random House; 2008. Annotated Edition.

Poff NL, Brown CM, Grantham TE, Matthews JH, Palmer MA, Spence CM, Wilby RL, Haasnoot M, Mendoza GG, Dominique KC, Baeza A. Sustainable water management under future uncertainty with eco-engineering decision scaling. Nat Clim Chang 2016;6(1):25–34.

Richey AS, Thomas BF, Lo M, Reager JT, Famiglietti JS, Voss K, Swenson S, Rodell M. Quantifying renewable groundwater stress with GRACE. Water Resour Res 2015;51(7):5217–38.

Richter BD, et al. Lost in development's shadow: the downstream human consequences of dams. Water Altern 2010;3:14–42.

Rockström J, Steffen W, Noone K, Persson Å, Chapin III FS, Lambin E, Lenton TM, Scheffer M, Folke C, Schellnhuber H, Nykvist B, De Wit CA, Hughes T, van der Leeuw S, Rodhe H, Sörlin S, Snyder PK, Costanza R, Svedin U, Falkenmark M, Karlberg L, Corell RW, Fabry VJ, Hansen J, Walker B, Liverman D, Richardson K, Crutzen P, Foley J. Planetary boundaries: exploring the safe operating space for humanity. Ecol Soc 2009;14(2):32.

Rosenzweig CJ, Elliott D, Deryng AC, Ruane C, Muller A, Arneth KJ, Boote C, Folberth M, Glotter N, Khabarov K. Assessing agricultural risks of climate change in the 21st century in a global gridded crop model intercomparison. Proc Natl Acad Sci 2014;111:3268–73.

Sebesvari Z, et al. Opportunities for considering green infrastructure and ecosystems in the Sendai framework monitor. Prog Disaster Sci 2019;2:100021.

Seddon S, Turner B, Berry P, Chausson A, Girardin CAJ. Grounding nature-based climate solutions in sound biodiversity science. Nat Clim Chang 2019;9:82–7.

Smith DM, Matthews JH, Bharati L, Borgomeo E, McCartney M, Mauroner A, Nicol A, Rodriguez D, Sadoff C, Suhardiman D, Timboe I, Amarnath G, Anish N. Adaptation's thirst: accelerating the convergence of water and climate action. In: Background paper prepared for the 2019 report of the global commission on adaptation. Rotterdam and Washington, D.C; 2019, www.gca.org.

Spinoni J, Naumann G, Vogt JV. Pan-European seasonal trends and recent changes of drought frequency and severity. Global Planet Change 2017;148:113–30.

Srinivasan V, Lambin EF, Gorelick SM, Thompson BH, Rozelle S. The nature and causes of the global water crisis: syndromes from a meta-analysis of coupled human-water studies. Water Resour Res 2012;48, W10516.

Srinivasan V, Konar M, Sivapalan M. A dynamic framework for water security. Water Secur 2017;1:12–20.

Staddon C, Scott CA. Putting water security to work: addressing global challenges. Water Int 2018;43(8):1017–25.

Steffen W, Richardson K, Rockstrom J, Cornell SE, Fetzer I, Bennett EM, Biggs R, Carpenter SR, de Vries W, de Wit CA, Folke C, Gerten D, Heinke J, Mace GM, Persson LM, Ramanathan V, Reyers B, Sorlin S. Planetary boundaries: guiding human development on a changing planet. Science 2015;347:1259855.

Sun S, Wang Y, Engel BA, Wu P. Effects of virtual water flow on regional water resources stress: a case study of grain in China. Sci Total Environ 2016;550:871–9.

UN-Water. Water security poster, https://www.unwater.org/publications/water-security-infographic/; 2013. accessed on 8–25-19.

UNHLP (United Nations High Level Panel on Water). Water Infrastructure and Investment, Sustainable Development. accessed on September 5, 2019 https://sustainabledevelopment.un.org/content/documents/hlpwater/08-WaterInfrastInvest.pdf; 2015.

UNISDR (United Nations Office for Disaster Risk Reduction). The human cost of weather related disasters: 1995–2015. Geneva, Switzerland: United Nations Office for Disaster Risk Reduction and Centre for Research on the Epidemiology of Disasters (CRED); 2016.

Vorosmarty CJ, Meybeck M, Pastore CL. Impair-then-repair: a brief history and global-scale hypothesis regarding human-water interactions in the Anthropocene. Daedalus 2015;144(3):94–109.

Vörösmarty CJ, et al. Ecosystem-based water security and the sustainable development goals (SDGs). Ecohydrol Hydrobiol 2018;18:317–33.

Wada Y, Florke M, Hanasaki N, Eisner S, Fischer G, Tramberend S, Satoh Y, van Vliet MTH, Yillia P, Ringler C, Burek P, Wiberg D. Modeling global water use for the 21st century: the water futures and solutions (WFaS) initiative and its approaches. Geosci Model Dev 2016;9:175–222.

WAP (United Nations World Water Assessment Programme). The united national world water development report 2017: wastewater: the untapped resource. Paris: UNESCO; 2017.

Williams AP, Seager R, Abatzoglou JT, Cook BI, Smerdon JE, Cook ER. Contributions of anthropogenic warming to California drought during 2012–2014. Geophys Res Lett 2015;42:6819–28.

Wood SLR, et al. Distilling the role of ecosystem services in the sustainable development goals. Ecosyst Serv 2018;29:70–82.

World Bank. High and dry: climate change, water and the economy. Washington, D.D: World Bank; 2016.

World Economic Forum. Water security: the water-food- energy-climate nexus. Washington, DC: Island Press; 2011.

World Economic Forum, editor. The global risks report 2019. 14th ed. Switzerland: World Economic Forum. Geneva; 2019.

WWAP (United Nations World Water Assessment Programme). The united National World Water Development Report 2012: managing water under uncertainty and risk. Paris: UNESCO; 2012.

WWAP (United Nations World Water Assessment Programme). The united National World Water Development Report 2015: Water for a sustainable world. Paris: UNESCO; 2015.

WWAP (United Nations World Water Assessment Programme). The United National world water development report 2016: water and jobs. Paris: UNESCO; 2016.

WWAP (United Nations World Water Assessment Programme). The united National World Water Development Report 2018: Nature-based solutions for water. Paris: UNESCO; 2018.

CHAPTER 2

History and development of nature-based solutions: Concepts and practice

Jan Cassin

Forest Trends, Washington, DC, United States

Cleary the problem of man and nature is not one of providing a decorative background for the human play, or even ameliorating the grim city: it is the necessity of sustaining nature as source of life, milieu, teacher, sanctum, challenge and, most of all, of rediscovering nature's corollary of the unknown in the self, the source of meaning.

Design with Nature, McHarg (1969).

Introduction

The term *nature-based solutions* (NBS) entered the lexicon in the late 2000s as a useful "umbrella" concept for approaches that rely on natural systems or natural processes to address societal challenges (Cohen-Shacham et al., 2016; Nesshover et al., 2017). Over the past 5–10 years, interest in NBS for water security, disaster risk reduction, climate mitigation and adaptation, resilient cities, health and sustainable development has grown exponentially (Fisher et al., 2019; Kabisch et al., 2017; Renaud et al., 2013; Seddon et al., 2020; UNDRR, 2020; WWAP, 2018). There are now NBS initiatives within individual businesses and business consortia (Ofosu-Amaah et al., 2021), individual municipalities and utilities as well as national governmental agencies (Cassin, 2021; Coxon et al., 2021; USACE, 2013), supra-national governments (EU, EC), nongovernmental organizations, academic departments, and international development institutions (Matthews and de la Cruz, 2020). Recognizing that both theory and practice are rapidly evolving, this chapter provides a brief history and overview of NBS and related concepts, particularly with respect to water security, discusses the utility of NBS as a unifying concept, and explores some of the ways that NBS is evolving.

The long (mostly hidden) history of nature-based solutions

As others have pointed out, while "nature-based solutions" is a new term, the concepts and practices embodied in NBS are not necessarily new (Cohen-Shacham et al., 2016; Matthews and de la Cruz, 2020; Nesshover et al., 2017). Yet, there has been limited treatment of the

Nature-Based Solutions and Water Security. https://doi.org/10.1016/B978-0-12-819871-1.00018-X

19

antecedents to the current expression of these ideas. In fact, many Indigenous peoples have rich traditions of land and water management based on nature (Berkes, 2018; Cassin and Ochoa-Tocachi, 2021). Revitalization of ancestral technologies and biocultural restoration initiatives are bringing these old nature-based practices into new prominence in places such as Peru and Hawaii (Bremer et al., 2021; Ochoa-Tocachi et al., 2019). Traditional agro-ecological systems and newer manifestations such as regenerative agriculture rely on and support natural processes to improve yields, reduce pests and diseases, improve resilience to drought, minimize soil erosion, maintain biodiversity, protect water quality, and regulate water flows (Altieri, 1995; Perfecto and Vandermeer, 2010).

Urban planning and landscape architecture, at least since the 18th century, have included traditions that build on nature or incorporate nature into the design of more healthful and sustainable cities. The Garden Cities movement in the United Kingdom and the City Beautiful Movement in the United States are just two examples (Hardy, 1991). Frederick Law Olmsted's emerald necklace in Boston, a 1100 acre chain of "blue-green" infrastructure of parks and waterways was designed to provide a sanctuary from the pressures of city life by bringing nature into the city: "the occasional contemplation of natural scenes of an impressive character, particularly if this contemplation occurs in connection with relief from ordinary cares, change of air and change of habits, is favourable to the health and vigor of man" (1865; in Olmsted, 1952). The seminal works of Ian McHarg contain some of the best expressions of "designing with nature" that preceded and foreshadowed NBS: "The natural restraints to flooding and drought are mainly the presence and distribution of vegetation, particularly, on the uplands and their steep slopes. Vegetation absorbs and utilizes considerable quantities of water; the surface roots, trunks of trees, stems of shrubs and plants, the litter of forest floor mechanically retard the movement of water, facilitating percolation, increasing evaporation opportunity....In short, diminishing the frequency and intensity of oscillation between flood and drought" (McHarg, 1964).

While current conceptions of NBS have been influenced by and can still learn from these related traditions, the emerging conceptual framework around NBS has particular strengths for addressing linked societal challenges in the 21st century.

The emergence of nature-based solutions

In the 1990s and early 2000s the limitations of traditional conservation approaches became evident, along with the growing concern that human development was fast approaching or broaching ecological limits and planetary boundaries (Rockström et al., 2009; Steffen et al., 2015). The economic, environmental, and social impacts and limitations of technological solutions, including the dominant approach to managing water resources with gray infrastructure, became ever more apparent in the last decades of the 20th century (Palmer et al., 2015; Rosenbloom, 2018). The state of dilapidation of existing water infrastructure in developed countries and its vulnerability to threats, including deteriorating source water quality and climate change, raised greater awareness of the need for new approaches (Cassin, 2021; NIAC, 2016). The 1990s and 2000s saw an explosion of interest in nature-based "green infrastructure," low impact development and sustainable urban drainage systems to address increasingly devastating and expensive urban flooding from inadequate stormwater management infrastructure.

At the same time, major global policy initiatives around biodiversity, climate change, sustainable development, and disaster risk management began to emphasize the role that ecosystems play in human well-being and the need to incorporate the value of ecosystems in development and economic decision-making (Costanza and Daly, 1992; MEA, 2005). Ecosystem-based management, the ecosystem approach, ecosystem-based adaptation and disaster risk management, environmental flows, and ecological engineering concepts gained prominence as more proactive, science-based, sustainable approaches for conserving biodiversity, managing terrestrial and coastal/marine resources, addressing climate impacts, and restoring the health of managed river systems (Cohen-Shacham et al., 2019; Poff et al., 1997). From this period two developments in particular contributed to the emergence of the NBS concept and are closely related to it: natural capital and ecosystem services (see Table 2.1).

As the services that people receive from ecosystems became more widely appreciated (Daily, 1997), it also became clear that natural systems are a distinct type of capital within the mix of human, social, and financial capital that underpin sustainable development (Costanza and Daly, 1992). National governments began to embrace natural capital and natural capital accounting based not only on older ideas of resource values but also on the basis of the flow of services natural capital provides to people. The government of the United Kingdom defines natural capital as: "The elements of nature that directly and indirectly produce value or benefits to people, including ecosystems, species, freshwater, land, minerals, the air and oceans, *as well as natural processes and functions*" (emphasis added) (UK Natural Capital Committee, 2014). In Canada, valuing and managing these stocks as a new kind of asset is being embraced by municipalities and municipal natural assets are defined as: "The stocks of natural resources or ecosystems that contribute to the provision of one or more services required for the health, well-being, and long-term sustainability of a community and its residents" (MNAI, 2019).

Viewing natural systems as a type of asset or capital also aligned with emerging ideas—particularly in the water sector—of "green" or "natural" infrastructure. Natural systems provide many of the same services that "gray" or built infrastructure systems provide, regulating water flows, filtering and purifying water, and protecting against hazards such as droughts and floods, often more cost-effectively and with significant co-benefits (Browder et al., 2019; Ozment et al., 2016).

Table 2.1: Nature-based solutions and related concepts.

Term/Concept	Source	Definition and description
		Nature-based solutions
Nature-based Solutions	UN World Water Assessment Program 2018	NBS are inspired and supported by nature and use, or mimic, natural processes to contribute to the improved management of water. An NBS can involve conserving or rehabilitating natural ecosystems and/or the enhancement or creation of natural processes in modified or artificial ecosystems. They can be applied at microscales (e.g., a dry toilet) or macroscales (e.g., watershed restoration). (WWAP, 2018)
Nature-based Solutions	IUCN	Actions to protect, sustainably manage, and restore natural or modified ecosystems that address societal challenges effectively and adaptively, simultaneously providing human well-being and biodiversity benefits. IUCN's NBS concept includes a broad range of solutions or approaches for water security as well as challenges such as conservation, climate adaptation, disaster risk reduction, and food security. IUCN's definition considers green infrastructure and natural infrastructure as subsets of NBS—the NBS that are explicitly delivering the same services as built or engineered infrastructure. (Cohen-Shacham et al., 2016)
Nature-based Solutions	Climate Bonds Initiative	NBS are the explicit, planned, intentional use of ecosystems to meet human needs; "nature-based infrastructure" uses ecological assets or "ecosystem-based features, processes, and functions" to address water needs. Under the umbrella of NBS, they distinguish: — natural features (protected or restored ecosystems); and — nature-based features (or features based on nature) that use or mimic the characteristics of ecosystems and natural processes but use human design, engineering, and construction to deliver or enhance a (usually single) service In distinguishing natural features from nature-based features, CBI follows US Army Corps of Engineers practice, which is also similar to the MNAI distinction of natural assets, enhanced assets, and engineered assets. (Climate Bonds Initiative, 2018)
Nature-based Solutions	European Commission	Living solutions inspired by, continuously supported by, and using nature designed to address various societal challenges in a resource-efficient and adaptable manner and to provide economic, social, and environmental benefits simultaneously. (Maes and Jacobs, 2017)

Nature-based Infrastructure	US Army Corps of Engineers	The US Army Corps of Engineers practice distinguishes between natural features and nature-based features, both of which deliver services that provide specific benefits to people. Natural features are: "created through the action of physical, geological, biological and chemical processes over time." Nature-based features are: "created by human design, engineering, and construction (in concert with natural processes) to provide specific services such as coastal risk reduction and other ecosystem services (e.g., habitat for fish and wildlife)." Nature-based features typically require human intervention to sustain the functions and services for which they were built. (Bridges, 2015)
Natural Infrastructure and Solutions Based on Nature	Natural Infrastructure for Water Security project, Peru	Natural infrastructure is an interconnected system of ecosystem components (water, soil, subsoil, vegetation, biodiversity) that performs one or more functions that provide services or benefits to people, such as hydrological regulation, carbon sequestration, flood mitigation, climate regulation, or erosion control. Solutions based on nature ensure or recover the functions of the natural infrastructure. Thus, a landscape (with its topography, soils, or climate) is composed of various ecosystems or natural infrastructure (such as a wetland, native forests, or an agricultural field), which host ecosystem functions and processes (such as evapotranspiration, water infiltration into soils, or primary biomass production). The functions will create ecosystem services (e.g., conservation of dry season river flows, reduction of soil erosion, or food production), which will generate benefits to society (e.g., nutrition, water security, or income), valued in hydrological, economic, social, and cultural terms.

Natural capital/natural assets

Natural Assets	Municipal Natural Asset Initiative (MNAI)	The stocks of natural resources or ecosystems that contribute to the provision of one or more services required for the health, well-being, and long-term sustainability of a community and its residents. Working definition of the MNAI. MNAI considers the general concept of green infrastructure as including natural assets (wetlands, forests, aquifers, lakes) as well as enhanced or engineered assets (raid gardens, bioswales, green roofs). MNAI, therefore, deals with the natural assets and not the enhanced or engineered assets. (MNAI, 2019)
Natural Capital/Natural Assets	World Forum on Natural Capital	The world's stocks of natural assets, which include geology, soil, air, water, and all living things. It is from this natural capital that humans derive a wide range of services, often called ecosystem services, which make human life possible. The focus on natural capital approaches has been on making the value of this capital more visible in decision-making, and on maintaining or restoring these stocks to ensure the flow of services is maintained. https://naturalcapitalforum.com/about/

Continued

Table 2.1: Nature-based solutions and related concepts—cont'd

Term/Concept	Source	Definition and description
Natural Capital	Natural Capital Finance Alliance	Natural capital comprises Earth's natural assets (soil, air, water, flora, and fauna), and the ecosystem services resulting from them, which make human life possible. Ecosystem goods and services from natural capital underpin productivity and the global economy. They provide services worth trillions of US dollars per year in equivalent terms and constitute food, fiber, water, health, energy, climate security, and other essential services for everyone. Neither these services nor the stock of natural capital that provides them are adequately valued in terms comparable to manufactured and financial capital. http://www.naturalcapitalfinancealliance.org/declaration-full-text
Natural Capital	Natural Capital Coalition	The stock of renewable and nonrenewable resources (e.g., plants, animals, air, water, soils, minerals) that combine to yield a flow of benefits to people. http://naturalcapitalcoalition.org/
Green infrastructure		
Green Infrastructure	European Commission	A strategically planned network of natural and semi-natural areas with other environmental features designed and managed to deliver a wide range of ecosystem services. http://ec.europa.eu/environment/nature/ecosystems/index_en.htm
Green Infrastructure	Benedict and McMahon (2000)	An interconnected network of protected land and water that supports native species, maintains natural ecological processes, sustains air and water resources, and contributes to the health and quality of life for America's communities and people. This conception of green infrastructure is similar to the narrower sense in which green infrastructure is used (often in the US and EU) to denote urban green infrastructure that is designed to address challenges around urban flooding and drainage, heat island, air quality, and/or recreation opportunities in highly developed urban areas.
Green-Blue Infrastructure and Natural Infrastructure	UNISDR (2017)	A strategically planned network of natural and semi-natural areas with other environmental features designed and managed to deliver a wide range of ecosystem services such as water purification, air quality, space for recreation, climate mitigation and adaptation, and management of wet weather impacts that provides many community benefits.
Ecosystem-based adaptation, disaster risk reduction, ecological engineering		
Ecosystem Services Approach	Martin-Ortega et al. (2015), Beaumont et al. (2017)	A way of understanding the complex relationships between nature and humans to support decision-making, with the aim of reversing the declining status of ecosystems and ensuring the sustainable use/management/conservation/ of resources. Has the following elements: (1) a focus on the status of ecosystems and the recognition of effects on human well-being; (2) understanding of the biophysical underpinning of ecosystems in terms of service delivery; (3) integration of natural and social sciences and other strands of knowledge for a comprehensive understanding of the service delivery process; and (4) incorporation of ecosystem services into environmental policy and management decisions.

Ecosystem-based Adaptation (EBA)	Convention on Biological Diversity	Ecosystem-based adaptation is the use of biodiversity and ecosystem services as part of an overall adaptation strategy to help people adapt to the adverse effects of climate change. (CBD, 2009)
Ecosystem-based Disaster Risk Reduction (ECO-DRR)	UNDRR	The sustainable management, conservation, and restoration of ecosystems to reduce disaster risk, with the aim to achieve sustainable and resilient development. (Estrella and Saalismaa, 2013)
Ecosystem-based Mitigation (EBM)	Sudmeier-Rieux et al. (2019)	EBM is the use of ecosystems for their carbon storage and sequestration service to aid climate change mitigation. (Sudmeier-Rieux et al., 2019)
Ecological Engineering	Mitsch (2012)	The design of sustainable ecosystems that integrate human society with its natural environment for the benefit of both (Mitsch, 2012). An example is "Building with Nature," a comprehensive engineering approach that seeks to enhance the use of natural ecological processes to achieve efficient and sustainable hydraulic infrastructural designs.

Nature-based solutions for water security: Reimagining infrastructure

NBS represent a new family of approaches that encompass many of the elements of existing approaches, capturing related concepts under a convenient unifying umbrella framework (Table 2.1) (Cohen-Shacham et al., 2016; Eggermont et al., 2015). As defined by the United Nations, NBS for water are "inspired and supported by nature and use, or mimic, natural processes to contribute to the improved management of water. An NBS can involve conserving or rehabilitating natural ecosystems and/or the enhancement or creation of natural processes in modified or artificial ecosystems. They can be applied at micro- (e.g., riparian buffers) or macro- (e.g., watershed restoration) scales" (UN-Water, 2018).

Although NBS can address a range of societal challenges, a critical element for managing resilient water systems is that NBS provide for an explicit focus on including the services from ecological processes and functions within infrastructure management systems (Matthews and de la Cruz, 2020). This can include recognizing and protecting existing natural infrastructure that is already providing an infrastructure service, such as a forested watershed that protects drinking water quality (Cassin, 2021). For example, in Peru in 2018, leaders of 23 water utilities signed the Piuray Declaration, committing to protecting the natural environments in source watersheds as critical elements in the country's water infrastructure (https://www.forest-trends. org/pressroom/at-perus-inaugural-national-water-summit-dozens-of-utilities-endorse-piuray- declaration-on-natural-infrastructure/). NBS can also use combined built and natural systems— hybrid systems that make use of natural processes to deliver services, such as the pre-Incan managed aquifer recharge *amuna-mamanteo* system in Peru (Ochoa-Tocachi et al., 2019).

Nature-based solutions as a unifying concept

While NBS is not an entirely new concept and encompasses many related concepts, it has become a useful term for a family of practices that share common elements. These include practices that recognize ecosystems themselves (natural assets, natural features, natural infrastructure) and practices (NBS) that are based on or use natural features, processes, and functions to meet human needs. A continuum of practices and assets exists based on the degree of management or intervention in natural processes, including natural ecosystems not actively or minimally managed, enhanced or actively managed assets, and hybrid natural- engineered assets (e.g., urban green infrastructure such as green roofs) (Fig. 2.1). These practices all aim to protect, maintain, and/or manage the ability of ecosystems to provide the ecosystem services that help address societal challenges.

While there is a continuum from minimal to intensive intervention or management, NBS are distinguished from gray infrastructure by the reliance on ecosystem structures, processes, and functions to deliver outcomes. For example, using native vegetation and soils to enhance infiltration and groundwater recharge to store water for later use would be considered an

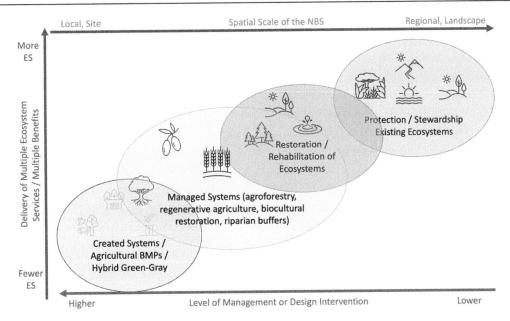

Fig. 2.1
A continuum of nature-based solutions.

NBS; while using a storage reservoir to store water for later use or using injection wells to artificially recharge groundwater do not use ecological functions or processes and would not be considered an NBS.

NBS may be designed for a specific priority benefit (e.g., improving water quality) but are by nature multifunctional and can simultaneously address multiple societal goals (e.g., biodiversity conservation, climate mitigation, food security, health). Finally, emerging best practice requires that the design and implementation of NBS need to consider the target benefit, mitigate trade-offs, and incorporate principles of equity, inclusion, sustainability, and resilience (Cassin and Locatelli, 2020; IUCN, 2020).

Evolving theory and practice

The theory and practice of NBS have been evolving rapidly over the past few years as NBS are explored or implemented in different sectors and for different purposes, such as water security, climate mitigation and adaptation, urban resilience, natural hazard mitigation, and sustainable development (Albert et al., 2020; Almenar et al., 2021; Bremer et al., 2021; Dushkova and Haase, 2020; Matthews et al., 2019; Seddon et al., 2020; Smith et al., 2019; WWAP (United Nations World Water Assessment Programme), 2018). Ongoing developments in two areas will be important for refining concepts, improving practice and evidence, and scaling implementation: standards and principles; and common typologies and frameworks.

Standards and principles

While NBS is useful as an umbrella concept, the size of the umbrella can create some conceptual fuzziness and increase the risk that "everything" becomes an NBS. At the same time, NBS can involve trade-offs across benefits and beneficiaries, result in unforeseen negative impacts, or be designed and implemented in a way that reduces their function and viability. NBS can sometimes result in "disservices," for example, urban trees by contributing pollen and volatile organic compounds can contribute to increased incidence of allergies, asthma, and respiratory health issues (Eisenman et al., 2019). If potential negative impacts are not addressed, the overall result of NBS may be negative. For example, much attention has been paid to planting trees to mitigate climate change, but large-scale tree plantations or afforestation in areas not previously forested can have significant negative effects on water yield (Alvarez-Garreton et al., 2019).

The need for more rigorous definitions of NBS and establishment of criteria or standards for what can be considered good practice for NBS has driven the development of principles and standards to guide design and implementation. The International Union for the Conservation of Nature (IUCN) has developed a global standard to help increase the scale and the impact of NBS, prevent misuse of the concept or negative outcomes, and provide guidance for funders, practitioners, and policy makers on evaluating the effectiveness of NBS (Box 2.1). Developed through a global consultation and review process, the standard is designed to provide clear parameters for defining NBS and a framework for benchmarking progress (IUCN, 2020).

As NBS are designed and implemented by a variety of actors and in a range of contexts, other principles and criteria are being developed to address specific applications or sectoral needs. Examples include principles for applying NBS in landscape planning (Albert et al., 2020) and

---

Box 2.1   IUCN global standard for NBS

IUCNs Global Standard defines eight criteria to guide the design, verification, and scaling up of NBS.
Criterion 1. NBS effectively address societal challenges.
Criterion 2. Design of NBS is informed by scale.
Criterion 3. NBS result in a net gain to biodiversity and ecosystem integrity.
Criterion 4. NBS are economically viable.
Criterion 5. NBS are based on inclusive, transparent, and empowering governance processes.
Criterion 6. NBS equitably balance trade-offs between achievement of their primary goal(s) and the continued provision of multiple benefits.
Criterion 7. NBS are managed adaptively, based on evidence.
Criterion 8. NBS are sustainable and mainstreamed within an appropriate jurisdictional context.
The standard provides indicators for evaluating these criteria and guidance for improving practice.

guidance for operationalizing NBS within the set of tools used in multilateral development (Matthews and de la Cruz, 2020). One place-based example is the *Guide for the Evaluation of Natural Infrastructure Interventions for Water Security: Effectiveness, Equity and Sustainability Scales* (Guía para la Evaluación de Intervenciones en Infraestructura Natural para la Seguridad Hídrica: Escala de Efectividad, Equidad y Sostenibilidad) developed specifically to guide natural infrastructure projects in Peru (Cassin and Locatelli, 2020).

The approach taken by the Peru guide is to evaluate NBS projects and initiatives by defining a scale from "business as usual" to "transformative practice" along three axes or criteria: effectiveness, equitability, and sustainability. Further, subdimensions are defined for each of the axes (Box 2.2). By defining steps in progressing from current practice to "best practice" NBS, the guide can be applied at different stages: from projects that are under design to those that have concluded their intervention. This allows assessing and documenting the extent to which the intervention addresses each of the subdimensions and recommendations for what can be considered in order to move into good or transformative practice. For each of the three criteria, the guide identifies attributes that describe what basic to transformative practices look like, provides questions to evaluate where in the scale a particular project fits with respect to each of the three criteria, and offers suggestions for verifying the evaluation.

Box 2.2   Guide for effective, equitable, and sustainable nature-based solutions

This approach is based on three dimensions and eight broad principles.
Principles

- **Focused on results.** Natural infrastructure provides water security and social, economic, and environmental benefits using clearly defined objectives, targets, and indicators.
- **Ecosystem-based.** Natural infrastructure uses ecosystem functions through protection, restoration, and enhancement for human well-being. It includes a full range of ecosystem types, from natural to modified.
- **Place-based.** Natural infrastructure is sensitive to and determined by site-specific natural and cultural contexts.
- **Sustainable and resilient.** Natural infrastructure maintains ecological integrity and biological and cultural diversity to enable long-term sustainable use and management of land and maintain the capacity of ecosystems to evolve over time; maintains and/or enhances the resilience of natural and social systems.
- **Integrated management for water security.** Natural infrastructure can be implemented alone but is more often part of an integrated approach in combination with "gray" or engineered approaches to water security.
- **Evidence-based.** Natural infrastructure encompasses evidence-based design using local, traditional, and scientific knowledge.

Continued

Box 2.2   Guide for effective, equitable, and sustainable nature-based solutions—cont'd

– **Gender equity.** Natural infrastructure delivers benefits in a fair and equitable manner, considering gender, class, and generational equity and explicitly assessing impacts and trade-offs. Includes full participation of women (and men) in water resources management (including natural infrastructure solutions) so that the different roles and knowledge of women and men are equally considered in decisions.
– **Participatory and transparent.** Promotes transparency and uses participatory processes in planning, design, and implementation.

Three dimensions

Effectiveness

The intervention effectiveness dimension allows an assessment, for example, of the extent to which improvements in the reliability of adequate water supplies of acceptable quality are achieved, while supporting nonhydrological benefits, such as contributions to livelihoods, through appropriately designed interventions and under adaptive management schemes and trade-offs. This dimension addresses the following subdimensions:

– hydrological objectives;
– co-benefit objectives;
– evidence-based results;
– monitoring and learning; and.
– negative impacts and trade-offs.

Equity

The aim is to ensure that interventions and their benefits include all stakeholders on an equal opportunity basis, addressing the needs of vulnerable groups and revaluing cultural and knowledge diversity. Equity has procedural, recognitional, and distributional aspects. The subdimensions that are part of the equity dimension are:

– stakeholder inclusion and engagement;
– gender equality;
– ethnic and cultural diversity;
– cultural context and local knowledge; and.
– benefit sharing.

Sustainability

Sustainability addresses those elements associated with the economic, social, and ecological aspects that will contribute to the viability, resilience, and durability of the NBS intervention and its outcomes. Sustainability has three subdimensions:

– economic sustainability;
– social sustainability; and.
– ecological sustainability.

Source: Cassin and Locatelli (2020)

Typologies and frameworks

As interest in NBS has grown, so have the number of frameworks and typologies to guide research, planning, design, and evaluation of NBS projects. These classify individual NBS practices by a number of attributes, including degree of engineering or intervention involved, number of ecosystem services or co-benefits provided, degree of maximizing a targeted service, scale of the intervention or scale of the impact, societal challenge addressed, type of practice (e.g., protection vs. restoration), ecosystem type, and disservices (Almenar et al., 2021; Cohen-Shacham et al., 2016; Eggermont et al., 2015; Gómez Martín et al., 2020; Trémolet and Karres, 2020). Scaling of NBS will be facilitated by having a common classification that allows decision-makers across different sectors and disciplines—urban planners, water utilities, development banks—to identify shared solutions to common challenges.

While frameworks and classifications need to be simple enough to be operational, there are some gaps in existing typologies that would improve their utility for decision-makers. For the most part, typologies do not integrate NBS and gray infrastructure options in ways that allow for assessing mixed portfolios (but see UNDEP-DHI, 2014). For example, integrated typologies could support planning for diversity and redundancy in reducing flood risks by combining NBS and built infrastructure options that together provide several layers of protection against flood risks.

Typologies could help link, but also clearly differentiate, NBS practices or interventions from supporting policies. For example, implementing most NBS requires a combination of management activities or interventions (e.g., restoring floodplain connectivity, planting riparian buffers) and enabling policies or regulations (e.g., conservation easements, land use zoning) and funding mechanisms (e.g., payments for ecosystem services). While NBS are most easily visualized, and most frequently discussed, as a set of conservation, restoration, land management, or hybrid green-gray practices or actions, these practices, when most effective, are supported by a set of enabling policies, institutional frameworks, governance and funding models, and technical tools and evidence that support the scoping, planning, design, funding, implementation, maintenance, and evaluation of NBS (see for example Cassin, 2021). Refining existing typologies could help decision-makers better understand how, for NBS to deliver desired outcomes, these more easily visualized NBS practices must be embedded in larger policy and governance systems for sustainability, equity, and resilience (Matthews et al., 2019).

NBS provide a new generation of integrative approaches that can fundamentally transform how we view and manage ecohydrological landscapes. As practice continues to evolve, clarity in definitions and criteria for good practice, frameworks, and typologies that support applied research, monitoring, and adaptive management to improve implementation and get to scale with NBS will be critical.

References

Albert C, Brillinger M, Guerrero P, Gottwalkd S, Henze J, Schmidt S, Ott E, Schroter B. Planning nature-based solutions: principles, steps, and insights. Ambio 2020. https://doi.org/10.1007/s13280-020-01365-1.

Almenar JB, Elliot T, Rugani B, Philippe B, Gutierrez TN, Sonnemann G, Geneletti D. Nexus between nature-based solutions, ecosystem services, and urban challenges. Land Use Policy 2021;100. https://doi.org/10.1016/j.landusepol.2020.104898.

Altieri MA. Agroecology: the science of sustainable agriculture. Boulder: Westview Press; 1995.

Alvarez-Garreton C, Lara A, Boisier JP, Galleguillos M. The impacts of native forests and forest plantations on water supply in Chile. Forests 2019;10:473. https://doi.org/10.3390/f10060473.

Beaumont NJ, Mongruel R, Hooper T. Practical application of the ecosystem service approach (ESA): lessons learned and recommendations for the future. Int J Biodivers Sci Manage 2017;13(3):68–78. https://doi.org/10.1080/21513732.2018.1425222.

Benedict MA, McMahon ET. Green infrastructure: smart conservation for the 21st century. Recreation 2000;37:4–7. https://doi.org/10.4135/9781412973816.n70.

Berkes F. Sacred ecology. 4th ed. New York: Routledge; 2018.

Bremer LL, Keeler B, Pascua P, Walker R, Sterling E. Nature-based Solutions, Sustainable Development and Equity. In: Cassin J, Matthews JH, Lopez-Gunn E, editors. Nature-based Solutions and Water Security: An Action Agenda for the 21st Century. Elsevier; 2021.

Bridges T. Use of natural and nature-based features (NNBF) for coastal resilience. US Army Corps for Engineers, Engineer Research and Development Center; 2015. https://erdc-library.erdc.dren.mil/jspui/handle/11681/4769.

Browder G, Ozment S, Bescos I, Gartner T, Lange G-M. Integrating green and gray: creating next generation infrastructure. Washington, DC: World Bank and World Resources Institute; 2019. https://files.wri.org/s3fs-public/integrating-green-gray-executive-summary.pdf.

Cassin J. Nature-based Solutions for source water protection in North America. In: Cassin J, Matthews JH, Lopez-Gunn E, editors. Nature-based solutions and water security: an action Agenda for the 21st Century. Elsevier; 2021.

Cassin J, Locatelli B. Guía para la Evaluación de Intervenciones en Infraestructura Natural para la Seguridad Hídrica. In: Escala de Efectividad, Equidad y Sostenibilidad. Forest Trends; 2020. Available at: https://www.forest-trends.org/publications/guia-ees/.

Cassin J, Ochoa-Tocachi BF. Learning from indigenous and local knowledge: the deep history of nature-based solutions. In: Cassin J, Matthews JH, Lopez-Gunn E, editors. Nature-based Solutions and Water Security: An Action Agenda for the 21st Century. Elsevier; 2021.

CBD. Connecting Biodiversity and Climate Change Mitigation and Adaptation; 2009. Report of the Second Ad Hoc Technical Expert Group on Biodiversity and Climate Change (No. Technical Series No. 41). Secretariat of the Convention of Biological Diversity, Montreal, Quebec, Canada.

Climate Bonds Initiative. Water infrastructure criteria under the climate bonds standard criteria document, https://www.climatebonds.net/standard/water; 2018.

Cohen-Shacham E, Walters G, Janzen C, Maginnis S, editors. Nature-based solutions to address global societal challenges. Gland, Switzerland: IUCN; 2016. https://doi.org/10.2305/IUCN.CH.2016.13.en.

Cohen-Shacham E, Andrade A, Dalton J, Dudley N, Jones M, Kumar C, Maginnis S, Maynard S, Nelson CR, Renaud FG, Welling R, Walters G. Core principles for successfully implementing and upscaling nature-based solutions. Environ Sci Policy 2019;98:20–9.

Costanza R, Daly HE. Natural capital and sustainable development. Conserv Biol 1992;6(1):37–46.

Coxon C, Gammie G, Cassin J. Mobilizing Funding for Nature-based Solutions: Peru's Drinking Water Tariff. In: Cassin J, Matthews JH, Lopez-Gunn E, editors. Nature-based Solutions and Water Security: An Action Agenda for the 21st Century. Elsevier; 2021.

Daily G, editor. Nature's services: societal dependence on natural ecosystems. Washington, DC; Covelo, CA: Island Press; 1997.

Dushkova D, Haase D. Not simply green: nature-based solutions as a concept and practical approach for sustainability studies and planning agendas in cities. Landarzt 2020;9(19). https://doi.org/10.3390/land9010019.

Eggermont H, Balian E, Azevedo JMN, Beumer V, Brodin T, Claudet J, et al. Nature-based solutions: new influence for environmental management and research in Europe. GAIA— Ecol Perspect Sci Soc 2015;24:243–8.

Eisenman TS, Churkina G, Jariwala SP, Kumar P, Lovasi GS, Pataki DE, Weinberger KR, Whitlow TH. Urban trees, air quality, and asthma: an interdisciplinary review. Landsc Urban Plan 2019;187:47–59.

Estrella M, Saalismaa N, editors. Ecosystembased disaster risk reduction (eco-DRR): an overview. In: The role of ecosystems in disaster risk reduction. Shibuya-ku, Tokyo: United Nations University Press; 2013.

Fisher B, Herrera D, Adams D, Fox HE, Gallagher L, Gerkey D, Gill D, Golden CD, Hole D, Johnson K, Mulligan M, Myers SS, Naidoo R, Pfaff A, Rasolofoson R, Selig ER, Tickner D, Treuer T, Ricketts T. Can nature deliver on the sustainable development goals? Lancet Planet Health 2019;3(3):e112–3. https://doi.org/10.1016/S2542-5196(18)30281-X.

Gómez Martín E, Máñez Costa M, Schwerdtner Máñez K. An operationalized classification of nature-based solutions for water-related hazards: from theory to practice. Ecol Econ 2020;167, 106460.

Hardy D. From garden cities to new towns: campaigning for town and country planning, 1899–1946. London: Chapman & Hall; 1991.

IUCN. IUCN global standard for nature-based solutions: a user-friendly framework for the verification, design, and scaling up of NBS. 1st ed. Gland, Switzerland: IUCN; 2020. https://portals.iucn.org/library/sites/library/files/documents/2020-020-En.pdf.

Kabisch N, Korn H, Stadler J, Bonn A, editors. Nature-based solutions to climate change adaptation in urban areas: linkages between science, policy, and practice. Cham, Switzerland: Springer; 2017.

Maes J, Jacobs S. Nature-based solutions for Europe's sustainable development. Conserv Lett 2017;10(1):121–4. https://doi.org/10.1111/conl.12216.

Martin-Ortega J, Ferrier RC, Gordon IJ, Khan S, editors. Water ecosystem services: a global perspective. Cambridge (UK): Cambridge University Press; 2015. 978-1- 107-10037-4.

Matthews J, de la Cruz EC. A Practitioner's guide to nature-based solutions. Protecting and investing in natural Capital in Asia and the Pacific. In: ADB technical assistant Consultant's report, project number: 50159-001. October 2020. Asian Development Bank; 2020.

Matthews JH, Matthews N, Simmons E, Vigerstol K. Wellspring: Sources water resilience and climate adaptation. Arlington, VA: The Nature Conservancy; 2019. https://www.nature.org/content/dam/tnc/nature/en/documents/Wellspring_FULL_Report_2019.pdf.

McHarg I. The place of nature in the city of man. Ann Am Acad Pol Soc Sci 1964;352(1). https://doi.org/10.1177/000271626435200102.

McHarg I. Design with nature. Garden City, New York: Natural History Press; 1969.

MEA (Millenium Ecosystem Assessment). Ecosystems and human well-being. Washington, D.C.: Island Press; 2005.

Mitsch WJ. What is ecological engineering? Ecol Eng 2012;45:5–12.

MNAI (Municipal Natural Assets Initiative). Advancing and integrating municipal natural asset management through asset management planning in Ontario. December 2019. Accessed on December 22, 2020 at https://mnai.ca/media/2020/01/MNAI_MNAPOntario.pdf; 2019.

Nesshover C, Assmuth T, Irvine KN, Rusch GM, Waylen KA, Delbaere B, Haase D, Jones-Walters L, Keune H, Kovacs S, Krauze K, Kulvik M, Rey F, van Dijk J, Vistad OI, Wilkinson ME, Wittmer H. The science, policy and practice of nature-based solutions: an interdisciplinary perspective. Sci Total Environ 2017;579:1215–27.

NIAC (National Infrastructure Advisory Council). Water Sector Resilience Final Report and Recommendations; 2016.

Ochoa-Tocachi BF, Bardales JD, Antiporta J, Perez K, Acosta L, Mao F, Zulkafli Z, Gil-Rios J, Angulo O, Grainger S, Gammie G, DeBievre B, Buytaert W. Potential contributions from pre-Inca infiltration infrastructure to Andean water security. Nature Sustain 2019;2:584–93.

Ofosu-Amaah N, Abell R, Fabre J, Fleming P, Matosich M, Morrison J, Varghese T. Nature-based solutions and corporate water stewardship. In: Cassin J, Matthews JH, Lopez-Gunn E, editors. Nature-based Solutions and Water Security: An Action Agenda for the 21st Century. Elsevier; 2021.

Olmsted FL. The Yosemite Valley and the mariposa big tree: a preliminary report 1865. Landsc Archit 1952;43(1):12–25.

Ozment S, Gartner T, Huber-Stearns H, Difrancesco K, Lichten N, Tognetti S. Protecting drinking water at the source lessons from watershed investment programs in the United States; 2016.

Palmer MA, Liu J, Matthews JH, Mumba M, D'Odorico P. Manage water in a green way. Science 2015;349(6248):584–5.

Perfecto I, Vandermeer J. The agroecological matrix as alternative to the land-sparing/agriculture intensification model. Proc Natl Acad Sci 2010;107:5786–91. https://doi.org/10.1073/pnas.0905455107.

Poff NL, Allan JD, Bain MB, Karr JR, Prestegaard KL, Richter BD, Sparks RE, Stromberg JC. The natural flow regime. Bioscience 1997;47(11):769–84.

Renaud FG, Sudmeier-Rieux K, Estrella M. The role of ecosystems in disaster risk reduction. New York, London, Tokyo: United Nations University Press; 2013.

Rockström J, Steffen W, Noone K, Persson Å, Chapin III FS, Lambin E, Lenton TM, Scheffer M, Folke C, Schellnhuber H, Nykvist B, De Wit CA, Hughes T, van der Leeuw S, Rodhe H, Sörlin S, Snyder PK, Costanza R, Svedin U, Falkenmark M, Karlberg L, Corell RW, Fabry VJ, Hansen J, Walker B, Liverman D, Richardson K, Crutzen P, Foley J. Planetary boundaries: exploring the safe operating space for humanity. Ecol Soc 2009;14(2):32.

Rosenbloom J. Fifty shades of gray infrastructure: land use and the failure to create resilient cities. In: 93 Washington law review; 2018. p. 317. Available at: https://digitalcommons.law.uw.edu/wlr/vol93/iss1/7.

Seddon N, Daniels E, Davis R, Chausson A, Harris R, Hou-Jones X, Hug S, Kapos V, Mace GM, Rizvi AR, Reid H, Roe D, Turner B, Wicander S. Global recognition of the importance of nature-based solutions to the impacts of climate change. Glob Sustain 2020;3:1–12. e15.

Smith DM, Matthews JH, Bharati L, Borgomeo E, McCartney M, Mauroner A, Nicol A, Rodriguez D, Sadoff C, Suhardiman D, Timboe I, Amarnath G, Anish N. Adaptation's thirst: Accelerating the convergence of water and climate action. In: Background paper prepared for the 2019 report of the global commission on adaptation. Rotterdam and Washington, D.C; 2019. , www.gca.org.

Steffen W, Richardson K, Rockstrom J, Cornell SE, Fetzer I, Bennett EM, Biggs R, Carpenter SR, de Vries W, de Wit CA, Folke C, Gerten D, Heinke J, Mace GM, Persson LM, Ramanathan V, Reyers B, Sorlin S. Planetary boundaries: guiding human development on a changing planet. Science 2015;347:1259855.

Sudmeier-Rieux K, Nehren U, Sandholz S, Doswald N. Disasters and ecosystems: resilience in a changing climate—source book. Geneva, Switzerland: UNEP and Cologne: TH Köln—University of Applied Sciences; 2019.

Trémolet S, Karres N. Resilient European Cities: Nature-based Solutions for Clean Water. London, United Kingdom: The Nature Conservancy; 2020.

UK Natural Capital Committee. Towards a framework for defining and measuring changes in natural capital. In: Working Paper 1, March 2014. Natural Capital Committee; 2014. https://assets.publishing.service.gov.uk/government/uploads/system/uploads/attachment_data/file/516946/ncc-working-paper-measuring-framework.pdf.

UNDRR. Ecosystem-based disaster risk reduction: Implementing nature-based solutions for resilience. Bangkok, Thailand: United Nations Office for disaster risk Reduction—Regional Office for Asia and the Pacific; 2020.

UNEP-DHI (United Nations Environment Programme). Green Infrastructure Guide for Water Management: Ecosystem-based management approaches for water-related infrastructure projects ISBN: 978–92–807-3404-1 Job Number: DEP/1827/NA; 2014.

UNISDR. Essential 5: Safeguard natural buffers to enhance the protective functions offered by natural ecosystems, in: How to make cities more resilient: A handbook for local government leaders (2017 edition). Geneva, Switzerland: United Nations Office for Disaster Risk Reduction; 2017. p. 54–7.

US Army Corps of Engineers. Coastal risk reduction and resilience. CWTS 2013-3, Washington: Directorate of Civil Works, US Army Corps of Engineers; 2013.

WWAP (United Nations World Water Assessment Programme). The united National World Water Development Report 2018: Nature-based solutions for water. Paris: UNESCO; 2018.

State of science and knowledge

CHAPTER 3

Addressing water security through nature-based solutions

Kari Vigerstol[a], Robin Abell[b], Kate Brauman[c], Wouter Buytaert[d], and Adrian Vogl[e]

[a]The Nature Conservancy, Arlington, VA, United States, [b]Conservation International, Arlington, VA, United States, [c]University of Minnesota, Minneapolis, MN, United States, [d]Imperial College London, London, United Kingdom, [e]Natural Capital Project, Stanford University, Stanford, CA, United States

Key messages

- The ability of NBS to deliver water security outcomes is extremely context-specific, dependent on the local biophysical setting, the NBS selected, the placement and scale of implementation, potential management of the NBS, and the time horizon since implementation.
- Selection of water security metrics should be based on their connection to the delivery of outcomes of interest for various stakeholder groups.
- A foundational understanding of hydrologic processes can be helpful in predicting the potential impact of activities on water storage and flows in a watershed.
- Directional and magnitudinal changes in water storage and flows are easier to predict when comparing protected natural areas to impervious degraded systems than in more nuanced changes in land cover and management.
- More evidence is available on the impact of NBS on water quality (vs water quality), including the ability of vegetation to filter out pollutants in overland flow such as excess sediment and nutrients.
- Monitoring outcomes and impacts of NBS is critical to inform adaptive management and to continue to inform our understanding and build the evidence base on the ability of NBS to deliver on water security outcomes.

Introduction

Nature-based solutions (NBS) can address multiple aspects of water security while providing co-benefits to communities, businesses, and nature (Abell et al., 2017; WWAP/UN-Water, 2018). Most NBS for water involve one or more of the following approaches: protecting or conserving existing natural systems (e.g., vegetation, soils); restoring or enhancing existing

Nature-Based Solutions and Water Security. https://doi.org/10.1016/B978-0-12-819871-1.00004-X

natural or managed systems to improve functionality; or adding natural elements to modified or created systems. Although the biophysical mechanisms by which NBS impact water quantity and quality are generally understood, the impacts of NBS on water security at the times, places, and scales of interest across geographic contexts are less certain.

There is a paucity of direct, rigorous studies of the effectiveness of NBS for water impacts, in large part because those impacts take time to manifest, and most projects are young. Much of our current understanding about the impact of NBS comes from comparing healthy and degraded watersheds (often with a focus on forests), from evaluations of agricultural best management practices or "climate smart" farming or ranching (e.g., effectiveness of conservation payments under the US Farm Bill), and from limited studies of ecological restoration projects (e.g., wetland restoration). Given growing interest in NBS, an emerging body of evidence derived directly from NBS projects should, over time, allow for improved understanding of the effectiveness of specific NBS in delivering targeted hydrological outcomes. This chapter aims to clarify—given the current state of the evidence—the confidence with which we might expect different NBS to deliver water quality, water quantity, and natural disaster mitigation benefits. This is followed by an exploration into how this knowledge is applied on the ground and how gaps in the knowledge base can be addressed.

NBS: Building resilience in a changing world

The terrestrial water cycle is affected by factors operating at multiple scales, from localized land-use changes to global climate change. At the same time, human society is highly dynamic, and its water needs and exposure to water-related risks are in constant flux (Milly et al., 2008). NBS can, depending on the context, help mitigate these risks and build resilience to current challenges and future change.

At the local scale, deforestation and other conversions of natural land cover tend to accelerate the movement of water, increasing peak flows and decreasing low flows. Conversion of land to water-hungry land uses, such as irrigated agriculture, can also impact the hydrologic cycle by increasing water withdrawals and further decreasing low flows and groundwater levels. Deteriorating water quality derives from many of these same land use changes as pollutants ranging from excess nutrients to chemicals and metals are added to the system (Van Vliet et al., 2017).

Global climate change will often add further pressure. Although the local scale impacts of climate change are highly variable, climate models suggest a global increase in extreme weather events, resulting in more frequent and intense droughts and floods (IPCC, 2014). Ambient temperature increases are also likely to negatively affect water resources by driving an increase in evapotranspiration, leading to reduced river baseflow and groundwater recharge, which may induce or intensify drought risk (Van Loon et al., 2016).

Climate change will also affect ecosystems and thereby have secondary effects on the water cycle. Ecosystem degradation, including vegetation changes and die-off, affects water-related ecosystem services, altering the timing and volume of water flow. Increased water temperature caused by climate change may exacerbate water quality issues such as eutrophication (IPCC, 2014).

Simultaneously, societal trends are increasing human vulnerability and exposure to water risks. Population growth and increasing urbanization put pressure on water resources and decrease per capita water availability and drought resilience, even if the supply remains stable (Graham et al., 2020; Vörösmarty et al., 2000). These same trends can lead to increased exposure to flood risk, as floodplains are encroached by settlements, often in conjunction with land use changes that have reduced the ability of the landscape to mitigate flood risk by infiltrating or slowing down overland water runoff. In one recent example, a nongovernmental organization-led study found that destruction of 1500 ha of wetlands in Phnom Penh for urban development is placing more than one million people at increased risk of flooding (LICADHO et al., 2020).

NBS can be a powerful tool to mitigate water security pressures (Matthews et al., 2019; WWAP/UN-Water, 2018). Many NBS, including (re)forestation, wetland creation, and floodplain reconnection, have a buffering effect on river flow and enhance groundwater storage. As such, they may be able to moderate the impact of increasingly variable rainfall patterns. NBS may also be instrumental in maintaining ecosystem integrity and health (e.g., through watershed restoration and conservation), which can increase ecosystem resilience to climate change. NBS that protect natural vegetation from future conversion are maintaining hydrologic function that might otherwise be lost. Further, some interventions classed as NBS reduce the exposure of people to water-related risks, for example, by zoning to restrict development and maintain vegetation in floodplains. The effectiveness of these types of NBS are driven as much by reducing exposure to flood risk as by avoiding changes in vegetation characteristics that could alter hydrologic flows.

Although many of the protective functions of NBS can be fulfilled by built infrastructure, the high flexibility and adaptive capacity of NBS are specific advantages. Uncertainties created by climate change and an unpredictable future render static solutions with high sunk costs and low adaptive capacity, such as large reservoirs and similar gray infrastructure, increasingly risky investments. NBS are often smaller, more affordable, more flexible, and more multipurpose than conventional interventions (Ochoa-Tocachi et al., 2019). As such, they are more compatible with the type of adaptive solutions and no-regret strategies that are advocated in the context of uncertain future change (Folke et al., 2005). NBS will not replace all reservoirs, piped water systems, and other critical gray infrastructure, but they serve an important role in building a flexible, robust water system to serve the multiple needs of communities in a changing world (Matthews et al., 2019).

Table 3.1: Nature-based solutions for water resource management.

Water management issue (Primary service to be provided)		Green Infrastructure solution	Location				Corresponding Grey Infrastructure solution (at the primar service level)
			Watershed	Floodplain	Urban	Coastal	
Water supply regulation (including drought mitigation)		Re/afforestation and forest conservation	▪	▪			Dams and groundwater pumping Water distribution systems
		Reconnecting rivers to floodplains		▪			
		Wetlands restoration/conservation	▪	▪			
		Constructing wetlands	▪	▪	▪		
		Water harvesting*			▪		
		Green spaces (bioretention and infiltration)			▪		
		Permeable pavements*			▪		
Water quality regulation	Water purification	Re/afforestation and forest conservation	▪				Water treatment plant
		Riparian buffers	▪	▪			
		Reconnecting rivers to floodplains		▪			
		Wetlands restoration/conservation	▪	▪			
		Constructing wetlands	▪	▪	▪		
		Green spaces (bioretention and infiltration)			▪		
		Permeable pavements*			▪		
	Erosion control	Re/afforestation and forest conservation	▪				Reinforcement of slopes
		Riparian buffers	▪	▪			
		Reconnecting rivers to floodplains		▪			
	Biological control	Re/afforestation and forest conservation	▪				Water treatment plant
		Riparian buffers	▪	▪			
		Reconnecting rivers to floodplains		▪			
		Wetlands restoration/conservation	▪	▪			
		Constructing wetlands	▪	▪			
	Water temperature control	Re/afforestation and forest conservation	▪				Dams
		Riparian buffers	▪	▪			
		Reconnecting rivers to floodplains		▪			
		Wetlands restoration/conservation	▪	▪	▪		
		Constructing wetlands	▪	▪	▪		
		Green spaces (shading of water ways)			▪		
Moderation of extreme events (floods)	Riverine flood control	Re/afforestation and forest conservation	▪				Dams and levees
		Riparian buffers		▪			
		Reconnecting rivers to floodplains		▪			
		Wetlands restoration/conservation	▪	▪			
		Constructing wetlands	▪	▪			
		Establishing flood bypasses		▪	▪		
	Urban stormwater runoff	Green roofs			▪		Urban stormwater infrastructure
		Green spaces (bioretention and infiltration)			▪		
		Water harvesting*	▪		▪		
		Permeable pavements*			▪		
	Coastal flood (storm) control	Protecting/restoring mangroves, coastal marshes and dunes				▪	Sea walls
		Protecting/restoring reefs (coral/oyster)				▪	

*Could be considered "grey" elements that enhance the ability of green infrastructure to deliver ecosystem services.
From UNEP-DHI. Green infrastructure guide for water management: ecosystem-based management approaches for water-related infrastructure projects. Produced jointly by UNEP, UNEP-DHI, IUCN and TNC; 2014.

A UNEP-DHI (2014) guide provides a useful introduction to NBS that can address specific water quantity, water quality, and water-related natural disaster challenges (Table 3.1). The guide includes NBS that may be implemented in upland watersheds, in floodplains, in urban areas and along coasts, although it is not an exhaustive inventory of NBS options. NBS are grouped by specific water management issues, with reference to comparable gray infrastructure solutions.

Elements of water security

Water security is defined by United Nations Water as "The capacity of a population to safeguard sustainable access to adequate quantities of acceptable quality water for sustaining livelihoods, human well-being, and socio-economic development, for ensuring protection against water-borne pollution and water-related disasters, and for preserving ecosystems in a climate of peace and political stability." This concept comprises four elements: drinking water and human well-being, ecosystems, water-related hazards and climate change, and economic activities and development. In this chapter we focus on how NBS can impact the biophysical aspects of water security, which touch on all of these elements in some way.

Perceptions of water security, and values placed on water security outcomes, can vary across stakeholder groups. For example, a rural community that accesses water directly from a surface water source may be interested in relatively consistent water levels or flows throughout the year and high water quality across a range of parameters. In contrast, a hydropower company with a large storage reservoir might be interested in annual water yield and low sediment loads but less concerned about flow at a specific time of year or other water quality issues. The water risks that these two groups face are quite different in nature and impact, but they all fall under the term water security.

These differing outcomes of interest can be translated into commonly measured water metrics, such as baseflow, turbidity, groundwater level, and soil moisture content (see Box 3.1). It is these types of metrics that the following sections will focus on as they explore what is known about the ability of NBS to affect individual metrics, and hence support water security outcomes, and where uncertainty still exists.

NBS and water quantity

Introduction

NBS affect where and when water is available in a watershed by modifying components of the hydrological cycle. NBS alter the landscape by adding or improving existing natural elements that modify water flows through four main levers, as shown in Fig. 3.1: (1) changes in above-ground vegetation characteristics, (2) subsurface vegetation characteristics, (3) land surface characteristics, and (4) soil characteristics.

Box 3.1   Key water security metrics

There are a variety of metrics typically used to describe the multiple aspects and desired outcomes of water security (see Table 3.2). This chapter links NBS to some common metrics, which are defined here.

Annual water yield is the total volume of water flowing out of a watershed in a given year, including both surface and subsurface flows. The net annual water yield is the difference between the precipitation that falls in that watershed and total evapotranspiration, assuming that there is no net storage in vegetation or soils over the course of a year.

Baseflow (also called "dry season flow") refers to water seeping into streams from groundwater and is often the only source of flow during the dry season when direct runoff (surface flow generated by rainfall) has stopped. Baseflow is represented by various indicators, such as the annual minimum of the 7-day flow, the mean of the 10% lowest flows or the baseflow fraction of total annual flow.

Flood peak is the highest stream water level during a storm event and reflects the volume of rainfall over time and the ability of the landscape to intercept and infiltrate precipitation and overland flow.

Infiltration is the portion of precipitation that soaks into the soil and subsurface via pores within the soil and channels along plant roots.

Interception is the portion of precipitation that is captured on plant surfaces, such as leaves within the canopy, stems, and/or the trunks of trees.

Groundwater level is the depth to groundwater at any given time. The groundwater level reflects the balance between water infiltrated into the aquifer over time and any losses, for example, by seepage into local streams or springs or water use via groundwater pumping. In unconfined aquifers, the water table is affected by infiltration in the connected landscape above, while confined aquifers typically have an infiltration source area disconnected from the above land surface.

Groundwater recharge is the process by which water moves down through the soil and subsurface to the water table; the process by which water enters aquifers.

Overland runoff is the portion of precipitation that does not infiltrate the soil but runs off the surface.

Turbidity describes the level of transparency of water as a result of suspended particles, such as sediment. Turbidity increases as sediment runs off land into streams. The amount of sediment ending up in a stream depends on the erosivity of local soils and streambanks, intensity of rain events and flood peaks, vegetative cover, and presence or absence of riparian buffers.

Pollutant concentration describes the presence or absence of a range of pollutants in a body of water. The concentration of nonpoint source pollutants, which come from diffuse sources in comparison to concentrated inputs from a waste disposal point, depends on the loading of pollutants on the landscape and the presence or absence of a vegetative cover, including riparian buffers.

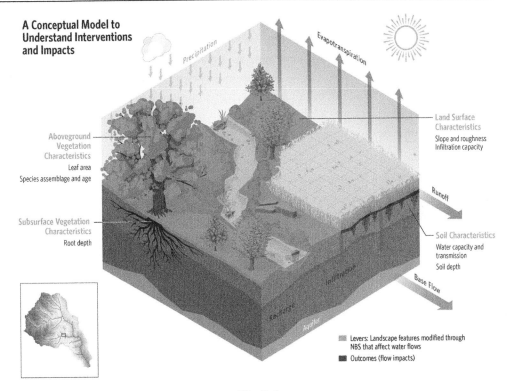

Fig. 3.1
A conceptual model to understand NBS interventions and potential impacts.

Many of the changes in above- and below-ground vegetation, land surface, and soil characteristics interact in complex ways. The ultimate impact that NBS will have on water flows depends on how these changes interact with one another—for example, the impact on water quantity of planting trees on former agricultural land will depend on whether the species of trees planted have a greater or lower water demand than the crops formerly grown there, whether they have deeper roots, and whether they are able to access deeper groundwater once they are mature.

The spatial and temporal scales at which the impact of NBS are assessed matters. Local impacts (e.g., improved infiltration, reduced runoff, improved soil water holding capacity, and soil water availability) typically manifest rapidly. Larger-scale impacts, such as groundwater recharge and downstream flow alteration, take more time to manifest and are much more dependent on the context within which NBS are embedded—whether the changes in vegetation, surface, and soil properties are qualitatively big enough and cover a large enough area to significantly alter landscape-scale hydrologic functioning.

Vegetation growth and soil development take time, as does water movement in the subsurface. Further, the impacts of NBS are often nonlinear and highly dependent on the complex interactions of soils, landscape configuration, and size of treatment. For example, the immediate soil compaction associated with the heavy equipment often used for deforestation can take decades to be reversed even if new trees are planted immediately. The effects of intervention may also change over time as vegetation grows and responds to interannual variation in precipitation and temperature and extreme events such as floods, drought, and wildfire. Therefore, understanding the hydrologic baseline and implementing ongoing monitoring before and after any NBS intervention is crucial to track impacts.

Many NBS are implemented with the expectation of more water overall (increased annual **water yield**, in general or during droughts), more water during the dry season (increased **base flows** and soil moisture, in general or during droughts), and less water during wet times (reduced **flood peaks**). Out of the eight outcomes identified in Table 3.2, four relate directly to water quantity: water supply for consumers, maintaining water levels and base flows, water availability for irrigation, and reducing or avoiding damages due to droughts and floods.

The underlying hydrologic functions that control whether these desired outcomes are achieved may be grouped into two general categories: (1) partitioning, that is, how much rainfall ends up as evaporation, surface runoff, or subsurface runoff, with implications for water supply reliability, maintaining base flows, and soil moisture; and (2) storage, which allows the landscape and vegetation to help smooth out extremes of high and low streamflows, thereby helping to reduce damages from floods and dry periods. The evidence for NBS effectiveness is discussed here in relation to these two categories of impact.

NBS effectiveness for water quantity

Maintaining reliable flows

Baseflow is generated by the complex interactions of vegetation, soil properties, topography, and the underlying geology of a region that influence how water infiltrates into the soil and subsequently moves through the subsurface (as shown in Fig. 3.1). Deforestation alters the balance between precipitation, rainfall interception by vegetation, transpiration (return of water to the atmosphere), runoff, and infiltration, thereby influencing the subsurface flow that provides baseflow during dry intervals. Depending upon the pre- and postintervention vegetation types and their relative rooting depths and water demand, the change can result in either increased or decreased annual and seasonal flows. In areas with adequate reservoir storage, changes in annual flows may be of greater concern (at least for some stakeholders) than seasonal flows. Maintaining a reliable water supply—either ensuring a predictable minimum flow year-to-year or a minimum level of consistent flow during the dry season—is one of the most crucial services demanded by people; therefore, it is one of the most common goals cited by programs that invest in NBS for water (Bennett and Carroll, 2014; Bremer et al., 2016). The evidence for the impacts of NBS on baseflow, however, is mixed.

Table 3.2: Example of the relationships between water security metrics used in this chapter and a range of stakeholder outcomes of interest.

Metric	Drinking water supply	Maintain stream levels	Water available for crops	Water quality for aquatic species	Reduced water treatment costs	Maintain water temperature within normal ranges	Reduced/avoided flood damages	Dam/reservoir longevity
Annual water yield	X	X	X					
Base flow	X	X	X	X	X	X		
Flood peak		X	X		X	X	X	X
Groundwater level	X	X	X					
Soil moisture content			X				X	
Turbidity	X		X	X	X		X	
Pollutant concentration	X		X	X	X		X	X

Studies of watershed degradation provide the clearest evidence that healthy vegetative cover has positive impacts on baseflow during the dry season, although the evidence is somewhat mixed. Fog capture contributes to baseflow in many tropical montane forests, which is another factor leading to a drop in baseflow when those forests are lost (Bruijnzeel et al., 2011). Conversely, increased dry season flow in land cleared for pasture, in areas of low-density development, and in secondary forest, as compared to forested watersheds, has been demonstrated in tropical systems (Bruijnzeel, 1989; Ogden et al., 2013) and in subtropical forests in North America (Price et al., 2011).

Forest restoration has mixed but often negative effects. A recent review showed that out of 53 cases where changes in baseflow were measured following forest restoration, afforestation, or forest expansion, 73% showed a decrease in baseflow, and only 21% recorded an increase (Filoso et al., 2017). While the exact mechanisms behind these different outcomes are still being actively studied, the causes have been variously attributed to soil infiltration and soil moisture capacity (Bruijnzeel, 1989) or water demand and access to water of pre- and postintervention vegetation relative to mean annual rainfall (Farley et al., 2005; Van Dijk et al., 2012). However, Kadykalo and Findlay (2016) found that, on average, wetlands can augment low flows.

The evidence for impacts of NBS in croplands and rangelands on downstream seasonal flows is limited to site-scale studies that measure change in infiltration, runoff, and short-term streamflow and use this information to infer changes in longer-term hydrologic response. In many cases, NBS in working landscapes, including edge-of-field or riparian buffer strips, conservation tillage, and rotational grazing have been shown to reduce soil compaction, improve infiltration and soil water holding capacity, and reduce short-term runoff. However, the impacts on baseflow and dry season flows over longer time periods and at larger scales are not well demonstrated, primarily due to a mismatch in study designs (i.e., most observations were reported at plot-scale and for short-term changes), and the scale at which watershed-level metrics of interest change (i.e., base flow at a watershed scale and over years) (Van Dijk et al., 2012).

Regardless of the magnitude and direction of changes in low flows, NBS that increase the existing canopy cover and, thus, evaporative demand from vegetation and those with deep roots that provide more access to water (e.g., reforestation, afforestation) typically decrease the total **annual water yield** (Farley et al., 2005; Zhang et al., 2017), and this effect is amplified as the intervention area grows larger relative to the watershed area (Brown et al., 2005; Dennedy-Frank, 2018; Filoso et al., 2017). However, if the natural vegetation has an overall lower water demand than whatever it is replacing, for example, when converting from water-intensive, irrigated agriculture, then reforestation would tend to increase water yield. Conversely, an initial increase in annual water yield following deforestation has been demonstrated in tropical forests, with flows tending to return to baseline over time (Bruijnzeel, 1989).

The effects of NBS on **recharge of deep groundwater** aquifers is a major gap in the published literature. When it comes to percolation and **shallow groundwater flow**, the evidence is somewhat better. However, as with baseflow, there is no clear directionality as the impacts depend strongly on the specific NBS interventions chosen, how the resulting changes in vegetation, land surface, and subsurface characteristics interact to influence infiltration and deep percolation and how these changes are mediated by connectivity between surface and subsurface aquifers. The same mechanisms that cause reduced baseflow following reforestation or afforestation can also reduce the potential for groundwater recharge, except in cases where flow through the soil and rooting zone is very rapid (Kim and Jackson, 2012). In cloud forests, however, where fog capture is significant, maintaining or improving vegetation cover can increase recharge (Brauman et al., 2012, 2014).

NBS such as crop rotation has been shown to either increase or decrease groundwater recharge, depending on context (Dakhlalla et al., 2016). Conservation tillage can decrease overland flow and increase recharge by focusing water in areas where it can flow below the root zone of crops (Leduc et al., 2001). A recent study of 1400-year-old infiltration enhancement ditches in the Andean highlands of Peru showed that such systems effectively enhance hillslope infiltration to recharge downslope springs (Ochoa-Tocachi et al., 2019).

In summary, the ability of NBS to maintain or enhance reliability of flows remains highly uncertain and depends heavily on the context, the type of intervention(s) chosen, and the spatial scale. Climate, topography, soil types, geology, and pre- and postintervention vegetation all play a role in determining whether a change in land cover will have a positive or negative impact on flows, relative to stakeholders' needs. The evidence that protecting healthy vegetation cover and functioning wetlands has positive impacts on baseflow is more straightforward. There is a sound theoretical basis and growing evidence that deforestation reduces low flows, due to the intensive soil compaction and increases in impervious cover normally associated with development (Price et al., 2011). However, the evidence for benefits of NBS that change the existing landscape elements (e.g., restoring vegetation, adding natural elements, etc.) is mixed.

Mitigating extremes: Floods and droughts

NBS are increasingly attractive strategies for **mitigating flood risks** because they are relatively less expensive than engineered—or "gray infrastructure"—solutions. Further, commonly used flood mitigation methods of channelizing rivers, armoring streambanks, and building dikes may mitigate local flooding while increasing flood risk downstream. The timing and magnitude of flooding in a watershed depends on three main pathways: (1) how much runoff is generated from hillslopes, (2) the timing and amount of runoff that accumulates in stream channels, and (3) how flow from multiple hillslopes and subwatersheds combine and interact in the network of channels, floodplains, and wetlands to generate downstream flooding (Dadson et al., 2017). NBS can influence each of these pathways, although the effects are typically more pronounced at local scales and for small flood events (e.g., pathway 1) and less significant as flow is integrated from many hillslopes and channels into larger watersheds (e.g., pathway 3).

NBS impact flooding via the first pathway by altering or maintaining the water retention capacity of hillslopes by increasing permeability and porosity of soil and associated infiltration. This is achieved by increasing the extent and types of vegetation with deep roots that increase macropore formation (Roberts et al., 2012) and by increasing surface roughness (Dadson et al., 2017). This leads to pathway number two, where the reduced volume and slower transit time of overland flow in upland areas alters the timing of flow from headwaters, so that water is less likely to arrive simultaneously at a given downstream point, reducing the potential flood peaks by spreading out the arrival of water in time (Dadson et al., 2017).

Further, increasing vegetation in and around channels can slow the flow of water there, delaying the arrival of floodwaters downstream. However, this effect diminishes as the area of the watershed increases, as the relative importance of hillslope and channel network routing in hydrograph formation is more pronounced at larger scales (Beven and Wood, 1993; Botter and Rinaldo, 2003; D'Odorico and Rigon, 2003; Robinson et al., 1995). The third pathway for NBS to impact flooding is by reconnecting and restoring floodplains and wetlands, thereby increasing total floodplain and channel storage (Dadson et al., 2017).

The evidence for the role of wetlands in flooding, however, is mixed. The direction and degree of impact will depend on the type of wetland, its surrounding landscape context, soil characteristics, management, its source of water (rain-fed, groundwater-fed, or river-fed), and its hydrological connectivity to the drainage network. Upland rain-fed wetlands (as opposed to those downstream in lowland areas adjacent to the floodplain) have been shown to have mixed effects, where in 41% of studies they actually enhanced flooding (Bullock and Acreman, 2003). While these topographic depressions can store water on hillslopes up to a certain point, once they "overflow" into the wider watershed drainage network, they become rapid runoff-response areas that can exacerbate larger flood events (Acreman and Holden, 2013). On the other hand, around 80% of studies in the same review suggested that downstream, floodplain wetlands reduce flooding (Bullock and Acreman, 2003). A more recent meta-analysis found that, on average, wetlands reduced frequency and magnitude of flooding and increased flooding return period (the likelihood that a given size flood will occur in any year), while also noting large variability in the magnitude of the impact (Kadykalo and Findlay, 2016).

NBS have been shown to reduce the frequency and depth of flooding only up to a point; because the landscape becomes saturated following heavy rain, there is little evidence that it can significantly impact the magnitude of the largest flood events and flood peaks in large watersheds (Dadson et al., 2017; Dennedy-Frank, 2018; O'Connell et al., 2007; Van Dijk et al., 2009). For example, Dadson et al. (2017) found that forested watersheds reduced flow peaks per unit area for storms less than 20% of the mean annual flood but found no significant difference between flow peaks during high flows. By far the biggest potential for NBS to mitigate flood risk is through protection or restoration of floodplains, which provides for natural storage and diversion in regularly flooded areas and reduces the exposure of lives and assets to risk when development in floodplains is permitted.

In terms of **drought**, NBS can impact the occurrence and severity of impacts in at least two ways: (1) by building or maintaining system resilience at the local level; and (2) by maintaining stability in the regional hydrologic cycle. NBS can help reduce the severity of impacts at a local scale via some of the same mechanisms previously described, such as through maintenance of base flow or maintenance of soil moisture holding capacity. The ability of the landscape to hold water and release it more slowly over time helps support water availability longer into the dry season, including during a drought. Some agricultural NBS, especially those that aim to increase soil organic matter, can increase soil moisture holding capacity, which can help with crop productivity longer into a drought and reduce the need for additional irrigation water (Bhadha et al., 2017). However, the ability of NBS to mitigate medium to longer term droughts is limited since NBS do not fundamentally alter the water balance; therefore, it is unlikely that they can prevent the worst impacts of droughts, or that they can balance the water budget where uses (including human, vegetation, and wildlife) exceed the long-term supply from precipitation.

At larger scales, the role of vegetation in regional weather patterns is clear. As they transpire, areas with more extensive vegetation cover cool the surrounding area and recharge atmospheric moisture, playing a key role in generating rainfall and driving regional climate patterns (Ellison et al., 2017). In areas that are highly dependent on rainfall generated from existing forests upwind, the effects of widespread deforestation, if replaced by vegetation with substantially less evapotranspiration, could cause increasing variability and changes in the seasonality of rainfall, potentially triggering an irreversible switch from a wetter to a drier climate (Sheil and Murdiyarso, 2009). For example, Keys et al. (2018) found that for eight of the world's megacities, over half of the rainfall that arrives in their source watersheds comes from upwind land areas, including Beijing, Buenos Aires, Chongqing, Karachi, Kinshasa, Lagos, São Paulo, and Wuhan. Widespread loss of forests in these upwind areas could have significant impacts on the water supplies for well over 100 million people.

NBS and water quality

Introduction

Globally, water quality is declining, largely driven by human activities that result in excessive sediments, industrial pollutants, nutrients, pathogens and toxins in water bodies (UNEP, 2016). This affects human health, food security, commerce, and quality of life, leading to growing demand for diverse solutions for clean water. While large inputs of pollutants, particularly heavy metals and manufactured chemicals, may overwhelm the ability of NBS to filter or break down pollutants, there is significant potential for NBS to improve water quality. NBS such as wetlands and other water retention features may be deployed throughout the landscape to mitigate broadly distributed (nonpoint) sources of pollutants such as sediment and nutrients, often originating on agricultural lands, and they may also be used to treat wastewater or to "spot-treat" pollution downstream of wastewater treatment plants. Whether a change in water quality is considered beneficial depends on the suite of desired uses of water (Keeler et al., 2012). For example, mussels remove suspended solids, bacteria, and phytoplankton from the water column, which is frequently interpreted as a benefit, but invasive zebra mussels in North America do so to the extent that waters become very clear and cannot support fish or other aquatic life (Macisaac, 1996).

Ecosystems are most often involved in improvement of water quality through the processing, uptake, and sequestration of nutrients, including nitrogen and phosphorus, as well as pesticides, herbicides, petroleum, and metals (Arora et al., 2010; Krutz et al., 2005; Williams, 2002). Vegetation can also keep soils in place so they are not mobilized as sediment in waterways, and vegetation may entrain sediments once they are mobilized. Ecosystems remove pollutants dissolved or entrained in water through physical (deposition, infiltration), geochemical (sorption, precipitation, occlusion), and biological (plant or microbial uptake) retention (Roberts et al., 2012).

These processes work in tandem; above-ground vegetation physically reduces surface flow speed, enabling sediment deposition, sorption, and infiltration (Arora et al., 2010; Sweeney and Newbold, 2014). Dense root systems increase permeability and porosity of soil and thereby increase infiltration (Roberts et al., 2012). Groundwater entering an ecosystem laterally or via infiltration interacts with soils and the rhizosphere, where constituents in it can be taken up by plants or denitrified by microbes (Roberts et al., 2012; Sweeney and Newbold, 2014). Certain species of plants may be selectively planted to uptake certain chemicals from soil and soil water, a process called phytoremediation (Mirza et al., 2014; Salt et al., 1998).

NBS effectiveness for water quality

NBS can help prevent nonpoint source pollutants from entering water in several ways. Continuous vegetative cover can reduce erosion, a source of sediment pollution, by improving both hillslope (Zuazo and Pleguezuelo, 2009) and bank stability (Fox and Wilson, 2010); root mass is also critical for soil stability (Gyssels et al., 2005). In tree plantations, intercropping with an herbaceous cover crop can reduce erosion substantially (Sidle et al., 2006). NBS that provide herbaceous cover can also reduce nutrient export, even following deforestation (Tomasella et al., 2009), and in some cases even when grasses are invasive (Wilcox et al., 2012).

For temperate and tropical forests, it is well established that a nutrient and sediment pulse follows deforestation and that forest regrowth quickly moderates this impact

(Tomasella et al., 2009). Timber harvesting increases sedimentation, which buffers can help reduce (Croke and Hairsine, 2006). Because intense fire is a substantial source of water pollution, NBS that reduce the frequency and intensity of fire can also reduce the export of sediment, nutrient, and other pollutants (Smith et al., 2011). The Rio Grande water fund in Northern New Mexico, for example, has invested in NBS to reduce the risk of catastrophic fire and associated water pollutants (Stone et al., 2017).

Once unwanted constituents have gotten into overland flow, NBS can often remove some but not all of these pollutants (see Table 3.3). Removal of dissolved pollutants, such as nitrogen, in surface flow by riparian buffers is moderate (Mayer et al., 2007). Removal efficiency for pesticides in solution is similar (Arora et al., 2010). Buffers also remove pollutants from surface flow by trapping sediments and the pollutants that are sorbed to them. As with absorption of dissolved nutrients and chemicals, the effectiveness of buffers in retaining sediment and the pollutants such as phosphorus or pesticides that attach to sediments, varies considerably (Gumiere et al., 2011; Liu et al., 2008). Buffers at the edge of forestry operations are also effective at reducing sediment delivery to water bodies (Norris, 1993). High overland water flows are probably the most important factor in reducing buffer strip effectiveness (Sweeney and Newbold, 2014).

Buffers also remove pollutants from water below ground. Removal of dissolved pollutants such as nitrogen in subsurface flow through a buffer is quite effective (Mayer et al., 2007). Nitrogen removal can even occur below the rooting zone, and these soil processes do not seem to be altered by above-ground disturbance (Sudduth et al., 2013).

Table 3.3: Summary of effectiveness of NBS for removal or filtering of water pollutants.

Pollutant	Effectiveness
Nitrate—surface flow over buffer	40% removal effectiveness (Mayer et al., 2007)
Nitrate—subsurface	55% (range: 26%–64%) for buffer widths < 40 m and 89% (range: 27%–99%) for buffer widths > 40 m (Sweeney and Newbold, 2014)
Nitrate—in stream	Between 0% and 50% of nitrogen can be processed in stream (Sudduth et al., 2013)
Pesticides in solution	Average of 45% with a range from 0% to 100% retention (Arora et al., 2010)
Sediment	Depends on buffer width, efficiency ranging from 54% to 100% (Liu et al., 2008). On average, buffers 10 m wide trap about 65% of sediments delivered by overland flow, while 30-m buffers trap about 85% of sediments (Sweeney and Newbold, 2014).
Phosphorus (attached to sediment)	41%–95% of sorbed P is, on average, removed by buffers (Roberts et al., 2012)
Pesticides (attached to sediment)	76% of pesticides sorbed to sediments are removed by buffers, with a range of 2%–100%

The mixed efficiency of buffers at pollution removal reflects a large number of factors, including soils, slope, precipitation patterns, and size of buffer area (Polyakov et al., 2005). Experiments done in controlled conditions with very low flows have frequently shown little additional water quality benefit from buffers exceeding 10 m wide, but experiments focused on more realistic situations, including heterogeneous flows and high flows, found that removal efficiency increases with width, which can help make up for selective flow paths through buffers (Sweeney and Newbold, 2014). Because geographic conditions play such a large role in buffer effectiveness, there is no clear consensus on the magnitude of effects at a watershed scale (Sweeney and Newbold, 2014).

Wetlands and floodplains are effective at removing pollutants (Hoffmann et al., 2009), and the interface between water and soil at a streambank is critical for processing nutrients (Lawrence et al., 2013). Worldwide, wetlands remove about 17% of anthropogenic reactive nitrate inputs (Jordan et al., 2011). The effectiveness of wetlands for treating wastewater has been recognized for at least a century; this has been measured and codified to the point that constructed wetlands are now a widely recognized and certified water treatment solution (Vymazal, 2011; Williams, 2002).

Within streams, vegetation improves water quality by stabilizing the channel as well as entraining and sequestering pollutants (Montakhab et al., 2012). In-stream processing of pollutants is varied and a function of loading, but between 0% and 50% of nitrogen can be processed in stream (Sudduth et al., 2013).

NBS may also provide direct additions of organic matter to water, which can ameliorate certain water quality problems. Vegetation contributes leaf litter and bulk debris to streams (Helmers et al., 2005) and provides dissolved organic matter through soil water (Leenheer and Croué, 2003). These inputs are critical to functioning, healthy river systems, providing nutrients and carbon at the base of the food web. In river systems starved of sediment, usually by upstream dams, coastal deltas can erode, as is happening in Venice, Italy, and New Orleans, United States. Over time, the vegetative control over sediment moving into and within rivers actually shapes the form of rivers in ways that can help support healthy riverine ecosystems (Statzner, 2012).

Although NBS generally require a much lower level of maintenance than gray infrastructure solutions, NBS are not necessarily maintenance-free. For example, vegetation that sequesters but does not break down toxic materials may need to be harvested; sequestered compounds can then be recycled or waste more easily disposed of (Mirza et al., 2014). Ecosystems that are taking up pollutants may saturate, losing their effectiveness and potentially remobilizing once-sequestered pollutants (Hoffmann et al., 2009; Roberts et al., 2012). Many types of pollutant removal by ecosystems are a function of loading, meaning that more removal occurs when more pollutants are present, up to a point when the pollutant is impacting the health of the ecosystem and its subsequent ability to filter runoff (Bouwman et al., 2005; Smith et al., 2003).

Key design and management considerations

Given the complexity of the systems within which NBS are applied, their success in delivering on objectives is dependent on a number of factors. A survey of academic literature and case studies of NBS implementation surfaces a few common themes about planning, design, and management of NBS that can maximize the potential for success for water security outcomes (Cohen-Shacham et al., 2019). Some of these themes arose in the previous sections on water quantity and quality and are described in more detail here:

Clarity on intended outcomes and a localized theory of change

At the outset of an NBS project, key stakeholders should clearly state and agree upon the intended outcomes. With this clarity, selection and design of NBS should be based on a full understanding of how each potential solution may deliver on the outcomes, taking into account local context and the possible trade-offs in selecting one solution over another.

Embracing the complexity

The relationships among NBS, the biophysical processes they affect, and the outcomes they are intended to deliver is complex, but not a black box. We have a solid understanding of hydrologic and ecological processes that underpin NBS and the uncertainties and assumptions that are part of this system. Getting clear early on in a project on the intersection between the greatest unknowns and areas of potential failure helps inform: (1) what should be monitored to reduce any gaps in knowledge, and (2) what should be most closely managed.

Getting to scale in the right places

Two of the most important factors determining the success of NBS interventions are placement and scale of implementation. For ecosystems to improve water quality, placement is critical, as potential ecosystem pollutant sinks must be in a flow path of contaminated water (Rittenburg et al., 2015). One common example of the importance of placement of NBS is the effectiveness of a riparian buffer in removing sediment or pollutants from overland flow before it reaches a water body, as compared to planting trees or other vegetation higher up in the watershed. A riparian buffer has the opportunity to improve the quality of water that has flowed from the top of the watershed all the way to the edge of the water body, as a sort of a last-step filter.

However, scale may trump location. If, for example, increased erosion due to deforestation is the primary challenge in a watershed, this problem will only truly be addressed when forest protection is implemented across a large enough swath to see a measurable reduction in sediment runoff and loading. A 10-ha project site in a 1000-ha watershed may make a difference locally, but it will have little impact on sediment loading to a reservoir downstream of the 1000 ha. For water quantity, evidence from paired-watershed studies suggests that flow changes cannot be statistically discriminated when forest cover is changed on less than 20% of the total watershed area (Dennedy-Frank and Gorelick, 2019; Van Dijk et al., 2012).

Understanding timing of benefits

An important factor to consider in determining the perceived success of an NBS project or program is the timing of measured outcomes. Water security outcomes can take years to appear, for at least two reasons: (1) getting to the desired scale of implementation may take some time due to a variety of factors, and (2) given that many NBS interventions include planting or management of vegetation, it takes time even after full-scale implementation for the NBS to reach their full potential.

Correcting geographic bias in experimental data

Like hydrologic science more generally, the evidence demonstrating the effectiveness of NBS comes disproportionately from the United States and Europe, and much remains unknown about both the performance of NBS and the basic hydrology of the tropics (Hamel et al., 2017; Ponette-González et al., 2014). Most studies of water quality impacts have focused on agricultural NBS in relatively flat landscapes, with some attention paid to forests, particularly in the context of timber harvest (Brauman, 2015). There are, therefore, some serious data gaps in our understanding of NBS performance in areas of the world where NBS are potentially of greatest interest.

Adaptive management

The impacts of NBS will change over time, as the canopy rainfall interception, water demand, and pollutant removal efficiency changes as vegetation is established, matures, and is (or is not) managed. Therefore, NBS should be considered a dynamic, rather than a static, strategy that requires adaptive management to respond to changing conditions and changing pressures on the broader landscape. Monitoring for intended impacts is critical to ensure they are being delivered and to inform potential changes to the NBS implementation to maximize effectiveness.

Monitoring and evaluation of NBS

Continuing to build the NBS evidence and knowledge base is impossible without effective monitoring and evaluation (M&E) of existing and new NBS programs and interventions. This is a challenge for a at least a few reasons:

- Measurable NBS impacts at the level desired often take quite a long time (10 + years), which many programs are not set up to accommodate for and most funders do not have the patience for.
- Parsing out the specific impact of the NBS interventions from other ongoing changes in a watershed can often be extremely difficult given the complexity of the biophysical system.

However, there are some guidelines that can help address some of these challenges in designing and implementing M&E systems for NBS:

- Create as much clarity as possible about the objectives of the NBS intervention(s) to ensure M&E is focused on the key outcomes of interest.
- Monitoring design should take into account spatial and temporal complexities. For example, programs should include monitoring of local impacts in the short term while tracking changes in the long-term outcomes at the downstream point of interest.

- Identify and track any factors of high uncertainty that could have an impact on the ability to deliver on outcomes.
- Track trends outside of the areas of intervention to account for changes that may overwhelm impacts from the interventions.
- If possible, monitoring should include a control watershed or subwatershed to demonstrate more clearly what happened over time with and without NBS program(s) or intervention(s).

Tools and guidance

Growing interest in the potential of NBS to effectively address water security concerns is being met with a growing body of knowledge and resources. To support the design and implementation of NBS programs, organizations, and institutions have developed helpful tools and guidance and have established focused areas of work around NBS for water security. Table 3.4 captures a sample of these tools and resources. For example, for readers looking to further explore the evidence base and specific literature on specific NBS and water security outcomes, the Nature-Based Solutions Evidence Platform from the University of Oxford offers a valuable resource.

The path ahead

Long before the term "nature-based solutions" was coined, people were using nature and natural processes to address and reduce water-related risks. For instance, since pre-Colombian times, small central Andean communities have utilized *mamanteo*, a nature-based stream diversion system applied in the mountains to improve downstream dry season stream flows (Cassin and Ochoa-Tocachi, 2021; Ochoa-Tocachi et al., 2019). Ancient Chinese and Egyptian cultures recognized the value of wetlands in filtering pollutants, and more recently, constructing wetlands for that purpose has been documented as far back as the turn of the 20th century (Brix, 1994). While NBS may currently be experiencing a surge of attention, the practice is anything but new.

What we know from this long history of NBS use, and from more recent scientific explorations that add to traditional knowledge, is that NBS has strong potential to help ensure water security in various contexts. We can often predict with greater certainty water quality versus water quantity benefits, and in general, there is more certainty around local as compared to downstream impacts. Predictions of NBS impacts on surface water are currently more accurate than those for predominantly groundwater-driven systems. In many instances, NBS applied at larger scales will be more effective than those applied over small areas, but location and design can be equally important, and even the largest NBS applications may be limited in their ability to mitigate the impact of catastrophic events like massive floods, droughts, and landslides.

Table 3.4: Additional tools and guidance on NBS evidence and modeling.

Resource	Type	Developed by	Example application/link
AgEvidence	Online database	The Nature Conservancy	https://www.agevidence.org/
Nature-Based Solutions Evidence Platform	Online database	University of Oxford's Nature-based Solutions Initiative	https://www.naturebasedsolutionsevidence.info/
UN World Water Development Report: Nature-Based Solutions	Report	UNESCO	http://www.unesco.org/new/en/natural-sciences/environment/water/wwap/wwdr/2018-nature-based-solutions/
Beyond the Source: The Environmental, Economic and Community Benefits of Source Water Protection	Report	The Nature Conservancy, Natural Capital Project, Forest Trends, the Inter-American Development Bank and the Latin American Water Funds Partnership	https://www.nature.org/en-us/what-we-do/our-insights/perspectives/a-natural-solution-to-water-security/
Putting Nature to Work: Integrating Green and Grey Infrastructure for Water Security and Climate Resilience	Report	World Bank; World Resources Institute	https://www.worldbank.org/en/news/feature/2019/03/21/green-and-gray
Nature-Based Solutions for Agricultural Water Management and Food Security	Discussion paper	Food and Agriculture Organization of the United Nations	http://www.fao.org/3/CA2525EN/ca2525en.pdf
Nature-Based Solutions for Water Utilities and Regulators	Technical brief	International Water Association; The Nature Conservancy	https://iwa-network.org/publications/nature-based-solutions-for-water-utilities-and-regulators/
Benefit Accounting for Nature-Based Solutions for Watersheds	Landscape assessment; guide; tool	CEO WM; Pacific Institute; The Nature Conservancy; LimnoTech; Danone	https://ceowatermandate.org/nbs/
Urban Water Blueprint	Website and report	The Nature Conservancy	http://water.nature.org/waterblueprint/#/intro=true
The Role of Natural Infrastructure in Water Resource Development	Website (blogs, book chapters, and photo stories)	Water, Land and Ecosystems (CGIAR)	https://wle.cgiar.org/natural-infrastructure-water
Ecosystem Based Adaptation and Water Security	Summary document	USAID	https://www.climatelinks.org/resources/ecosystem-based-adaptation-and-water-security
Guía para la Evaluación de Intervenciones en Infraestructura Natural para la Seguridad Hídrica: Escala de Efectividad, Equidad y Sostenibilidad	Guide	Forest Trends	https://www.forest-trends.org/publications/guia-ees/

Table 3.4: Additional tools and guidance on NBS evidence and modeling—cont'd

Resource	Type	Developed by	Example application/link
Guía HIRO—Herramienta de Identificación Rápida de Oportunidades para la Infraestructura Natural en la Gestión del Riesgo de Desastres	Guide	Forest Trends	https://www.forest-trends.org/publications/guia-hiro/
Metodologías CUBHIC: Restauración y Protección de Humedales	Guide	Forest Trends; Kieser & Associates	https://www.forest-trends.org/publications/metodologias-cubhic-restauracion-y-proteccion-de-humedales/
Metodologías CUBHIC: Qochas	Guide	Forest Trends; Kieser & Associates	https://www.forest-trends.org/publications/metodologias-cubhic-qochas/
InVEST	Modeling tool	Natural Capital Project	https://www.cabdirect.org/cabdirect/abstract/20133222272
SWAT	Modeling tool	US Department of Agriculture	https://www.sciencedirect.com/science/article/pii/S0022169416000524
Links to other surface water modeling tools	Website	US Environmental Protection Agency	https://www.epa.gov/ceam/surface-water-models-assess-exposures

Much of our knowledge about the functions of large-scale NBS comes not from intentional NBS applications but from the study of naturally occurring intact systems. Retaining intact ecosystems will almost always be more effective and affordable than restoring or rehabilitating ecosystems after they have been degraded or lost. The strongest evidence for watershed-level impacts points to protecting existing natural vegetation as the most reliable way to maintain hydrologic services. Typically, neither active restoration nor passive recovery results in a full restoration of services (Jones et al., 2018), underscoring the importance of maintaining watersheds that are in good hydrologic condition.

NBS exist along a spectrum, and most will be similar to engineered, gray infrastructure in requiring thoughtful design and management to meet water-related objectives now and into the future. There is rarely a one-size-fits-all nature-based solution, especially if there are potential trade-offs to be balanced. For instance, a wetland constructed to filter pollutants may, depending on location, size, soil type, and topography, lead to greater downstream flood risk. Or there may be trade-offs among water-related benefits and other objectives, like food security if agricultural or pastoral land are restored to forest. Very little research to date has explored trade-offs of NBS, and while in many instances the risks will be low, practitioners should be mindful of unintended consequences.

NBS practitioners are often "learning while doing," which underscores the importance of building robust M&E elements into existing and future NBS programs. Strong M&E programs are critical to the success of individual projects, as they can provide information for adaptive management. Demonstrated success of NBS via M&E can also make the case for increased funding or sustainable financing of projects. More broadly, there is a need to grow the knowledge base to continually improve the design and management of NBS to deliver on water security outcomes. Only when actors who have historically relied on gray approaches to water security feel confident in the application of NBS will these approaches become mainstream.

Research: in addition to the tools and resources provided in Table 3.4, there are a number of efforts and programs underway to further our understanding of how NBS can provide water security outcomes and support effective planning and implementation. Box 3.2 provides an example of a program that looks at the context-specific outcomes of NBS application.

Box 3.2   Blue Forest Conservation case study

In an effort to better understand the nuances of water quantity impacts of forest fuel treatments, such as forest thinning and prescribed burns, on watersheds, the University of California at Merced partnered with Blue Forest Conservation and the Sierra Adaptive Management Project to conduct an extensive study in the central and southern Sierra Nevada Mountains. Researchers gathered and analyzed data over multiple years and used hydrologic modeling to assess the potential impacts over a larger area.

What they found is that even when similar treatments are applied across the landscape, local biophysical context alters the potential impact. For example, decreasing vegetation by 8% in the American River watershed resulted in an average 12% water yield increase, while similar interventions near Yosemite National Park only increased annual runoff by 3%. The difference in outcomes is most likely due in large part to different precipitation patterns in the two watersheds, as the remaining trees in the Yosemite area watersheds used all the extra water that was made available from the treatment, while the American River watershed was less water-limited and excess water was measured as additional runoff. Modeled simulations of forest fires in these watersheds showed similar differences in the water quantity outcomes in the two study areas. A reduction in vegetation of 40% due to a simulated forest fire resulted in a 75% difference in increased water yield between the two sites.

What was even more interesting is that when less intensive forest fuel treatments were applied, it appeared that the extent of forest canopy had a greater impact on water quantity than the total biomass across the landscape, although both were important factors. Forest canopy determines how much sunlight hits the forest floor, which in turn affects temperature, snowmelt, and evapotranspiration.

One of the primary learnings from this study is that water quantity impacts of forest fuel treatments that are measured in one location are not necessarily transferable to other locations that differ in weather, soil characteristics, and vegetative cover. However, this study highlights some of the key factors that determine water yield changes in a watershed under forest fuel interventions and demonstrates an approach that can be applied in other watersheds to deepen the understanding of changes in vegetation and water quantity outcomes.

(Source: https://snri.ucmerced.edu/news/2019/water-yield-forest-thinning-depends-how-where-and-how-much.)

References

Abell R, et al. Beyond the source: the environmental, economic and community benefits of source water protection. Arlington, VA, USA: The Nature Conservancy; 2017.

Acreman M, Holden J. How wetlands affect floods. Wetlands 2013;33(5):773–86. https://doi.org/10.1007/s13157-013-0473-2.

Arora K, Mickelson SK, Helmers MJ, Baker JL. Review of pesticide retention processes occurring in buffer strips receiving agricultural Runoff1. J Am Water Resour Assoc 2010;46(3):618–47.

Bennett G, Carroll N. Gaining depth: state of watershed investment 2014. Washington, DC: Forest Trends' Ecosystem Marketplace; 2014.

Beven KJ, Wood EF. Flow routing and the hydrological response of channel networks. Channel Netw Hydrol. Chichester: John Wiley; 1993. p. 99–128.

Bhadha JH, Capasso JM, Khatiwada R, Swanson S, LaBorde C. Raising soil organic matter content to improve water holding capacity; 2017. Document SL447 from the Soil and Water Science Department, UF/IFAS extension.

Botter G, Rinaldo A. Scale effect on geomorphologic and kinematic dispersion. Water Resour Res 2003;39. https://doi.org/10.1029/2003WR002154.

Bouwman AF, Van Drecht G, Knoop JM, Beusen AHW, Meinardi CR. Exploring changes in river nitrogen export to the world's oceans. Global Biogeochem Cycles 2005;19(1).

Brauman KA. Hydrologic ecosystem services: linking ecohydrologic processes to human well-being in water research and watershed management. Wiley Interdiscip Rev Water 2015;2(4):345–58.

Brauman KA, Freyberg DL, Daily GC. Land cover effects on groundwater recharge in the tropics: ecohydrologic mechanisms. Ecohydrology 2012;5(4):435–44. https://doi.org/10.1002/eco.236.

Brauman KA, Freyberg DL, Daily GC. Impacts of land-use change on groundwater supply: ecosystem services assessment in Kona, Hawaii. J Water Resour Plan Manag 2014;141(5):1–11. https://doi.org/10.1061/(ASCE)WR.1943-5452.0000495.

Bremer LL, Auerbach DA, Goldstein JH, Vogl AL, Shemie D, Kroeger T, et al. One size does not fit all: natural infrastructure investments within the Latin American Water Funds Partnership. Ecosyst Serv 2016;17:217–36. https://doi.org/10.1016/j.ecoser.2015.12.006.

Brix H. Use of constructed wetlands in water-pollution control—historical development, present status, and future perspectives. Water Sci Technol 1994;30(8):209–23. https://doi.org/10.2166/wst.1994.0413.

Brown AE, Zhang L, McMahon TA, Western AW, Vertessy RA. A review of paired catchment studies for determining changes in water yield resulting from alterations in vegetation. J Hydrol 2005;310(1):28–61. https://doi.org/10.1016/j.jhydrol.2004.12.010.

Bruijnzeel LA. (De)forestation and dry season flow in the tropics: a closer look. J Trop For Sci 1989;1(3):229–43.

Bruijnzeel LA, Mulligan M, Scatena FN. Hydrometeorology of tropical montane cloud forests: emerging patterns. Hydrol Process 2011;25:465–98. https://doi-org.stanford.idm.oclc.org/10.1002/hyp.7974.

Bullock A, Acreman M. The role of wetlands in the hydrological cycle. Hydrol Earth Syst Sci 2003;7(3):358–89. https://doi.org/10.5194/hess-7-358-2003.

Cambodian League for the Promotion and Defense of Human Rights (LICADHO), Cambodian Youth Network (CYN), Equitable Cambodia (EC), Sahmakum Teang Tnaut (STT). Smoke on the water: a social and human rights impact assessment of the destruction of the Tompoun/Cheung Ek wetlands, https://teangtnaut.org/wp-content/uploads/2020/07/Smoke-on-the-Water_Eng_Final..pdf; 2020.

Cassin J, Ochoa-Tocachi B. Learning from indigenous and local knowledge: the deep history of nature-based solutions. Nature-Based Solutions and Water Security: An Action Agenda for the 21st Century. Elsevier; 2021.

Cohen-Shacham, et al. Core principles for successfully implementing and upscaling nature-based solutions. Environ Sci Policy 2019;98:2019. https://doi.org/10.1016/j.envsci.2019.04.014.

Croke JC, Hairsine PB. Sediment delivery in managed forests: a review. Environ Rev 2006;14(1):59–87.

Dadson SJ, Hall JW, Murgatroyd A, Acreman M, Bates P, Beven K, et al. A restatement of the natural science evidence concerning catchment-based 'natural' flood management in the UK. Proc R Soc A 2017;473(2199). https://doi.org/10.1098/rspa.2016.0706, 20160706.

Dakhlalla AO, Parajuli PB, Ouyang Y, Schmitz DW. Evaluating the impacts of crop rotations on groundwater storage and recharge in an agricultural watershed. Agric Water Manag 2016;163:332–43. https://doi.org/10.1016/j.agwat.2015.10.001.

Dennedy-Frank PJ. Effects of land-cover change on streamflow: analysis of watershed simulations from around the world. Stanford, CA: Stanford University; 2018. Retrieved from http://purl.stanford.edu/qf950rf6314.

Dennedy-Frank PJ, Gorelick SM. Insights from watershed simulations around the world: watershed service-based restoration does not significantly enhance streamflow. Glob Environ Change 2019;58. https://doi.org/10.1016/j.gloenvcha.2019.101938.

D'Odorico P, Rigon R. Hillslope and channel contributions to the hydrologic response. Water Resour Res 2003;39. https://doi.org/10.1029/2002WR001708.

Ellison D, et al. Trees, forests and water: cool insights for a hot world. Glob Environ Change 2017;43:51–61. https://doi.org/10.1016/j.gloenvcha.2017.01.002.

Farley KA, Jobbagy EG, Jackson RB. Effects of afforestation on water yield: a global synthesis with implications for policy. Global Change Biol 2005;11(10):1565–76.

Filoso S, Bezerra MO, Weiss KCB, Palmer MA. Impacts of forest restoration on water yield: a systematic review. PLoS ONE 2017;12(8). https://doi.org/10.1371/journal.pone.0183210, e0183210.

Folke C, Hahn T, Olsson P, Norberg J. Adaptive governance of social-ecological systems. Annu Rev Env Resour 2005;30(1):441–73.

Fox GA, Wilson GV. The role of subsurface flow in hillslope and stream bank erosion: a review. Soil Sci Soc Am J 2010;74(3):717–33.

Graham NT, et al. Humans drive future water scarcity changes across all shared socioeconomic pathways. Environ Res Lett 2020. https://doi.org/10.1088/1748-9326/ab639b.

Gumiere SJ, Le Bissonnais Y, Raclot D, Cheviron B. Vegetated filter effects on sedimentological connectivity of agricultural catchments in erosion modelling: a review. Earth Surf Process Landf 2011;36(1):3–19.

Gyssels G, Poesen J, Bochet E, Li Y. Impact of plant roots on the resistance of soils to erosion by water: a review. Prog Phys Geogr Earth Environ 2005;29(2):189–217.

Hamel P, Riveros-Iregui D, Ballari D, Browning T, Célleri R, Chandler D, Chun KP, Destouni G, Jacobs S, Jasechko S, Johnson M, Krishnaswamy J, Poca M, Pompeu PV, Rocha H. Watershed services in the humid tropics: opportunities from recent advances in ecohydrology. Ecohydrology 2017;11(3), e1921.

Helmers MJ, Eisenhauer DE, Dosskey MG, Franti TG, Brothers JM, McCullough MC. Flow pathways and sediment trapping in a field-scale vegetative filter. Trans ASAE 2005;48(3):955–68.

Hoffmann CC, Kjaergaard C, Uusi-Kämppä J, Hansen HCB, Kronvang B. Phosphorus retention in riparian buffers: review of their efficiency. J Environ Qual 2009;38(5):1942–55.

IPCC. Climate change 2014: impacts, adaptation, and vulnerability. Part A: global and sectoral aspects. In: Field CB, Barros VR, Dokken DJ, Mach KJ, Mastrandrea MD, Bilir TE, et al., editors. Contribution of working group II to the fifth assessment report of the intergovernmental panel on climate change. Cambridge, United Kingdom and New York: Cambridge University Press; 2014.

Jones, et al. Restoration and repair of Earth's damaged ecosystems. Proc R Soc B Biol Sci 2018;285(1873):20172577. http://rspb.royalsocietypublishing.org/lookup/doi/10.1098/rspb.2017.2577.

Jordan SJ, Stoffer J, Nestlerode JA. Wetlands as sinks for reactive nitrogen at continental and global scales: a meta-analysis. Ecosystems 2011;14(1):144–55.

Kadykalo AN, Findlay CS. The flow regulation services of wetlands. Ecosyst Serv 2016;20(C):91–103. Elsevier.

Keeler BL, Polasky S, Brauman KA, Johnson KA, Finlay JC, O'Neill A, Kovacs K, Dalzell B. Linking water quality and well-being for improved assessment and valuation of ecosystem services. Proc Natl Acad Sci U S A 2012;109(45):18619–24.

Keys PW, Wang-Erlandsson L, Gordon LJ. Megacity precipitationsheds reveal tele-connected water security challenges. PLoS One 2018;13(3). https://doi.org/10.1371/journal.pone.0194311, e0194311.

Kim JH, Jackson RB. A global analysis of groundwater recharge for vegetation, climate, and soils. Vadose Zone J 2012;11:1.

Krutz LJ, Senseman SA, Zablotowicz RM, Matocha MA. Reducing herbicide runoff from agricultural fields with vegetative filter strips: a review. Weed Sci 2005;53(3):353–67.

Lawrence JE, Skold ME, Hussain FA, Silverman DR, Resh VH, Sedlak DL, Luthy RG, McCray JE. Hyporheic zone in urban streams: a review and opportunities for enhancing water quality and improving aquatic habitat by active management. Environ Eng Sci 2013;30(8):480–501.

Leduc C, Favreau G, Schroeter P. Long-term rise in a Sahelian water-table: the continental terminal in south-West Niger. J Hydrol 2001;243(1):43–54. https://doi.org/10.1016/S0022-1694(00)00403-0.

Leenheer JA, Croué J-P. Peer reviewed: characterizing aquatic dissolved organic matter. Environ Sci Technol 2003;37(1):18A–26A.

Liu X, Zhang X, Zhang M. Major factors influencing the efficacy of vegetated buffers on sediment trapping: a review and analysis all rights reserved. No part of this periodical may be reproduced or transmitted in any form or by any means, electronic or mechanical, including photocopying, recording, or any information storage and retrieval system, without permission in writing from the publisher. J Environ Qual 2008;37(5):1667–74.

Macisaac HJ. Potential abiotic and biotic impacts of Zebra mussels on the inland waters of north America1. Am Zool 1996;36(3):287–99.

Matthews J, Matthews N, Simmons E, Vigerstol K. Wellspring: source water resilience and climate adaptation. Arlington, VA: The Nature Conservancy; 2019. https://www.nature.org/content/dam/tnc/nature/en/documents/Wellspring_FULL_Report_2019.pdf.

Mayer PM, Reynolds SK, McCutchen MD, Canfield TJ. Meta-analysis of nitrogen removal in riparian buffers. J Environ Qual 2007;36(4):1172–80.

Milly PCD, Betancourt J, Falkenmark M, Hirsch RM, Kundzewicz ZW, Lettenmaier DP, Stouffer RJ. Stationarity is dead: whither water management? Science 2008;317:573–4.

Mirza N, Mahmood Q, Shah MM, Pervez A, Sultan S. Plants as useful vectors to reduce environmental toxic arsenic content. Sci World J 2014;2014. https://doi.org/10.1155/2014/921581, 921581.

Montakhab A, Yusuf B, Ghazali AH, Mohamed TA. Flow and sediment transport in vegetated waterways: a review. Rev Environ Sci Biotechnol 2012;11(3):275–87.

Norris V. The use of buffer zones to protect water quality: a review. Water Resour Manag 1993;7(4):257–72.

Ochoa-Tocachi BF, Bardales JD, Antiporta J, Pérez K, Acosta L, Mao F, Zulkafli Z, et al. Potential contributions of pre-Inca infiltration infrastructure to Andean water security. Nat Sustain 2019;2:584–93. https://doi.org/10.1038/s41893-019-0307-1.

Ogden FL, Crouch TD, Stallard RF, Hall JS. Effect of land cover and use on dry season river runoff, runoff efficiency, and peak storm runoff in the seasonal tropics of Central Panama. Water Resour Res 2013;49(12):8443–62. https://doi.org/10.1002/2013WR013956.

O'Connell PE, Ewen J, O'Donnell G, Quinn P. Is there a link between agricultural land-use management and flooding? Hydrol Earth Syst Sci 2007;11:96–107. https://doi.org/10.5194/hess-11-96-2007.

Polyakov V, Fares A, Ryder MH. Precision riparian buffers for the control of nonpoint source pollutant loading into surface water: a review. Environ Rev 2005;13(3):129–44.

Ponette-González AG, Marín-Spiotta E, Brauman KA, Farley KA, Weathers KC, Young KR. Hydrologic connectivity in the high-elevation tropics: heterogeneous responses to land change. Bioscience 2014;64(2):92–104.

Price K, Jackson CR, Parker AJ, Reitan T, Dowd J, Hall JS. Effects of watershed land use and geomorphology on stream low flows during severe drought conditions in the southern Blue Ridge Mountains, Georgia and North Carolina, United States. Water Resour Res 2011;47. https://doi.org/10.1029/2010WR009340, W02516.

Rittenburg RA, Squires AL, Boll J, Brooks ES, Easton ZM, Steenhuis TS. Agricultural BMP effectiveness and dominant hydrological flow paths: concepts and a review. J Am Water Resour Assoc 2015;51(2):305–29.

Roberts WM, Stutter MI, Haygarth PM. Phosphorus retention and remobilization in vegetated buffer strips: a review. J Environ Qual 2012;41(2):389–99.

Robinson JS, Sivapalan M, Snell JD. On the relative roles of hillslope processes, channel routing, and network geomorphology in the hydrologic response of natural catchments. Water Resour Res 1995;31(12):3089–101. https://doi.org/10.1029/95WR01948.

Salt DE, Smith RD, Raskin I. Phytoremediation. Annu Rev Plant Physiol Plant Mol Biol 1998;49(1):643–68.

Sheil D, Murdiyarso D. How forests attract rain: an examination of a new hypothesis. Bioscience 2009;59:341–7. https://doi.org/10.1525/bio.2009.59.4.12.

Sidle RC, Ziegler AD, Negishi JN, Nik AR, Siew R, Turkelboom F. Erosion processes in steep terrain—truths, myths, and uncertainties related to forest management in Southeast Asia. For Ecol Manage 2006;224(1–2):199–225.

Smith SV, Swaney DP, Talaue-Mcmanus L, Bartley JD, Sandhei PT, McLaughlin CJ, Dupra VC, Crossland CJ, Buddemeier RW, Maxwell BA, Wulff F. Humans, hydrology, and the distribution of inorganic nutrient loading to the ocean. Bioscience 2003;53(3):235–45.

Smith HG, Sheridan GJ, Lane PNJ, Nyman P, Haydon S. Wildfire effects on water quality in forest catchments: a review with implications for water supply. J Hydrol 2011;**396**(1):170–92.

Statzner B. Geomorphological implications of engineering bed sediments by lotic animals. Geomorphology 2012;157–158:49–65.

Stone M, Afrin Z, Gregory A. An investigation into the potential impacts of watershed restoration and wildfire on water yields and water supply resilience in the Rio Grande water fund project area; 2017. Prepared for the Middle Rio Grande Conservancy District.

Sudduth EB, Perakis SS, Bernhardt ES. Nitrate in watersheds: straight from soils to streams? J Geophys Res Biogeosci 2013;118(1):291–302.

Sweeney BW, Newbold JD. Streamside forest buffer width needed to protect stream water quality, habitat, and organisms: a literature review. J Am Water Resour Assoc 2014;50(3):560–84.

Tomasella J, Neill C, Figueiredo R, Nobre AD. Water and chemical budgets at the catchment scale including nutrient exports from intact forests and disturbed landscapes. In: Keller M, Bustamante M, Gash J, Dias PS, editors. Amazonia and global change. Geophysical Monograph Series 186. American Geophysical Union; 2009. p. 505–24.

UNEP. A snapshot of the world's water quality: towards a global assessment. Nairobi, Kenya: United Nations Environment Programme; 2016. p. 162.

UNEP-DHI. Green infrastructure guide for water management: ecosystem-based management approaches for water-related infrastructure projects; 2014. Produced jointly by UNEP, UNEP-DHI, IUCN and TNC.

Van Dijk AIJM, Van Noordwijk M, Calder IR, Bruijnzeel SLA, Schellekens J, Chappell NA. Forest–flood relation still tenuous—comment on 'global evidence that deforestation amplifies flood risk and severity in the developing world' by C. J. A. Bradshaw, N.S. Sodi, K. S.-H. Peh and B.W. Brook. Global Change Biol 2009;15(1):110–5.

Van Dijk AIJM, Peña-Arancibia JL, Bruijnzeel LA. Land cover and water yield: inference problems when comparing catchments with mixed land cover. Hydrol Earth Syst Sci 2012;16(9):3461–73.

Van Loon AF, Gleeson T, Clark J, Van Dijk AIJM, Stahl K, Hannaford J, et al. Drought in the Anthropocene. Nat Geosci 2016;9(2):89–91.

Van Vliet MTH, Florke M, Wada Y. Quality matters for water scarcity. Nat Geosci 2017;10(11):800–2.

Vörösmarty CJ, Green P, Salisbury J, Lammers RB. Global water resources: vulnerability from climate change and population growth. Science 2000;289:284–8.

Vymazal J. Constructed wetlands for wastewater treatment: five decades of experience. Environ Sci Technol 2011;45(1):61–9.

Wilcox BP, Turnbull L, Young MH, Williams CJ, Ravi S, Seyfried MS, Bowling DR, Scott RL, Germino MJ, Caldwell TG, Wainwright J. Invasion of shrublands by exotic grasses: ecohydrological consequences in cold versus warm deserts. Ecohydrology 2012;5(2):160–73.

Williams JB. Phytoremediation in wetland ecosystems: progress, problems, and potential. Crit Rev Plant Sci 2002;21(6):607–35.

WWAP/UN-Water. Nature-based solutions for water. Paris: UNESCO; 2018.

Zhang M, et al. A global review on hydrological responses to forest change across multiple spatial scales: importance of scale, climate, forest type and hydrological regime. J Hydrol 2017;546:44–59. https://doi.org/10.1016/j.jhydrol.2016.12.040.

Zuazo VCHD, Pleguezuelo CROR. Soil-erosion and runoff prevention by plant covers: a review. In: Lichtfouse E, Navarrete M, Debaeke P, Véronique S, Alberola C, editors. Sustainable agriculture. Dordrecht, Netherlands: Springer; 2009. p. 785–811.

CHAPTER 4

Nature-based solutions, water security and climate change: Issues and opportunities

Jan Cassin[a] and John H. Matthews[b]
[a]Forest Trends, Washington, DC, United States, [b]Alliance for Global Water Adaptation, Corvallis, OR, United States

Introduction

For the past 10 years, the World Economic Forum Global Risks reports have noted water supply crises and water hazards related to extreme weather as the most significant global risks in terms of likelihood and severity of impact, with biodiversity loss and ecosystem degradation consistently in the top risks (World Economic Forum, 2019). In both 2020 and 2021, failure to mitigate and adapt to climate change had the highest likelihood and risk ratings to the global economy and peoples' well-being (World Economic Forum, 2020, 2021). Climate change is an additional factor that interacts with other stressors on earth systems from human actions such as pollution of water sources, biodiversity loss, habitat conversion and fragmentation, and unsustainable resource exploitation (Watson et al., 2018).

Water and climate in particular are closely linked. People will feel the effects of climate change mostly through water. From more and longer or deeper droughts impacting food and energy production, loss of life and damages from more extreme storms and flooding, or the unpredictability of rainfall that can lead farmers to abandon land and can spur internal and international migration—fundamentally, climate is water (Nyugen and Hens, 2019; Jiménez Cisneros et al., 2014). Water provision is critical for economic development, underpinning agricultural production and fisheries, supporting biodiversity, energy generation, and water-based transportation, meaning that the water crisis is a systemic risk that affects all major global challenges (WWAP, 2016; World Economic Forum, 2011, 2019). There is an increasingly urgent need to identify and implement robust climate adaptation measures that can help build greater resilience to a range of possible future conditions.

While hard (engineered gray infrastructure) and soft approaches, such as better early warning systems for hazards, have been the most prevalent adaptation approaches to date, there is growing interest in the potential for nature-based solutions (NBS) as cost-effective, flexible solutions that can complement gray infrastructure in climate adaptation (Chausson et al., 2020; Global Commission on Adaptation, 2019; Jones et al., 2012; Seddon et al., 2020a).

Nature-Based Solutions and Water Security. https://doi.org/10.1016/B978-0-12-819871-1.00017-8

Importantly, because NBS are inherently multifunctional, they can address the interdependent challenges of ecosystem degradation, biodiversity loss, human well-being, and climate change together. Furthermore, because water is an enabler of adaptation in other sectors (Smith et al., 2019), NBS focused on water risks are particularly well-suited in providing systemic, cross-sectoral, and integrative adaptation measures.

In this chapter we provide an introductory guide to the issues and opportunities for addressing climate change with NBS, particularly NBS for climate-resilient water management. Recognizing that this topic is complex—combining the complexities of water security, climate change and social-ecological systems—and is a rapidly evolving area of research and practice, we hope to highlight some important concepts and the potential and limitations for using NBS to address climate change.

Nature-based solutions for climate adaptation

NBS are a useful umbrella concept for other established approaches, such as ecosystem-based adaptation, which work with nature to address societal challenges (Table 4.1) (Cassin and Matthews, 2021a; Cohen-Shacham et al., 2016). NBS have become increasingly important in climate change policy, as reflected in the recognition that NBS have an important role to play in Nationally determined contributions (NDCs) and that a majority of signatories to the Paris Agreement have included NBS in some form in their NDCs (Seddon et al., 2020b; Timboe et al., 2019). However, systematic and targeted implementation at scale is still in the distance. In most of these NDCs there are general goals or visions for including NBS in adaptation or mitigation, but fewer cases of specific nature-based actions, and fewer still of measurable targets for these actions (Seddon et al., 2020b).

Table 4.1: Related terms and definitions for approaches to societal challenges based on nature.

Term/concept	Source	Definition and description
Nature-Based Solutions		
Nature-based Solutions	UN World Water Assessment Program 2018	Nature-based solutions (NBS) are inspired and supported by nature and use, or mimic, natural processes to contribute to the improved management of water. An NBS can involve conserving or rehabilitating natural ecosystems and/or the enhancement or creation of natural processes in modified or artificial ecosystems. They can be applied at microscales (e.g., a dry toilet) or macroscales (e.g., watershed restoration) (WWAP, 2018).
Nature-based Solutions	IUCN	Actions to protect, sustainably manage, and restore natural or modified ecosystems that address societal challenges effectively and adaptively, simultaneously providing human well-being and biodiversity benefits. IUCŃs NBS concept includes a broad range of solutions or approaches for water security and challenges such as conservation, climate adaptation, disaster risk reduction, and food security. IUCŃs definition considers green infrastructure and natural infrastructure as subsets of NBS—the NBS that are explicitly delivering the same services as built or engineered infrastructure (Cohen-Shacham et al., 2016).

Table 4.1: Related terms and definitions for approaches to societal challenges based on nature—cont'd

Term/concept	Source	Definition and description
Nature-based Solutions	Climate Bonds Initiative	NBS are the explicit, planned, intentional use of ecosystems to meet human needs; "nature-based infrastructure" uses ecological assets or "ecosystem-based features, processes, and functions" to address water needs. Under the umbrella of NBS, they distinguish: — natural features (protected or restored ecosystems); and — nature-based features (or features based on nature) that use or mimic the characteristics of ecosystems and natural processes but use human design, engineering, and construction to deliver or enhance a (usually single) service. In distinguishing natural features from nature-based features, CBI follows US Army Corps of Engineers practice, which is also similar to the MNAI distinction of natural assets, enhanced assets, and engineered assets (Climate Bonds Initiative, 2018).
Nature-based Solutions	European Commission	Living solutions inspired by, continuously supported by, and using nature designed to address various societal challenges in a resource efficient and adaptable manner and to provide simultaneously economic, social, and environmental benefits (Maes and Jacobs, 2017).
Green Infrastructure		
Green Infrastructure	European Commission	A strategically planned network of natural and seminatural areas with other environmental features designed and managed to deliver a wide range of ecosystem services (http://ec.europa.eu/environment/nature/ecosystems/index_en.htm).
Green Infrastructure	Benedict and McMahon 2002	An interconnected network of protected land and water that supports native species, maintains natural ecological processes, sustains air and water resources, and contributes to the health and quality of life for America's communities and people. This conception of green infrastructure is similar to the narrower sense in which green infrastructure is used (often in the United States and European Union) to denote urban green infrastructure that is designed to address challenges around urban flooding and drainage, heat island, air quality, and/or recreation opportunities in highly developed urban areas (Benedict and McMahon, 2000).
Green-Blue Infrastructure and Natural Infrastructure	UNISDR	A strategically planned network of natural and seminatural areas with other environmental features designed and managed to deliver a wide range of ecosystem services such as water purification, air quality, space for recreation, climate mitigation and adaptation, and management of wet weather impacts that provides many community benefits (UNISDR, 2017).
Ecosystem-based Adaptation, Disaster Risk Reduction, and Mitigation		
Ecosystem-based Adaptation (EBA)	Convention on Biological Diversity	Ecosystem-based adaptation is the use of biodiversity and ecosystem services as part of an overall adaptation strategy to help people adapt to the adverse effects of climate change (CBD, 2009).

Continued

Table 4.1: Related terms and definitions for approaches to societal challenges based on nature—cont'd

Term/concept	Source	Definition and description
Ecosystem-based Disaster Risk Reduction (ECO-DRR) Ecosystem-based Mitigation (EBM) Ecological Engineering	UNDRR	The sustainable management, conservation, and restoration of ecosystems to reduce disaster risk, with the aim to achieve sustainable and resilient development (Estrella and Saalismaa, 2013). EBM is the use of ecosystems for their carbon storage and sequestration service to aid climate change mitigation (Sudmeier-Rieux et al., 2019). The design of sustainable ecosystems that integrate human society with its natural environment for the benefit of both (Mitsch, 2012). An example is "Building with Nature" (Box 2 and p. 16), a comprehensive engineering approach that seeks to enhance the use of natural ecological processes to achieve efficient and sustainable hydraulic infrastructural designs.

Most of the discussion of NBS for climate has focused on the potential for mitigation through the sequestration and storage of carbon in ecosystems—primarily in terrestrial vegetation and soils (Fargione et al., 2018; Griscom et al., 2020). While NBS can play a significant mitigation role in carbon capture and storage, some conclusions about the mitigation potential may be overestimated (Veldman et al., 2019), and there are potential downsides if not done well (e.g., massive afforestation, nonnative plantations). These downsides include: taking attention away from the need for aggressive emissions reductions; impacts to biodiversity; displacing Indigenous communities; displacing lands vital for food production; and negatively impacting water resources (Anderson et al., 2019; Seddon et al., 2019; Smith et al., 2017; Veldman et al., 2015). We focus here primarily on NBS for adaptation, recognizing that many NBS for adaptation will also have mitigation benefits.

Adaptation, resilience, and nature-based solutions

Resilience, adaptation, and vulnerability are closely linked concepts (see Box 4.1). Resilience, in particular, is a term that has been used in different (often vague) ways across different disciplines, including engineering, psychology, and ecology (Olsson et al., 2015). A common understanding of resilience is the ability to cope with stress and/or return to a prior normal condition following a period of stress or "an ability to recover from or adjust easily to misfortune or change" (Merriam Webster, https://www.merriam-webster.com/dictionary). Although resilience has been used in a wide variety of senses, the term "resilience" comes from the Latin *resilire*, which means to spring or bounce back. This meaning is aligned with traditional engineering concepts of resilience and early ecological concepts and refers the

Box 4.1   Risk, resilience, and adaptation.

Adaptation: The UNFCCC defines adaptation as "adjustments in ecological, social, or economic systems in response to actual or expected climatic stimuli and their effects or impacts. It refers to changes in processes, practices, and structures to moderate potential damages or to benefit from opportunities associated with climate change...to respond to the impacts of climate change that are already happening, as well as prepare for future impacts." Or the specific actions taken to respond to or prepare for particular climate impacts.

Adaptive capacity: Adaptive capacity is the ability of an individual, social, or environmental system or a natural or built asset to adjust to a hazard or changing conditions, cope with change, and or respond to new opportunities. The ability to adjust or innovate in response to changing conditions.

Exposure: The presence of people, assets, and ecosystems in places where they could be adversely affected by hazards; the extent to which these are subject to potential impacts of climate change.

Resilience: Encompasses the ability to absorb or withstand shocks and maintain function, including through reorganization, adaptive capacity, learning, and transformation (Folke, 2006). Resilience includes aspects of resistance to change, recovery from change (bouncing back), and adapting in the face of change to a new state (bounce back with transformation).

Risk: Risk can be defined as a function of threat and vulnerability.

Sensitivity: The degree to which a system, population, or resource is or might be affected by hazards. The degree to which a system is affected by or responsive to climate impacts.

Threat: The actions that can negatively impact an individual, community, or system.

Vulnerability: The degree of potential damage, influenced by conditions determined by physical, social, economic, and environmental factors or processes, which increase the susceptibility to an individual or community to the impact of hazards. Vulnerability is influenced by sensitivity, exposure, and adaptive capacity.

From: C2ES (Center for Climate and Energy Solutions). What is climate resilience and why does it matter? Clim Essent 2019; ETH (Swiss Federal Institute of Technology Zurich). Expressions of resilience: from 'bounce back' to adaptation. Factsheet, 3RG report. Zurich: Risk and Resilience Research Group, Center for Security Studies (CSS) ETH; 2012; Folke C. Resilience: The Emergence of a Perspective for Social-Ecological Systems Analyses. Glob Environ Chang 2006;16(3):253–67. Pergamon. doi:https://doi.org/10.1016/j.gloenvcha.2006.04.002; IPCC. Climate change 2014: impacts, adaptation, and vulnerability. In: Field CB, Barros VR, Dokken DJ, Mach KJ, Mastrandrea MD, Bilir TE, Chatterjee M, Ebi KL, Estrada YO, Genova RC, Girma B, Kissel ES, Levy AN, MacCracken S, Mastrandrea PR, White LL, editors. Part A: Global and sectoral aspects. Contribution of working group II to the fifth assessment report of the intergovernmental panel on climate change. Cambridge and New York, NY: Cambridge University Press; 2014. 1132 pp.; Matthews JH, Matthews N, Simmons E, Vigerstol K. Wellspring: Sources water resilience and climate adaptation. Arlington, VA: The Nature Conservancy; 2019. https://www.nature.org/content/dam/tnc/nature/en/documents/Wellspring_FULL_Report_2019.pdf; UNDRR. Report of the open-ended intergovernmental expert working group on indicators and terminology relating to disaster risk reduction. UNGA 2016; 2017.

ability of a system to return to its normal functioning state following a disturbance (ETH, 2012). In both disciplines, this sense of resilience was tied to a view of equilibrium systems operating under a narrow set of well-defined, historical boundary conditions where change largely happens within those boundaries.

Olsson et al. (2015) describe two different concepts of resilience in the scientific literature: bounce back (closer to the engineering concept of resilience); and bounce back with transformation (current ecological concepts of resilience). Bounce back with transformation acknowledges that change can be extreme and can create conditions that are outside the "normal" or historical boundary conditions. Resilience in these cases requires a transformation or adaptation to the new conditions, resulting in a new state that still maintains basic functionality but may look very different from the original system (Folke et al., 2010). Both ecosystems and human societies have exhibited this type of resilience—recovery with transformation, avoiding collapse by transforming in a way that allows them to thrive in a new state (Haldon et al., 2018; Linner and Wibeck, 2019; Mundy, 2015).

Climate resilience for water security

Given the rapid pace of the alteration of earth systems over the past half-century (Rockström et al., 2009; Steffen et al., 2015), in both engineering and ecology disciplines the conception of resilience has been evolving into one in which resilience is seen as including the ability to be flexible in the face of change, to "change with change." As the pace of climate change and other social-ecological stressors is increasing, the resilience of linked social-ecological systems will need to increasingly stress adaptation or "bouncing back with transformation" (Folke, 2006; Matthews et al., 2019; Poff et al., 2016). Climate-resilient water security will require this concept of resilience—being able to adjust to change, recover from it, and when recovery is not possible, reorganize in new ways that allow society to meet needs for well-being (Burgess et al., 2019; Matthews et al., 2019; Rodina, 2018; Smith et al., 2019).

There are several factors that make preparing for climate-resilient water systems particularly challenging. Hydrological systems are inherently variable, but societies have been able to manage this variability based on historical patterns that defined typical ranges of variation (e.g., return periods for floods of a particular magnitude). We have designed our water infrastructure—mostly gray—to be robust within the range of variation that we defined from past conditions, and we could be reasonably sure that conditions would not exceed these boundaries, except perhaps with a very low probability that is accepted as a reasonable risk. However, past history is no longer a reliable guide for how hydrological systems will change under future climates (and are already changing) making it very difficult to identify robust water solutions (Milly et al., 2008).

In addition, societies must now make decisions about water management under "deep uncertainty" (Lempert et al., 2003; Smith et al., 2019). Deep uncertainty arises when stakeholders and decision-makers cannot agree on or cannot know how a particular system will work, how likely different possible future states of the system are, or how important different outcomes might be (Marchau et al., 2019). Under deep uncertainty, many plausible future climate conditions and resulting hydrological systems can be identified, but it is not possible to rank them in terms of likelihood of occurrence or outcome. We are not able to really predict future climate and hydrological conditions or their impacts, and we are not even able to predict what the range of possible future conditions will be (Smith et al., 2019).

As Smith et al. (2019) point out, this means that if we are to ensure water security (access to reliable supplies of adequate quality, mitigation of hazards) under future climate conditions, we need to identify solutions that can balance robustness (function over a range of conditions) and flexibility (capable of change/adaptation). Robust measures are those that, with a high degree of confidence, will perform well under a range of future conditions (Wilby and Dessai, 2010). Flexible approaches are capable of adjusting strategies or changing course when there are no credible choices between possible alternative futures (we just do not know) or future changes result in impacts that were not anticipated. Flexibility means not closing off options and being able to choose a new path in the future as conditions change or results of changing conditions become obvious. In other words, flexibility helps increase the chances that we will choose adaptation pathways that do not lead to dead ends (Hallegatte et al., 2012).

Importance of integrating NBS and hard adaptation approaches

As Smith et al. (2019) argue, NBS in combination with gray infrastructure is a key strategy for balancing robustness and flexibility in developing climate-resilient water infrastructure. Incorporating NBS in water resilience strategies provides diversity and redundancy—both traits that contribute to flexibility and adaptive capacity. For example, sponge cities or Copenhagen's blue-green cloudburst strategies that combine green infrastructure—urban wetlands, forests, green roofs, and green streets—with traditional stormwater management, such as detention ponds and conveyance tunnels, results in a diversity of potential responses to urban flooding. The green infrastructure absorbs and slows down flood waters, preventing or greatly reducing damage from most precipitation events but may be overwhelmed by very large events. The gray infrastructure provides a backstop to cope with larger events by storing large amounts of water on the surface or conveying flood waters downstream. Using both natural and engineered water storage to withstand times of scarcity can combine both robustness and flexibility. For example, maintaining forested natural areas or restoring ancestral NBS that enhance groundwater recharge and storage, can complement surface storage in reservoirs providing redundant systems if one source of supply becomes threatened or fails (Cassin and Ochoa-Tocachi, 2021; Ochoa-Tocachi et al., 2019). Relying only on NBS or gray infrastructure is not likely to be a successful adaptation strategy, and importantly, reliance only on expensive, long-lasting, and inflexible gray infrastructure solutions can make adopting NBS in the future more difficult or expensive (see Tellman et al., 2018 for an example).

Jones et al. (2012) similarly identify three contexts in which NBS should be more central to adaptation actions. First, by complementing gray infrastructure, NBS can increase the overall adaptive capacity, balancing robustness and flexibility. They cite the example of Yangtze River flood dikes along the river that have increased flooding in some areas, eliminated fisheries, degraded floodplain wetlands, and exacerbated water quality in lakes cut off from the river. A new system of restoring environmental flows and reconnecting floodplain wetlands in combination with setting levees back further away from the river, provides additional flood storage and improving water quality, fisheries, and agriculture while the relocated levees provide a backstop in the case of very large floods (Yu et al., 2009).

Second, in many cases NBS can be a more cost-effective alternative, for example, protecting source watershed forests and wetlands to maintain water quality for drinking water supplies can reduce treatment costs and/or avoid the need for new treatment plants, while also providing a suite of cobenefits (see review in Cassin, 2021; Matthews et al., 2019). Finally, in contexts where gray infrastructure solutions are not possible or have already resulted in maladaptation, NBS are the only adaptation option. For years the city of New Orleans relied on flood control levees to manage riverine and coastal flooding from large storms, but during Hurricane Katrina, the levees actually held floodwaters in, causing additional damage (Day et al., 2007). Relocating levees and restoring coastal marshes are providing more protection to New Orleans while addressing some of the negative impacts of the original levee system.

Adaptation services from nature

The ability of human societies to thrive, especially under change, is dependent on the delivery of ecosystem services, which underpin all aspects of human well-being (Wood et al., 2018). Climate adaptation measures have generally been categorized in two ways: hard approaches that use gray infrastructure-based engineering measures such as levees, seawalls, or dams; and soft approaches that use information, capacity building, institutions, and governance (Hale et al., 2011; Parry et al., 2009). NBS for adaptation are a third approach that shares aspects of both hard and soft measures (Jones et al., 2012). Nature-based or ecosystem-based adaptation uses the capacity of natural systems to "buffer people against the adverse impacts of climate change through the sustainable delivery of ecosystem services" (Jones et al., 2012). Adaptation approaches based on nature and on maintaining natural infrastructure intact can help reduce risks of natural disasters, improve food security, maintain reliable flows of water, support diversified economies, and regulate local and regional climates (Lavorel et al., 2015).

Lavorel et al. (2015) define adaptation services of nature as "the benefits to people from increased social ability to respond to change, provided by the capacity of ecosystems to moderate and adapt to climate change and variability." Adaptation services therefore include the ability of ecosystems to provide buffering against change (resistance) and the ability of ecosystems to recover from change through adapting and transforming in the face of change—both the bounce back and bounce back with transformation aspects of resilience. Buffering capacity in the face of change helps ensure that the current suite of ecosystem services that are providing services to people will continue to be delivered, including services that may not be valued now but provide insurance or option values in the future. For example, if water is currently abundant, the function of mountain forests in recharging local aquifers may not be valued now, but it will become valuable under future conditions of more variable and unpredictable precipitation.

Adaptive capacity of ecosystems

Ecosystems support not only adaptation services for people but also the delivery of these services, and one of the strengths of nature-based approaches for adaptation arises from the inherent adaptive capacity of ecosystems themselves, their ability to evolve and adapt to change. Importantly, ecosystem attributes that enable the persistence of current ecosystem services also support the transformative capacity of ecological systems. Attributes such as functional diversity and connectivity can provide both robustness and flexibility in adaptive responses, so that continued delivery of ecosystem services can be robust under a variety of possible future conditions (Lavorel et al., 2019; Oliver et al., 2015).

Ecosystems will be impacted by climate change but also have the capacity to adapt and transform as conditions change, resulting in new suites of ecosystem services that provide new benefits to people. For example, permanent wetlands that provide little flood mitigation under current conditions may transform to intermittent wetlands under drier future conditions, but because intermittent wetlands are not continually saturated, they have the capacity to absorb more runoff and provide some flood mitigation. Understanding how ecosystems may transform under climate change can provide clues for choosing which NBS to implement to enhance adaptation and social-ecological resilience as conditions change.

The option values of ecosystems for the future are also important for adaptation services. Lavorel et al. (2019) argue that many ecosystem services that are not highly valued now may become critically important under new climate conditions. For example, local climate regulation services, such as shade from tree canopies that moderate air and water temperatures and maintain soil moisture, may become much more important in providing resilience to increased drought conditions under warmer temperature regimes. The full suite of ecosystem services provided by natural systems, even those not currently of obvious value, are the basis for maintaining flexible, open adaptation pathways for the future (Lavorel et al., 2019; Tellman et al., 2018).

What makes ecosystems resilient?

If ecosystem resilience underpins the delivery of ecosystem services and our capacity to adapt to climate change, what makes ecosystems resilient? There is a broad consensus on common resilience attributes of both terrestrial and freshwater systems (Table 4.2) (Biggs et al., 2020;

Table 4.2: Ecosystem attributes that contribute to adaptive capacity and social-ecological resilience.

Ecosystem attribute	Contribution to resilience
Landscape Level Mechanisms	
Connectivity	Connectivity of terrestrial habitats or upstream/downstream and river/floodplain connectivity in freshwater systems influence resilience in a number of ways. Connectivity supports population persistence: through the "rescue effect" or the ability to recolonize an area following a disturbance; allowing movement between habitats critical for different life stages (e.g., salmon access to spawning in headwaters streams as well as ocean waters for adult stages); allowing movement to new areas as current habitats become unsuitable with climate change (e.g., room for salt marshes to migrate inland as sea levels rise).
Spatial Heterogeneity	Spatial variation contributes to functional and species diversity (a greater variety of habitats that can support different species) while also ensuring the complexity and diversity of ecological processes and functions (which deliver ecosystem services) are maintained. Channelizing rivers and disconnecting floodplains reduces the complexity of river systems and eliminates processes of over bank flooding and flood storage, water purification and filtration, and groundwater recharge.
Temporal Variability	In freshwater systems especially, temporal variation in the natural flow regime, the timing and magnitude of seasonal flow patterns, for example, are the basis for maintaining biodiversity and functional diversity.
Size of Habitat Area	Connectivity also contributes to maintaining large, unfragmented habitat areas, which in turn are related to maintaining larger population sizes, which contributes to greater genetic diversity. Habitat fragmentation that leads to small population sizes reduces genetic diversity and the capacity for adaptive evolution.
Community Level Mechanisms	
Functional Redundancy	Many different species may perform the same ecological functions, resulting in functional redundancy within communities. For example, many different types of insects, birds, and mammals function as pollinators. This redundancy is sometimes referred to as the "insurance effect of biodiversity" because it makes it possible for functions and ecosystem services to be maintained even if some species are lost. Functional redundancy among riparian tree and shrub species has been found to be important in maintaining nutrient cycling functions in riparian buffers under changing nutrient loads (Bruno et al., 2016). Functional redundancy is generally increased by species diversity, genetic diversity, spatial heterogeneity, and trait (response) diversity.
Species or Population Level Mechanisms	
Sensitivity to Environmental Change/ Response Traits Intrinsic Rate of Population Growth	Species vary in their behavioral or physiological traits that mediate sensitivity to environmental conditions or the ability to resist change. For example, different species of trees have different capacities to tolerate drought conditions, which makes some more resistant to drought conditions or to have a low vulnerability to drought. The greater the response diversity across species, the wider the set of environmental conditions that can be tolerated. Recovery from disturbances can be facilitated by high growth rates as populations recover to predisturbance levels more quickly. However, response traits and growth rates are frequently "bundled" in ways that intertwine resistance and recovery. For example, trees that are more resistant to drought may also be slower growing, meaning that their ability to recover from disturbance is limited, so a trait that may be resilient when droughts are intermittent, relatively infrequent, and not long duration, could mean reduced resilience when drought periods become more intense, frequent, and long-lasting. If growth rates are slow, recovery time will not be sufficient to balance resistance to short, more minor droughts.

Table 4.2: Ecosystem attributes that contribute to adaptive capacity and social-ecological resilience—cont'd

Ecosystem attribute	Contribution to resilience
Genetic Diversity (both short and long-term responses to change)	Greater genetic diversity in a population increases the chances that some individuals will have traits that are resistant to disturbances or new conditions. This in turn increases the chance that some members of the population will persist in the face of disturbance and the functions and ecosystem services they support will also persist. Genetic diversity within a species can increase the chances that traits will evolve as conditions change, allowing species to adjust to new conditions. Although some trait evolution can be rapid, the very rapid pace of anthropogenic environmental change will likely outpace evolutionary adaptation for many species.

Grill et al., 2019; Grantham et al., 2019; Lavorel et al., 2015; Oliver et al., 2015; Poff, 2018; Walker, 2020). Many of these attributes work by supporting diversity at multiple levels, including habitat diversity, trait diversity, functional diversity, genetic diversity, and species diversity.

What are the implications for nature-based adaptation?

The emerging evidence base, remaining knowledge gaps, and recommendations for implementing NBS for adaptation have been the subject of several recent reviews re (see for example, Chausson et al., 2020; Kapos et al., 2019; Malhi et al., 2020; Morecroft et al., 2019; Seddon, 2018). The policy perspective and integrating NBS in national and international climate adaptation policies and frameworks have also been addressed recently (see OECD, 2020; UNDRR, 2020). Here we discuss additional implications that need to be taken into consideration in planning, designing, and managing NBS for adaptation.

Nature-based solutions' vulnerability to climate impacts

While there is uncertainty about specific climate-induced changes in natural systems in particular locations, there is consensus that climate change will impact those attributes of ecosystems that contribute to resistance, recovery, and adaptive capacity (Hoegh-Guldberg et al., 2018). For example, more diverse plant communities resist and recover from droughts more quickly than monocultures or species poor communities (Chapin and Diaz, 2020; Hutchison et al., 2018). Plant diversity is likely to decline in much of the world under warmer, drier conditions and more frequent droughts (Harrison et al., 2020). Nature-based adaptation strategies, such as protecting or restoring forests and grasslands (e.g., high altitude paramos) for their role in maintaining reliable water flows, maintaining soil moisture, and reducing risks from moderate droughts, may not be able to provide these services under warmer, drier conditions in the future. Designs for NBS that explicitly include steps to protect existing biodiversity or enhance biodiversity in restored or managed systems are more likely to continue providing adaptation services as the climate changes.

The increasing frequency and severity of extreme events that are already being seen with climate change will make NBS even more vulnerable than they will be to more gradual climate changes (Johnstone et al., 2016). There is growing evidence that extreme conditions (frequent severe floods, deep and long droughts), which are becoming more frequent, are taxing the ability of many ecosystems to recover from repeated disturbances. Under climate change new disturbance regimes are likely to be radically different from those under which ecosystems have evolved. Examples include more frequent and intense wildfires in forests in the western United States, frequent episodes of coral bleaching in the Great Barrier Reef, and deeper droughts in the Amazon basin (Hughes et al., 2019; Johnstone et al., 2016). Rather than recover to some prior state, these ecosystems may be transforming into new states—novel ecosystems characterized by new sets of species and a new suite of ecosystem services (Hobbs et al., 2009).

Abrupt ecological change

The design and implementation of NBS (ideally) is based on the best available ecological science. Ecologists have a long tradition of studying how ecosystems change over time, but these changes, for the most part, have been more gradual than the accelerating pace of climate change. The phenomenon of abrupt change in ecological systems (ACES) is becoming more common, and these changes are increasing in frequency (Ratajczak et al., 2018). Abrupt changes are leading to tipping points or rapid shifts into new systems, for example from forest to savanna or grassland (Halofsky et al., 2018; Kitzberger et al., 2016). Turner et al. (2020) suggest several insights about abrupt changes: they are more likely triggered by extreme events rather than mean trends in climate; they are often the result of interactions among multiple drivers (climate, land use change, pollution); and they depend on contingencies, such as the sequence and frequency of disturbances or ecological memory. Ecological memory refers to the legacy of past disturbances and the individuals of different species, adaptations, and materials that remain after a disturbance. This postdisturbance legacy then determines the future trajectory—or path dependency—of the system under new disturbances.

How can NBS practitioners address these challenges?

While climate change and other anthropogenic stressors will influence how effective nature-based adaptation approaches will be, there are ways that practitioners and decision-makers can address these challenges.

- Existing natural systems, especially those that are intact and healthy, are already providing adaptation services and should be protected.

- When designing and implementing NBS based on restoration or managed systems (e.g., agroforestry), manage for resilience:
  - Enhance and manage for connectivity in the landscape in which NBS are implemented.
  - Enhance and manage for diversity (species, response traits, functional diversity, redundancy).
  - Design NBS that cover a larger area (more likely to persist and deliver desired adaptation services at scale).
- Combine hard, soft, and nature-based measures in diverse portfolios that are more likely to be robust and flexible.
- Understand how interacting stressors may affect the performance and viability of the NBS over time.
- Avoid NBS that may be maladaptive (e.g., nonnative tree plantations; afforestation that can affect water yield, and resilience to future climate change).
- Embrace the likely need for some form of ongoing management and adaptive management of NBS.

NBS have great potential to enhance social-ecological resilience but are not a silver bullet, and implementing successful solutions will require careful planning, commitment to learning, and adjusting approaches both as we learn more and as conditions change.

References

Anderson CM, DeFries RS, Litterman R, Matson PA, Nepstad DC, Pacala S, Schlesinger WH, Shaw MR, Smith P, Weber C, Field CB. Natural climate solutions are not enough: decarbonizing the economy must remain a critical priority. Science 2019;363(6430):933–4.

Benedict MA, McMahon ET. Green infrastructure: smart conservation for the 21st century. Recreation 2000;37:4–7. https://doi.org/10.4135/9781412973816.n70.

Biggs CR, Yeager LA, Bolser DG, Bonsell C, Dichiera AM, Hou Z, Keyser SR, Khursigara AJ, Lu K, Muth AF, Negrete B, Erisman BE. Does functional redundancy affect ecological stability and resilience? A review and meta-analysis. Ecosphere 2020;11(7):e03184.

Bruno D, Gutiérrez-Cánovas C, Sánchez-Fernández D, Velasco J, Nilsson C, et al. Impacts of environmental filters on functional redundancy in riparian vegetation. J Appl Ecol 2016;53:846–55.

Burgess R, Horbatuck K, Beruvides M. From mosaic to systemic redux: the conceptual foundation of resilience and its operational implications for water resource management. Systems 2019;7(3):32. https://doi.org/10.3390/systems7030032.

Cassin J. Nature-based solutions for source water protection in North America. In: Cassin J, Matthews JH, Lopez-Gunn E, editors. Nature-based solutions and water security: An action agenda for the 21st century. Elsevier; 2021.

Cassin J, Matthews JH. Setting the scene: nature-based solutions and water security. In: Cassin J, Matthews JH, Lopez-Gunn E, editors. Nature-based solutions and water security: An action agenda for the 21st century. Elsevier; 2021a.

Cassin J, Ochoa-Tocachi B. Learning from indigenous and local knowledge: the deep history of nature-based solutions. In: Cassin Jan, Matthews John H, Lopez-Gunn Elena, editors. Nature-Based Solutions and Water Security: An Action Agenda for the 21st Century. Elsevier; 2021.

CBD. Connecting biodiversity and climate change mitigation and adaptation: Report of the second ad hoc technical expert group on biodiversity and climate change. Technical series no. 41, Montreal, QC: Secretariat of the Convention of Biological Diversity; 2009.

Chapin S, Diaz S. Interactions between changing climate and biodiversity: shaping humanity's future. PNAS 2020;117(12):6295–6.

Chausson A, Turner B, Seddon D, Chabaneix N, Girardin CAJ, Kapos V, Key I, Roe D, Smith A, Woroniecki S, Seddon N. Mapping the effectiveness of nature-based solutions for climate change adaptation. Glob Chang Biol 2020;26(11):6134–55.

Climate Bonds Initiative. Water infrastructure criteria under the climate bonds standard criteria document, https://www.climatebonds.net/standard/water; 2018.

Cohen-Shacham E, Walters G, Janzen C, Maginnis S, editors. Nature-based solutions to address global societal challenges. IUCN International Union for Conservation of Nature; 2016. https://doi.org/10.2305/IUCN.CH.2016.13.en.

Day Jr JW, et al. Restoration of the Mississippi Delta: lessons from hurricanes Katrina and Rita. Science 2007;315:1679–84.

Estrella M, Saalismaa N, editors. Ecosystem-based disaster risk reduction (Eco-DRR): an overview. In: The role of ecosystems in disaster risk reduction. Shibuya-ku, Tokyo: United Nations University Press; 2013.

ETH (Swiss Federal Institute of Technology Zurich). Expressions of resilience: from 'bounce back' to adaptation. In: Factsheet, 3RG report. Zurich: Risk and Resilience Research Group, Center for Security Studies (CSS) ETH; 2012.

Fargione JE, et al. Natural climate solutions for the United States. Sci Adv 2018;4:eaat1869. Accessed through FIU Digital Commons https://digitalcommons.fiu.edu/fce_lter_journal_articles/494.

Folke C. Resilience: the emergence of a perspective for social-ecological systems analyses. Glob Environ Chang 2006;16(3):253–67. Pergamon https://doi.org/10.1016/j.gloenvcha.2006.04.002.

Folke C, Carpenter SR, Walker B, Scheffer M, Chapin T, Rockström J. Resilience thinking: integrating resilience, adaptability and transformability. Ecol Soc 2010;15(4):20. https://doi.org/10.5751/ES-03610-150420.

Global Commission on Adaptation. Adapt now: A global call for leadership on climate resilience. Rotterdam and Washington, DC: Global Center on Adaptation and World Resources Institute; 2019.

Grantham TE, Matthews JH, Bledsoe BP. Managing freshwater ecosystems for ecological resilience in a changing climate. Water Secur 2019;8:100049.

Grill G, Lehner B, Thieme M, Geenen B, Tickner D, Antonelli F, Babu S, Borrelli P, Cheng L, Crochetiere H, Macedo HE. Mapping the world's free-flowing rivers. Nature 2019;569(7755):215–21. https://doi.org/10.1038/s41586-019-1111-9.

Griscom BW, Busch J, Cook-Patton SC, Ellis PW, Funk J, Leavitt SM, et al. National mitigation potential from natural climate solutions in the tropics. Philos Trans R Soc B 2020;375(1794):20190126.

Haldon J, Mordechai L, Newfield TP, Chase AF, Izdebski A, Guzowski P, Labuhn I, Roberts N. History meets palaeoscience: consilience and collaboration in studying past societal responses to environmental change. PNAS 2018;115(13):3210–8.

Hale LC, Newkirk S, Beck M. Helping coastal communities adapt to climate change. Solutions 2011;2:84–5.

Hallegatte S, Shah A, Lempert R, Brown C, Gill S. Investment decision making under deep uncertainty—Application to climate change. Washington, DC: World Bank; 2012.

Halofsky JS, Donato DC, Franklin JF, Halofsky JE, Peterson DL, Harvey BJ. The nature of the beast: examining climate adaptation options in forests with stand-replacing fire regimes. Ecosphere 2018;9(3):e02140. https://doi.org/10.1002/ecs2.2140.

Harrison S, Spasojevic MJ, Li D. Climate and plant community diversity in space and time. Proc Natl Acad Sci U S A 2020;117:4464–70.

Hobbs RJ, Higgs E, Harris JA. Novel ecosystems: implications for conservation and restoration. Trends Ecol Evol 2009;24(11):599–605.

Hoegh-Guldberg O, Jacob D, Taylor M, Bindi M, Brown S, Camilloni I, Diedhiou A, Djalante R, Ebi KL, Engelbrecht F, Guiot J, Hijioka Y, Mehrotra S, Payne A, Seneviratne SI, Thomas A, Warren R, Zhou G.

Impacts of 1.5°C global warming on natural and human systems. In: Masson-Delmotte V, Zhai P, Pörtner H-O, Roberts D, Skea J, Shukla PR, Waterfield T, editors. Global warming of 1.5°C. An IPCC special report on the impacts of global warming of 1.5°C above pre-industrial levels and related global greenhouse gas emission pathways, in the context of strengthening the global response to the threat of climate change, sustainable development, and efforts to eradicate poverty. IPCC (Intergovernmental Panel on Climate Change); 2018.

Hughes TP, Kerry JT, Conolly SR, Baird AH, Eakin CM, Heron SF, Hoey AS, Hoogenboom MO, Jaconbson M, Liu G, Pratchett MS, Skirving W, Torda G. Ecological memory modifies the cumulative impact of recurrent climate extremes. Nat Clim Chang 2019;9:40–3.

Hutchison C, Gravel D, Guichard F, Potvin C. Effect of diversity on growth, mortality, and loss of resilience to extreme climate events in a tropical planted forest experiment. Sci Rep 2018;8:15443.

Jiménez Cisneros BE, Oki T, Arnell NW, Benito G, Cogley JG, Döll P, Jiang T, Mwakalila SS, Barros VR, Dokken DJ, Mach KJ, Mastrandrea MD, Bilir TE, Chatterjee M, Ebi KL, Estrada YO, Genova RC, Girma B, Kissel ES, Levy AN, MacCracken S, Mastrandrea PR, White LL. Freshwater resources. In: Field CB, editor. Climate change 2014: Impacts, adaptation, and vulnerability. Part A: Global and sectoral aspects. Contribution of working group II to the fifth assessment report of the intergovernmental panel on climate change. Cambridge and New York, NY: Cambridge University Press; 2014. p. 229–69.

Johnstone JF, Allen CD, Franklin JF, Frelich LE, Harvey BJ, Higuera PE, Mack MC, Meentemeyer RK, Metz MR, Perry GLW, Schoennagel T, Turner MG. Changing disturbance regimes, ecological memory, and forest resilience. Front Ecol Environ 2016;14(7):369–78.

Jones HP, Hole DG, Zavaleta ES. Harnessing nature to help people adapt to climate change. Nat Clim Chang 2012;2:504–9.

Kapos V, Wicander S, Salvaterra T, Dawkins K, Hicks C. The role of the natural environment in adaptation, background paper for the global commission on adaptation. Rotterdam and Washington, DC: Global Commission on Adaptation; 2019.

Kitzberger T, Perry GLW, Paritsis J, Gowda JH, Tepley AJ, Holz A, Veblen TT. Fire-vegetation feedbacks and alternative states: common mechanisms of temperate forest vulnerability to fire in southern South America and New Zealand. N Z J Bot 2016;54:247–72. https://doi.org/10.1080/0028825X. 2016.1151903.

Lavorel S, et al. Ecological mechanisms underpinning climate adaptation services. Glob Chang Biol 2015;21:12–31. https://doi.org/10.1111/gcb.12689.

Lavorel S, et al. Mustering the power of ecosystems for adaptation to climate change. Environ Sci Policy 2019;92:87–97.

Lempert RJ, Popper S, Bankes S. Shaping the next one hundred years: New methods for quantitative, long-term policy analysis. Santa Monica, CA: RAND; 2003.

Linner B, Wibeck V. Sustainability transformations: Agents and drivers across societies. Cambridge: Cambridge University Press; 2019.

Maes J, Jacobs S. Nature-based solutions for Europe's sustainable development. Conserv Lett 2017;10(1):121–4. https://doi.org/10.1111/conl.12216.

Malhi Y, Franklin J, Seddon N, Solan M, Turner MG, Field CB, Knowlton N. Climate change and ecosystems: threats, opportunities and solutions. Philos Trans R Soc B 2020;375. https://doi.org/10.1098/rstb.2019.0104. 2019.0104.

Marchau VAWJ, Walker WE, Bloemen PJTM, Popper SW, editors. Decision making under deep uncertainty: From theory to practice. Springer; 2019. https://doi.org/10.1007/978-3-030-05252-2.

Matthews JH, Matthews N, Simmons E, Vigerstol K. Wellspring: Sources water resilience and climate adaptation. Arlington, VA: The Nature Conservancy; 2019. https://www.nature.org/content/dam/tnc/nature/en/documents/ Wellspring_FULL_Report_2019.pdf.

Milly PCD, Betancourt J, Falkenmark M, Hirsch RM, Kundzewicz ZW, Lettenmaier DP, Stouffer RJ. Stationarity is dead: whither water management? Science 2008;319:5863. https://doi.org/10.1126/science.1151915.

Mitsch WJ. What is ecological engineering? Ecol Eng 2012;45:5–12. https://doi.org/10.1016/j. ecoleng.2012.04.013.

Morecroft MD, Duffield S, Harley M, Pearce-Higgins JW, Stevens N, Watts O, Whitaker J. Measuring the success of climate change adaptation and mitigation in terrestrial ecosystems. Science 2019;366:e1329.

Mundy B. The death of Aztec Tenochtitlan the life of Mexico City. Austin, TX: University of Texas Press; 2015.

Nyugen AR, Hens L. Human ecology of climate change hazards in Vietnam: Risks for nature and humans in lowland and upland areas. Cham: Springer Climate; 2019.

Ochoa-Tocachi BF, Bardales JD, Antiporta J, Perez K, Acosta L, Mao F, Zulkafli Z, Gil-Rios J, Angulo O, Grainger S, Gammie G, DeBievre B, Buytaert W. Potential contributions from pre-Inca infiltration infrastructure to Andean water security. Nat Sustain 2019;2:584–93.

OECD. Nature-based solutions for adapting to water-related climate risks: Policy perspectives. OECD Environment Policy Paper No. 21, OECD; 2020.

Oliver TH, et al. Biodiversity and resilience of ecosystem functions: review. Trends Ecol Evol 2015;30(11):673–84.

Olsson L, Jerneck A, Thoren H, Persson J, O'Byrne D. Why resilience is unappealing to social science: theoretical and empirical investigations of the scientific use of resilience. Sci Adv 2015;1:e1400217.

Parry M, et al. Assessing the costs of adaptation to climate change: A review of the UNFCCC and other recent estimates. IIED and Grantham Institute for Climate Change; 2009.

Poff NL. Beyond the natural flow regime? Broadening the hydro-ecological foundation to meet environmental flows challenges in a non-stationary world. Freshw Biol 2018;63(8):1011–21.

Poff NL, Brown CM, Grantham TE, Matthews JH, Palmer MA, Spence CM, Wilby RL, Haasnoot M, Mendoza GG, Dominique KC, Baeza A. Sustainable water management under future uncertainty with eco-engineering decision scaling. Nat Clim Chang 2016;6(1):25–34.

Ratajczak Z, Carpenter SR, Ives AR, Kucharik CJ, Ramiadantsoa T, Stegner MA, Williams JW, Zhang J, Turner MG. Abrupt change in ecological systems: inference and diagnosis. Trends Ecol Evol 2018;33:513–26. https://doi.org/10.1016/j.tree.2018.04.013.

Rockström J, Steffen W, Noone K, Persson Å, Chapin III FS, Lambin E, Lenton TM, Scheffer M, Folke C, Schellnhuber H, Nykvist B, De Wit CA, Hughes T, van der Leeuw S, Rodhe H, Sörlin S, Snyder PK, Costanza R, Svedin U, Falkenmark M, Karlberg L, Corell RW, Fabry VJ, Hansen J, Walker B, Liverman D, Richardson K, Crutzen P, Foley J. Planetary boundaries: exploring the safe operating space for humanity. Ecol Soc 2009;14(2):32.

Rodina L. Defining 'water resilience': debates, concepts, approaches, and gaps. Wiley Interdiscip Rev Water 2018;6(2). https://doi.org/10.1002/wat2.1334. e1334–18. John Wiley & Sons, Ltd.

Seddon N. Nature-based solutions: Delivering national-level adaptation and global goals. IIED Briefing Paper, International Institute for Environment and Development; 2018. https://pubs.iied.org/pdfs/17484IIED.pdf.

Seddon N, Turner B, Berry P, Chausson A, Girardin C. Grounding nature-based climate solutions in sound biodiversity science. Nat Clim Chang 2019;9:84–7. https://doi.org/10.1038/s41558-019-0405-0.

Seddon N, Chausson A, Berry P, Gerardin CAJ, Smith A, Turner B. Understanding the value and limits of nature-based solutions to climate change and other global challenges. Philos Trans R Soc B 2020a;375:20190120.

Seddon N, Daniels E, Davis R, Chausson A, Harris R, Hou-Jones X, Hug S, Kapos V, Mace GM, Rizvi AR, Reid H, Roe D, Turner B, Wicander S. Global recognition of the importance of nature-based solutions to the impacts of climate change. Glob Sustain 2020b;3(e15):1–12.

Smith DM, Matthews JH, Bharati L, Borgomeo E, McCartney M, Mauroner A, et al. Adaptation's thirst: accelerating the convergence of water and climate action. In: Background paper prepared for the 2019 report of the global commission on adaptation. Global Commission on Adaptation; 2019. Rotterdam and Washington, DC www.gca.org.

Smith A, et al. How natural capital delivers ecosystem services: a typology derived from a systematic review. Ecosyst Serv 2017;26:111–26. https://doi.org/10.1016/j.ecoser.2017.06.006.

Steffen W, Richardson K, Rockstrom J, Cornell SE, Fetzer I, Bennett EM, Biggs R, Carpenter SR, de Vries W, de Wit CA, Folke C, Gerten D, Heinke J, Mace GM, Persson LM, Ramanathan V, Reyers B, Sorlin S. Planetary boundaries: guiding human development on a changing planet. Science 2015;347:1259855.

Sudmeier-Rieux K, Nehren U, Sandholz S, Doswald N. Disasters and ecosystems: Resilience in a changing climate—Source book. Geneva: UNEP and Cologne: TH Köln—University of Applied Sciences; 2019.

Tellman B, Bausch JC, Eakin H, Anderies JM, Mazari-Hiriart M, Manuel-Navarete D, Redmand CL. Adaptive pathways and coupled infrastructure: seven centuries of adaptation to water risk and the production of vulnerability in Mexico City. Ecol Soc 2018;23(1):1. https://doi.org/10.5751/ES-09712-230101.

Timboe I, Pharr K, Matthews JH. Watering the NDCs: National climate planning for 2020—How water-aware climate policies can strengthen climate change mitigation & adaptation goals. Corvallis, OR: Alliance for Global Water Adaptation (AGWA); 2019. Nov. 2019 https://www.wateringthendcs.org/.

Turner MG, Calder WJ, Cumming GS, Hughes TP, Jentsch A, LaDeau SL, Lenton TM, Shuman BN, Turetshy MR, Ratajczak Z, Williams JH, Williams AP, Carpenter SR. Climate change, ecosystems and abrupt change: science priorities. Philos Trans R Soc B 2020;375:20190105.

UNDRR. Ecosystem-based disaster risk reduction: Implementing nature-based Solutions for resilience, United Nations Office for disaster risk. Bangkok: Reduction—Regional Office for Asia and the Pacific; 2020.

UNISDR. Essential 5: safeguard natural buffers to enhance the protective functions offered by natural ecosystems. In: How to make cities more resilient: A handbook for local government leaders. 2017 ed. Geneva: United Nations Office for Disaster Risk Reduction; 2017. p. 54–7.

Veldman JW, et al. Where tree planting and forest expansion are bad for biodiversity and ecosystem services. Bioscience 2015;65:1011–8. https://doi.org/10.1093/biosci/biv118.

Veldman JW, et al. Comment on "the global tree restoration potential". Science 2019. https://doi.org/10.1126/science.aay7976.

Walker BH. Resilience: what it *is* and is *not*. Ecol Soc 2020;25(2):11. https://doi.org/10.5751/ES-11647-250211.

Watson JEM, et al. The exceptional value of intact forest ecosystems. Nat Ecol Evol Perspect 2018;2:599–610.

Wilby RL, Dessai S. Robust adaptation to climate change. Weather 2010;65(7):180–5.

Wood SLR, Jones SK, Johnson JA, Brauman KS, Chaplin-Kramer R, Fremier A, Girvetz E, Gordon LJ, Kappel CV, Mandle L, Mulligan M, O'Farrell P, Smith WP, Willemen L, Zhang W, DeClerck FA. Distilling the role of ecosystem services in the sustainable development goals. Ecosyst Serv 2018;2212-0416. 29(Part A):70–82. https://doi.org/10.1016/j.ecoser.2017.10.010.

World Economic Forum. Water security: The water-food-energy-climate nexus. Washington, DC: Island Press; 2011.

World Economic Forum, editor. The global risks report 2019. 14th ed. Geneva: World Economic Forum; 2019.

World Economic Forum, editor. The global risks report 2020. 15th ed. Geneva: World Economic Forum; 2020.

World Economic Forum, editor. The global risks report 2021. 16th ed. Geneva: World Economic Forum; 2021.

WWAP (United Nations World Water Assessment Programme). The United Nations World Water Development Report 2016: water and jobs. Paris: UNESCO; 2016.

WWAP (United Nations World Water Assessment Programme). The United Nations World Water Development Report 2018: Nature-based solutions for water. Paris: UNESCO; 2018.

Yu X, et al. Freshwater management and climate change adaptation: experiences from the central Yangtze in China. Clim Dev 2009;1:241–8.

CHAPTER 5

Nature-based solutions, sustainable development, and equity

Leah L. Bremer[a,*], **Bonnie Keeler**[b,†], **Pua'ala Pascua**[c,†], **Rebecca Walker**[b,†], and **Eleanor Sterling**[c,†]

aUniversity of Hawai'i at Mānoa, Honolulu, HI, United States, bUniversity of Minnesota, Minneapolis, MN, United States, cCenter for Biodiversity and Conservation, American Museum of Natural History, New York, NY, United States

Key messages

- Healthy ecosystems and human well-being are closely linked.
- Water security is central to human well-being; water is essential for physical and mental health, food security, poverty reduction, energy production, resilient cities, ecosystem health, cultural values, and economic opportunity.
- Nature-based solutions (NBS) have been recognized as a key tool for water security (Sustainable Development Goal 6), and importantly, they are multifunctional—NBS for water can help achieve multiple other Sustainable Development Goals (SDGs).
- NBS can be implemented in a variety of ways, including through urban green infrastructure, payments for ecosystem services, and biocultural approaches.
- NBS must be designed and implemented to explicitly address issues of equity and social justice.
- Emerging best practices for NBS include being attentive to legacies of inequity and social injustice, consideration of potential trade-offs, and explicit consideration of distributional, procedural, and recognitional equity. These best practices are key to successful and equitable NBS for water security.

Water, well-being, nature-based solutions and equity

Water is central to sustainable development and critical to all dimensions of ecosystem and human well-being, including economic opportunity, food and energy production, and human physical and mental health. In 2015, the United Nations explicitly recognized the links between human and environmental health by adopting 17 SDGs aiming to "protect the planet from degradation…so it can support the needs of present and future generations." SDG 6 (ensure availability and sustainable management of water and sanitation for all) focuses specifically on

---

* Lead author.
† Co-authors.

Nature-Based Solutions and Water Security. https://doi.org/10.1016/B978-0-12-819871-1.00016-6

water. This is a critical and ambitious goal, as it is currently estimated that 2.1 million people do not have access to safe drinking water sources and 4.5 million people lack safely managed sanitation services (WHO/UNICEF, 2017). Progress toward this goal is also critical for all other SDGs (see Table 5.1). For example, achieving SDG 1 (end poverty in all its forms everywhere), requires equitable access to clean water for drinking, sanitation, and other uses.

Table 5.1: Key SDGs related to NBS for water along with example targets and their relevance to SDGs and equity.

Goal	Example target	Relevance to NBS and equity
Goal 1: No poverty	By 2030, eradicate extreme poverty for all people everywhere, currently measured as people living on $1.25 a day	Clean water is essential for livelihoods. It is important to ensure that NBS does not reduce access to land and livelihoods
Goal 2: Zero hunger	By 2030, ensure sustainable food production systems and implement resilient agricultural practices that increase productivity and production that help maintain ecosystems, strengthen capacity for adaptation to climate change, extreme weather, drought, flooding and other disasters, and progressively improve land and soil quality	NBS can contribute to sustainable water sources for agriculture. NBS (e.g., agroforestry) can be sustainable agriculture. It is important to ensure that NBS strengthens or enables subsistence and other agricultural systems for vulnerable populations
Goal 3: Good health and well-being	By 2030, end preventable deaths of newborns under 5 years of age	There is potential for NBS to contribute to reduced diarrheal disease, particularly when combined with gray infrastructure, behavioral change, and policy reform. It is important to ensure that NBS does not create unequal reductions in access to land for food and other resources important for health and well-being
Goal 5: Gender equality	Ensure women's full and effective participation and equal opportunities for leadership at all levels of decision-making in political, economic, and public life	Women must be involved in decision making of NBS projects, and gender-specific outcomes must be taken into account
Goal 6: Clean water and sanitation	By 2020, protect and restore water-related ecosystems, including mountains, forests, wetlands, aquifers, and lakes. Support and strengthen the participation of local communities in improving water and sanitation management	Done with care, NBS can contribute to clean water especially in access to safe drinking water and protecting and restoring water related to ecosystems
Goal 8: Decent work and economic growth	By 2030, devise and implement policies to promote sustainable tourism that creates jobs and promotes local culture and products	It is important to ensure that NBS create rather than reduce employment and economic opportunity and incorporate meaningful and culturally relevant jobs

Table 5.1: Key SDGs related to NBS for water along with example targets and their relevance to SDGs and equity—cont'd

Goal	Example target	Relevance to NBS and equity
Goal 10: Reduce inequality within and among countries	By 2030, progressively achieve and sustain income growth of the bottom 40% at a rate higher than the national average	Support NBS that increase income and reduce income inequality (i.e., by supporting incomes of more marginalized or low-income communities, ideally, and at least do not decrease it)
Goal 11: Sustainable cities and communities	By 2030, provide universal access to safe, inclusive, and accessible, green and public spaces, in particular for women and children, older persons and persons with disabilities; by 2030, ensure access for all to adequate, safe, and affordable housing and basic services and upgrade slums	NBS can increase green space and provide wastewater services. It is important to explicitly pay attention to equity and combine with gray infrastructure approaches (e.g., ensure access to clean piped water)
Goal 13: Take urgent action to combat climate change and its impacts	Promote mechanisms for raising capacity for effective climate change-related planning and management in least developed countries and small island developing States, including focusing on women, youth, and local and marginalized communities	NBS can function as climate adaptation and mitigation measures, but must be done in a broader context of addressing past and current inequities
Goal 14: Conserve and sustainably use the oceans, seas, and marine resources for sustainable development	By 2025, prevent and significantly reduce marine pollution of all kinds, in particular form land-based activities, including marine debris and nutrient pollution	NBS can help reduce land-based pollutants to culturally and economically important marine ecosystems
Goal 15: Protect, restore, and promote sustainable use of terrestrial ecosystems, sustainably manage forests, combat desertification, and halt and reverse land degradation and halt biodiversity loss	By 2020, ensure the conservation, restoration, and sustainable use of terrestrial and inland freshwater ecosystems and their services, in particular forests, wetlands, mountains, and drylands in line with obligations under international agreements	NBS can directly help achieve these protection, restoration, and sustainable management goals. It is important to ensure fair agreements that do not threaten land or livelihood rights
Goal 16: Promote peaceful and inclusive societies for sustainable development, provide access and build effective, accountable and inclusive institutions at all levels	Ensure responsive, inclusive, participatory, and representative decision-making at all levels; broaden and strengthen the participation of developing countries in the institutions of global governance	NBS should be designed in line with equity and inclusion principles
Goal 17: Strengthen the means of implementation, and revitalize the Global Partnership for Sustainable Development	Mobilize additional financial resources for developing countries from multiple sources	NBS for water can be financed broadly for multiple benefits

NBS are living solutions inspired by and supporting nature. They are designed to provide economic, social, and environmental benefits (European Commission, 2015; Kabisch et al., 2016). NBS for water are inspired by and utilize nature to improve water quality, reduce peak flows, and increase base flows and groundwater recharge. As such, NBS offer a critical tool for achieving SDG 6 and other SDGs underpinned by clean and ample water supplies. This will be the case particularly where NBS are combined with gray infrastructure solutions (e.g., point-of-source filters, improved pipes) to address existing inequities in water availability and access.

As a starting point for discussion of equity and NBS for water, it is important to acknowledge that while improving over time, access to ample and safe water for drinking, sanitation, and hygiene via primarily gray infrastructure is highly unequal (UNICEF and WHO, 2019). For example, at the global scale in 2017, 9 out of 10 of the 785 million people living without access to safe drinking water live in sub-Saharan Africa (400 billion), East and Southeast Asia (161 billion), and Central and South Asia (145 billion), whereas it is estimated that > 99% of the population of the highest income countries have access to at least basic drinking water services (UNICEF and WHO, 2019). Across the world, disparities in water access are shrinking but continue among urban and rural populations (UNICEF and WHO, 2019). Even in areas where access to basic water services is considered high, there are continued inequalities along racial, ethnic, and class lines (Adams et al., 2016; Keeler et al., 2020).

NBS for water efforts that aim to achieve SDG 6 and other SDGs in an equitable way must be aware of and address legacies and root causes of inequity and social justice (directly addressed in SDGs 5, 10, and 16; Table 5.1; Kharas et al., 2020). For example, the lack of clean reliable water for drinking, cooking, and personal hygiene is often less about physical water availability and more about policy failure by local or regional governments to ensure equitable access and distribution of water. Often insufficient investments in infrastructure and lack of oversight, regulation, or resources to protect and treat water supplies interact with environmental degradation to put vulnerable communities at risk (Pérez-Foguet and Garriga, 2011). Likewise, achieving multiple SDGs in an equitable way requires careful consideration and mitigation of potential negative trade-offs associated with reaching one goal at the expense of another. For example, without careful attention to the local context and equity concerns, policies around SDG 15 (protect, restore and promote sustainable use of terrestrial ecosystems) could reduce access to important water sources. In contrast, deliberate participatory planning processes are more likely to produce just outcomes and areas of synergy among goals, such as cases where investments in forest protection and sustainable management can contribute to clean drinking water for rural communities.

We argue that the potential for NBS to contribute to multiple SDGs requires a concerted focus on equity. In this chapter we consider three dimensions of equity: the distribution of the costs, benefits, burdens, and rights of NBS projects among different groups of people (distributional

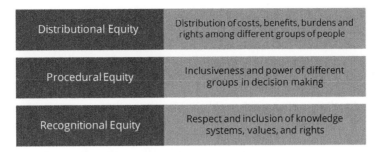

Fig. 5.1
The three dimensions of equity.

equity); the degree of inclusiveness in decision making around NBS projects (procedural equity); and respect for knowledge systems, values, and rights of stakeholders in the design and implementation of NBS (recognitional equity) (McDermott et al., 2013; Pascual et al., 2014) (Fig. 5.1). Equity concerns include understanding how multiple factors, including race, gender, and socio-economic status, influence how groups participate in and are affected by NBS projects. Equally important is considering the equitable inclusion of diverse social-cultural worldviews, ontologies, and values, especially those of communities impacted by NBS projects, in NBS project planning and decision-making processes.

Organization of the chapter

This chapter is organized into three main sections. First, we review emerging evidence on the links between land management strategies used in NBS and the SDGs, including human health, nutrition, sustainable urban areas, and local livelihoods. Second, we provide several concrete examples of policies, programs, and approaches to NBS solutions and demonstrate the critical importance of attention to equity and justice in their design and implementation. Specifically, we explore urban green and gray infrastructure, investments in watershed services through payments for ecosystem services, and biocultural approaches to sustainable development. Finally, we conclude with a discussion of emerging best practices in improving social equity in NBS for water.

Nature-based solutions are an important tool for achieving multiple sustainable development goals

Chapter 3 addressed the ways that NBS for water can address key water supply risks through improved water quality and regulation. NBS also have the potential to deliver multiple additional benefits beyond direct water outcomes, including human health, food security and local livelihoods, and climate mitigation and adaptation. In this section, we review prominent examples of evidence of the linkages between NBS land management strategies and SDGs related to health, food security, and sustainable urban areas and economies. Table 5.1 provides additional examples of linkages between NBS for water and a range of SDGs.

Nature-based solutions for water and health

SDG 3: Ensure healthy lives and promote well-being for all at all ages

There is growing evidence that how we manage landscapes has direct and indirect impacts on human health and well-being and that well-designed NBS can be a "strategic health partner" (Gaffikin, 2013). While links between environmental and human health have long been recognized, increased availability of data and development of new analytical methods have improved understanding of how, when, and where investments in NBS provide human health benefits (Bratman et al., 2012, 2015; Herrera et al., 2017; Myers et al., 2013). For example, in an evaluation of demographic and health survey data from 35 low-income countries, researchers identified correlations between watershed condition (tree cover) and lower incidence of diarrheal disease in children in rural areas (Fisher et al., 2019; Herrera et al., 2017).

Growing research also points to greater emergence and transmission of infectious disease as a result of ecosystem loss and degradation (Guo et al., 2019; Patz et al., 2004, 2008). Approximately 60% of all infectious diseases and 75% of all emerging infectious diseases come from wild or domestic animals (i.e., zoonotic). This includes the Ebola outbreak, linked to forest loss and greater human-wildlife interaction (UNEP, 2016), and the Nipah virus, linked to deforestation and displacement of fruit bats into agricultural areas, which infected pigs and then humans. Other zoonotic diseases include swine flu, avian flu, Zika, and COVID-19, among many others (Ahmad et al., 2020; Bonilla-Aldana et al., 2020). Deforestation and biodiversity loss have also been associated with increased risk of infection of a number of vector-borne (e.g., spread through mosquitoes and other vectors) diseases including malaria, West Nile, Dengue, and Lyme disease (MacDonald and Mordecai, 2019; Myers et al., 2013). Though disease transmission and infection are complex and variable, this research points to the potential of NBS focused on landscape-scale ecosystem protection to contribute to human health outcomes through infectious disease regulation. To achieve positive health and equity outcomes, NBS interventions need to be combined with policy and behavioral solutions (e.g., mosquito nets for malaria; rapid response to emerging infections; safe water storage and hygiene, point-of-source filters), and also must be carefully designed to reduce burdens and trade-offs associated with any losses in access to lands important for livelihoods, food security, and cultural practices.

NBS management strategies can also have important links to mental health. Recent psychological studies have found that nature-based experiences are connected to decreased depression, improved cognitive functioning, and reduced stress (Bratman et al., 2012, 2015; Hough, 2014). For example, research on the Japanese tradition of Shirin-Yoku (forest bathing) has demonstrated that time in nature improves concentration and mood, reduces blood glucose levels, and boosts immune functioning (Lee et al., 2011; Li et al., 2008; Morita et al., 2007; Ohtsuka et al., 1998). A growing body of research also demonstrates that time spent caring for or stewarding land can be an important source of connection to place and to community in many societies (Bremer et al., 2018b; Chan et al., 2016, 2018; Pascua et al., 2017). Where NBS can be designed with communities and to incorporate Indigenous and local knowledge and practices (procedural and recognitional equity), the potential for positive human well-being benefits are high.

Nature-based solutions for water and food security

SDG 2: End hunger, achieve food security and improved nutrition, and promote sustainable agriculture

NBS management strategies can also contribute to sustainable agricultural production, food security, and reduced hunger. Most directly, NBS landscapes can include both agricultural lands and natural ecosystems with wild foods used for hunting and gathering. Examples of agricultural NBS systems are *lo'i kalo* (wetland and flooded field taro systems) in Hawai'i, which have been shown to retain sediment (Bremer et al., 2018a), while producing a staple Hawaiian crop with important cultural, community, and family health benefits (Aikau et al., 2016; McGregor, 2007). Agroforestry systems, a spectrum of practices that integrate trees with tended and harvested plants or animals (Hastings et al., 2020), are another type of NBS system that can provide important water-related ecosystem services (sediment retention, flood mitigation; Jose, 2009) and support diverse diets high in micronutrients (vitamin A, iron, folate, etc.; Ickowitz et al., 2016). The diversity of crops from agroforests, similar to wild food harvesting, can help combat "hidden hunger" or critical micronutrient deficiencies, which impact nearly two billion people and cause serious health problems, particularly among women and children in low-income countries (IFPRI, 2016). NBS can also take the form of home, rooftop, and urban gardens, which can provide important urban hydrologic services, while also contributing to healthy food supplies.

NBS management strategies can also support agricultural production and food security by protecting the pollinator habitat on which many of the crops that provide essential micronutrients rely, including fruit and vegetables like pumpkins, melons, and tropical fruits (Chaplin-Kramer et al., 2014; Eilers et al., 2011). An analysis of the benefits of source watershed protection by Abell et al. (2017) found that a loss of pollination, which could occur if landscapes surrounding agricultural areas fall below a 30% threshold of forest and other pollinator habitat, could decrease the availability of vitamin A by 40% or more. Many of the areas most impacted by loss of pollinator habitat overlap with areas where deficiencies and associated health impacts (e.g., reduced immunity and night blindness) are already high. As with other links between NBS and health, the actual health impact of changes in micronutrient availability will depend on access to food imports as well as dietary habits (Ellis et al., 2015).

Nature-based solutions for water and inclusive, resilient cities

SDG 11: Make cities and human settlements inclusive, safe, resilient, and sustainable

Nature in cities has been shown to benefit water quality and quantity in numerous ways, including through improved retention of stormwater, reduced peak flows and urban flooding, enhanced aquifer recharge, and improved water quality in urban lakes, rivers, and streams (Keeler et al., 2019). NBS are often presented as win-wins for cities by improving urban welfare, enhancing resilience and adaptation to climate change, and supporting habitat for urban biodiversity. However, recent work has also highlighted the potential negative consequences of implementing NBS in the form of ecosystem "disservices" such as degraded air quality and pollen production from tree canopy, nutrient loading to urban lakes from decomposing leaf litter, and infrastructure damage from tree roots and limbs. Particularly in cities, the benefits of NBS are mediated by complex social, ecological, and technical factors that impact where, when, and for whom NBS deliver benefits or costs (Keeler et al., 2019; McPhearson et al., 2015).

Nature-based solutions for water and sustainable economies

SDG 8: Promote sustained, inclusive, and sustainable economic growth; full and productive employment; and decent work for all

Another important area of potential synergy between NBS management strategies and the SDGs is through job creation and economic opportunity. While some NBS programs have been criticized as activity-reducing and paying people to do nothing (Pattanayak et al., 2010), well-designed NBS projects can create jobs and income opportunities. For example, the Guandu Water Producer Project (Água e Floresta) in Brazil created jobs associated with reforestation (Bremer et al., 2016a) and many other programs will need growth in the NBS green jobs sector. An analysis of the "restoration economy" in the United States found that regulation-driven environmental protection, mitigation, and restoration activities were directly responsible for 126,000 jobs and about $9.5 billion in economic output annually (BenDor et al., 2015). An additional 95,000 jobs and $15 billion were due to indirect business-to-business activities and household spending. Together, employment from the restoration economy represents more workers than directly employed in the United States in coal mining, oil and gas extraction, or steel production.

Job creation must both acknowledge and respect fair wages/compensation in addition to locally and culturally informed pathways for development and sustainability. For instance, "productive employment" (as specified in the SDGs) should also encompass traditional occupations like subsistence fishing and farming (Sterling et al., 2017). These occupations foundationally rely on and build upon Indigenous and local knowledge to support food security and food sovereignty, while also promoting community social cohesion and the maintenance and application of place-based subsistence practices (Pascua et al., 2017).

Nature-based solutions: Toward social justice and equity in practice
Nature-based solutions must consider current and historical legacies of injustice that influence existing inequities

While NBS projects are often framed in a narrative in which all groups share the benefits, sustainability policies and projects often play out on landscapes shaped by historical legacies of colonialism and discrimination based on gender, race, ethnicity, and class. As a result, the effects of NBS projects are often experienced unevenly among different groups (Nelson et al., 2020; Park and Pellow, 2013). For example, where NBS create or support programs that restrict access to or use of upstream ecosystems without a careful consideration of social and livelihood trade-offs, NBS may deliver clean water to a downstream city, while exacerbating inequities upstream (lacking distributional equity). Similarly, green infrastructure in cities can improve water quality while increasing the cost of living, which can displace low-income residents (Haase et al., 2017). Thus, it is critical to consider "equity for whom," meaning a careful consideration of the intersectional impacts of NBS projects by gender, socio-economic status, race, ethnicity, and other dimensions of identity.

For example, although the United States is estimated to have > 99% access to safe water, 1.46 million people continue to lack plumbed access to safe water, much of which traces to historical legacies of discrimination and marginalization. In an analysis of the spatial pattern of plumbed access to potable water in the United States, Deitz and Meehan (2019) observed that households that lack plumbing are concentrated in poor, rural, communities of color. In particular, this study found that within census tracts with incomplete plumbing, American Indian or Native Alaskan households were 3.7 times more likely and Black and Hispanic households 1.2 times more likely than White households to lack complete plumbing (Deitz and Meehan, 2019). The authors attribute this pattern to declines in investments in water infrastructure for these households over the 20th century. Insufficient water infrastructure was linked to income, race, and ethnicity, and rural-urban status, demonstrating how political and economic inequality, marginalization, and exclusion play a key role in shaping clean water access in the United States (Deitz and Meehan, 2019).

Inequality in gray water infrastructure has several important implications for NBS for water. First, without acknowledging and actively confronting current and historical legacies of structural racism and resulting racial and economic inequality across multiple scales, efforts to implement NBS will likely only serve to reproduce unequal landscapes of environmental privilege and exclusion (Keeler et al., 2020). Second, NBS can fill an infrastructure gap left by inadequate policy and infrastructure. Where access to water treatment and piped distribution systems is limited, protecting or restoring the natural environment may be an important way to protect water supplies, underscoring the critical role of NBS in locations of unequal investment in water infrastructure. Working with communities to codevelop plans to protect these natural resources and the ecosystem services they provide is essential to ensuring just outcomes and long-term sustainability. Paying attention to procedural and recognitional equity (i.e., ensuring that decision-making processes are inclusive) can also help to increase the potential for NBS to provide cobenefits in addition to water that matter to relevant communities (e.g., access to local food and other valued plants; reduced temperatures). However, it is also critical to acknowledge the limitations of NBS approaches in meeting all infrastructural needs—while NBS have the potential to help fill the infrastructure gap, NBS will often need to be combined with gray infrastructure to truly address water inequalities.

Cases: Nature-based solutions for water in the context of equity

We now consider three approaches to NBS for water in the context of equity: green infrastructure in cities, investments in watershed services through payments for ecosystem services (PES), and biocultural restoration. We argue that a more comprehensive understanding of the existing landscapes of inequality and the historical contexts that gave rise to those inequities is essential for ensuring NBS are designed, targeted, and implemented in ways that promote just and equitable outcomes. Throughout our case studies, we consider multiple dimensions of equity outlined in the introduction of this chapter: distributional equity (distribution of the costs, benefits, burdens, and rights of NBS projects among different groups of people); procedural equity (the degree of inclusiveness in decision making around NBS projects); and recognitional equity (respect for knowledge systems, values, and rights of stakeholders in the design and implementation of NBS).

Case study 1: Urban NBS for water

Cities are increasingly turning to NBS to improve water quality and mitigate flood risk, issues that disproportionately impact low-income and minority communities (Douglas et al., 2012; Maantay and Maroko, 2009). Urban NBS—such as greenspaces, bioswales, and rain gardens—have an important role to play in addressing environmental injustice in cities (distributional equity), provided they are designed and implemented in inclusive ways that are sensitive to the social, economic, and political contexts of their communities (procedural and recognitional equity).

However, without adequate attention to social and equity outcomes, investments in green infrastructure NBS risk deepening existing inequalities. For example, NBS in the form of urban parks or river restoration can drive up property values and rents, displacing low-income communities. This limits access to the benefits of greening efforts to wealthier residents and ultimately deepens inequities (Angelo, 2019; Gould and Lewis, 2018). This phenomenon, known as "green gentrification" (Gould and Lewis, 2016), "environmental gentrification" (Checker, 2011; Pearsall and Pierce, 2010), or "eco-gentrification" (Quastel, 2009), is the process through which greening increases the desirability and thus the cost of an area, pricing out socially vulnerable residents (Angelo, 2019). Green gentrification highlights the need to forefront equity in the decision-making phase and evaluation of NBS success in order to ensure equitable outcomes.

Green gentrification was originally recognized in the context of brownfield remediation and environmental justice projects (Pearsall and Pierce, 2010) but is now understood to include a wide range of sustainability initiatives in the urban environment, including the creation of parks, bike lanes, urban agriculture, and green infrastructure (Anguelovski et al., 2018). Pearsall and Pierce (2010) describe the effect of brownfield remediation on low-income communities in New York City. Brownfields are former industrial sites that pose threats to environmental and human health, normally via the contamination of soil and water. In 2007, New York City moved to clean up its brownfields, located disproportionately near minority, low-income, and elderly populations. However, within 3 years of clean-up, surrounding property values increased between 100% and 200% (Pearsall and Pierce, 2010). Rising costs forced vulnerable communities to leave their neighborhoods, often moving further from jobs and community support systems or to locations where they faced new environmental hazards. Accordingly, benefits of brownfield clean-up were enjoyed by White, affluent populations, leading to low distributional equity.

Displacement via green gentrification (i.e., failure to achieve equity in the distribution of NBS benefits) highlights the consequences of inequity in decision-making (i.e., recognitional and procedural equity) around NBS. The emphasis on environmental sustainability serves to narrow the planning scope and detach urban NBS projects from equity goals (Finn and Mccormick, 2011; Rigolon and Németh, 2018). By limiting the discussion to the technical aspects of stormwater retention or water quality impacts, the voices of nonexperts are excluded from planning conversations. This problem can be exacerbated by the outsourcing of NBS implementation by cities to environmental NGOs or private developers, removing decision-making power over the project from the democratic process (Checker, 2011; Rigolon and Németh, 2018). Removing decision-making power from communities serves to segregate the environmental goals of NBS projects from their effects on the communities in which they are implemented. Projects implemented without community engagement are likely to fail to reflect the needs, values, and identities of the people they are intended to serve and ultimately result in displacement (Kabisch and Haase, 2014).

Several strategies have emerged to address inequity in NBS projects in cities. When implemented thoughtfully, NBS can improve water quality and mitigate flood risk—issues that disproportionately impact poorer, historically marginalized communities both in the United States (Balazs et al., 2012; Butler et al., 2016; Douglas et al., 2012; Maantay and Maroko, 2009) and around the world (Adams et al., 2016; De Risi et al., 2013; Luh et al., 2013; Sakijege et al., 2012). Increasing equity in access to urban green space is essential to building more just cities, as high-quality green spaces can have positive impacts on physical and mental health (van den Berg et al., 2015). In addition, shifting urban spaces from industrial spaces toward well-maintained and protected green spaces can have positive impacts on water and air quality while increasing groundwater recharge and reducing flood risks to downstream communities.

Critically, early and sustained community engagement is needed to ensure concerns are heard, understood, and integrated (Rigolon and Christensen, 2019; Wolch et al., 2014). For this engagement to be meaningful and effective, the community needs substantive decision-making power over the project plan and implementation. Additionally, to ensure distributional equity, the impact of greening projects on housing should be considered at the onset of the project, including engaging with city governments and developers to explore policies like community benefits agreements, inclusionary zoning, and other interventions that build wealth in surrounding communities while allowing residents to stay in place. Finally, equity outcomes and indicators need to be a central part of success metrics in the evaluation of any NBS project (Cassin, 2021). Bringing equity concerns to the forefront is essential to ensuring that the benefits of these projects are shared by all populations.

Case study 2: Source watershed protection through investments in watershed services

Investments in watershed services, including through PES and other incentive programs such as water funds, are a growing approach to implement NBS (Bremer et al., 2016a; Salzman et al., 2018). These programs involve transfers of resources from water users to upstream communities and farmers to implement source watershed protection activities believed to sustain or enhance water quality and quantity. Some of the most dominant voices in neoclassical PES theory have argued that social goals should remain secondary to environmental effectiveness goals (Pagiola et al., 2008; Wunder, 2008; Wunder et al., 2018). However, it is increasingly recognized that social goals and equity are important in their own right and are also important for long-term environmental effectiveness (Chan et al., 2018; Corbera et al., 2020; Pascual et al., 2014; Shapiro-Garza et al., 2019). Moreover, few, if any, PES programs are market-mechanisms, but rather can be better conceptualized as development pathways that have the potential to be adapted by local actors, including through the incorporation of local notions of equity and social justice (McElwee et al., 2020; Muradian et al., 2010; Shapiro-Garza et al., 2019; VonHedemann, 2019).

A key component of distributional equity in PES is who is willing, able, and eligible to participate. While some national-scale PES programs (e.g., Ecuador's SocioBosque program) have substantial participation among small and low-income farmers and rural communities (Bremer et al., 2014a), others have found that programs tend to favor large landowners (Lansing, 2014; Zbinden and Lee, 2005). In some cases, PES exclude those without formal land tenure, which excludes both the landless and many of the most vulnerable landowners (Wunder, 2008). Even when insecure tenure is not an issue, others have noted that marginalized communities and small farmers may lack the social networks or financial means necessary to complete entry requirements or may not be aware of the program (Bremer et al., 2014a; Zbinden and Lee, 2005). PES may also be less desirable to small farmers than larger landowners due to low-incentive payments for small landholdings, a reliance on available land for subsistence food security, or a fear of land expropriation (Grieg-Gran et al., 2005; Wunder, 2008; Zbinden and Lee, 2005).

In response to these obstacles, some PES programs have introduced measures to increase access to programs. These measures include relaxed tenure requirements, inclusion of poverty criteria in prioritization models, and decreasing price structures with increasing area enrolled (Bremer et al., 2014a). Some PES schemes have also moved toward incorporating "working landscapes," which are more likely to attract marginalized rural farmers who rely on their land for subsistence and who are unable to "lock up" productive land (Rosa et al., 2004). For example, Costa Rica's national watershed PES program added in agroforestry systems as part of incentivized landscapes, which successfully increased participation by smaller farmers and helped to boost food production (Cole, 2010). Similarly, the *Agua por la Vida y Sostenibilidad* (Water for Life and Sustainability) water fund in the Cauca Valley, Colombia, focuses much of its activities on NBS management around improving crop and livestock production (e.g., agroforestry and silviculture systems) with the explicit goal of helping to diversify livelihoods while also retaining sediment and increasing infiltration (Bremer et al., 2016a).

Distributional equity, in terms of the distribution of benefits and burdens, depends on the local context and the design of the program. For example, China's Sloping Lands Conservation Program (SLCP) has generally provided greater relative benefits for low-income households and has been found to decrease income inequality (Groom and Palmer, 2012; Li et al., 2011). However, there have been few studies evaluating outcomes of this program on multiple dimensions of human well-being beyond income, such as social cohesion and cultural values, leaving open questions about broader impacts. In contrast, Costa Rica's national PES program has generally reported greater income benefits for larger, wealthier landowners than for smaller landholders, though efforts such as the agroforestry project previously described have helped to increase benefits for low-income farmers (Ina, 2013; Cole, 2010). Many studies have found that perceived nonmonetary benefits (e.g., ability to protect valued forests; social networks) motivate participation in PES even where financial benefits are small to null (Arriagada et al., 2015; Bremer et al., 2014b, 2018b; Grillos, 2017). The importance of cobenefits demonstrates the importance of considering multiple and locally meaningful dimensions of well-being in planning for and assessing equity in PES.

PES can open up access to important funding to invest in collective and individual development (Loft et al., 2017; Murtinho and Hayes, 2017). In Ecuador, for example, well-organized rural communities have utilized PES payments to finance carefully planned community projects, including rural banks, sustainable agricultural production, and sanitation projects (Bremer et al., 2014b). However, in cases where programs are inserted into areas where power relations and land rights are highly uneven, PES can also increase existing inequalities and marginalize values and rights of the most vulnerable people (Corbera et al., 2020; Rodríguez de Francisco et al., 2013; Shapiro-Garza et al., 2019). For instance, in contexts of uneven land tenure security, which is common throughout much of the world, PES may benefit wealthy large landowners with secure tenure, while inadvertently reducing access to landless people or informal users. Likewise, where communities enroll collectively held land, there is a risk of elite capture of benefits, and/or uneven livelihood risks and losses that exacerbate existing inequalities (Hayes et al., 2019). This is the case in collectively held land in the Andes, for example, where community members with the least economic resources tend to be most dependent on highland grasslands or páramo grasslands and have the most to lose if these areas are enrolled into PES programs, which restrict grazing without adequate compensation (Bremer et al., 2014b).

There are several important strategies that can be utilized to improve distributional equity in PES arrangements. First, it is critical that PES program designers take the time to understand the historical and local contexts in order to understand power dynamics as well as likely winners and losers with PES incentives and land use changes. This requires time, relationship building, and working with local communities to design program structure, incentives, and rules in line with local perceptions of fairness. Programs can also be designed to focus specifically on improving livelihoods of landholders and community members who are most dependent on land affected by PES agreements. For example, in the Andes, several PES programs have focused specifically on working with communities to strengthen agricultural production in lower elevation areas or develop more sustainable activities within areas enrolled in PES targeted at those who utilized these areas prior to program onset. In one case, a community described how PES helped the community achieve a broad community goal of reducing heavy cattle grazing and burning of the páramo (high Andean grasslands), which the community relied on for their water supply, while also increasing the livelihoods of women in the community through alpaca husbandry and creation of a womens textile cooperative (Farley et al., 2011).

Gender equity (SDG 5) is a critical dimension of equity in PES that has been little explored in watershed PES (Richards, 2013). Changes in land management practices related to PES will differentially affect men and women, and these impacts need to be taken into account in project design and implementation. A recent study of early outcomes of REDD + programs (Reducing Emissions from Deforestation and Forest Degradation—a type of carbon-focused PES), across six countries, found that perceived well-being decreased in REDD + intervention villages compared to control villages and that women specifically fared worse (Larson et al., 2018). This is partially attributed to a lack of meaningful attention to gender in program design, including that women perceived empowerment, social and community unity, and having their own source of income as critical to their well-being (Larson et al., 2018). Thus, aggregated accounts of the outcomes of PES for households from the perspective of male heads of households fail to tell the full story. As with other dimensions of equity, it is essential for PES programs to understand existing gender inequities and evaluate how programs will affect men and women differently across socio-economic, cultural, and racial backgrounds. Designing programs that contribute to greater gender equity (in line with local conceptualizations of fairness and justice), must involve women in planning and decision-making processes. It is critical to pay attention to power dynamics where female and other voices may not be heard in group discussions and planning sessions and seek ways for meaningful inclusion, such as working directly with womeńs cooperatives and other groups.

Procedural equity (inclusive decision-making processes) and recognitional equity (respect for local knowledge, values, norms, and rights in decision making) are important dimensions of equity in their own right and also increase the likelihood of fair distribution of costs, benefits, and burdens. It is widely agreed that increasing participation of small farmers, rural communities, women, and other vulnerable groups in PES design and decision-making processes can help increase long-standing participation of these groups and lead to more socially just outcomes (Bétrisey et al., 2016, 2018; Wilburn et al., 2017). The most successful and equitable programs will be those emerging from or codesigned with local communities. For example, research on a water fund in Colombia demonstrated the importance of long-term relationships between program staff and indigenous communities and that participation was contingent on the project being designed by the community to meet locally defined goals and objectives (Nelson et al., 2020).

There has been a push by some PES theorists and high-level practitioners to "get the science right" (Naeem et al., 2015), including through better targeting protection and restoration activities to reach downstream ecosystem service goals. Yet, without long-term engagement with upstream actors, these targeting exercises risk obscuring the goals, values, and livelihoods that characterize diverse upstream communities and deepening existing inequities. Targeting efforts and program design need to consider power relations, political context, and social goals alongside hydrologic ecosystem service goals. Otherwise, ecosystem service modeling efforts risk marginalizing the values of those living in the watershed at the expense of (often higher income and more powerful) downstream interests (Nelson et al., 2020). PES programs are most likely to be effective and equitable when they work with communities and individual landholders to understand their goals and values and see where synergies can be found with downstream goals (Kolinjivadi et al., 2015, 2017).

Attention to procedural and recognitional equity can also help design more environmentally effective NBS projects. A prominent example comes from Lima, Perú, where a water fund and partners have worked with a local community to restore ancient infiltration channels (*amuñas*). This project served to elevate ancestral knowledge, build trust, and increase interest in watershed protection (Bremer et al., 2016b). A team of hydrologists recently documented that this technology also offers significant benefits for Lima's water supply and is one of the most promising source watershed protection activities for Lima to invest in given additional funding available through recent federal legislation (Ochoa-Tocachi et al., 2019). Such a discovery never would have happened through a top-down, externally driven approach.

Finally, PES programs are also increasingly recognizing the importance of addressing systemic inequities, including access to clean water, through a combination of NBS and gray infrastructure. For example, the FONAG water fund in Quito, Ecuador, has moved away from focusing only on protecting highland ecosystems toward working with outskirt peri-urban communities to address water security by extending the municipal water service to these areas. Once communities have clean water, conversations about ecosystem conservation begin. This shift in approach has important equity benefits. Similarly, smaller municipal water funds in Bolivia run through *Fundación Natura* have worked with rural communities to obtain clean drinking water (secured through combined NBS and gray infrastructure approaches; Grillos, 2017). While NBS management activities can provide a suite of benefits, these cases demonstrate the importance of systemic approaches that also utilize gray infrastructure to address legacies of environmental injustice.

Case study 3: Biocultural restoration

Biocultural conservation and restoration efforts start with, and are based on, locally and culturally meaningful knowledge, processes, perspectives, and worldviews that have long been practiced by Indigenous peoples and local communities (IPLCs) around the world (Chang et al., 2019; Gavin et al., 2015; Morishige et al., 2018; Pascua et al., 2017). Using a biocultural approach in conservation and restoration means building from local perspectives, values, and priorities and recognizing interactions between human and environmental well-being and resilience (Sterling et al., 2017). Given this foundation, biocultural approaches also carry important implications on social justice and equity when highlighting perspectives from groups who are typically underrepresented in decision-making. Biocultural approaches, collectively referred to as the biocultural paradigm, are social-ecological, systemic, and recognized for their significant contributions to local and global sustainability through both policy and on-the-ground action (Merçon et al., 2019). While variations of a biocultural approach are broadly applicable, these approaches are particularly relevant for NBS projects working in collaboration with IPLCs and other groups with generational ties to a place. Biocultural approaches often focus on decolonizing land management and revitalizing Indigenous practices around care for resources (Kealiikanakaoleohaililani et al., 2018; Kimmerer, 2013). Kimmerer (2011, p. 258), director of the Center of Native Peoples and the Environment at SUNY College of Environmental Science and Forestry and a leader in biocultural approaches, refers to "reciprocal restoration" as the "the mutually reinforcing restoration of land and culture such that repair of ecosystem services contributes to cultural revitalization, and renewal of culture promotes restoration of ecological integrity." IPLCs manage or exercise tenure rights over 25% of the Earth's land surface, including 40% of the world's land-based protected areas (Garnett et al., 2018). Beyond existing tenure rights, the vast majority of land on Earth (and thus where NBS are located) are on and around Indigenous lands and traditional territories (Garnett et al., 2018). Thus, there are clear responsibilities and opportunities for such projects to be both more effective and more equitable through greater attention to procedural and recognitional equity and by incorporating place-based cultural perspectives into NBS design.

Biocultural restoration approaches are on the rise in many areas, including in Hawai'i where a recent special issue highlighted the diversity and extent of these efforts, many of which can be classified as NBS for water (Chang et al., 2019). For example, ongoing efforts to restore Indigenous agricultural systems including *lo'i kalo* (wetland and flooded field taro systems) and *loko i'a* (Hawaiian fishponds) have been shown to provide important downstream water quality benefits while also supporting linked social, cultural, and ecological benefits (Bremer et al., 2018a; Winter et al., 2020). These restoration efforts stand in stark contrast with NBS projects focused only on green infrastructure approaches without considering cultural practices; for instance, projects that may impede traditional and customary gathering/harvest or inhibit access to natural resources necessary to perpetuate customary practices. Moreover, in places like Hawai'i where traditional and customary practices are protected by law, projects that inadvertently ignore cultural practices may be in direct violation of existing legal mandates (e.g., maintaining in-stream flow standards necessary to support native tenants' appurtenant water rights for Indigenous agriculture practices) (Sproat, 2009).

Reviving traditional and customary resource governance and management systems using a biocultural approach to restoration provides a suite of water and other benefits (Chang et al., 2019). However, through an added emphasis on place-based and participatory design, biocultural approaches to restoration also offer the potential to improve distributional, procedural, and recognitional equity. Biocultural approaches could be combined with incentive mechanisms such as PES to provide water-related benefits while also contributing to more equitable conservation and development pathways.

Emerging best practices and guidance

Based on our review, we offer the following conclusions related to distributional equity, procedural equity, and recognitional equity in NBS, along with a set of emerging best practices for NBS that contribute to multiple SDGs in a fair and just manner.

Distributional equity

It is critical to acknowledge that NBS can have both positive and negative outcomes, and these outcomes may be distributed unequally (e.g., changes in property values, lost livelihoods, changes in nonmaterial values and well-being). There needs to be a careful consideration of ecological and social benefits, but also costs, including losses to livelihoods and ways of life where land use restrictions or change affect local communities. PES and other incentive programs need to be rooted in a rights-based approach, where programs strengthen capacity and sovereignty over land and resources.

Access to participation in an NBS program and associated benefits should be available across and within communities, regardless of race, ethnicity, class, identity, and so forth. Engagement in the planning process increases the likelihood of NBS success by building trust and understanding and by creating more culturally relevant NBS programs (Innes, 2018). PES programs have worked to increase equity in access through relaxing formal land tenure requirements, incorporating working landscapes, and creating tiered payment schemes that favor smaller landholders. Biocultural approaches may also increase the relevance and desirability of NBS, particularly in communities that feel marginalized by conventional green infrastructure approaches.

Procedural and recognitional equity

NBS projects need to use diverse methodologies and approaches to meaningfully include stakeholder and rightsholder communities in planning and decision-making and to ensure that NBS do not prioritize certain values or knowledge systems over others. PES programs that incorporate community values in a meaningful way have been shown to be more effective and socially acceptable, and in many places reciprocal framings of these programs will be more appropriate than economic transaction framings. Biocultural approaches offer potential to align NBS with processes and actions that are attuned to local and cultural contexts.

Emerging best practices for equitable NBS for water

1. **Account for the historical context.** Understanding legacies of discriminatory policies, colonization, and institutional and power structures helps explain how we got to where we are. This historical and societal context is key to understanding a place and its people.
2. **Assess existing inequalities and the distribution of environmental benefits and burdens.** Any future implementation of NBS should be aware of inequities across groups that would affect implementation and changes in flows of benefits.
3. **Analyze the distributional impacts of any proposed change produced by NBS** to better understand how a given policy or proposal or restoration activity may affect the flow of benefits and harms to different groups of people.
4. **Design policy implementation to be inclusive of diverse and underrepresented perspectives**, keeping in mind power imbalances and capacity constraints.
5. **Pay attention to diverse value formations and uses of NBS** including cultural uses, nontimber forest products, and relational nonmaterial values. These may not be evident, especially from the vantage point of technical elites, policy makers, and engineers.
6. **Give communities a central role in ongoing management and maintenance of NBS—cultivate a sense of shared investment and ownership.** Sufficiently resource these efforts so they are likely to succeed. Use tools like community land trusts, cooperative ownership models, and inclusionary zoning, to make sure that benefits flow to communities and allow them to stay in place.

Acknowledgments

We thank Lauren Fisher of the University of Minnesota for assistance with editing.

References

Abell R, Asquith N, Boccaletti G, Bremer L, Chapin E, Erickson-Quiroz A, Higgins J, Kang S, Karres N, Lehner B, McDonald R, Shemie D, Simmons E, Sridhar A, Vigerstol K, Vogl A, Wood S. Beyond the source: the environmental, economic and community benefits of source water protection. Arlington, VA, USA; 2017.

Adams EA, Boateng GO, Amoyaw JA. Socioeconomic and demographic predictors of potable water and sanitation access in Ghana. Soc Indic Res 2016;126(2):673–87. https://doi.org/10.1007/s11205-015-0912-y.

Ahmad T, Khan M, Haroon, Musa TH, Nasir S, Hui J, Bonilla-Aldana DK, Rodriguez-Morales AJ. COVID-19: zoonotic aspects. Travel Med Infect Dis 2020;, 101607. Advance online publication https://doi.org/10.1016/j.tmaid.2020.101607.

Aikau HK, Ann D, Camvel K. Cultural traditions and food: Kānaka maoli and the production of poi in the heʻeʻia wetland. Food Cult Soc 2016;19(3):539–61. https://doi.org/10.1080/15528014.2016.1208340.

Angelo H. Added value? Denaturalizing the "good" of urban greening. Geogr Compass 2019;13(8). https://doi.org/10.1111/gec3.12459.

Anguelovski I, Connolly JJT, Masip L, Pearsall H. Assessing green gentrification in historically disenfranchised neighborhoods: a longitudinal and spatial analysis of Barcelona. Urban Geogr 2018;39(3):458–91. https://doi.org/10.1080/02723638.2017.1349987.

Arriagada RA, Sills EO, Ferraro PJ, Pattanayak SK. Do payments pay off? Evidence from participation in Costa Rica's PES program. PLoS One 2015;10(7):1–17. https://doi.org/10.1371/journal.pone.0131544.

Balazs CL, Morello-Frosch R, Hubbard AE, Ray I. Environmental justice implications of arsenic contamination in California's San Joaquin Valley: a cross-sectional, cluster-design examining exposure and compliance in community drinking water systems. Environ Health 2012;11(1):84. https://doi.org/10.1186/1476-069X-11-84.

BenDor T, Lester TW, Livengood A, Davis A, Yonavjak L. Estimating the size and impact of the ecological restoration economy. PLoS One 2015;10(6). https://doi.org/10.1371/journal.pone.0128339, e0128339.

Bétrisey F, Mager C, Rist S. Local views and structural determinants of poverty alleviation through payments for environmental services: Bolivian insights. World Dev Perspect 2016;1:6–11. https://doi.org/10.1016/j.wdp.2016.05.001.

Bétrisey F, Bastiaensen J, Mager C. Payments for ecosystem services and social justice: using recognition theories to assess the Bolivian Acuerdos Recíprocos por el Agua. Geoforum 2018;92:134–43. https://doi.org/10.1016/j.geoforum.2018.04.001.

Bonilla-Aldana DK, Dhama K, Rodriguez-Morales AJ. Revisiting the one health approach in the context of COVID-19: a look into the ecology of this emerging disease. Adv Anim Vet Sci 2020;8(3):234–7. https://doi.org/10.17582/journal.aavs/2020/8.3.234.237.

Bratman GN, Hamilton JP, Daily GC. The impacts of nature experience on human cognitive function and mental health: nature experience, cognitive function, and mental health. Ann N Y Acad Sci 2012;1249(1):118–36. https://doi.org/10.1111/j.1749-6632.2011.06400.x.

Bratman GN, Daily GC, Levy BJ, Gross JJ. The benefits of nature experience: improved affect and cognition. Landsc Urban Plan 2015;138:41–50. https://doi.org/10.1016/j.landurbplan.2015.02.005.

Bremer LL, Farley KA, Lopez-Carr D. What factors influence participation in payment for ecosystem services programs? An evaluation of Ecuador's SocioPáramo program. Land Use Policy 2014a;36:122–33. https://doi.org/10.1016/j.landusepol.2013.08.002.

Bremer LL, Farley KA, Lopez-carr D, Romero J. Conservation and livelihood outcomes of payment for ecosystem services in the Ecuadorian Andes : what is the potential for 'win—win' ? Ecosyst Serv 2014b;1–18. https://doi.org/10.1016/j.ecoser.2014.03.007.

Bremer LL, Auerbach DA, Goldstein JH, Vogl AL, Shemie D, Kroeger T, Nelson JL, Benítez SP, Calvache A, Guimarães J, Herron C, Higgins J, Klemz C, León J, Sebastián Lozano J, Moreno PH, Nuñez F, Veiga F, Tiepolo G. One size does not fit all: natural infrastructure investments within the Latin American Water Funds Partnership. Ecosyst Serv 2016a;17:217–36. https://doi.org/10.1016/j.ecoser.2015.12.006.

Bremer LL, Gammie G, Maldonado O. Participatory social impact assessment of water funds: a case study from Lima, Perú. Forest Trends; 2016b. https://www.forest-trends.org/wp-content/uploads/imported/for183-sia-report-english-16-0701-web-pdf.pdf.

Bremer LL, Falinski K, Ching C, Wada CA, Burnett KM, Kukea-Shultz K, Reppun N, Chun G, Oleson KLL, Ticktin T. Biocultural restoration of traditional agriculture: cultural, environmental, and economic outcomes of Loʻi Kalo restoration in Heʻeia, Oʻahu. Sustainability 2018a;10(12):4502. https://doi.org/10.3390/su10124502.

Bremer LL, Brauman KA, Nelson S, Prado KM, Wilburn E, Fiorini ACO. Relational values in evaluations of upstream social outcomes of watershed payment for ecosystem services : a review. Curr Opin Environ Sustain 2018b;35:116–23. https://doi.org/10.1016/j.cosust.2018.10.024.

Butler LJ, Scammell MK, Benson EB. The Flint, Michigan, water crisis: a case study in regulatory failure and environmental injustice. Environ Justice 2016;9(4):93–7. https://doi.org/10.1089/env.2016.0014.

Cassin J. History and development of nature-based solutions: concepts and practice. In: Cassin J, Matthews JH, Lopez-Gunn E, editors. Nature-based solutions and water security: an action agenda for the 21st century. Elsevier; 2021.

Chan KMA, Balvanera P, Benessaiah K, Chapman M, Díaz S, Gómez-Baggethun E, Gould R, Hannahs N, Jax K, Klain S, Luck GW, Martín-López B, Muraca B, Norton B, Ott K, Pascual U, Satterfield T, Tadaki M, Taggart J, Turner N. Why protect nature? Rethinking values and the environment. Proc Natl Acad Sci U S A 2016;113(6):1462–5. https://doi.org/10.1073/pnas.1525002113.

Chan C-S, Si FH, Marafa LM. Indicator development for sustainable urban park management in Hong Kong. Urban For Urban Green 2018;31:1–14. https://doi.org/10.1016/j.ufug.2018.01.025.

Chang K, Winter KB, Lincoln NK. Hawaiʻi in focus: navigating pathways in global biocultural leadership. Sustainability 2019;11(1):283. https://doi.org/10.3390/su11010283.

Chaplin-Kramer R, Dombeck E, Gerber J, Knuth KA, Mueller ND, Mueller M, Ziv G, Klein AM. Global malnutrition overlaps with pollinator-dependent micronutrient production. Proc R Soc B Biol Sci 2014;281(1794). https://doi.org/10.1098/rspb.2014.1799.

Checker M. Wiped out by the "Greenwave": environmental gentrification and the paradoxical politics of urban sustainability. City Soc 2011;23(2):210–29. https://doi.org/10.1111/j.1548-744X.2011.01063.x.

Cole RJ. Social and environmental impacts of payments for environmental services for agroforestry on small-scale farms in southern Costa Rica. Int J Sust Dev World 2010;17(3):208–16. https://doi.org/10.1080/13504501003729085.

Corbera E, Costedoat S, Ezzine-de-Blas D, Van Hecken G. Troubled encounters: payments for ecosystem services in Chiapas, Mexico. Dev Change 2020;51(1):167–95.

De Risi R, Jalayer F, De Paola F, Iervolino I, Giugni M, Topa ME, Mbuya E, Kyessi A, Manfredi G, Gasparini P. Flood risk assessment for informal settlements. Nat Hazards 2013;69(1):1003–32. https://doi.org/10.1007/s11069-013-0749-0.

Deitz S, Meehan K. Plumbing poverty: mapping hot spots of racial and geographic inequality in U.S. household water insecurity. Ann Am Assoc Geogr 2019;109(4):1092–109. https://doi.org/10.1080/24694452.2018.1530587.

Douglas EM, Kirshen PH, Paolisso M, Watson C, Wiggin J, Enrici A, Ruth M. Coastal flooding, climate change and environmental justice: identifying obstacles and incentives for adaptation in two metropolitan Boston Massachusetts communities. Mitig Adapt Strat Glob Chang 2012;17(5):537–62. https://doi.org/10.1007/s11027-011-9340-8.

Eilers EJ, Kremen C, Smith Greenleaf S, Garber AK, Klein A-M. Contribution of pollinator-mediated crops to nutrients in the human food supply. PLoS One 2011;6(6). https://doi.org/10.1371/journal.pone.0021363, e21363.

Ellis AM, Myers SS, Ricketts TH. Do pollinators contribute to nutritional health? PLoS One 2015;10(1):1–17. https://doi.org/10.1371/journal.pone.0114805.

European Commission, Directorate-General for Research and Innovation. Towards an EU research and innovation policy agenda for nature-based solutions & re-naturing cities: final report of the Horizon 2020 expert group on 'Nature-based solutions and re-naturing cities': (full version). Publications Office of the European Union; 2015. http://dx.publications.europa.eu/10.2777/765301.

Farley KA, Anderson WG, Bremer LL, Harden CP. Compensation for ecosystem services: an evaluation of efforts to achieve conservation and development in Ecuadorian páramo grasslands. Environ Conserv 2011;38(4):393–405.

Finn D, Mccormick L. Urban climate change plans: how holistic? Local Environ 2011;16(4):397–416. https://doi.org/10.1080/13549839.2011.579091.

Fisher B, Herrera D, Adams D, Fox HE, Gallagher L, Gerkey D, Gill D, Golden CD, Hole D, Johnson K, Mulligan M, Myers SS, Naidoo R, Pfaff A, Rasolofoson R, Selig ER, Tickner D, Treuer T, Ricketts T. Can nature deliver on the sustainable development goals? Lancet Planet Health 2019;3(3):e112–3. https://doi.org/10.1016/S2542-5196(18)30281-X.

Gaffikin L. The environment as a strategic healthcare partner. Curr Opin Obstet Gynecol 2013;25(6):494–9. https://doi.org/10.1097/GCO.0000000000000021.

Garnett ST, Burgess ND, Fa JE, Fernández-Llamazares Á, Molnár Z, Robinson CJ, et al. A spatial overview of the global importance of indigenous lands for conservation. Nat Sustain 2018;1(7):369–74. https://doi.org/10.1038/s41893-018-0100-6.

Gavin MC, McCarter J, Mead A, Berkes F, Stepp JR, Peterson D, Tang R. Defining biocultural approaches to conservation. Trends Ecol Evol 2015;30(3):140–5. https://doi.org/10.1016/j.tree.2014.12.005.

Gould KA, Lewis TL. Green gentrification: urban sustainability and the struggle for environmental justice. Routledge; 2016.

Gould KA, Lewis TL. From green gentrification to resilience gentrification: an example from Brooklyn1: from green gentrification to resilience gentrification. City Community 2018;17(1):12–5. https://doi.org/10.1111/cico.12283.

Grieg-Gran M, Porras I, Wunder S. How can market mechanisms for forest environmental services help the poor? Preliminary lessons from Latin America. World Dev 2005;33(9):1511–27. https://doi.org/10.1016/j.worlddev.2005.05.002.

Grillos T. Economic vs non-material incentives for participation in an in-kind payments for ecosystem services program in Bolivia. Ecol Econ 2017;131:178–90. https://doi.org/10.1016/j.ecolecon.2016.08.010.

Groom B, Palmer C. REDD + and rural livelihoods. Biol Conserv 2012;154:42–52. https://doi.org/10.1016/j.biocon.2012.03.002.

Guo Q, Fei S, Potter KM, Liebhold AM, Wen J. Tree diversity regulates forest pest invasion. Proc Natl Acad Sci 2019;116(15):7382–6.

Haase D, Kabisch S, Haase A, Andersson E, Banzhaf E, Baró F, Brenck M, Fischer LK, Frantzeskaki N, Kabisch N, Krellenberg K, Kremer P, Kronenberg J, Larondelle N, Mathey J, Pauleit S, Ring I, Rink D, Schwarz N, Wolff M. Greening cities—to be socially inclusive? About the alleged paradox of society and ecology in cities. Habitat Int 2017;64:41–8. https://doi.org/10.1016/j.habitatint.2017.04.005.

Hastings Z, Ticktin T, Botelho M, Reppun N, Kukea-Shultz K, Wong M, et al. Integrating co-production and functional trait approaches for inclusive and scalable restoration solutions. Conserv Sci Pract 2020;2(9), e250.

Hayes T, Grillos T, Bremer LL, Murtinho F, Shapiro E. Collective PES: more than the sum of individual incentives. Environ Sci Policy 2019;102(April):1–8. https://doi.org/10.1016/j.envsci.2019.09.010.

Herrera D, Ellis A, Fisher B, Golden CD, Johnson K, Mulligan M, Pfaff A, Treuer T, Ricketts TH. Upstream watershed condition predicts rural children's health across 35 developing countries. Nat Commun 2017;8(1):811. https://doi.org/10.1038/s41467-017-00775-2.

Hough RL. Biodiversity and human health: evidence for causality? Biodivers Conserv 2014;23(2):267–88. https://doi.org/10.1007/s10531-013-0614-1.

Ickowitz A, Rowland D, Powell B, Salim MA, Sunderland T. Forests, trees, and micronutrient-rich food consumption in Indonesia. PLoS One 2016;11(5). https://doi.org/10.1371/journal.pone.0154139, e0154139.

IFPRI. 2015 annual report. International Food Policy Research Institute (IFPRI); 2016. http://ebrary.ifpri.org/cdm/ref/collection/p15738coll2/id/130442.

Ina P. Payments for environmental services: lessons from the Costa Rican PES programme, https://mpra.ub.uni-muenchen.de/47186/; 2013.

Innes J. Planning through consensus building: a new view of the comprehensive planning ideal, https://www.taylorfrancis.com/books/e/9781351179522/chapters/10.4324%2F9781351179522-13; 2018.

Jose S. Agroforestry for ecosystem services and environmental benefits: an overview. Agrofor Syst 2009;76. https://doi.org/10.1007/s10457-009-9229-7.

Kabisch N, Haase D. Green justice or just green? Provision of urban green spaces in Berlin, Germany. Landsc Urban Plan 2014;122:129–39. https://doi.org/10.1016/j.landurbplan.2013.11.016.

Kabisch N, Frantzeskaki N, Pauleit S, Naumann S, Davis M, Artmann M, Haase D, Knapp S, Korn H, Stadler J, Zaunberger K, Bonn A. Nature-based solutions to climate change mitigation and adaptation in urban areas: perspectives on indicators, knowledge gaps, barriers, and opportunities for action. Ecol Soc 2016;21(2). JSTOR https://www.jstor.org/stable/26270403.

Kealiikanakaoleohaililani K, Kurashima N, Francisco KS, Giardina CP, Louis RP, McMillen H, Asing CK, Asing K, Block TA, Browning M, Camara K, Camara L, Dudley ML, Frazier M, Gomes N, Gordon AE, Gordon M, Heu L, Irvine A, et al. Ritual + sustainability science? A portal into the science of aloha. Sustainability 2018;10(10):3478. https://doi.org/10.3390/su10103478.

Keeler BL, Hamel P, McPhearson T, Hamann MH, Donahue ML, Meza Prado KA, Arkema KK, Bratman GN, Brauman KA, Finlay JC, Guerry AD, Hobbie SE, Johnson JA, MacDonald GK, McDonald RI, Neverisky N, Wood SA. Social-ecological and technological factors moderate the value of urban nature. Nat Sustain 2019;2(1):29–38. https://doi.org/10.1038/s41893-018-0202-1.

Keeler BL, Derickson KD, Waters H, Walker R. Advancing water equity demands new approaches to sustainability science. One Earth 2020;2:211–3. https://doi.org/10.1016/j.oneear.2020.03.003.

Kharas H, McArthur JW, Ohno I, editors. Leave no one behind: time for specifics on the sustainable development goals. Washington, DC: The Brookings Institution; 2020.

Kimmerer R. Restoration and reciprocity: the contributions of traditional ecological knowledge. In: Egan D, Hjerpe EE, Abrams J, editors. Human dimensions of ecological restoration. Island Press/Center for Resource Economics; 2011. p. 257–76. https://doi.org/10.5822/978-1-61091-039-2_18.

Kimmerer R. Braiding sweetgrass: indigenous wisdom, scientific knowledge and the teachings of plants. Milkweed Editions; 2013.

Kolinjivadi V, Grant A, Adamowski J, Kosoy N. Geoforum juggling multiple dimensions in a complex socio-ecosystem: the issue of targeting in payments for ecosystem services. Geoforum 2015;58:1–13. https://doi.org/10.1016/j.geoforum.2014.10.004.

Kolinjivadi V, Van Hecken G, Rodríguez de Francisco JC, Pelenc J, Kosoy N. As a lock to a key? Why science is more than just an instrument to pay for nature's services. Curr Opin Environ Sustain 2017;26–27:1–6. https://doi.org/10.1016/j.cosust.2016.12.004.

Lansing DM. Unequal access to payments for ecosystem services: the case of Costa Rica: unequal access to payments for ecosystem services. Dev Change 2014;45(6):1310–31. https://doi.org/10.1111/dech.12134.

Larson A, et al. Gender lessons for climate initiatives: a comparative study of REDD + impacts on subjective wellbeing. World Dev 2018;108:86–102.

Lee J, Park BJ, Tsunetsugu Y, Ohira T, Kagawa T, Miyazaki Y. Effect of forest bathing on physiological and psychological responses in young Japanese male subjects. Public Health 2011;125(2):93–100. https://doi.org/10.1016/j.puhe.2010.09.005.

Li Q, Morimoto K, Kobayashi M, Inagaki H, Katsumata M, Hirata Y, Hirata K, Shimizu T, Li YJ, Wakayama Y, Kawada T, Ohira T, Takayama N, Kagawa T, Miyazaki Y. A forest bathing trip increases human natural killer activity and expression of anti-cancer proteins in female subjects. J Biol Regul Homeost Agents 2008;22(1):45–55.

Li J, Feldman MW, Li S, Daily GC. Rural household income and inequality under the Sloping Land Conversion Program in western China. Proc Natl Acad Sci 2011;108(19):7721–6. https://doi.org/10.1073/pnas.1101018108.

Loft L, Le DN, Pham TT, Yang AL, Tjajadi JS, Wong GY. Whose equity matters? National to local equity perceptions in Vietnam's payments for forest ecosystem services scheme. Ecol Econ 2017;135:164–75. https://doi.org/10.1016/j.ecolecon.2017.01.016.

Luh J, Baum R, Bartram J. Equity in water and sanitation: developing an index to measure progressive realization of the human right. Int J Hyg Environ Health 2013;216(6):662–71. https://doi.org/10.1016/j.ijheh.2012.12.007.

Maantay J, Maroko A. Mapping urban risk: flood hazards, race, & environmental justice in New York. Appl Geogr 2009;29(1):111–24. https://doi.org/10.1016/j.apgeog.2008.08.002.

MacDonald AJ, Mordecai EA. Amazon deforestation drives malaria transmission, and malaria burden reduces forest clearing. Proc Natl Acad Sci 2019;116(44):22212–8. https://doi.org/10.1073/pnas.1905315116.

McDermott M, Mahanty S, Schreckenberg K. Examining equity: a multidimensional framework for assessing equity in payments for ecosystem services. Environ Sci Policy 2013;33:416–27. https://doi.org/10.1016/j.envsci.2012.10.006.

McElwee P, Huber B, Nguyễn THV. Hybrid outcomes of payments for ecosystem services policies in Vietnam: between theory and practice. Dev Chang 2020;51(1):253–80. https://doi.org/10.1111/dech.12548.

McGregor D. Nā Kua'āina: living Hawaiian culture. University of Hawai'i Press; 2007.

McPhearson T, Andersson E, Elmqvist T, Frantzeskaki N. Resilience of and through urban ecosystem services. Ecosyst Serv 2015;12:152–6.

Merçon J, Vetter S, Tengö M, Cocks M, Balvanera P, Rosell JA, Ayala-Orozco B. From local landscapes to international policy: contributions of the biocultural paradigm to global sustainability. Glob Sustain 2019;2. https://doi.org/10.1017/sus.2019.4, e7.

Morishige K, Andrade P, Pascua P, Steward K, Cadiz E, Kapono L, Chong U. Nā Kilo 'Āina: visions of biocultural restoration through indigenous relationships between people and place. Sustainability 2018;10(10):3368. https://doi.org/10.3390/su10103368.

Morita E, Fukuda S, Nagano J, Hamajima N, Yamamoto H, Iwai Y, Nakashima T, Ohira H, Shirakawa T. Psychological effects of forest environments on healthy adults: Shinrin-yoku (forest-air bathing, walking) as a possible method of stress reduction. Public Health 2007;121(1):54–63. https://doi.org/10.1016/j.puhe.2006.05.024.

Muradian R, Corbera E, Pascual U, Kosoy N, May PH. Reconciling theory and practice: an alternative conceptual framework for understanding payments for environmental services. Ecol Econ 2010;69(6):1202–8. https://doi.org/10.1016/j.ecolecon.2009.11.006.

Murtinho F, Hayes T. Communal participation in payment for environmental services (PES): unpacking the collective decision to enroll. Environ Manag 2017;59(6):939–55. https://doi.org/10.1007/s00267-017-0838-z.

Myers SS, Gaffikin L, Golden CD, Ostfeld RS, Redford KH, Ricketts TH, Turner WR, Osofsky SA. Human health impacts of ecosystem alteration. Proc Natl Acad Sci 2013;*110*(47):18753–60. https://doi.org/10.1073/pnas.1218656110.

Naeem S, Ingram JC, Varga A, Agardy T, Barten P, Bennett G, Bloomgarden E, Bremer LL, Burkill P, Cattau M, Ching C, Colby M, Cook DC, Costanza R, DeClerck F, Freund C, Gartner T, Goldman-Benner R, Gunderson J, et al. Get the science right when paying for nature's services. Science 2015;347(6227):1206–7. https://doi.org/10.1126/science.aaa1403.

Nelson SH, Bremer LL, Meza Prado K, Brauman KA. The political life of natural infrastructure: water funds and alternative histories of payments for ecosystem services in Valle del Cauca, Colombia. Dev Chang 2020;51(1):26–50. https://doi.org/10.1111/dech.12544.

Ochoa-Tocachi BF, Bardales JD, Antiporta J, Pérez K, Acosta L, Mao F, Zulkafli Z, Gil-Ríos J, Angulo O, Grainger S, Gammie G, De Bièvre B, Buytaert W. Potential contributions of pre-Inca infiltration infrastructure to Andean water security. Nat Sustain 2019;2(7):584–93. https://doi.org/10.1038/s41893-019-0307-1.

Ohtsuka Y, Yabunaka N, Takayama S. Shinrin-yoku (forest-air bathing and walking) effectively decreases blood glucose levels in diabetic patients. Int J Biometeorol 1998;41(3):125–7. https://doi.org/10.1007/s004840050064.

Pagiola S, Rios AR, Arcenas A. Can the poor participate in payments for environmental services? Lessons from the Silvopastoral Project in Nicaragua. Environ Dev Econ 2008;13(3):299–325. https://doi.org/10.1017/S1355770X08004270.

Park LS-H, Pellow DN. The slums of Aspen: immigrants vs. the environment in America's Eden. New York, NY: NYU Press; 2013.

Pascua P, McMillen H, Ticktin T, Vaughan M, Winter KB. Beyond services: a process and framework to incorporate cultural, genealogical, place-based, and indigenous relationships in ecosystem service assessments. Ecosyst Serv 2017;26:465–75. https://doi.org/10.1016/j.ecoser.2017.03.012.

Pascual U, Phelps J, Garmendia E, Brown K, Corbera E, Martin A, Gomez-Baggethun E, Muradian R. Social equity matters in payments for ecosystem services. Bioscience 2014;64(11):1027–36. https://doi.org/10.1093/biosci/biu146.

Pattanayak SK, Wunder S, Ferraro PJ. Show me the money: do payments supply environmental services in developing countries? Rev Environ Econ Policy 2010;4(2):254–74. https://doi.org/10.1093/reep/req006.

Patz JA, Daszak P, Tabor GM, Aguirre AA, Pearl M, Epstein J, et al. Unhealthy landscapes: policy recommendations on land use change and infectious disease emergence. Environ Health Perspect 2004;112(10):1092–8.

Patz JA, Olson SH, Uejio CK, Gibbs HK. Disease emergence from global climate and land use change. Med Clin North Am 2008;92(6):1473–91.

Pearsall H, Pierce J. Urban sustainability and environmental justice: evaluating the linkages in public planning/policy discourse. Local Environ 2010;15(6):569–80. https://doi.org/10.1080/13549839.2010.487528.

Pérez-Foguet A, Garriga RG. Analyzing water poverty in basins. Water Resour Manag 2011;25(14):3595–612. https://doi.org/10.1007/s11269-011-9872-4.

Quastel N. Political ecologies of gentrification. Urban Geogr 2009;30(7):694–725. https://doi.org/10.2747/0272-3638.30.7.694.

Richards M. What do we know about gender and other social impacts of IWS projects?: A literature review. Forest Trends; 2013. https://www.forest-trends.org/publications/what-do-we-know-about-gender-and-other-social-impacts-of-iws-projects/.

Rigolon A, Christensen J. Greening without Gentrification: learning from parks-related anti-displacement strategies nationwide. University of California Institute of the Environment and Sustainability; 2019. https://www.ioes.ucla.edu/project/prads/.

Rigolon A, Németh J. "We're not in the business of housing:" environmental gentrification and the nonprofitization of green infrastructure projects. Cities 2018;81:71–80. https://doi.org/10.1016/j.cities.2018.03.016.

Rodríguez de Francisco JC, Budds J, Boelens R. Payment for environmental services and unequal resource control in Pimampiro, Ecuador. Soc Nat Resour 2013;26(10):1217–33. https://doi.org/10.1080/08941920.2013.825037.

Rosa LP, dos Santos MA, Matvienko B, dos Santos EO, Sikar E. Greenhouse gas emissions from hydroelectric reservoirs in tropical regions. Clim Change 2004;66(1/2):9–21. https://doi.org/10.1023/B:CLIM.0000043158.52222.ee.

Sakijege T, Lupala J, Sheuya S. Flooding, flood risks and coping strategies in urban informal residential areas: the case of Keko Machungwa, Dar es Salaam, Tanzania. Jàmbá: J Disaster Risk Stud 2012;4(1):10. https://doi.org/10.4102/jamba.v4i1.46.

Salzman J, Bennett G, Carroll N, Goldstein A, Jenkins M. The global status and trends of payments for ecosystem services. Nat Sustain 2018;1(3):136–44. https://doi.org/10.1038/s41893-018-0033-0.

Shapiro-Garza E, McElwee P, Van Hecken G, Corbera E. Beyond market logics: payments for ecosystem services as alternative development practices in the global south. Dev Change 2019;51(1):3–25. https://doi.org/10.1111/dech.12546.

Sproat DK. Ola I Ka Wai: a legal primer for water use and management in Hawai'i. Ka Huli Ao & Office of Hawaiian Affairs; 2009. https://www.law.hawaii.edu/sites/www.law.hawaii.edu/files/content/news/18470/WaterPrimer.pdf.

Sterling EJ, Filardi C, Toomey A, Sigouin A, Betley E, Gazit N, Newell J, Albert S, Alvira D, Bergamini N, Blair M, Boseto D, Burrows K, Bynum N, Caillon S, Caselle JE, Claudet J, Cullman G, Dacks R, et al. Biocultural approaches to well-being and sustainability indicators across scales. Nat Ecol Evol 2017;1(12):1798–806. https://doi.org/10.1038/s41559-017-0349-6.

UNEP (United Nations Environment Programme). UNEP frontiers 2016 report: emerging issues of environmental concern. Nairobi, Kenya: United Nations Environment Programme; 2016.

United Nations Children's Fund (UNICEF), World Health Organization. Progress on household drinking water, sanitation and hygiene 2000–2017. Special focus on inequalities, https://www.unicef.org/media/55276/file/Progress%20on%20drinking%20water,%20sanitation%20and%20hygiene%202019%20.pdf; 2019.

van den Berg M, Wendel-Vos W, van Poppel M, Kemper H, van Mechelen W, Maas J. Health benefits of green spaces in the living environment: a systematic review of epidemiological studies. Urban For Urban Green 2015;14(4):806–16. https://doi.org/10.1016/j.ufug.2015.07.008.

VonHedemann N. Incentives, livelihoods, and forest ecology: payments for ecosystem services in Guatemala's Western Highlands. University of Arizona; 2019. Retrieved from https://repository.arizona.edu/handle/10150/631878. [Accessed 2 March 2020].

Wilburn E, Bremer L, Brauman K, Gould R, Seehusen S, Prado K, Hamel P. Voices from the field: participant perspective on growth and sustainability opportunities for water producer projects, https://naturalcapitalproject.stanford.edu/sites/g/files/sbiybj9321/f/publications/voicesfromthefield_eng.pdf; 2017.

Winter K, Lincoln N, Berkes F, Alegado R, Kurashima N, Frank K, et al. Ecomimicry in indigenous resource management: optimizing ecosystem services to achieve resource abundance, with examples from Hawai'i. Ecol Soc 2020;25(2).

Wolch JR, Byrne J, Newell JP. Urban green space, public health, and environmental justice: the challenge of making cities "just green enough.". Landsc Urban Plan 2014;125:234–44. https://doi.org/10.1016/j.landurbplan.2014.01.017.

World Health Organization (WHO), the United Nations Children's Fund (UNICEF). Progress on drinking water, sanitation and hygiene 2017 update and SDG baselines, https://www.who.int/mediacentre/news/releases/2017/launch-version-report-jmp-water-sanitation-hygiene.pdf; 2017.

Wunder S. Payments for environmental services and the poor: concepts and preliminary evidence. Environ Dev Econ 2008;13(3):279–97. https://doi.org/10.1017/S1355770X08004282.

Wunder S, Brouwer R, Engel S, Ezzine-de-Blas D, Muradian R, Pascual U, Pinto R. From principles to practice in paying for nature's services. Nat Sustain 2018;1(3):145–50. https://doi.org/10.1038/s41893-018-0036-x.

Zbinden S, Lee DR. Paying for environmental services: an analysis of participation in Costa Rica's PSA program. World Dev 2005;33(2):255–72. https://doi.org/10.1016/j.worlddev.2004.07.012.

State of policy and governance

CHAPTER 6

Why governments embrace nature-based solutions: The policy rationale

Kathleen Dominique[a,*], Nathanial Matthews[b], Lisa Danielson[c,†], and John H. Matthews[d]

[a]OECD, Paris, France, [b]Global Resilience Partnership, Stockholm, Sweden, [c]Government of Canada, Ottawa, ON, Canada, [d]Alliance for Global Water Adaptation, Corvallis, OR, United States

Nature-based solutions: A rising policy priority

Resilient management of water resources is fundamental for the broader achievement of the Sustainable Development Goals (SDGs), many of which are contingent on the achievement of SDG 6. Water management is especially relevant for food security, healthy lives, clean energy, sustainable cities, marine and terrestrial biodiversity, and ecosystems. The sound management of water resources and provision of water services is critical for economic development, underpinning agricultural production and fisheries, supporting biodiversity, energy generation, and water-based transportation (Cassin and Matthews, 2021).

Our economies, health, and social systems are deeply intertwined with Earth systems that are rapidly transforming. As a result, we are now beginning to see interconnected financial and economic crises coupled with extreme weather events exacerbated by climate change, such as droughts, floods, and disease outbreaks that have repercussions from local to global scales (LaPorte, 2007). As the need for a fundamental shift from trying to control change and maintain stability toward living with turbulence and uncertainty becomes increasingly acute, the importance of resilient infrastructure and nature-based solutions (NBS) has grown. This has led many to call for the development of novel approaches to infrastructure to be undertaken in a way that enhances resilience in a world of growing uncertainty and depleting resources. However, existing policy frameworks, institutional arrangements, and financing approaches provide a more conducive enabling environment for traditional gray

---

* The views expressed in the chapter are the author's own and do not necessarily reflect the views of OECD member countries.

† The views expressed in the chapter are the author's own and do not necessarily reflect the views of the Government of Canada.

infrastructure, and there is broad scope to increase the systematic consideration of NBS as a part of efforts to improve water management and the delivery of water-related services.

This chapter explores the rationale for governments to promote NBS as a complement or alternative to gray infrastructure. It also examines the distinctive features of NBS compared to gray infrastructure that may pose challenges for their implementation at scale. Finally, the chapter discusses several levers for policy change that would serve to encourage the more systematic uptake of NBS, alongside gray infrastructure. In conclusion, the chapter ends with select reflections on political economy considerations and priorities for future policy action.

An attractive alternative or complement to gray infrastructure to improve resilience

Limitations of traditional large-scale gray infrastructure is driving interest in NBS to enhance resilience

Water infrastructure is critical to development. While many high-income countries have well-developed water infrastructure, the majority of low- and middle-income countries have a water infrastructure deficit that has been shown to significantly impede development. For these countries, water infrastructure (often taking the form of traditional gray water infrastructure) is seen as a basic step in the development process or its hydraulic mission. Hydraulic missions are typically characterized by engineering-led, large-scale gray infrastructure projects such as dams and irrigation projects. The focus within hydraulic missions is to secure water and seek to control and exploit it as a resource. Phases of hydraulic missions spawn hydraulic bureaucracies, sometimes called hydrocracies, which are often headed by influential decision makers (Molle et al., 2009). Such hydrocracies are often a mix of top-down decision-making driven by private and state actors, including ministries and government departments responsible for irrigation, energy, and water, as well as politicians, land elites, development, state and private banks, and construction companies (Molle et al., 2009). Countries in their hydraulic missions will be in various stages of the process of making decisions on basic infrastructure development, such as water and sanitation systems, roads and storm water drainage, modern agricultural technologies, and large hydropower dams.

The majority of the decisions regarding water infrastructure are predicated on engineering modeling and thinking that typically base projections of future needs, and thus plans, on historical patterns of water availability and use. This includes designs and plans based on the assumption that future climatic and hydrological conditions will be broadly similar to those of the past. This thinking has been at the foundation of many of the world's largest water infrastructure projects including the Three Gorges Dam, the High Aswan Dam, the Hoover Dam, the Great Man-Made River, and the Kaleshwaram Lift Irrigation Project. These projects have been critical for national development, providing income, generating electricity, reducing floods, and providing irrigation for agriculture and food security and water for human consumption. They have also come at enormous costs due to their impacts on people and ecosystems. Large-scale water infrastructure projects that have limited or weak environmental mitigation plans, have displaced millions of people, altered the flow, temperature, water quality, and sediment loads of rivers and had profound impacts on the ecosystems that provide the livelihood foundation for hundreds of millions of people (Matthews, 2016; WCD, 2000).

In the past, the negative impacts of gray, large-scale water infrastructure on people and ecosystems were significant, but today they are even more profound and interconnected. The world is now quite different than it was even 50 years ago. The world's population has more than doubled from 3.6 billion people in 1970 to 7.8 billion in 2020. This enormous population increase has coincided with massive development in technology and economic activity that has resulted in humans now modifying climate, weather, land, ecosystems, oceans, and the cryosphere in fundamental and often enduring ways. As Steffen et al. (2015) have noted, our human alterations are accelerating change across these systems in an unprecedented manner, and this is reshaping the architecture of risk and its implications for infrastructure (Keys et al., 2019). Despite these new realities and the changing landscape of risk, many state and nonstate actors in developing countries are still pushing ahead with hydraulic missions that are dominated by traditional planning and gray infrastructure that ignores or downplays system connections and impacts across ecosystems, health, and economies. Prominent examples include, but are not limited to, the rapid hydropower development along the Mekong and Amazon Basins and the Grand Ethiopian Renaissance Dam (Matthews et al., 2012; Middleton and Matthews, 2014). The COVID-19 pandemic confirmed that to keep our economic systems healthy, we must also protect our socio-ecological systems. It has further confirmed that many of our current economic and social systems are fragile to shocks and stresses and lack resilience.

From control to resilience

Several principles have been suggested that enable water resilience, based on the capacity of freshwater ecosystems to respond to climate shifts now and in the past (Le Quesne et al., 2010; Poff, 2017), summarized in Box 6.1.

These key resilience principles reflect the enabling conditions necessary for water systems to compensate and adapt on their own to climate shifts, as they have largely done throughout history and prehistory. In the context of intensive human dependence and management of these water systems, high uncertainty about emerging conditions emphasizes the need for systematically incorporating climate impacts into water management governance, including policies, practices, and investment approaches (Schmidt and Matthews, 2017; Wolf et al., 2020). Indeed, natural processes may now need more active human management in order to build, restore, and enhance resilience in water systems (Matthews et al., 2019). This new reality has called into question the costs and risks associated with large-scale gray water infrastructure projects versus their benefits and suitability to enhance resilience in an

---

Box 6.1   Principles to enable water resilience.

Strengthening the resilience of water systems can be promoted through several principles that enhance the capacity of freshwater ecosystems to respond to changing conditions:

- *Managing for hydrologic connectivity.* Connectivity includes maintaining and/or restoring functional connections within and between ecosystems and habitats for the movement of nutrients, organisms, and ecological processes and functions. Grill et al. (2019) recently defined a promising set of metrics around connectivity.
- *Managing for temporal environmental variability.* For freshwater ecosystems, temporal variability is most often associated with the disturbance regime, also known as the natural flow regime or the annual hydrograph (Poff et al., 1997). The seasonal flow patterns in aquatic systems and water quality play a critical role in maintaining ecosystem health. Distortions of the natural flow regime from shifts in climate or operational decisions can trigger rippling and profound effects across ecosystems (Poff, 2017).
- *Managing for spatial heterogeneity.* Spatial variation within freshwater ecosystems, sometimes also referred to as spatial heterogeneity, which broadly refers to ensuring that: (1) hydrological and hydraulic aspects of freshwater systems are maintained and driven by ecological and hydrological processes rather than human modification intended to supersede or simplify natural functions (e.g., natural meanders in contrast to "channelizing" a river, which reduces complexity); (2) functional diversity can persist across the food web; and (3) organisms are not trapped in one habitat, region, or ecological population or community.
- *Managing freshwater ecosystems at the basin scale.* Effective governance is critical for implementation, including the ability to develop a basin-wide "shared vision" of what resilient water systems can and should look like in practice.

From Le Quesne T, Matthews JH, Von der Heyden C, Wickel AJ, Wilby R, Hartmann J, Pegram G, et al. Flowing forward: Freshwater ecosystem adaptation to climate change in water resources management and biodiversity conservation. Washington, DC: World Bank; 2010; Poff NL. Beyond the natural flow regime? Broadening the hydro-ecological foundation to meet environmental flows challenges in a non-stationary world. Freshw Biol 2017;12:1011–21. https://doi.org/10.1111/fwb.13038.

---

ever-changing environment. Notably, NBS can assist in managing uncertainty related to climate change by avoiding or delaying lock-in to capital-intensive gray infrastructure, allowing for flexibility to adapt to changing circumstances (OECD, 2013).

Developing resilient water infrastructure recognizes that hydrology, ecosystems, water management, and climate are intertwined and that the integrity of ecosystems is necessary for the integrity of communities—and vice versa. Many regions are already facing serious threats from climate impacts, and these threats are likely to intensify and grow, overcoming decades and centuries. Restoring the functions of the landscapes and ecosystems that provide for and regulate our natural water systems is one of the most reliable means society has for continuing to thrive and develop while alleviating poverty and promoting sustainable growth through a more effective vision of resource management (Poff et al., 2016).

NBS have been proposed as a suite of ecosystem-related approaches to address pressing challenges facing humanity and our natural systems, including climate change and water resilience (see Cassin and Matthews, 2021, for further discussion). NBS can vary in terms of how "natural" or engineered a solution is, from protecting a fully intact ecosystem such as an old-growth forest to implementing an engineered wetland. The US Army Corps of Engineers, for instance, distinguishes between natural features (which are existing ecological processes and ecosystems) and NBS (which are somewhat broader and include designed and reconstructed and ecological analog approaches, such as "new" wetlands on brownfield sites) (Bridges et al., 2015). What all NBS have in common is that they seek to maximize the ability of nature to provide ecosystem services that help address a human challenge, such as climate change adaptation or disaster risk reduction.

NBS are especially useful to improve water quantity and quality through filtration and flow regulation. They can be used alone or in conjunction with other solutions to produce the desired outcomes. For example, NBS such as protection or restoration of natural vegetation in the watershed can be combined with a smaller-scale gray infrastructure water treatment facility at the point of diversion to deliver clean, reliable water at a reasonable cost (see also Cassin, 2021; Vigerstol et al., 2021).

Advantages of NBS: Cost-effectiveness and cobenefits

NBS have been shown to provide diverse benefits to communities, ecosystems, and cities, including carbon sequestration, habitat for local fauna, rich soils for sustainable agriculture and clean water for industry, energy, and consumption. Upstream forest protection, reforestation, and improved agricultural practices could improve water quality for four out of five large cities around the world and can deliver these benefits cost-effectively: one in six cities can pay for natural solutions solely through savings in water treatment costs (Cities4Forests, 2020). The diversity and affordability of benefits underscores the urgency to mobilize the power of nature to meet water security challenges in a sustainable way.

While acknowledging that NBS (and green infrastructure) is not a replacement for all large-scale gray infrastructure and that a diverse range of approaches to climate change adaptation and mitigation are needed, NBS nevertheless have an important role in providing resilient and cost-effective infrastructure (Culwick and Bobbins, 2016: Thiele et al., 2020). They often require less capital expenditure than traditional gray infrastructure, while providing multiple benefits (Culwick and Bobbins, 2016; Kithiia and Lyth, 2011). Culwick and Bobbins (2016) point out that, "green infrastructure can provide infrastructure alternatives where the cost of traditional gray infrastructure is prohibitively high." A notable example is the Diepsloot informal settlement in Johannesburg, South Africa. Diepsloot does not have a formalized storm water infrastructure system due to its high cost; however, a community-based organization called WASSUP (Water, Amenities, Sanitation Services

Upgrading Programme), together with academics from the University of the Witwatersrand, have explored opportunities to develop low-cost green infrastructure to address the risks of both standing water and flooding (Culwick and Bobbins, 2016). Initial findings indicate that such opportunities that use available resources and local skills exist (e.g., a pilot soak-away garden has been developed by WASSUP to help in absorbing excess surface water) (Culwick and Bobbins, 2016).

Distinct challenges posed by nature-based solutions from a policy perspective

While there is a clear and compelling case for the use of NBS for water management in many contexts, policy challenges arise in moving from pilot projects to upscaling and broader use (Wolf et al., 2020). To date, many of the NBS examples for water management are the result of specific pilot programs, dedicated funding efforts, and advocacy by specific actors such as NGOs or specific departments within governments or development finance institutions. The broader mainstreaming of NBS remains limited to date, due in part to the distinctive features of NBS, which generate certain "mismatches" with the existing policy framework and institutional arrangements. Many of the same characteristics that make NBS so attractive in the first place, such as cobenefits, modularity, and dynamism (as highlighted in the previous section) can lead to specific challenges that limit or prevent the operationalization of NBS through existing institutions, policies, and financial instruments. Some of the challenges facing the widespread use of NBS can be understood as being intrinsic to the characteristics of NBS themselves. This section summarizes these distinctive features.

Difficulties in quantifying and monetizing benefits

In addition to their primary purpose, such as improving water quality, NBS generate additional social, economic, and environmental cobenefits that can be difficult to quantify and, consequently, to monetize. While NBS with clear market benefits, such as improved water supply, increased aquaculture yield, or tourism revenue can be more easily valued, many of the benefits of NBS face quantification and, therefore, monetization challenges. One challenge is that many of the benefits of NBS are in terms of avoided costs, which can be difficult to measure in their own right, given the stochastic nature of damaging events and the public good benefits of protection (OECD, 2019a). In the case of flooding, for example, the difficulties in translating findings between unique sites is compounded by the inherent low frequency of flood events, with each event characterized by a number of variables (Wingfield et al., 2019). Finally, the existing methods for assessing, valuing, and monitoring cobenefits not traded on the market (such as improved mental health, due to access to nature, or improved biodiversity) are often underdeveloped or challenging to apply (Trémolet et al., 2019). Without a clear understanding of benefits, it becomes challenging both to get buy-in for the use of NBS and to secure revenue streams and finance.

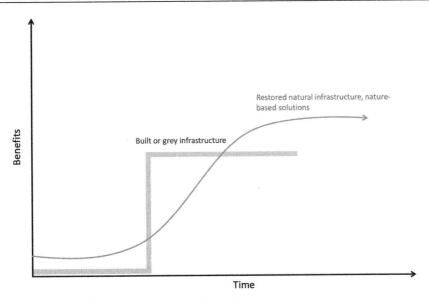

Fig. 6.1

Time lags in realizing NBS benefits. Existing natural infrastructure is already providing water-related benefits, but the full benefits take time to develop following restoration and/or management, compared to gray infrastructure. Gray infrastructure is fully functional once construction is completed. NBS are multifunctional and come with cobenefits that go beyond the specific targeted water outcome. Gray infrastructure has a design life, and once it reaches the end of this life, functionality is lost and/or requires significant new investment to recover functionality (e.g., siltation of reservoirs that eliminates storage). While natural infrastructure may need protection or active management, it is self-sustaining in a way that gray infrastructure is not and can continue to provide benefits over a longer period of time.

Time lag between investment and realization of benefits

Some NBS, especially those involving the restoration of badly degraded ecosystems, can be slow to develop their adaptation benefits or deliver potential cobenefits in full. The growth rate of the living components of an NBS and stage of maturity can substantially affect their effectiveness—for example, a young forest will not have the same water retention benefits as a mature one (OECD, 2020). Compared to conventional gray infrastructure, the timeframe for benefits to be fully realized follows a distinctly different pattern (see Fig. 6.1 for illustration, modified from Browder et al. (2019)).

The time lag between investment and realization of benefits can be a barrier to politicians seeking quick results or investors seeking short- or medium-term returns (Browder et al., 2019), and NBS can be poorly suited for quick response to a crisis. At the same time, the slow growth of NBS can be an advantage, as they can appreciate in value over time with the regeneration of nature and its associated ecosystem services, as opposed to the high depreciation costs associated with gray infrastructure (OECD, 2020).

Uncertainty related to performance and lack of performance track records

Ecosystems are not static, as they are made of living components that evolve over time. As previously noted, this can be a benefit, as it means NBS can adapt to changing environmental and risk conditions, thereby potentially exceeding the design lifetime of gray infrastructure (Browder et al., 2019). However, the dynamism of NBS also introduces new sources of uncertainty, which increases the difficulty in developing solid predictions about the likely value of working with NBS.

Ecosystem responses to climate change and the resilience of the ecosystem to withstand climate change impacts are a major source of uncertainty. For example, coral reefs can provide natural protection against storm surges; however, the IPCC Special Report on Global Warming of 1.5°C found that between 70% and 90% of coral reefs would be lost if the temperature increased by 1.5°C, and more than 99% if temperature increased by 2°C (Masson-Delmotte et al., 2018). Research gaps remain in understanding complex, adaptive system dynamics, including feedback and the potential for climate change and other major disruptions to affect ecosystems and the future provision of associated services (Guerry et al., 2015). In addition, the efficacy of NBS in response to different intensity and frequency of threats contribute to the uncertainty related to performance of such approaches.

Novel sources of uncertainty can result in policy makers, regulators, and/or permitting agencies prioritizing gray infrastructure over NBS because it is familiar and easily understood with respect to existing policy and regulatory standards (OECD, 2020). For example, a review of local governments in the United Kingdom found that natural flood management techniques were often avoided based on perceived regulatory hurdles (Huq and Stubbings, 2015).

High transaction costs due to specificity of ecosystem and climatic dynamics, and the multiple parties engaged

The planning of NBS must consider ecosystem dynamics at various spatial scales. For example, modifying amounts of sediment upstream will influence downstream coastline stability and, therefore, can determine the effectiveness of interventions. While the planning of gray infrastructure takes these dynamics into account, for gray infrastructure, it is a matter of optimizing functionality and efficiency and avoiding unforeseen effects. For NBS, these dynamics are fundamental for the success of the intervention. This means that NBS often require coordinated landscape-scale interventions, which involve multiple stakeholders across jurisdictions (World Bank, 2017). The cost of engaging and negotiating with multiple stakeholders, working across regulatory jurisdictions, and collaborating with dispersed landowners can be time consuming and costly. In addition, those responsible for providing the desired service of the NBS (such as a water authority) may not have the capacity or legal legitimacy to engage with necessary stakeholders. Finally, the specificity of ecosystem dynamics means that no two NBS are the same and each project requires tailored input.

In addition, the reliance on a broader range of stakeholders needed to deliver benefits and performance over time can complicate the maintenance of NBS. While the long-term operations and maintenance of gray infrastructure is typically the direct responsibility of the service provider, the implementation and maintenance of NBS often require the active support of local citizens and landowners (through responsibilities such as replanting trees or maintaining natural water retention sites) (Browder et al., 2019). In many cases, responsibilities for the ongoing maintenance of an NBS project remains unspecified, which poses a risk to the continuity of delivering the desired socioeconomic and environmental benefits in the long term (Kabisch et al., 2016).

In addition to challenges linked to the features that make NBS unique from gray infrastructure, many challenges arise from the rigidity of existing policy and institutional arrangements (discussed further in this chapter). Institutional norms and path dependency, whereby decision makers chose familiar options, can be a daunting barrier to widespread NBS use. Furthermore, infrastructural approaches are deeply ingrained in certain professional contexts (such as water engineers) and can shape institutional practices. These biases are compounded by cognitive barriers, which can prevent action on climate change adaptation in general, such as the discounting of climate risks and challenges to decision-making under uncertainty (OECD, 2020).

Limits to standardization

There is an inherent tension between the uniqueness of individual NBS projects and potential for replicability and scaling up. In general, NBS performance and benefits (for example, flood reduction) are context- and site-specific and fluctuate over time. This creates difficulty in designing standards, which imply a one-size-fits-all approach (Seddon et al., 2020). Due to the diversity of NBS across geographies and climates, there is an intrinsic limit to potential for standardization and replication. However, a globally recognized standard can help ensure the quality and credibility of NBS and support efforts for mainstreaming and scaling up. In 2020, IUCN launched a Global Standard for NBS, which comprises eight criteria and indicators, to support the effective operationalization of best-practice principle (IUCN, 2020).

Barriers to funding and financing nature-based solutions

Decision makers and stakeholders may support NBS, but funding and financing new types of approaches to deliver water management and infrastructure confronts a number of barriers. In the case of water service providers, these actors may be limited in their ability to fund NBS for a range of reasons, including accounting rules relative to asset capitalization, issues linked to land ownership, or constraints on the ability to finance activities outside of their service area, among others (Trémolet et al., 2019). Funding and financing for NBS are further discussed in Trémolet et al. (2021).

Levers for change to make the enabling environment more conducive to NBS

Overcoming rigidity of existing policy and institutional arrangements that favor the status quo

While NBS have demonstrated their potential to deliver a range of cobenefits and strengthen the resilience of economies, ecological systems, and communities, their more systematic application is unlikely to occur without concerted policy impetus. However, a status quo bias can result in reinforcement of past project, program, and investment design favoring conventional, gray infrastructure. While overall policy frameworks may generally be conducive to allowing for NBS, a shift in approach may be difficult to emerge due to a certain degree of "path dependency" related to approaches used in the past and general resistance to change (Trémolet et al., 2019). Thus, a need to overcome the rigidity of existing policy and institutional arrangements that favor the status quo should not be overlooked in efforts to scale up NBS for water management (see, for example, Coxon et al., 2021).

Integrating an explicit focus on NBS into strategic policy and planning processes

Beyond a focus on pilot projects and dedicated programs, NBS would benefit from explicit inclusion in strategic policy and planning processes and documents, such as National Water Strategies, National Adaptation Plans, National Determined Contributions related to the UN Framework Convention on Climate Change, among others. For example, in the case of cities, in order for local decision makers to adopt NBS as part of their water management approach, they generally need to be included within a specific plan or strategy or integrated into the overarching development plan (Kremer et al., 2016).

Recent studies document a significant number of countries including references to NBS in key policy-planning documents. According to a review by the OECD, roughly two-thirds of OECD countries[a] include explicit references to NBS as part of their National Adaptation Plans, of which managing water-related risks are a key element (OECD, 2020). In the European Union, the EU Floods Directive (2007) promotes nature and risk-based adaptation planning as a part of a holistic approach to flood risk management. A recent study found that this directive spurred the majority of EU member states (26 in total) to include NBS in some of all of their flood risks management plans (Trémolet et al., 2019).

Shifting from approach focused on delivering infrastructure to delivering services

Making the enabling environment more conducive to systematic consideration of NBS can be supported by shifting from an output-based approach focused on building infrastructure to an

---

[a] A total of 24 countries out of the 35 OECD member countries that have National Adaptation Plans.

outcome-based approach focused on delivering services, such as water flow regulation, flood prevention and control, water quality improvement, and so on. This promotes a shift from a focus from delivering infrastructure (typically conventional gray infrastructure) to delivering services, widening the range of possible solutions that can do so in a cost-effective way. This is an important consideration for policy makers in government, legislators, regulators, and development finance institutions. One example on how water regulators can contribute to such a shift is the adoption of a regulatory framework that directly and indirectly stimulates innovation. For example, an economic regulator can promote specific service quality targets through the use of performance indicators while allowing utilities the flexibility to reach those targets in the most cost-efficient manner. Regulators can allow companies to derogate from existing rules (e.g., allowing pilot projects and regulatory experiments) or use funds collected via tariffs to finance certain projects (Guerrini, 2020).

In addition to economic and environmental regulation, procurement systems often lead local governments, water managers, or utilities to define performance in narrowly focused or prescriptive technical terms (e.g., volume of water going through tertiary treatment) rather than in outcome terms (e.g., volume of water discharged back into the environment that meets environmental standards) (Trémolet et al., 2019). Adjusting procurement rules and processes to allow for outcome-based approaches would make the enabling environment more conducive to NBS.

Defining NBS as an asset class and allowing for NBS in the regulated asset base of service providers

Expanding the traditional understanding of what constitutes an "asset" in the realm of water management would help more broadly legitimize the use of NBS to deliver water services and expand their uptake (Cassin, 2021). This requires ensuring that the regulatory, legislative, and policy context recognize the services that NBS can deliver and allow for their use in the course of delivering regulated public services. Water managers, service providers, city managers, and urban planners, among others, face limitations in terms of what types of investments can be justified and included in the regulated asset base. Such distinctions also inform the tariff setting process and the investment costs that can be legally recovered from users.

The role of water sector regulators is especially crucial here. For example, when Sabesp in São Paulo, Brazil, one of the leading water and sewage service providers in Latin America, faced a crisis of water shortage during the historic drought of 2014–15, it had to drastically readjust investment planning to strengthen the system's resilience. The regulator's role in allowing for the inclusion of these investments to increase resilience in the regulated asset base was decisive (OECD, 2019a). More broadly, the Association of Latin American Water Regulators (ADERASA) developed guidance on green infrastructure focused on how utilities can design and implement watershed investments (Bennett and Ruef, 2016).

Another prominent example is that of California, where the Water Code was revised in 2016 to declare it to be state policy that source watersheds are recognized and defined as "integral components of California's water infrastructure" (State of California, 2016). The law allows forests and meadows to qualify as water infrastructure and ecosystem restoration and conservation activities to qualify as eligible maintenance and repair activities for water infrastructure. This enables available water infrastructure financing to be used to protect or restore landscapes that are used for water supply (Bennett and Ruef, 2016).

Removing jurisdictional constraints limiting implementation and funding of NBS

Regulatory frameworks and other rules governing the disbursement of public funds can limit options to fund and finance NBS for water management. Such restrictions can relate to what can be financed, who can receive financing, and cross-jurisdictional considerations. In terms of jurisdictional constraints, many government departments and agencies lack the authority to disburse funds outside of the jurisdictions. There is often an intrinsic spatial mismatch between the supply of ecosystem services and demand for these services, which require cross-boundary and cross-scale cooperation (Kremer et al., 2016). The relevant spatial scale of NBS projects are determined by ecosystem boundaries, not jurisdictional ones. For example, payments for ecosystem services, such as a forest restoration project designed to improve water quality can face constraints due to the project's location outside of the jurisdiction of the water operator servicing a particular city. Brazil has addressed this challenge by establishing laws to facilitate cross-jurisdictional, state-wide payments for ecosystem services (Browder et al., 2019).

Improving evidence base quantifying cobenefits and performance of NBS

One of the most practical obstacles for implementing NBS is the additional resource requirements presented by the new choices, options, and paths to increase resilience. Standard economic evaluation techniques are not well equipped to account fully for the range of nonmarket benefits delivered by NBS. This results in undervaluing ecosystem services and resilience benefits. Benefits may not accrue for years or even decades and might not guarantee a sufficient return on investment to be selected using tools that heavily discount uncertainty and risk. Assessing the attribution of impacts arising from a complex chain of causal links, such as those involved in the delivery of ecosystem services, presents significant challenges and is likely to remain so with emerging novel climate and ecosystem conditions. Such complexity regarding causation of impacts can complicate implementation of policies such as payments for ecosystems (Guerry et al., 2015).

NBS typically present significant challenges for the quantification of benefits, their monetization, and subsequently, barriers for financing (OECD, 2019b). Methods for tracking and valuing cobenefits from NBS (in terms of climate, biodiversity, or social cohesion) are insufficiently developed or not truly operational to drive investment decisions resulting in few water service providers systematically considering NBS in the context of investment plans (Trémolet et al., 2019).

Some well-established initiatives have made important strides in quantifying benefits related to NBS. For example, the Water Fund in Quito Ecuador (FONAG) has promoted catchment protection for 20 years. It devotes significant effort to monitoring the impact of the interventions. Impact monitoring entails the quantification of benefits in terms of water quality and quantity, with feedback on the design of the portfolio, and as an input for return-on-investment studies. These efforts are fundamental to promoting trust in FONAG and sustaining financial contributions that support its operations (De Bièvre and Coronel, 2021).

The use of economic evaluation approaches better suited to the assessment of NBS would allow for consideration of their distinctive features compared to gray infrastructure. For example, more widespread use of multicriteria analysis (MCA), which considers both monetary and nonmonetary benefits could support broader consideration of NBS. In practice, infrastructure planners, project developers, and financial institutions are constrained by evaluation techniques dictated by their institutions. Expanding the range of techniques that can be used and raising awareness about when and how they can be applied can ensure techniques more conducive to NBS are considered. One illustration is the Asian Development Bank's *Practitioner's Guide to Nature-Based Solutions*, which highlights provisions to support NBS and adaptation more generally. For instance, loan project officers have the option to invoke a 6% discount rate for NBS instead of the standard 9%. They can also opt to use MCA instead of a standard cost-benefit analysis (or a combination of both) in order to better capture cobenefits related to NBS (ADB, 2020).

Summary and reflections on future developments

Historically, decisions about water infrastructure have been driven by an engineering perspective strongly anchored to historical patterns of water availability and use. The negative impacts of large-scale, gray water infrastructure on people and ecosystems have been significant in the past, but today these are even more pronounced as populations and economies grow and pressure on natural resources becomes more acute. At the same time, increasing uncertainty about future climatic conditions and the related impact on the hydrological cycle has led to a call for the development of novel approaches to delivering the benefits of water management and water services in a way that strengthens resilience of ecosystems, communities, and economies in the context of changing conditions.

In this context, NBS have emerged as an attractive complement or alternative to gray infrastructure. NBS can assist in managing uncertainty related to changing conditions by avoiding or delaying lock-in to capital-intensive gray infrastructure while delivering a range of cobenefits cost-effectively. Despite the growing interest in NBS, their uptake has been mainly confined to distinct pilot programs and dedicated initiatives rather than emerging from mainstream decision-making related to infrastructure, water management, and urban planning. Barriers to broader scaling relate not only to some of the distinctive features of NBS (as discussed in "Distinct challenges posed by nature-based solutions from a policy perspective" section) but also to the status quo bias of existing policy frameworks, institutional arrangements and financing approaches, which provide a more conducive enabling environment for traditional gray infrastructure. The more systematic application of NBS is unlikely to occur without concerted policy impetus. Levers for policy change that can encourage the consideration of NBS alongside gray infrastructure (as discussed in "Levers for change to make the enabling environment more conducive to NBS" section), can provide useful ways forward for policy makers, project developers, and financial institutions seeking to support NBS.

While practical policy levers and measures to encourage the uptake of NBS point to promising opportunities for scaling, political economy considerations should not be overlooked. Prominent policy economy issues related to environmental policy decisions typically concern the distributional impacts, vested interests, and political and social acceptability (OECD, 2017). As NBS gain more attention and interest, policy economy concerns regarding who benefits and who does not from shifts in infrastructure and service delivery approaches are likely to become more visible and acute. Notably, the implementation of NBS requires significant shifts in land use and spatial planning, which can impact (in both positive and potential negative ways) on the agricultural sector, urban development, and property developers, among others. How benefits are articulated, coalitions are constructed, and negative impacts are compensated will also be important issues for governments to consider in the pursuit of NBS in the years to come.

References

Asian Development Bank (ADB). A practitioner's guide to nature-based solutions; 2020.

Bennett G, Ruef F. Alliances for green infrastructure: state of watershed investment 2016. Washington, DC: Forest Trends; 2016. Available online at: http://www, forest-trends.org/documents/files/doc_5463.pdf.

Bridges T, et al. Use of natural and nature-based features (NNBF) for coastal resilience. Washington, DC: US Army Corps of Engineers; 2015. ERDC SR-15-1.

Browder G, et al. Integrating green and gray, https://openknowledge.worldbank.org/handle/10986/31430; 2019. [Accessed 7 November 2019].

Cassin J. Nature-based solutions for source water protection in North America. In: Cassin J, Matthews JH, Lopez Gunn E, editors. Nature-based solutions and water security: An agenda for the 21st century. Elsevier; 2021 [forthcoming].

Cassin J, Matthews J. Setting the scene: nature-based solutions and water security. In: Cassin J, Matthews JH, Lopez Gunn E, editors. Nature-based solutions and water security: An agenda for the 21st century. Elsevier; 2021 [forthcoming].

Cities4Forests. 2020. Available at: https://cities4forests.com/wp-content/uploads/2020/06/C4F-Urban-Forests-for-Healthier-Cities.pdf [Accessed 23 October 2020].

Coxon C, Gammie G, Cassin J. Mobilizing funding for nature-based solutions: Peru's drinking water tariff. In: Cassin J, Matthews JH, Lopez Gunn E, editors. Nature-based solutions and water security: An agenda for the 21st century. Elsevier; 2021 [forthcoming].

Culwick C, Bobbins K. A framework for a green infrastructure planning approach in the Gauteng city-region. GCRO Research Report No. 04, Gauteng City-Region Observatory; 2016.

De Bièvre B, Coronel L. Investing in catchment protection: the water fund model. In: Leflaive X, Dominique K, Alaerts G, editors. Investing in water and growth: Recent developments and perspectives. Elsevier; 2021 [forthcoming].

Grill G, Lehner B, Thieme M, et al. Mapping the world's free-flowing rivers. Nature 2019;569:215–21. https://doi. org/10.1038/s41586-019-1111-9.

Guerrini A. Water regulatory trends to 2030. ARERA Board Member and President of the Network of European Water Regulators (WAREG); 2020. http://wareg.org/documents.php?q=view&id=2.

Guerry A, et al. Natural capital and ecosystem services informing decisions: from promise to practice. Proc Natl Acad Sci 2015. https://doi.org/10.1073/pnas.1503751112.

Huq N, Stubbings A. How is the role of ecosystem services considered in local level flood management policies: case study in Cumbria, England. JEAPM 2015;17(04):1550032.

IUCN. IUCN global standard for nature-based solutions; 2020. https://doi.org/10.2305/IUCN.CH.2020.08.en.

Kabisch N, et al. Nature-based solutions to climate change mitigation and adaptation in urban areas: perspectives on indicators, knowledge gaps, barriers, and opportunities for action. Ecol Soc 2016;21(2):39. https://doi. org/10.5751/ES-08373-210239.

Keys PW, Galaz V, Dyer M, Matthews N, Folke C, Nyström M, Cornell SE. Anthropocene risk. Nat Sustain 2019;2(8):667–73.

Kithiia J, Lyth A. Urban wildscapes and green spaces in Mombasa and their potential contribution to climate change adaptation and mitigation. Environ Urban 2011;23(1):251–65.

Kremer, et al. The value of urban ecosystem services in New York City: a spatially explicit multicriteria analysis of landscape scale valuation scenarios. Environ Sci Policy 2016. https://doi.org/10.1016/j.envsci.2016.04.012.

LaPorte TR. Critical infrastructure in the face of a predatory future: preparing for untoward surprise. J Conting Crisis Manag 2007;15(1):60–4.

Le Quesne T, Matthews JH, Von der Heyden C, Wickel AJ, Wilby R, Hartmann J, Pegram G, et al. Flowing forward: Freshwater ecosystem adaptation to climate change in water resources management and biodiversity conservation. Washington, DC: World Bank; 2010.

Masson-Delmotte V, et al. Summary for policy makers. In: Global warming of 1.5°C an IPCC special report on the impacts of global warming of 1.5°C above pre-industrial levels and related global greenhouse gas emission pathways, in the context of strengthening the global response to the threat of climate change, sustainable development, and efforts to eradicate poverty. IPCC; 2018. https://report.ipcc.ch/sr15/pdf/sr15_spm_final.pdf. [Accessed on 3 June 2019].

Matthews N. 'People and fresh water ecosystems: pressures, responses and resilience. Aquat Procedia 2016;6:99–105.

Matthews N, Nicol A, Seide WM. Constructing a new water future?: an analysis of Ethiopia's current hydropower development. In: Handbook of land and water grabs in Africa. Routledge; 2012. p. 311–23.

Matthews N, Simmons E, Vigerstol K, Matthews J. Wellspring: Source water resilience and climate adaptation, https://www.nature.org/content/dam/tnc/nature/en/documents/Wellspring_FULL_Report_2019.pdf; 2019. [Accessed 20 October 2020].

Middleton C, Matthews N. Whose risky business? Public–private partnerships, build-operate-transfer and large hydropower dams in the Mekong Region. In: Hydropower development in the Mekong Region. Routledge; 2014. p. 143–68.

OECD. Water and climate change adaptation: Policies to navigate uncharted waters. Paris: OECD Publishing; 2013.

OECD. The political economy of biodiversity policy reform. Paris: OECD Publishing; 2017. https://doi.org/10.178 7/9789264269545-en.

OECD. Responding to rising seas: comparing OECD countries' approaches to coastal adaptation. Paris: OECD Publishing; 2019a.

OECD. Summary of the 4th meeting of the roundtable on financing water, http://www.oecd.org/water/Summary-RTmeeting-26and27June.pdf; 2019b.

OECD. Nature-based solutions for adapting to water-related climate risks. OECD environment policy papers, No. 21, Paris: OECD Publishing; 2020. https://doi.org/10.1787/2257873d-en.

Poff NL. Beyond the natural flow regime? Broadening the hydro-ecological foundation to meet environmental flows challenges in a non-stationary world. Freshw Biol 2017;12:1011–21. https://doi.org/10.1111/fwb.13038.

Poff NL, Allan JD, Bain MB, Karr JR, Prestegaard KL, Richter B, Sparks R, Stromberg J. The natural flow regime: a new paradigm for riverine conservation and restoration. BioScience 1997;47:769–84.

Poff NL, Casey L, Brown M, Grantham TE, Matthews JH, Palmer M, Spence CM, Wilby R, et al. Sustainable water management under future uncertainty with eco-engineering decision scaling. Nat Clim Chang 2016;6:25–34.

Schmidt JJ, Matthews N. Global challenges in water governance: Environments, economies, societies. Springer; 2017.

Seddon N, et al. Understanding the value and limits of nature-based solutions to climate change and other global challenges. Philos Trans R Soc B 2020;375(1794):20190120. https://doi.org/10.1098/rstb.2019.0120.

State of California. Source watersheds: Financing. Assembly Bill-2480, vol. AB-2480; 2016. https://leginfo.legislature.ca.gov/faces/billTextClient.xhtml?bill_id=201520160AB2480.

Steffen W, Broadgate W, Deutsch L, Gaffney O, Ludwig C. The trajectory of the Anthropocene: the great acceleration. Anthropol Rev 2015;2(1):81–98.

Thiele T, Alleng G, Biermann A, Corwin E, Crooks S, Fieldhouse P, Herr D, Matthews N, Roth N, Shrivastava A, von Unger M, Zeitlberger. Blue infrastructure finance: A new approach, integrating naturebased solutions for coastal resilience. Gland: IUCN; 2020.

Trémolet S, et al. Investing in nature for Europe water security. The Nature Conservancy, Ecologic Institute and ICLEI; 2019.

Trémolet S, et al. Funding and financing to scale nature-based solutions for water security. In: Cassin J, Matthews JH, Lopez Gunn E, editors. Nature-based solutions and water security: An agenda for the 21st century. Elsevier; 2021 [forthcoming].

Vigerstol K, Abell R, Brauman K, Buytaert W, Vogl A. Addressing water security through nature-based solutions. In: Cassin J, Matthews JH, Lopez Gunn E, editors. Nature-based solutions and water security: An agenda for the 21st century. Elsevier; 2021 [forthcoming].

Wingfield T, et al. Natural flood management: beyond the evidence debate. Area 2019;51(4):743–51. https://doi.org/10.1111/area.12535.

Wolf S, Pham M, Matthews N, Bubeck P. Understanding the implementation gap: policy-makers' perceptions of ecosystem-based adaptation in Central Vietnam. Clim Dev 2020;13(39):1–14.

World Bank. Implementing nature-based flood protection: Principles and implementation guidance. World Bank; 2017. http://documents.worldbank.org/curated/en/739421509427698706/pdf/120735-REVISED-PUBLIC-Brochure-Implementing-nature-based-flood-protection-web.pdf. [Accessed 15 June 2018].

World Commission on Dams. Dams and development: a new framework for decision-making. A report of the World Commission on Dams. London: Earthscan; 2000. http://awsassets.panda.org/downloads/wcd_dams_final_report.pdf. [Accessed 5 September 2020].

Molle F, Mollinga P, Wester P. Hydraulic bureaucracies and the hydraulic mission: flows of water, flows of power. Water Altern 2009;2:328–49.

CHAPTER 7

Nature-based solutions in international policy instruments

Ingrid Timboe[a],* and Kathryn Pharr[b],†
[a]Alliance for Global Water Adaptation, Corvallis, OR, United States, [b]International Water Policy Consultant, London, United Kingdom

Introduction to NBS and policy: How can nature support our global goals?

The complex societal challenges facing policymakers today are deeply interlinked and require transformative, systems-level approaches to reduce risk and build resilience in the face of deep uncertainty about the future (OECD, 2020). For example, water security cannot be separated from climate change, natural hazards, food security, urbanization, poverty alleviation, and global pandemics (OECD, 2020). Each of these challenges has a water component, though none are solely water issues. Such challenges require resilient and effective responses that are sustainable for both communities and ecosystems. Responses must be flexible, robust, and based on a holistic management approach and cross-sectoral collaboration (IUCN, 2015). Policy can play a key role in a strong enabling environment for these approaches, yet current international policy frameworks are not designed or implemented in ways that promote integrated planning across sectors, across policy frameworks, or across borders in regions with shared natural resources. Therefore, policy makers must enact measures that address multiple risks simultaneously and, ideally, provide tangible benefits that often cross policy agendas and administrative, national, and sectoral boundaries.

In this environment of increasing policy complexity and greater ecological limitation, nature-based solutions (NBS) can play a productive role in providing interlinked, transformative solutions to complex challenges. Such solutions (e.g., restoring natural river drainage, mangroves, and wetlands) have already been deployed around the world to combat water pollution, cyclones and other natural disasters, urbanization, and groundwater depletion. Indigenous cultures have long recognized nature's essential role in human flourishing (Berkes, 2017). However, contemporary interest in NBS has flourished only over the past few decades as the negative impacts of unsustainable development practices have spurred national

---

* Lead author.
† Contributing author.

Nature-Based Solutions and Water Security. https://doi.org/10.1016/B978-0-12-819871-1.00015-4

governments and the international community to craft new policies aimed at protecting and restoring nature (Gomez-Baggethun et al., 2010). This chapter provides an introduction to the major global policy frameworks and identifies potential areas for targeted, evidence-based NBS to contribute to the global goals and thereby enhance societal and ecological resilience.

Advantages and limitations of NBS

A focus on resilience and adaptation as a key element for solving these complex societal challenges has highlighted the limitations of conventional engineered solutions—termed gray infrastructure—that were touted as the solutions to many water issues in the twentieth century. Traditional gray infrastructure is routinely built for a single purpose (e.g., a hydropower dam) with operation parameters based on historic records (such as a 100-year flood recurrence). Thus, gray infrastructure may be unable to cope with the novel or compounding shocks and stressors associated with climate change such as repeat floods, reduced annual precipitation, or extreme storms.

On the other hand, NBS such as green infrastructure, can provide the flexibility and robustness needed to cope with both known and unknown risks resulting from climate change, natural hazards, or other challenges. It can be adjusted or removed with greater ease than traditional gray assets. For example, water managers in the semi-arid Mexican state of Nuevo Leon have employed revegetation, reforesting, and soil conservation on degraded lands to reduce runoff and soil erosion from intense rainfall and recharge local aquifers in order to protect ecosystems and drinking water supplies and to store water for periods of drought (Hesselbach et al., 2019). These green infrastructure investments exist alongside traditional water infrastructure (such as conveyance pipes, dams, and water treatment facilities) and help improve the functionality of these assets over time by reducing stress from extreme weather events. They also provide benefits that extend beyond providing water, including improvements to biodiversity and food security.

This hybrid gray-green infrastructure can provide a powerful framework for how to envision long-term sustainable infrastructure planning, design, management, and investment in an era of ongoing climate change (Matthews et al., 2019). Research from the Netherlands has further demonstrated that using floodplain restoration (a nature-based solution) alongside dykes and levees is more cost-effective and provides a wider range of socio-economic and ecological benefits in the medium- to long-term (Brouwer and Van Ek, 2004) than implementing gray infrastructure. Hybrid infrastructure for mitigating storm surges, such as reef restoration alongside seawalls, are often more resilient over time than seawalls alone and are less expensive; in addition, they can provide important cobenefits for coastal biodiversity, climate adaptation, and fisheries (Ferrario et al., 2014).

NBS is still an evolving concept; thus, work to engage all stakeholders is particularly critical where stakeholders' understanding of both NBS and the advantages that ecosystems can provide to protect and sustain local populations is limited. Many residents are more familiar and comfortable with gray infrastructure (Sherren et al., 2019), and institutional structures have been developed to approve, commission, and monitor gray infrastructure (Rahman et al., 2019). The town of Truro, Nova Scotia, has proven an excellent case study on how to work with local and provincial stakeholders to restore the Onslow salt marsh, a tidal wetland, as the best method to protect the town against flooding (Rahman et al., 2019).

It is important to stress that NBS is not equally applicable in all contexts and can cause harm if misapplied. Interest in—but not clear understanding of—NBS has led to cases of misapplication (Seddon et al., 2020). For example, failure to consider water use can lead to restored ecosystems using too much water, reducing the amount of water available for other uses and putting pressure on local communities. NBS developed in isolation or at an inappropriate scale rarely delivers the needed benefit (IUCN, 2020). Policymakers must work with local stakeholders who will use and benefit from the NBS intervention as well as natural and social scientists to ensure that the correct NBS is chosen for the best fit to achieve the project's goals within a broader context, particularly looking at the whole ecosystem and sometimes the whole catchment.

Standardized, stepwise approaches to NBS implementation are critical to ensuring that these limitations are considered from the outset. To address the lack of standard indicators (Möhner, 2018), IUCN and its partners have recently created create a uniform set of evidence-based NBS principles and standards (see: Cohen-Shacham et al., 2019). This project is well under way, with the first edition of the Global Standard finalized in early 2020 (IUCN, 2020), subject to ongoing revision and amendment as initial projects are completed and assessed. It is hoped that such clear documentation of NBS for ecosystem services will help policy and decision makers operationalize NBS in their policies and projects.

Policy considerations

Water policy makers in particular are increasingly drawn to NBS's perceived ability to address a number of societal challenges at once and to bring water into broader conversations around societal resilience. For example, NBS such as coastal habitat protection or floodplain reconnection can provide "multiple risk, multiple benefit" options in regions facing current or anticipated water risks stemming from sea-level rise, storm surge, coastal population growth, and flood risk. As an added benefit, NBS may help provide some coherence between policy agendas, which, in practice, have the tendency to perpetuate institutional and sectoral silos (UNESCO and UN-Water, 2020). At the same time, not all water policy challenges will benefit equally from NBS; NBS should only be applied with care to ensure that any proposed solutions match the unique context of implementation.

National governments play a critical role in mainstreaming and scaling up the use of NBS by designing the institutional, financial, and regulatory environment necessary for both public and private sectors to adopt NBS and for mandating local participation in the NBS process. As countries set their national commitments and create implementation plans within global policy frameworks, there is a clear opportunity to better define, develop, and finance NBS to address these important challenges. It is estimated that global investment in new and upgraded infrastructure assets will need to top USD97 trillion by 2040 (Global Infrastructure Hub and Oxford Economics, 2017). Ensuring that these investments are resilient should be a top priority for national governments. While NBS can and should look different in each national plan, standardized metrics for measuring and monitoring NBS, such as the IUCN Global Standard, can help add robustness to national adaptation strategies and plans and attract new sources of finance, for example, green bonds (Trémolet and Karres, 2020).

Introduction to the global policy frameworks: Sustainable development, climate change, disaster risk reduction, and biodiversity

The year 2015 marked a watershed year for global policy with three major UN frameworks—the 2030 Agenda for Sustainable Development, the Paris Agreement on Climate Change, and the Sendai Framework for Disaster Risk Reduction (DRR)—all signed within the span of 12 months. Collectively referred to as the "post-2015 agenda for action," these agreements provide the foundation for sustainable, low-carbon, and resilient development under a changing climate. While none of these frameworks is legally binding, Parties to these agreements are requested to submit progressively ambitious plans detailing their intended national contributions to each agenda, based on their individual capacities and priorities, and to report on their progress over regular intervals. Much of the work of this post-2015 agenda advanced the 1990s work on the Convention on Biological Diversity (CBD) that sought to protect biodiversity and secure ecosystem services, which are critical to the success of NBS. Unlike the post-2015 frameworks, CBD is legally binding but similarly has its own regular reporting against common indicators, as shown in Table 7.1.

While all four of these international policy frameworks form a historic step toward addressing the world's most pressing challenges, they still suffer from some of the same problems that plagued earlier agreements: namely, they reinforce existing administrative and technocratic silos and include overlapping and sometimes contradictory targets (UNESCO and UN-Water, 2020). Each framework includes its own set of indicators and modes of implementation (see Table 7.1). This reality often results in a burdensome administrative load for national staff charged with designing, implementing, and reporting on national commitments. An attempt to overcoming these barriers can be seen in the CBD's indicators for its 2011–2020 Aichi targets; many of the indicators are aligned with Sustainable Development Goal (SDG) targets and marked as such.

Encouragingly, some effort has been made to improve the coherence and connectivity between the frameworks at the national level (Hollins, 2018; Seidler et al., 2018) and

Table 7.1: Modes of Implementation.

Policy framework	Number of indicators	Country-level commitments	Reporting interval
2030 Agenda	232	No uniform vehicle; each government determines its own strategy for addressing the 17 SDGs. Each year, countries are also encouraged to undertake a voluntary national review (VNR) of their progress.	Continuous
Paris Agreement	In development (no common national adaptation indicators)	Nationally determined contributions (NDCs); national adaptation plans (NAPs); national communications; long-term strategies	Every 5 years
Sendai Framework for DRR	38	National DRR strategies (Target E), including targets, indicators, and time frames	Continuous
Convention on Biological Diversity	144	National reports and national biodiversity strategies and action plans (NBSAP)	Every 4 years

to consider some of the critical trade-offs in moving toward various goals of different frameworks simultaneously at the national and international levels (Hirji et al., 2017; Parkinson et al., 2019). For example, research has highlighted the trade-offs of limited water availability when unsustainable overuse of groundwater is used to increase food production to meet the SDG 2 on reducing hunger (Perry et al., 2017). More generally, weak institutional coordination, lack of political mandates for integrated planning, and insufficient funding has complicated efforts to address systemic risks and integrate planning (Hirji et al., 2017).

Although NBS is not a silver bullet, it can be a vehicle for addressing many of these policy challenges. NBS can also be significantly cheaper than traditional gray infrastructure projects (Cooper, 2020) and are, by definition, robust and flexible solutions (Anisha et al., 2020; Mauroner et al., 2021). Furthermore, robust implementation of NBS requires ongoing coordination among stakeholders (Matthews et al., 2019), which can further promote cooperation among different agencies and ministries within these policy frameworks.

One key area of convergence between the frameworks is enhancing societal resilience. Given the fundamental role that ecosystems play in supporting human wellbeing, NBS can potentially provide benefits that address goals related to all three 2015 agendas as seen in Fig. 7.1 (UNFCCC, 2017). For example, NBS to reduce communities' disaster risks can also support multiple SDGs (including SDG 2: Zero Hunger, SDG 3: Good Health and Wellbeing, SDG 6: Clean Water and Sanitation, SDG 11: Sustainable Cities and Communities, and SDG 15: Life on Land), and climate mitigation and adaptation as well as CBD targets. Other types of NBS have a more limited scope and may only address a single societal challenge at a time. Regardless, both narrow and broad NBS interventions are potentially useful and should be considered when developing national climate, DRR, and sustainable development policies.

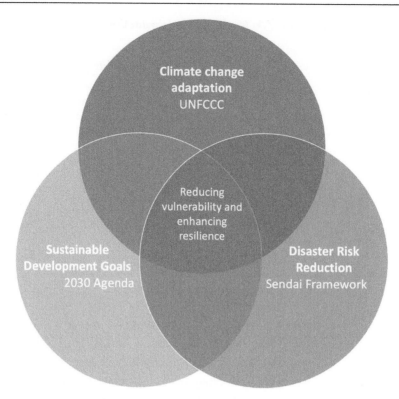

Fig. 7.1

Connecting Action Agenda 2030, the Sendai Framework and the Paris Agreement. *Courtesy: UNFCCC. Opportunities and options for integrating climate change adaptation with the Sustainable Development Goals and the Sendai Framework for Disaster Risk Reduction 2015–2030. Technical paper. Bonn, Germany: UNFCCC; 2017.*

While NBS can offer important solutions for bridging the agendas of all four frameworks, as previously noted, not all NBS will be equally relevant or suitable for each framework and its specific indicators. Policymakers must work with natural and social scientists to ensure each identified NBS project is evaluated for trade-offs and benefits among all stakeholders and is not implemented in a silo. Therefore, applied NBS must be compatible with a range of goals and must be adjustable as conditions on the ground change. For example, a constructed wetland to assist a water treatment facility can both withstand and adapt to climate impacts. If designed and implemented in coordination with relevant decision makers, this intervention could also benefit SDGs 6: Clean Water and Sanitation and 13: Climate Action; CBD Aichi targets 8: Pollution Reduced, 14: Ecosystem Services, and 15: Ecosystem Resilience and Restoration; and Sendai Target D: Substantially Reduce Damage to Critical Infrastructure and Service Disruptions.

The remainder of this section looks at each of the four international policy frameworks in turn, highlighting the ways in which NBS can support the global agendas. Six case studies from around the world demonstrating tangible integration of NBS into national DRR, climate, biodiversity, and sustainable development policy can be found in Boxes 7.1, 7.2, and 7.3.

NBS and the Sendai Framework for Disaster Risk Reduction 2015–2030

The Sendai Framework for DRR was adopted at the Third UN World Conference on DRR in Sendai, Japan, on March 18, 2015, with the aim of achieving the "substantial reduction of disaster risk and losses in lives, livelihoods and health and in the economic, physical, social, cultural and environmental assets of persons, businesses, communities and countries" (UNDRR, 2015). It includes four priorities for action to prevent new disaster risks and reduce existing ones before 2030:

1. Understanding disaster risk
2. Strengthening disaster risk governance to manage disaster risk
3. Investing in disaster reduction for resilience
4. Enhancing disaster preparedness for effective response, and to "Build Back Better" in recovery, rehabilitation and reconstruction (UNDRR, 2015)

Every 2 years, the UN Office for DRR (UNDRR) compiles a global assessment report, which details global progress toward these goals. Reporting on implementation is highly encouraged but voluntary, with member states self-reporting progress against a set of 38 indicators aligned with seven global targets.

The Sendai Framework is a shift in international policy from reacting to hazards to focusing on how to prevent them and minimize their impact to communities and ecosystems when they do occur. One of the chief purposes of the Sendai Framework is to promote active prevention and improved rebuilding (build back better) strategies aimed at increasing resilience and reducing long-term risk from both sudden and slow-onset hazards within and across sectors (Sendai Priorities 3 and 4). Prior to the enactment of the Sendai Framework, national disaster strategies largely focused on offering postdisaster relief, which has been shown to be more costly both in terms of dollars spent and lives lost (Multi-Hazard Mitigation Council, 2019). This shift toward prevention and preparedness remains an ongoing process mediated by complex interactions between several disaster drivers including climate change, inequality, demographic change, population distribution, and environmental degradation (Briceño, 2015).

For example, in the last 20 years, climate change has caused increased frequency of natural disasters, with three-quarters of them being water-related such as floods, droughts, and cyclones (UNESCO and UN-Water, 2020). The uncertainty surrounding climate change requires solutions that are resilient and proactive, which is also the aim of the Sendai Framework. As such, climate change is playing an increasingly profound role in disaster recovery, altering conceptions of what is meant by "returning to normal" as landscapes transform under a warming climate (Matthews et al., 2018). Increasingly strong storms, shifting precipitation patterns, and longer and more intense wildfire seasons mean that disaster response must shift from the goal of quickly returning to a predisaster state to a goal of creating resilient societies able to persist, adapt, or transform as necessary over the medium- to long-term.

NBS are uniquely suited to this challenge. For disaster risk prevention, NBS can offer salient opportunities to redefine "build back better" to incorporate a reduction in future hazard risks and support for new development pathways that promote ecosystem integrity and adaptive management. Water-dependent ecosystems such as forests, wetlands, and coastal systems can reduce human physical exposure to natural hazards by buffering damage, protecting critical infrastructure and property, and supporting faster recover (Cohen-Shacham et al., 2016; Tasnim and Irfanullah, 2020). Safeguarding vital ecosystem services can contribute to societal resilience in the face of change (see Box 7.1), and NBS can also significantly enhance the design and construction of critical infrastructure—both for reducing disaster risk and for postdisaster recovery. In the case of DRR interventions, failure has the potential to be catastrophic; thus, great care must be taken to identify and implement appropriate solutions collaboratively with a multisector approach.

Box 7.1   NBS for disaster risk reduction and management

Case study 1: Reducing urban flood risk in China

Rapid urbanization like the kind that China has experienced in the first quarter of the twenty-first century brings many challenges: ecological degradation, water pollution, heat-island effects, and urban flooding. Such urbanization—combined with changing rainfall patterns linked to climate change—increases flood risk and decreases urban resilience (Qiao et al., 2020). In 2013, the Chinese government responded with the national Sponge City Construction Project (SCCP), originally aimed at addressing stormwater management, which quickly grew into a much larger program (Qiao et al., 2020). The term "sponge city" refers to the city's new ability to respond to environmental changes and natural hazards by absorbing and purifying water during weather events and then releasing the stored water during times of scarcity (Qiao et al., 2020). As the director of water management at Arcadis China, Wen Mei Dubbelaar, explains, the purpose is to "give space back to the river...[instead of] fighting the water" (Li, 2019). With anticipated costs around USD 1.5 trillion (10 trillion Yuen) and covering 657 cities, the goal of the SCCP is to combine gray and green infrastructure to help reverse the effects of rapid urbanization through rainwater harvesting, ecological restoration, more effective flood control, and water quality improvements (Jia et al., 2017).

Despite national support, implementation of the SCCP has faced several challenges including: lack of understanding regarding NBS within provincial and local government offices, limited technical guidance for implementing combined gray and green infrastructure, lack of close coordination among local stakeholders and the national government, and financial hurdles (Jia et al., 2017; Chen and Guo, 2019). To enhance local coordination and overcome these challenges, Sponge City pilot cities have set up "Sponge City Offices" where all bureaus connected to urban water management are now represented, mirroring the close multiagency collaboration for the SCCP at the ministry level (Jia et al., 2017). Local authorities are also able to adapt SCCP projects to fit their needs. As of 2017, nearly 130 cities across China were creating plans to become sponge cities (Liu et al., 2017).

Continued

Box 7.1 NBS for disaster risk reduction and management—cont'd

The SCCP is a central part of China's DRR strategy and reflects all four priorities from the Sendai Framework: it understands disaster risk, strengthens the governance needed to manage that risk, invests in resilience, and enhances preparedness for a better response. In other words, it moves from reactive to proactive management of urban flooding, with an emphasis on flexible solutions designed to function for decades to come.

Case Study #2: Integrated DRR and climate planning in Vanuatu

The Pacific island of Vanuatu is highly vulnerable to extreme weather events, saltwater intrusion, and sea-level rise. Ocean acidification, overfishing, and invasive species have further led to the deterioration of many of Vanuatu's natural defense systems, including its coral reefs. In addition to increasing hazard risk, the loss of reefs has jeopardized the livelihood of local people, many of whom depend on fishing and tourism. In 2015, Vanuatu released an integrated national climate change adaptation and DRR plan that highlights the use of NBS. The plan prioritizes green over gray infrastructure for maintaining ecosystem functions such as using coastal revegetation instead of sea walls to minimize flooding concerns (SPC, 2015). The plan also concentrates on actions for adaptation and risk reduction actions that expand, integrate, and protect important ecosystems such as conservation areas, heritage sites, locally managed areas, carbon sinks, and vulnerable habitats (SPC, 2015).

One example of this integrated climate and disaster risk planning in action is a recent coral restoration partnership between Vanuatu's Nguna-Pele Marine and Land Protected Area Network, the Secretariat of the Pacific Community (SPC), and the Deutsche Gesellschaft für Internationale Zusammenarbeit (GIZ) GmbH. This NBS project included the establishment of new heat-tolerant coral varieties, which led to the return of coral-dependent fish and helped stabilize eroding coastlines. In an early test of this project, the new corals were able to withstand 2015's Cyclone Pam, which was one of the strongest cyclones ever to hit the island. The project leaders also focused on extensive community engagement throughout the project, including a concerted effort to engage women and girls. By the end, over 3000 community members participated in the 3-year program.

This NBS example connects all four frameworks discussed in this chapter: SDGs 5: Gender Equality, 13: Climate Action, 14: Oceans and Coastal Zones, and 17: Partnerships; CBD Aichi targets 1, 5, 6, 10, 14, 15, and 19; Sendai Framework Priorities 1 and 2; and Vanuatu's national adaptation priorities 2 (sustainable tourism development), 3 (community-based marine resource management), and 5 (DRR strategic adaptation priorities). With this work, Vanuatu models the valuable synergies possible among these policy frameworks at both the national and community levels and the role of NBS in establishing these connections.

NBS is valuable in all types of disasters. Until 2020, most policy conversations on disasters hinged on weather-related events, but the recent global COVID-19 pandemic has raised the question of resilience in the realm of global health challenges and risks. In 2020, the High-Level Experts and Leaders Panel on Water and Disasters (HELP) released a set of *Principles to Address Water-related DRR under the COVID-19 Pandemic* (HELP, 2020) aimed at addressing how areas facing natural hazards can avoid becoming virus hotspots, while ensuring that postpandemic economic responses are undertaken in a comprehensive, risk-sensitive manner. In such situations, NBS can potentially improve outcomes when it is applied for source water protection against disease and water-related disasters, which can exacerbate viral transmission. NBS to ensure water quality is particularly key in these situations where handwashing is needed to mitigate human-to-human virus transmissions (Adikari and Yoshitani, 2009).

This coordination of investments for NBS can be accomplished within the Sendai Framework when countries are setting their national DRR targets. Because national policy is often the primary driver of local action, incorporating NBS into these national targets and national disaster planning guidelines can help attract DRR finance and the enthusiasm to make these interventions mainstream across the country. At the same time, locally led NBS implementation can inform the monitoring, adjustment, and scaling up of these interventions over time.

NBS and the 2030 Agenda for Sustainable Development

During the September 2015 UN Sustainable Development Summit, world leaders joined together to adopt the 2030 Agenda for Sustainable Development, which includes 17 integrated SDGs, 169 targets, and over 280 indicators. UN Secretary General Ban Ki-moon described the framework as a "big, bold agenda…a roadmap to ending global poverty, building a life of dignity for all and leaving no one behind" (UN, 2015). Addressing a range of societal needs including poverty reduction, health, education, gender equality, and social protection—while tackling climate change and environmental protection—the 2030 Agenda is not legally binding, instead relying on countries to institute, implement, and monitor their own national plans and policies for achieving the SDGs by 2030. Countries are encouraged to submit progress reports as part of the voluntary national review (VNR) process, which are summarized into an annual synthesis document on global progress toward the 2030 Agenda.

The 17 SDGs are designed to be integrated together, with complimentary targets connecting the societal, economic, and environmental aspects of the agenda. In practice, however, the SDGs have been addressed largely in isolation with planners failing to account for the synergies and trade-offs between the goals (Moyer and Bohl, 2019; Seddon et al., 2020). This siloed approach has undermined progress. With less than 10 years left until 2030, the UN has called on countries to accelerate efforts to adopt multisectoral, integrated approaches to financing and implementing the SDGs, in particular noting the urgent need for action to reverse continued ecosystem degradation, which further undermines all the goals (High-Level Political Forum (HLPF), 2020).

Resilient and diverse ecosystems, which rely on clean and available water, underpin healthy communities and economies. As such, NBS can support nearly all the SDGs, not just those directly related to the biosphere or water resources. For example, SDG 2: Zero Hunger, SDG

7: Affordable and Clean Energy, SDG 11: Sustainable Cities and Communities, and SDG 12: Responsible Consumption and Production all rely on healthy ecosystems and reliable, good quality water. Because NBS often requires integrated planning among multiple sectors and stakeholders (van Ham and Klimmek, 2017), it can also help improve integrated planning and implementation of the SDGs. Standardized, step-wise approaches to NBS implementation, like the IUCN Global Standard, are critical to ensuring that these considerations are taken into account from the outset. Examples in Box 7.2 demonstrate some positive NBS water applications in different contexts.

Many nature-based SDG projects are already underway. In the Southern Caucus mountains of Azerbaijan, several hybrid green-gray infrastructure measures have been employed to address the significant pasture degradation that led to the loss of soil, livelihoods, and biodiversity and to increased risk of landslides and flooding. The national government of Azerbaijan, with funding from GIZ, has worked with local communities to pilot projects including riverbed stabilization, afforestation, and terraces as well as trainings for local communities

---

Box 7.2   NBS for the Sustainable Development Goals

Case study #3: Building with nature in Indonesia
On the northern Indonesian island of Java, coastal flooding and erosion hazards currently affect nearly 30 million people due to groundwater extraction, unsustainable coastal development, and the removal of mangroves for aquaculture production (WI, 2016). Demak is a town in northern Java where the sea-level rise along its shoreline is expected to cause flooding six kilometers inland by 2100, affecting over 70,000 people and covering 6000 ha of ponds currently used for aquaculture production (WI, 2016). A Wetlands International initiative "Building with Nature" seeks to address these issues through the implementation of locally appropriate hydraulic NBS infrastructure in Demak that involves recreating native mangrove forests (WI, 2016). Because the mangroves could not simply be replanted, permeable bamboo structures were placed along the shore for land reclamation before replanting the mangroves (Ecoshape, 2020).

   This project takes the economic needs of the community into account and ensures close community collaboration throughout the process so that mangrove restoration can take place alongside aquaculture cultivation (Ecoshape, 2020). In its 2019 VNR, Indonesia highlights the Building with Nature Program and its integrated risk management approach to combining sustainable development and hazard risk reduction (Republic of Indonesia, 2019). Once established, mangrove forests will provide flood protection for people and infrastructure (SDGs 9, 11 and 13) and sustainable aquaculture will help improve food security (SDG 2), livelihoods (SDG 8), and reduce water pollution (SDG 6). Mangroves also provide important nearshore habitat for aquatic life (SDG 14). According to the VNR, Building with Nature projects have "resulted in, among others, the improvement in the quality of the environment, the acknowledgement of community interest, elevation in community capacity, a rise in community income, a new found awareness in the environment and preparedness in facing the risks of disasters" (Republic of Indonesia, 2019).

Continued

Box 7.2   NBS for sustainable development goals—cont'd

Case study #4: Community-based landscape management in the Ecuadorean Andes
Forests and alpine wetlands in Ecuador—the primary source of livelihoods for most rural
Ecuadorean communities—are highly endangered by both climate and land use change. Ecuador
is one of the most biologically diverse countries in the world, yet it also has one of the highest
deforestation rates in South America (Oppla, 2020). Expanded aquaculture and agriculture, oil
exploration, and mining have damaged high-Andean moors known as paramos, which, along
with highland forests, form the primary source of water in the upper and lower Andes region
(Morales, 2011). Paramos are a particularly important type of natural infrastructure in this
region, acting like a giant sponge that absorbs rainwater and slowly releases it over time. This
is particularly important for agriculture in the valleys below (Morales, 2011). The loss of these
forests and wetlands have had negative impacts on human food and water security, flood and
landslide risk, local livelihoods, and biodiversity.

Begun in 2008 as an initiative of Ecuador's Ministry of the Environment, the Socio Bosque
Program funds landowners, particularly local and Indigenous communities, to protect the native
forests, paramos, and other native vegetation on their property with priority given to areas with
rapid land use change and those deemed critical to maintaining ecosystem services (Oppla,
2020; FAO, 2020). Within a decade, the Socio Bosque Program had 1.5 million hectares
under management (Alacron, 2018). This program is one of several focused on sustainable
management of natural resources that the government of Ecuador reported on under SDG 15:
Life on Land as part of its 2018 VNR (Ecuador 2018). Socio Bosque also connects to SDGs
8: Decent Work and Economic Growth and 13: Climate Action, contributing to both climate
change adaptation (reduced flood risk) and mitigation (carbon sequestration) and enhancing
biodiversity (Alacron, 2018). In 2018, a new initiative "Greening Ecuador" expanded the
work done with Socio Bosque with a focus on several NBS including land conservation, green
infrastructure for cities, the bioeconomy, environmental education, forest management, and
reforestation (Alacron, 2018).

---

on implementing and maintaining erosion control measures. This regional NBS project
simultaneously addresses several SDGs, including SDG 1: No Poverty, SDG 12: Responsible
Consumption and Production, SDG 13: Climate Action, and SDG 15: Life on Land. In
addition, the case studies from these pilot projects have been compiled in a handbook to be
distributed among regional partners and national ministries to facilitate knowledge sharing
and lessons learned (Koeppler, 2020).

NBS and the Paris Agreement on Climate Change

The final framework agreement to be signed in 2015 was the Paris Agreement on Climate
Change, signed in Paris during the 21st session of the Conference of the Parties (COP21) to
the United Nations Framework Convention on Climate Change (UNFCCC). The primary goal
of this agreement is to address climate change by limiting the global average temperature

increase to "well below" 2°C from preindustrial levels and to strengthen the ability of countries to manage climate impacts (UN, 2016). Starting in 2020, every 5 years Parties to the Paris Agreement are required to submit national strategies known as Nationally Determined Contributions (NDCs) demonstrating how they will contribute to reducing global emissions and adapt to climate change. Parties are also invited (although not required) to submit adaptation strategies as part of their contribution.

NBS has strong potential to contribute to the mitigation and adaptation goals of the Paris Agreement. The Agreement itself calls out the critical role of ecosystems in achieving its goals, noting the "importance of the conservation and enhancement, as appropriate, of sinks and reservoirs of the greenhouse gases referred to in the Convention…[including] biomass, forests and oceans as well as other terrestrial, coastal and marine ecosystems" (UN, 2016). In recognition of this, most Organization for Economic Co-operation and Development (OECD) countries now explicitly mention NBS in their national adaptation plans (NAPs) or strategies (OECD, 2020). Roughly 78% of all intended NDCs include some variation of NBS, primarily in relation to adaptation targets (Seddon et al., 2019), though details for NBS contributions are lacking in the NDCs and other detailed documents like the National Adaptation Plans (NAPs).

One of the primary reasons that NBS are so attractive is because they often provide important cobenefits. Nevertheless, very few existing NDCs account for the mitigation cobenefits of NBS for adaptation or the adaptation benefits of NBS for mitigation. National planners tasked with writing NDCs should be encouraged to take advantage of NBS for achieving multiple policy goals simultaneously—and for considering and weighing the potential benefits and trade-offs of different mitigation and adaptation actions across sectors.

In terms of adaptation benefits, ecosystems play a crucial role in building climate-resilient communities by helping attenuate climate impacts, like extreme heat and sea-level rise, and by improving the functionality of critical infrastructure despite an uncertain climate future. Climate change negatively impacts freshwater ecosystems by fundamentally altering streamflow and water quality (Timboe et al., 2019), thereby posing new risks to drinking water and sanitation and to the ecosystems themselves. Therefore, NBS must ensure ecosystems are resilient enough to continue to provide or expand ecosystem services.

NBS in the form of ecosystem-based mitigation can make a powerful contribution to reducing greenhouse gas (GHG) emissions by preventing the degradation and loss of ecosystems, with recent research suggesting that NBS could provide approximately 30% of the cost-effective mitigation that is needed to stabilize warming to below 2°C above preindustrial levels by 2030 through ecosystems like forests storing GHGs and CO_2 emissions (Griscom et al., 2017; Matthews and van Noordwijk, 2014; Anisha et al., 2020; IPCC, 2014). Wetlands— particularly coastal and estuarine wetlands—are able to store up to six times more carbon that forests, with peatlands alone storing roughly 30% of all terrestrial carbon (Parish et al., 2008). Conversely, destruction of wetlands releases those sequestered GHGs, diminishing the likelihood of meeting the Paris Agreement's targets. NBS projects, therefore, can protect and possibly expand carbon sinks like existing wetlands, but NDCs thus far have not been utilizing such strategies.

Countries now have an opportunity to strengthen their national climate plans and policies by adding specific, measurable, and robust NBS targets to the next round of NDCs and NAPs. As part of this enhancement process, national climate policy makers should explore opportunities to align NBS climate targets with other policy goals and assess potential trade-offs.

NBS and Convention on Biological Diversity

Predating the post-2015 action agenda, the CBD set the stage for, and continues to play a critical role in, safeguarding the Earth's biodiversity (IISD, 2019). The CBD currently governs the sustainable use and conservation of biodiversity under its Strategic Plan for Biodiversity 2011–2020 and the Aichi Biodiversity targets (CBD, 2020). Through concepts such as eco-DRR and ecosystem-based adaptation (EbA), NBS has played a strong supporting role in the 2011–2020 strategy and implementation of the Aichi targets. In 2018, the CBD Conference of the Parties (CBD COP) adopted CBD Technical Series 93 on EbA for climate change adaptation and DRR (PEDRR and FEBA, 2020). This document provides examples of successful water-related NBS projects, such as flood buffers and watershed monitoring in Thailand's Khon Kaen Province to address increased flood and drought conditions that link Aichi targets 14 and 15 with SDGs 6, 13, and 15 and Sendai Priorities 2 and 3 (CBD Secretariat, 2019).

Work under the CBD framework is integral to the societal resilience goals of the Sendai Framework, the 2030 Agenda, and the Paris Agreement (CBD et al., 2017). Biodiversity loss and climate change are mutually reinforcing; ecosystem degradation or fragmentation can turn carbon sinks into significant carbon emitters while loss of habitat can increase the severity of climate impacts such as drought. Inversely, biodiversity can enhance efforts toward DRR, sustainable development, and climate change mitigation and adaptation goals (IISD, 2019). A 2017 CBD Technical Note links the 17 SDGs to at least one (and often more) Aichi targets such as 8: pollution reduced, 14: ecosystem services, and 15: ecosystem restoration and resilience (CBD et al., 2017).

With the current CBD strategy ending in 2020 and the UN Biodiversity Conference (CBD COP 15) happening in 2021, focus has shifted toward adopting a new, post-2020 Global Biodiversity Framework (GBF) (IISD, 2020). With this new biodiversity framework and set of targets, there is a push to ensure that new indicators align with those of other frameworks such as the SDGs and the Paris Agreement (PEDRR and FEBA, 2020). Examples of NBS that include the CBD framework are Case Study 2: Vanuatu (Box 7.1) and Case Study 5: Tunisia (Box 7.3).

Box 7.3   NBS for the Paris Agreement on Climate Change

Case study #5: Combating desertification with traditional grassland management in Tunisia
Tunisia has focused extensively on NBS in their NDC, including a wide range of ecosystem-based interventions for mitigation and adaptation, including the restoration and sustainable management of degraded grasslands in central and southern Tunisia (Rankovic et al., 2017). Land clearing, overgrazing of livestock, and population growth in Tunisia, with additional pressure coming from increasing aridity due to climate change, have led to desertification and reduced water security among these important ecosystems (Belgacem et al., 2019, Abdellatif et al., 2017). Recognizing the need to reverse desertification, the Tunisian government has made this a priority area in its NDC.

Adaptation actions that are based in local traditions are often more resilient over time and can be a powerful method to spur behavior change among local populations. Traditional methods of managing rangelands in southern Tunisia, called Gdel, have been proven to be resilient to climate change impacts if anthropogenic pressures are concurrently addressed (Belgacem et al., 2019). Gdel involves community-led sanctions around when and where grazing can occur throughout the year to reduce grazing pressure on fragile rangelands (Gamoun et al., 2018). Additionally, this practice limits when esparto (*Stipa tenacissima*) can be harvested to ensure sustainable cultivation. This native grass has important economic value for local communities, which use it for basket weaving and the production of other goods (Gamoun et al., 2018), and is under pressure by increased desertification in the region. When southern Tunisia moved from a pastoral society to a largely agropastoral one, traditional practices like Gdel were often abandoned (Belgacem et al., 2019). Thanks to recent interventions by local communities and the national government, some historical practices are being rediscovered.

It is also important to note that this emphasis on NBS is not new for Tunisia, which has been working to improve ecosystems nationally for years under the CBD. It has done so by formalizing a series of environmental laws and policies such as creating 40 Ramsar sites for wetlands protection and starting a national council to combat desertification (Convention on Biological Diversity (CBD), 2020). The Tunisian government is now working to integrate their climate planning within the context of this existing body of work on biodiversity.

Case study #6: Enhancing climate resilience in Mozambique's port cities
Due in part to its geographic location and rapid urbanization, Mozambique is highly vulnerable to natural hazards including tropical cyclones and both coastal and riverine flooding. Climate change has already begun to exacerbate these risks. Port cities like Quelimane, Nacala, and Beira are currently dealing with multiple challenges, including insufficient stormwater infrastructure that has struggled to cope during cyclones and extended rainfall periods, alongside saltwater intrusion and loss of coastal protection from storm surge stemming from the destruction of coastal mangrove forests (CES and Inros Lackner SE., 2020).

Continued

Box 7.3   NBS for the Paris Agreement on climate change—cont'd

Beira, Mozambique's largest coastal city, is considered one of the African cities most at risk from climate impacts, including extreme heat, rising seas, groundwater depletion, coastal erosion, tropical cyclones, and inadequate drainage (CES and Inros Lackner SE., 2020). To address these challenges, the national government, in cooperation with GIZ, implemented a series of NBS for coastal climate resilience and risk reduction. The first phase restored the natural drainage capacity of the Chiveve River, a tidal river that flows through central Beira and is prone to flooding due to heavy siltation and floodplain loss (CES and Inros Lackner SE., 2020). As part of the Mozambique Cities and Climate Change Project (3CP), the World Bank and the Global Facility for Disaster Reduction & Recovery funded the next phase, which included mangrove and coastal habitat restoration and the creation of urban green spaces in Beira, Quelimane, and Nacala.

The utility of these preventative interventions has already been tested. When Cyclone Idai struck Beira in 2019, much of the city experienced limited impact, sparing 250,000 people from the worst of the floods thanks to improved drainage and measures for flood protection (Garcia and Nagar, 2020). To further strengthen Beira's disaster preparedness in the wake of Idai, the World Bank is funding new coastal protections and additional drainage rehabilitation in and around the city (Garcia and Nagar, 2020). Thanks in part to the success of these initial interventions, vulnerability assessments and NBS recommendations for the port cities of Quelimane and Nacala have been recently completed.

Toward implementation: NBS as a connector of national climate, development, biodiversity, and DRR policy

The overarching purpose of this chapter is to identify the ways in which water-centric NBS can align and support the diverse (and sometimes conflicting) agendas of the four global policy frameworks. It has been demonstrated how NBS can be applied to each framework to reach targets across sectors. NBS can also provide greater coherence between policy agendas as many NBS such as source water protection or soil conservation provide benefits—and illuminate trade-offs—across sustainable development, climate, biodiversity, and DRR agendas. NBS can also help national governments improve their interdepartmental coordination to effectively address complex, interrelated societal challenges such as water security.

Some countries have managed to link multiple policy agendas using NBS, while others have only managed to coordinate two of them. Case Study 2 in Box 7.1, Vanuatu manages to link all four: the CBD, the Paris Agreement (both mitigation and adaptation), the 2030 Agenda, and the Sendai Framework (SPC, 2015). Few existing NDCs make concrete connections to the Sendai Framework or the 2030 Agenda with a few exceptions such as Colombia, Ghana, and Guatemala (UNFCCC, 2017). Chile has integrated the SDGs and the Paris Agreement, linking progress of SDG 13: Climate Action to its national climate commitments in its 2019 VNR (Government of Chile, 2019). Chile's 2020 NDC explicitly connects its mitigation and adaptation commitments to relevant SDG targets and states that it will favor the application of

NBS across all of its national climate and commitments, which are framed by the 2030 Agenda (Government of Chile, 2020). The Costa Rican government is working to integrate its National Disaster Risk Management Policy with its NDC (Government of Costa Rica, 2016) and plans to launch a National EbA Strategy. With support from the German International Climate Initiative (IKI), Costa Rica has already completed several pilot NBS projects that included working with local Indigenous communities on topics such as strengthening local water governance (Reid et al., 2018). Other countries such as Vietnam, Burkina Faso, Kiribati, and Jordan are taking similar action, while Bolivia is working to integrate its national watershed plan (Plan Nacional de Cuencas de Bolivia) with both the SDGs and its 2020 NDC.

Implementing these solutions at the national level also requires a strong understanding of the relevant national political, governmental, and institutional frameworks at play and may require policy reforms to facilitate the necessary cross-sectoral coordination that robust NBS implementation requires. Interdepartmental NBS working groups or committees may be useful for overcoming jurisdictional silos. Dedicated, independent budgets and staff for NBS programs can improve the project financing, ensuring that no single department will be responsible for funding or administering these multistakeholder projects. Because NBS are largely implemented at the local level, coordination between national ministries and local policy and decision makers is also key.

Financing has proven a strong incentive for using NBS to collaborate across these frameworks. Public-private partnerships can also support the engagement and buy-in of local communities for NBS, especially if the cobenefits for their communities are clearly stated and understood (Droste et al., 2017). Framing DRR interventions in terms of their climate adaptation benefits may also improve access to climate finance, which often has extensive criteria for determining project finance eligibility. This is why, according to the UNFCCC's Adaptation Committee (2020), many countries are now working to explicitly align their NAPs and other national adaptation measures with the Sendai Framework and/or the 2030 Agenda. Still others have included innovative financial risk management and insurance measures in their adaptation plans, with Antigua and Barbuda planning to make an index-based crop insurance scheme available for farmers within the next few years (UNFCCC AC, 2020).

Transboundary considerations

Just as solutions to these complex challenges such as water security and climate change need to be cross-sectoral and implemented within the various policy frameworks discussed in this chapter, NBS need to be coordinated within the context of the larger ecosystem in which they are based. In the case of freshwater systems, this often requires taking a transboundary approach. In freshwater basins that transcend national borders, many countries have formed river basin organizations (RBOs) or other institutional or regulatory agreements that more effectively and efficiently comanage these resources (Blumstein et al., 2016).

The importance of transboundary work in the global policy frameworks discussed in this chapter can be seen in some formal documentation and in applied projects. While much of the reporting for these frameworks has been undertaken at the national level, work to meet the climate, DRR, biodiversity, and sustainable development targets have not stopped at national borders. For example, several NDCs such as Vietnam (2016), Palestine (2017), and Moldova (2020) highlight the value and importance of transboundary cooperation. Over the last few decades, the CBD has looked to transboundary water management as an important way to ensure water allocation and management for sustainable use and biodiversity conservation under Article 5 of the Convention (Brels et al., 2008).

Using this foundation, transboundary NBS projects have been applied to DRR, climate, sustainable development, and often in combination. The Lake Chad Basin Commission has participated in several sustainable development and climate adaptation NBS projects aimed at reviving surface water in the region by restoring local wadis and wetlands that have been affected by climate change (Ross, 2018; UNESCO, 2019). Elsewhere, in the lower Ganges-Brahmaputra-Meghna basin of South Asia, partners in Bangladesh, India, and Nepal are working to implement eco-DRR projects to address flooding and drought related to climate change, unsustainable development, and land use change (IUCN, 2018). The connection between these projects and national climate, DRR or sustainable development plans such as the NDCs remains tenuous at best; thus, there is an opportunity to better align transboundary NBS activities to national priorities and targets under the 2015 action agenda. Another example is the International Convention for the Protection of the Danube River (ICPDR), which implemented a Flood Action Program that included green and gray infrastructure for DRR and climate change adaptation. Within this action program, the Living Danube Project as a public-private partnership with the Coca-Cola Foundation, ICPDR, and the World Wildlife Fund has worked on riverine and wetland habitation restoration projects that have exceeded the initial goal of restoring 5300 ha (ICPDR, TCCF, and WWF, 2020). Further work to build evidence for the use of transboundary NBS is being undertaken by IUCN and the Partnership for Environment and DRR and Adaptation (PEDRR).

Funding for transboundary NBS projects remains a challenge, as the majority of development and climate funding is targeted toward national and local governments, but RBOs are starting to develop innovative funding models. The development of bankable transboundary NBS projects requires capacity, structure, and common vision to propose projects that share both the risks and benefits. One of the keys to the Living Danube Project's success was its significant blended, public-private finance. This allowed the basin organization greater freedom to pursue projects that were aligned with its mandate and the goals of its member states. Elsewhere, the Cubango-Okavango River Basin has partnered with several organization to create its own endowment fund that can finance resilient NBS projects (OKACOM, 2019) while the Niger Basin Authority has obtained funds from several large donors to implement green infrastructure investments such as flood plain reconstruction to address water scarcity (African Development Bank (AfDB), 2018). As these types of NBS projects become more urgent due to the transboundary nature of climate impacts, more work is needed to understand how best to structure and finance nature-based projects in international settings.

Conclusion

This chapter has reviewed the value that NBS bring to global policy frameworks—both as a solution to complex, societal issues and as a mechanism for addressing some of the coordination and implementation issues those frameworks face at local, national, and international level. Water-centric NBS projects have been applied at national and transboundary levels to meet targets for climate change adaptation and mitigation, disaster risk recovery, biodiversity conservation, and sustainable development. Through case studies from around the world, NBS for water security have met policy targets across frameworks and have promoted multisectoral collaboration.

References

Abdellatif BM, Neffati M, Belgacem AO. Restoration and rehabilitation of degraded Saharan communal rangelands in southern Tunisia. J New Sci 2017;25(6).

Adikari Y, Yoshitani J. Global trends in water-related disasters: an insight for policymakers. Paris, France: UNESCO, ICHARM; 2009.

African Development Bank (AfDB). Niger Basin authority will get US$134 million to spur development and climate change adaptation in the Niger Basin (PIDACC). On November 15, 2020, available at: https://www.afdb.org/en/news-and-events/niger-basin-authority-will-get-us-134-million-to-spur-development-and-climate-change-adaptation-in-the-niger-basin-pidacc-18654; 2018.

Alacron I. Greening the country is the promise of the Ecuadorian government. *El Comercio*. On November 1, 2020, available at: https://www.elcomercio.com/tendencias/reverdecer-pais-acciones-gobierno-ambiente.html#:~:text=Reverdecer%20Ecuador%20es%20la%20iniciativa,Macas%2C%20provincia%20de%20Morona%20Santiago; 2018.

Anisha NF, Mauroner A, Lovett G, Neher A, Servos M, Minayeva T, Schutten H, Minelli L. Locking carbon in wetlands: enhancing climate action by including wetlands in NDCs. Corvallis, Oregon and Wageningen, The Netherlands: Alliance for Global Water Adaptation and Wetlands International; 2020.

Belgacem AO, Salem FB, Gamoun M, Chibani R, Louhaichi M. Revival of traditional best practices for rangeland restoration under climate change in the dry areas. Int J Clim Change Strateg Manag 2019;11(5):643–59.

Berkes F. Sacred ecology. 4th ed. New York: Routledge; 2017.

Blumstein S, Pohl B, Tanzler D. Water and climate diplomacy: integrative approaches for adaptive action in transboundary River Basins. In: German Federal Foreign Office Climate Diplomacy Report. Berlin. Germany: adelphi research gemeinnützige GmbH; 2016.

Brels S, Coates D, Loures F. Transboundary water resources management: the role of international watercourse agreements in implementation of the CBD; 2008 [Secretariat of the Convention on Biological Diversity].

Briceño S. Looking back and beyond Sendai: 25 years of international policy experience on disaster risk reduction. Int J Disaster Risk Reduct 2015;6(1):1–7.

Brouwer R, Van Ek R. Integrated ecological, economic and social impact assessment of alternative flood control policies in the Netherlands. Ecol Econ 2004;50(1):1–21.

CBD. Processes and meetings. On November 5, 2020, available at: http://www.cbd.int/process; 2020.

CBD, FAO, WB, UNEP, and UNDP. Biodiversity and the 2030 agenda for sustainable development: technical note. On November 5, 2020, available at: https://www.cbd.int/development/doc/biodiversity-2030-agenda-technical-note-en.pdf; 2017.

CBD Secretariat. Voluntary guidelines for the design and effective implementation of ecosystem-based approaches to climate change adaptation and disaster risk reduction and supplementary information technical series no 93, https://www.cbd.int/doc/publications/cbd-ts-93-en.pdf; 2019.

CES & Inros Lackner SE. Upscaling nature-based flood protection in Mozambique's cities: knowledge note. World Bank On October 31, 2020, available at: https://reliefweb.int/sites/reliefweb.int/files/resources/Upscaling-Nature-Based-Flood-Protection-in-Mozambique-s-Cities-Knowledge-Note.pdf; 2020.

Chen F, Guo H. Disaster risks and response strategies in process of urbanization in China. In: Contributing paper to the UNDRR Global Assessment Report 2019. Geneva, Switzerland: UNDRR; 2019.

Cohen-Shacham E, Walters G, Janzen C, Maginnis S, editors. Nature-based solutions to address global societal challenges. Gland, Switzerland: IUCN; 2016.

Cohen-Shacham E, Andrade A, Dalton J, Dudley N, Jones M, Kumar C, Welling R. Core principles for successfully implementing and upscaling nature-based solutions. Environ Sci Policy 2019;98:20–9.

Convention on Biological Diversity (CBD). Tunisia-main details biodiversity facts. Convention on biological diversity. On October 31, 2021, available at: https://www.cbd.int/countries/profile/?country=tn; 2021.

Cooper R. Nature-based solutions and water security. K4D helpdesk report 813. Brighton, UK: Institute of Development Studies; 2020.

Droste N, Schroter-Schlaack C, Hansjurgens B, Zimmermann H. Implementing nature-based solutions in urban areas: financing and governance aspects. In: Kabisch N, Korn H, Stadler J, Bonn A, editors. Nature-based solutions to climate change adaptation in urban areas, theory and practice of urban sustainability transitions. Cham: Springer; 2017.

Ecoshape. "About this pilot." Building with nature Indonesia. On October 30, 2020, available at: https://www.ecoshape.org/en/pilots/building-with-nature-indonesia/; 2020.

FAO. Evaluación de los recursos forestales mundiales 2020 Informe Ecuador. On November 1, 2020, available at http://www.fao.org/3/cb0102es/cb0102es.pdf; 2020.

Ferrario F, Beck MW, Storlazzi CD, Micheli F, Shepard CC, Airoldi L. The effectiveness of coral reefs for coastal hazard risk reduction and adaptation. Nat Commun 2014;5(1):1–9.

Gamoun M, Werner J, Louhaichi M. Traditional grazing-management practice makes an impact in Southern Tunisia. ICARDA. On October 30, 2020, available at: https://www.icarda.org/media/drywire/traditional-grazing-management-practice-makes-impact-southern-tunisia; 2018.

Garcia AC, Nagar A. Building sustainable resilience for sub-Saharan Africa's urban era. World of Opportunity On October 31, 2020, available at: https://medium.com/world-of-opportunity/building-sustainable-resilience-for-sub-saharan-africas-urban-era-48ca329c04a6; 2020.

Global Infrastructure Hub and Oxford Economics. Global infrastructure outlook. Sydney, NSW, Australia: Global Infrastructure Hub; 2017. https://outlook.gihub.org/?utm_source=GIHub+Homepage&utm_medium=Project+tile&utm_campaign=Outlook+GIHub+Tile.

Gomez-Baggethun E, de Groot R, Lomas PL, Montes C. The history of ecosystem services in economic theory and practice: from early notions to markets and payment schemes. Ecol Econ 2010;69:1209–18.

Government of Chile. 2nd National voluntary report Chile 2019. On November 7, 2020, available at: https://sustainabledevelopment.un.org/content/documents/23507Informe_Nacional_Voluntario_CHILE_Junio_2019_final_1.pdf; 2019.

Government of Chile. Chile's nationally determined contribution (NDC). English translation. 2020 edition. available at: https://www4.unfccc.int/sites/ndcstaging/PublishedDocuments/Chile%20First/Chile%27s_NDC_2020_english.pdf; 2020.

Government of Costa Rica. Costa Rica's intended nationally determined contribution. English Translation. San José, Costa Rica. On November 7, 2020, available at: https://www4.unfccc.int/sites/ndcstaging/PublishedDocuments/Costa%20Rica%20First/INDC%20Costa%20Rica%20Version%202%200%20final%20ENG.pdf; 2016.

Griscom BW, Adams J, Ellis PW, Houghton RA, Lomax G, Miteva DA, Woodbury P. Natural climate solutions. Proc Natl Acad Sci 2017;114(44):11645–50.

Hesselbach MH, Sánchez de Llanos JA, Reyna-Sáenz F, García Moral FJ, León SJ, Torres-Origel F, Gondor A. Plan de Conservación del Fondo de Agua Metropolitano de Monterrey, México. Arlington, VA: The Nature Conservancy; 2019.

High-Level Experts and Leaders Panel on Water and Disasters (HELP). Principles to address water-related disaster risk reduction (DRR) under the COVID-19 pandemic. Tokyo, Japan: HELP; 2020.

High-Level Political Forum (HLPF). President's summary of the 2020 High-Level Political Forum on Sustainable Development. United Nations Sustainable Development Goals Knowledge Platform; 2020. https://sustainabledevelopment.un.org/content/documents/269252020_HLPF_Presidents_summary.pdf.

Hirji R, Nicol A, Davis R. South Asia climate change risks in water management. Washington, D.C.: World Bank. Colombo, Sri Lanka: International Water Management Institute (IWMI); 2017.

Hollins L. In: Update on relevant activities under the UNFCCC process. Presentation to the UN Economic Commission for Europe (UNECE) On November 5, 2020; 2018. available at: https://www.unece.org/fileadmin/DAM/stats/documents/ece/ces/ge.33/2018/mtg4/S3_2_unfccc.pdf.

ICPDR, The Coca-Cola Foundation, WWF. Cross-sectoral partnership for wetlands:lessons from the living danube partnership. Presentation, Stockholm World Water Week On November 5, 2020. Available at: https://www.worldwaterweek.org/event/9454-cross-sectoral-partnership-for-wetlands-lessons-from-the-living-danube-partnership; 2020.

IISD. Why biodiversity matters: mapping the linkages between biodiversity and the SDGs. On November 5, 2020, available at: https://sdg.iisd.org/commentary/policy-briefs/why-biodiversity-matters-mapping-the-linkages-between-biodiversity-and-the-sdgs/; 2019.

IISD. UN biodiversity conference (CBD COP 15). On November 5, 2020, available at: http://sdg.iisd.org/events/2020-un-biodiversity-conference/; 2020.

IPCC. Climate change 2014: synthesis report. In: Pachauri RK, Meyer LA, editors. Contribution of working groups I, II and III to the fifth assessment report of the intergovernmental panel on climate change ore writing team. Geneva, Switzerland: IPCC; 2014.

IUCN. Water and climate change: building climate change resilience through water management and ecosystems. Gland, Switzerland: IUCN Issues Brief; 2015. https://www.iucn.org/sites/dev/files/import/downloads/water_and_climate_change_issues_brief.pdf.

IUCN. Nature-based solutions and their application in river basin management. In: BRIDGE GBM CSO dialogue report. Bangkok, Thailand: IUCN; 2018.

IUCN. Global standard for nature-based solutions. A user-friendly framework for the verification, design and scaling up of NbS. 1st ed. Gland, Switzerland: IUCN; 2020.

Jia H, Wang Z, Zhen X, Clar M, Shaw LY. China's sponge city construction: a discussion on technical approaches. Front Environ Sci Eng 2017;11(4):18.

Koeppler M. Ecosystem-based erosion control in Azerbaijan PANORAMA—solutions for a healthy planet On November 5, 2020. available at: https://panorama.solutions/en/solution/ecosystem-based-erosion-control-azerbaijan; 2020.

Li J. Inside China's leading 'sponge city': Wuhan's war with water. The guardian On November 8, 2020. available at: https://www.thegvcuardian.com/cities/2019/jan/23/inside-chinas-leading-sponge-city-wuhans-war-with-water; 2019.

Matthews RB, van Noordwijk M. From euphoria to reality on efforts to reduce emissions from deforestation and forest degradation (REDD+). Mitig Adapt Strat Glob Chang 2014;19(6):615–20.

Matthews J, Timboe I, Amani A, Bhaduri A, Dalton J, Dominique K, Fletcher M, Gaillard-Picher D, Holmgren T, Leflaive X, McClune, K. … Yokota, T. Mastering disaster in a changing climate: reducing disaster risk through resilient water management. Global Water Forum; 2018. https://globalwaterforum.org/2018/12/02/mastering-disaster-in-a-changing-climate-reducing-disaster-risk-through-resilient-water-management/.

Matthews J, Matthews N, Simmons E, Vigerstol K. Wellspring: source water resilience and climate adaptation. Arlington, VA: The Nature Conservancy; 2019. https://www.nature.org/content/dam/tnc/nature/en/documents/Wellspring_FULL_Report_2019.pdf.

Möhner A. The evolution of adaptation metrics under the UNFCCC and its Paris Agreement. In: Christiansen L, Martinez G, Naswa P, editors. Adaptation metrics: perspectives on measuring, aggregating and comparing adaptation results; 2018. p. 14–27 [UNEP DTU Partnership].

Morales M. The case of Ecuador. In: Greiber T, Schiele S, editors. Governance of ecosystem services: lessons learnt from Cameroon, China, Costa Rica, and Ecuador. Gland, Switzerland: IUCN; 2011. p. 93–118.

Moyer JD, Bohl DK. Alternative pathways to human development: assessing trade-offs and synergies in achieving the sustainable development goals. Futures 2019;105:199–210.

Multi-Hazard Mitigation Council. Natural Hazard Mitigation Saves: 2019 Report. Principal Investigator Porter, K; Co-Principal Investigators Dash, N,, Huyck, C., Santos, J., Scawthorn, C.; Investigators: Eguchi, M., Eguchi, R., Ghosh., S., Isteita, M., Mickey, K., Rashed, T., Reeder, A.; Schneider, P.; and Yuan, J., Directors, MMC. Investigator Intern: Cohen-Porter, A, Washington, D.C.: National Institute of Building Sciences; 2019. https://www.nibs.org/page/mitigationsaves.

OECD. Nature-based solutions for adapting to water-related climate risks. OECD environment policy paper no. 21. OECD Publishing: Paris, France; 2020.

OKACOM. Factsheet: Cubango-Okavango River Basin fund. On November 1, 2020, available at https://www.okacom.org/sites/default/files/documents/CORB%20Fund%20Factsheet.pdf; 2019.

Oppla. Ecuador: the socio bosque program. On November 1, 2020, available at https://oppla.eu/casestudy/18372; 2020.

Parish F, Sirin A, Charman D, Joosten H, Minayeva T, Silvius M, Stringer L, editors. Assessment on Peatlands, biodiversity and climate change: main report. Kuala Lumpur, Malaysia, and Wageningen. The Netherlands: Global Environment Centre and Wetlands International; 2008.

Parkinson S, Krey V, Huppmann D, Kahil T, McCollum D, Fricko O, Raptis C. Balancing clean water-climate change mitigation trade-offs. Environ Res Lett 2019;14(1), 014009.

PEDRR & FEBA. Promoting nature-based solutions in the post-2020 global biodiversity framework. On November 5, 2020, available at https://www.iucn.org/sites/dev/files/promoting_nbs_in_the_post-2020_global_biodiversity_framework.pdf; 2020.

Perry C, Steduto P, Karajeh F. Does improved irrigation technology save water? A review of the evidence. In: Discussion paper on irrigation and sustainable water resources management in the Near East and North Africa. Cairo, Egypt: FAO; 2017. http://www.fao.org/3/I7090EN/i7090en.pdf.

Qiao XJ, Liao KH, Randrup TB. Sustainable stormwater management: a qualitative case study of the sponge cities initiative in China. Sustain Cities Soc 2020;53:101963.

Rahman HM, Sherren K, van Proosdij D. Institutional innovation for nature-based coastal adaptation: lessons from salt marsh restoration in Nova Scotia, Canada. Sustainability 2019;11(23):6735.

Rankovic A, Chan S, Laurans Y. Implementing nature-based solutions in climate policies: what's in it for biodiversity?—first lessons from Morocco and Tunisia, studies N°07/17. Paris, France: IDDRI; 2017.

Reid H, Pérez de Madrid M, Ramírez O. In: Ecosystem-based approaches to adaptation: strengthening the evidence and informing policy. Research results from the adaptation, vulnerability and ecosystems (AVE) project, Costa Rica and Panama. Project Report. IIED, London; 2018.

Republic of Indonesia. Voluntary national reviews (VNR): empowering people and ensuring inclusive and equality. Available at: https://sustainabledevelopment.un.org/content/documents/2380320190708_Final_VNR_2019_Indonesia_Rev3.pdf; 2019.

Ross W. Lake Chad: can the vanishing lake be saved? *BBC News*. On November 1, 2020, available at https://www.bbc.co.uk/news/world-africa-43500314; 2018.

Secretary of the Pacific Community (SPC). Vanuatu climate change and disaster risk reduction policy 2016-2030. Government of the Republic of Vanuatu. Suva, Fiji: SPC; 2015.

Seddon N, Sengupta S, García-Espinosa M, Hauler I, Herr D, Rizvi AR. Nature-based solutions in nationally determined contributions: synthesis and recommendations for enhancing climate ambition and action by 2020. Gland, Switzerland and Oxford, UK: IUCN and University of Oxford; 2019.

Seddon N, Chausson A, Berry P, Girardin CA, Smith A, Turner B. Understanding the value and limits of nature-based solutions to climate change and other global challenges. Philos Trans R Soc B 2020;375(1794):20190120.

Seidler R, Dietrich K, Schweizer S, Bawa KS, Chopde S, Zaman F, Khaling S. Progress on integrating climate change adaptation and disaster risk reduction for sustainable development pathways in South Asia: evidence from six research projects. Int J Disaster Risk Reduct 2018;31:92–101.

Sherren K, Bowron T, Graham JM, Rahman HMT, van Proosdij D. Coastal infrastructure realignment and salt marsh restoration in Nova Scotia, Canada. Chapter 5. In: Responding to rising seas: OECD country approaches to tackling coastal risks. Paris, France: OECD Publishing; 2019. p. 111–35.

Tasnim T, Irfanullah HM. Rethinking nature: a pathway towards sustainable development? UNEP Perspectives Issue no 38, http://www.icccad.net/wp-content/uploads/2020/09/Perspective-2020-AUG-D5.pdf; 2020.

Timboe I, Pharr K, Matthews JH. Watering the NDCs: national climate planning for 2020—how water-aware climate policies can strengthen climate change mitigation & adaptation goals. Corvallis, OR, USA: AGWA; 2019. https://alliance4water.org/wateringthendcs.

UNESCO. BIOsphere and Heritage of Lake Chad (BIOPALT) project. On November 1, 2020, available at https://en.unesco.org/biopalt/restoration; 2019.

UNESCO, UN-Water. United Nations World Water Development Report 2020: Water and Climate Change. Paris, France: UNESCO; 2020.

UNFCCC. Opportunities and options for integrating climate change adaptation with the Sustainable Development Goals and the Sendai Framework for Disaster Risk Reduction 2015-2030. Technical paper, Bonn, Germany: UNFCCC; 2017.

UNFCCC Adaptation Committee (AC). Synthesis report on how developing countries are addressing hazards, focusing on relevant lessons learned and good practices in the context of the recognition of adaptation efforts of developing countries. In: AC18/SREP/5D. Bonn, Germany: UNFCCC; 2020. https://unfccc.int/documents/254565.

United Nations (UN). Summit charts new era of sustainable development. On November 2, 2020, available at https://www.un.org/sustainabledevelopment/blog/2015/09/summit-charts-new-era-of-sustainable-development-world-leaders-to-gavel-universal-agenda-to-transform-our-world-for-people-and-planet/; 2015.

United Nations (UN). Paris Agreement. Paris: United Nations; 2016. p. 1–27.

United Nations Office for Disaster Risk Reduction (UNDRR). Sendai Framework for Disaster Risk Reduction 2015-2030. In: Geneva. Switzerland: UN; 2015.

van Ham C, Klimmek H. Partnerships for nature-based solutions in urban areas—showcasing successful examples. In: Kabisch N, Korn H, Stadler J, Bonn A, editors. Nature-based solutions to climate change adaptation in urban areas. Theory and practice of urban sustainability transitions. Cham: Springer; 2017.

Wetlands International (WI). Building with nature Indonesia. Wetlands International Case Study. On October 30, 2020, available at https://www.wetlands.org/casestudy/building-with-nature-indonesia/; 2016.

Ecuador. Voluntary national review 2018. Sustainable development goals knowledge platform, https://sustainabledevelopment.un.org/memberstates/ecuador; 2018. [Accessed 1 November 2020].

Liu H, Jia Y, Niu C. "Sponge city" concept helps solve China's urban water problems. Environ Earth Sci 2017;76(14):473.

Trémolet S, Karres N. Resilient European cities: nature-based solutions for clean water. London, UK: The Nature Conservancy; 2020.

Mauroner A, Anisha N, de la Cruz E, Barrios E, Guntoju SS. Operationalizing nature-based solutions in low- and middle-income countries: redefining and "greening" project development. In: Cassin J, Matthews JH, Lopez-Gunn E, editors. Nature-based solutions and water security: an agenda for the 21st century. Elsevier; 2021.

CHAPTER 8

Legible rivers, resilient rivers: Lessons for climate adaptation policy from the Wild and Scenic Rivers Act

Denielle M. Perry

School of Earth and Sustainability, Northern Arizona University, Flagstaff, AZ, United States

Key messages

- Proactive conservation is vital for riverine biodiversity and ecosystem service protection.
- Conservation grounded in adaptive management can serve as a nature-based climate adaptation policy.
- Conservation legibility acts can foster resilient riverine socio-ecological systems.
- Federal river conservation can be viewed as an act of eco-governmentality.
- Conservation-inclusive integrated water resource management (IWRM) strategies can address inequalities in water governance.

Introduction

Unlike technology-based infrastructure, ecosystem-based approaches to climate change adaptation that utilize natural capital offer measures that take an interconnected view of climate change, biodiversity, and sustainable resource management (Grantham et al., 2019; Munang et al., 2013; Poff et al., 2015; Vörösmarty et al., 2018; WWAP, 2018). A key tenet in this nature-based adaptation approach for resilient ecosystems is ecosystem protection through conservation policy strategies that monitor, assess, and adapt for current and future conditions (Grantham et al., 2019; Hermoso et al., 2016; Poff, 2018). Yet the creation of adaptation policies is often stymied by a lack of prioritization or obligation within governments and private entities (Henstra, 2015; Tuusa et al., 2013) and by legal and institutional barriers (Cosens et al., 2014; Palmer et al., 2009). Given the political limitations to creating new conservation policies and deploying them at the scale necessary for effective climate adaptation, Parenti (2015) suggests the state will be called upon to address the climate crisis by expanding upon the legibility practices it already conducts on the environment. Against that backdrop, this chapter is grounded in the concept of **legibility**—"a reductive process, geared explicitly toward representation of *what interests the state*, and is thus tied closely to the surveillance, regulation, and control of both people and environments" (Kirsch, 2002, p. 556).

Nature-Based Solutions and Water Security. https://doi.org/10.1016/B978-0-12-819871-1.00011-7

Taking into account that legibility produces nature in an abstract way for capital (Scott, 1998), legibility can be understood as the state's way of knowing its resources through surveying, inventorying, cataloging, and management exercises that serve its purposes.

This chapter exposes the historical intricacies of how the US government territorialized riverine resources and ecosystem services through legibility acts first to adapt to regional climates and later to mitigate ecological degradation. The state did this in part for capital accumulation but also for preservation purposes and to foster state legitimacy. From periods of territorial expansion to ones of economic contraction across the US landscape, dams and water diversions were subsidized by the state, making it an example of a hydraulic society (Meehan, 2012; Parenti, 2015; Worster, 1985). While economically stimulating and highly profitable for those directly benefiting from these projects, abstracting water by these means degraded biodiversity habitat and other ecosystem services that rivers provide such as clean water, recreational opportunities, and fisheries. Over time, and through social relations of production centered on resource conservation, the US government intervened in the trajectory of development it had set on course half a century before in the early 1900s. Archival evidence highlights how the state was driven by discourse of scarcity and ecological crisis to turn toward familiar legibility practices, pursuing a policy that could strike a balance between two forms of capital production reliant on rivers. Technical water resource development projects would ultimately be offset by the conservation of free-flowing rivers and their vital riverine ecosystem services through the Wild and Scenic Rivers Act of 1968 (WSRA).

Here the WSRA is understood as a standalone policy that at once melds together, is fortified by, and fortifies a complex of multiple unprecedented policies created during the Johnson administration shaped by legibility acts grounded on notions of scarcity and security (see Fig. 8.1). These policies ultimately reflect an era of evolving national environmental priorities. After threading these policies together, the legibility imperative that drove federal agencies to survey and inventory free-flowing rivers with high water quality, catalog outstandingly remarkable values (ORVs), and establish policy to protect and ultimately manage the rivers becomes apparent. In a federal system that devolves water rights to states, this legibility exercise can be considered an attempt to restructure governance of riverine resources deemed important to US society writ large.

Moreover, this chapter illustrates how the identification and conservation of ORVs in the 1960s environmental era reflects the emergence of a new paradigm in understanding socio-ecological systems —that of ecosystem services and their resilience. Drawing on Foucault's rhetoric of discourse and biopower, this chapter then makes a case for situating climate adaptation language within a riverine ecosystem service protection framing in the present day, as such management is increasingly important across scales, especially regarding water resources (Poff et al., 2015). Adopting adaptation language can serve to transform state practices of resource governance through linguistic practices that facilitate incorporation of adaptation strategies into policy (Kendrick, 2012; Rose et al., 2006; Sharp and Richardson,

Fig. 8.1

Timeline (1964–68) of federal water-related policies created during the Johnson administration that culminated in the WSRA.

2001). Such a transition is necessary because the lack of knowledge about adaptation and its relevance to ecosystem services stands as a barrier to implementing effective policies (Urwin and Jordan, 2008; Wilson, 2006). Thus, this framing can link adaptation relevance to river management strategies around the globe for sustainable riverine ecosystem governance.

Background and theoretical context

Territorializing rivers through legibility: From scarcity to resilience

The United States has a long history of dam building. Geographers have shown this history of development is inextricably linked to the history of nation-building through the territorialization of water (Evenden, 2009; Graf, 1999; Meehan, 2012; Vogel, 2012). Developers rationalized projects on abstracted scientific notions of the hydrologic cycle by quantifying availability, codifying laws to govern resources, and devising management regimes, at last rendering water legible to the state and those who would exploit its resources

(Linton, 2014). Legibility practices, as Scott (1998) suggests, are central to the state territorial project for capital accumulation through natural resource development. Laws and policies authorize, create, and reconfigure nature's role in capital (Collard and Dempsey, 2017). At the core of the capitalist state is the management, mediation, delivery, and production of the environment within its defined territory (Parenti, 2015). Water development projects that shape, move, control, and employ water exemplify "environment making" legibility acts conducted for the benefit of the state. Yet, critical analyses of legibility show that through reductionist measurements that seek to simplify resources for legibility purposes, states can unintentionally produce scarcity conditions (Perramond, 2013; Scott, 1998). Such is the case with the national policy of dam building in distinct regions of the country.

Take, for instance, the case of the arid Western United States. Here, despite compelling recommendations by Major John Wesley Powell to divide territory according to physiographic characteristics based on water availability —no less a proposal justified in exhaustive government surveys of the region —Congress elected instead to pursue a simplified system of property allocation based on geometry and the rectangular land survey (Kirsch, 2002). This territorial project, lacking concern for environmental limitations, set in motion state-produced water scarcity that manifested in limited mining potential outside of streambeds and on farms with little to no water to support crops and livestock. After homesteading settlers and industrialists realized the constraints these environmental limitations affected on production in these once frontier lands, they called on the government for intervention. The resultant Prior Appropriation Doctrine was adopted as a temporal water rights system devised to govern the development and allocation of waters far removed from real property (Benson, 2012; Gates et al., 1993; Wilkinson, 1992). Launched by the Bureau of Reclamation and the Federal Reclamation Act of 1904, a marriage of policy and agency soon came to focus on facilitating economic expansion and federal state building in the West. Spurred by development-minded interest groups who, according to Polanyi (1944), call upon the state when it serves to advantage them, an "iron triangle" of Congressional committees and federal agencies imposed a system of state-funded dams, reservoirs, and irrigation canals in a largely arid region—in the process reconfiguring flows and consolidating federal power in Western territory (Lawrence, 2011; McCool, 1987; Meehan, 2012; Worster, 1985).

In quick succession, as the United States reeled from the Great Depression, capital found new ways to tap rivers for production. New Deal economic stimulus policies and the Congressional Authorization of the Tennessee Valley Authority in 1933 promoted public works projects, further advancing the state territorial project. Increased dam development brought power to poverty-stricken rural areas of the Southeast and the Pacific Northwest and worked to reclaim flood-prone lands for development. In so doing, the state produced nature to combat depressed economics, to protect assets in the built environment, and later to fuel the military industrial complex for national security purposes (Evenden, 2009). Dams are such a pillar of development in the United States that today somewhere between 75,000 (Graf, 1999) and 90,500 dams are used for energy production, agriculture, municipal use, navigation, flood control, and recreation (ASCE, 2017). The great wealth and security generated through water resource development did not, however, come without a price.

As Scott (1998) demonstrates in his examination of state schemes to govern resources, legibility projects, no matter how well intentioned, can have ramifications that ripple out from project nuclei to negatively impact society and the environment in profound and unforeseen ways. Following Marx, capital's "mindless exploitation" of nature lacks foresight to consider remote externalities of production because its focus rests on reaping immediate accumulation successes (Harvey, 2001, p. 53). In this sense reminiscent of the First Law of Ecological Bloodymindedness (Brown, 2005), as the state capitalized on dams, these structures simultaneously trapped water and sediment, altered habitat, and cut off migration corridors for aquatic and terrestrial species, ultimately degrading the nation's river ecosystems (Auerbach et al., 2014; Strayer and Dudgeon, 2010). Moreover, mining, manufacturing, and agriculture sectors stimulated by these projects produced polluting effluent problems that poisoned rivers, thereby compounding ecosystem impairment (Bernhardt and Palmer, 2011; Harden et al., 2014; Strayer and Dudgeon, 2010; Wohl, 2005). Dams also displaced people culturally connected with rivers (WCD, 2001). Thus, fueled by desires to preserve nature, conserve vital water resources, protect homelands, and capitalize on recreation opportunities after decades of building monolithic nature-modifying structures, a movement took shape to interrupt the state trajectory of dam construction. A major, yet understudied policy outcome of this movement 50-plus years ago was the WSRA, a legibility act created to identify and protect free-flowing river corridors and their ecosystem services known in policy terms as ORVs.

Today, climate change portends a whole host of threats to riverine communities and ecosystem services from floods and droughts to species invasions and extirpations (Darwall et al., 2018; Parry et al., 2007; Palmer et al., 2009). With mounting temperatures and hydrologic changes bringing a resurgence of path-dependent big infrastructure, new projects are being ushered in around the world to adapt to water resource management challenges. New and augmented dams and levees are focused on channeling river resources for renewable energy production (UNGCF, 2017; World Water Assessment Programme (WWAP), 2009), irrigated agriculture (Perry and Praskievicz, 2017), flood control (IPCC, 2008), and municipal uses (Finley, 2016; Weiser, 2016). Such gray infrastructure development can lead to increased habitat fragmentation (Grill et al., 2019), water temperature changes, water quality degradation, and exacerbation of species invasions (Darwall et al., 2018). With such pressures, there are increasing calls to protect biodiversity and other riverine ecosystem services to foster resilience (Darwall et al., 2018; Henstra, 2015; IPCC, 2008; Palmer et al., 2014; Perry, 2017; Thompson, 2015; United Nations, 2009; Vörösmarty et al., 2010). Resilient rivers are likely to maintain their current conditions in the face of perturbation, preserving socio-ecological values important to society (Lake, 2013; Poff, 2018).

In riverine ecosystems, biophysical climate resilience may be engendered by free and consistent flow unencumbered by dams (Arthington et al., 2010; Palmer et al., 2008), intact riparian forests and floodplains (Malhi et al., 2008), habitat refugia (Naiman et al., 2000; Thrush et al., 2013), and habitat connectivity (Sundstrom et al., 2012; Thrush et al., 2013; Van Looy et al., 2019). Both longitudinal and lateral connectivity facilitate recovery after perturbation by allowing the movement of water, sediment, organic matter, and organisms within and between the channel and floodplain (Bernhardt and Leslie, 2013; Churchill et al., 2013; Gerisch et al., 2012; Norden et al., 2009; Thompson et al., 2017). Meanwhile, species diversity enhances the stability of biotic communities when faced with variable conditions (Stachowicz et al., 2008). Thus, protection of free-flowing rivers may promote riverine ecosystem resilience by maintaining both terrestrial and aquatic corridors important for the movement of both species and materials during disturbances.

With nearly every major river basin in the US fragment by dams (Graf, 1999) and less than half of large rivers remaining free-flowing globally (Grill et al., 2019), river conservation is vital for freshwater biodiversity and other ecosystem service protection and to promote resilience (Tickner et al., 2020). Implementing such protections will depend on the creation and/or expansion of conservation policies designed to protect riverine ecosystems in an integrated way, taking into consideration both the aquatic and terrestrial components (Leal et al., 2020; Abell and Harrison, 2020). Inspired by Parenti's (2015) call to improve upon the legibility acts the state already conducts on the environment to contend with the socio-ecological crises of climate change, this research uncovers the rationale for creating the first riverine ecosystem conservation policy in the world by answering the following questions: How did legibility, changing environmental values, scientific discourse, and notions of rational resource use factor in shaping the WSRA? How, in turn, did this complex rationale guide the reconfiguration of the state's water resource policy from one of development to one of conservation? By answering these questions, the benefits of legibility acts are reconsidered while lessons emerge that can assist in today's shaping of nature-based climate adaptation policies to promote resilient riverine ecosystems.

Methodology

Investigation for this project proceeded largely through extensive archival research at the Lyndon Baines Johnson (LBJ) Presidential Library in Austin, Texas, and the National Archives at Denver. White House files related to legislation, programs, budgets, and personal communications concerned with water and other natural resource policies during the LBJ administrations were brought to light. In total, 134 boxes, many of which had yet to be breached since the administration's end, were examined over the course of three 40-hour weeks. Given the limited temporal scope (1963–68) captured in these materials, 17 boxes of files from the Department of Interior were then examined at the National Archives at

Denver over the course of 3 days. These boxes contained similar materials from previous and subsequent presidential administrations.

Archival materials were photographed and later converted to PDF files for qualitative data analysis utilizing NVivo QDA software. The purpose of collecting archival materials was twofold. Because "it is important to show how discourse, or changes in discourse, make a difference to what happens in policy processes or in society more broadly" (Sharp and Richardson, 2001, p. 196), the first objective was to expose empirical details related to how the WSRA was conceptualized. The second goal was to uncover the lesser known, yet inherently linked, environmental policies, programs, and values factoring into the policy design. By situating a critical eye on taken-for-granted events and knowledge, archival documents were coded and analyzed to reveal common themes and then arranged within six broad categories: (1) environmental values/ideologies; (2) location; (3) stakeholders; (4) policy; (5) capacity; and (6) science. This discourse analysis revealed the political-ecological trends elaborated on in the following sections.

Analysis and discussion

Making biopower legible in the landscape through conservation

> *It is hereby declared to be the policy of the U.S. that certain selected rivers of the Nation which, with their immediate environments, possess outstandingly remarkable scenic, recreational, geologic, fish and wildlife, historic, cultural or other similar values, shall be preserved in free-flowing condition, and that they and their immediate environments shall be protected for the benefit and enjoyment of present and future generations. The Congress declares that the established national policy of dams and other construction at appropriate sections of the rivers of the U.S. needs to be complemented by a policy that would preserve other selected rivers or sections thereof in their free-flowing condition to protect the water quality of such rivers and to fulfill other vital national conservation purposes.*
>
> ***President Lyndon B. Johnson, October 2, 1968***

As President Johnson proclaimed, the WSRA was created to fulfill conservation needs at a time when rapid degradation from water development projects threatened the nation's river heritage. This first of its kind, which celebrated its 50th anniversary in 2018, the policy signified a shifting focus from a national policy of engineering projects for economic growth and human development to one of protecting rivers for posterity and recognizing the intrinsic existence value of nature. Generally, the WSRA is considered a manifestation of attempts to reconcile roughly 30 years of tensions between the national preservation movement and regional interests centered on dam development, as the caricature in Fig. 8.2 depicts (Burce, 2008; Daniels, 2009; Palmer, 1993, 2017). Yet, probing deeper into the archives reveals LBJ's words also signal an evolution in national environmental values related to water resources. Initially this transformation stemmed from three interconnected socio-economic phenomena:

Fig. 8.2
LBJ considering the creation of a wild and scenic rivers policy as dam development interests look on.
Courtesy of Brinkman. White House social files alpha box 1755. Austin, TX: LBJ Presidential Library; 1965.

urbanization, population growth, and the rise of the outdoor recreation industry. It turns out discourse on these trends, along with the management of finite resources, was driving environmental policies to govern what today are referred to as ecosystem services, or the goods and benefits people derive from ecosystems (Costanza et al., 2017; Finlayson et al., 2005), and their role in capital accumulation.

In the post–World War II baby boom era, the United States, like much of the world, underwent exponential population growth and rapid rates of urbanization. From 1950 to 1960, the US population grew by 19%, increasing from 150.7 to 179.3 million (Hobbs and Stoops, 2002). From 1940 to 1970, as the economy shifted from primary agricultural production to secondary manufacturing, people moved to the cities en masse (Platt et al., 2013). On the heels of this transition, outdoor recreation became capital's new environmental focus as rural areas once regarded as sources of raw materials for extraction became new sites of accumulation through tourism. Shifting environmental priorities began with recreation as a new precedence for land management. In economic terms, burgeoning industries of technical recreational equipment viewed nature preservation as "good business" and supported conservation policies to advance their interests (Rittenhouse, 1965). In social terms, recreation served as a partial solution to problems attributed to rapid rates of urbanization and increasing leisure time. These points were made clear in the following archive excerpt culled from the Bureau of Outdoor Recreation's (now defunct) declaration of intent.

> *There needs to be public understanding that recreation is not only a renewing experience but also serious business. It is serious national business both because of its economic impact and its beneficial effect on the physical, cultural, social and moral well-being of the American People. It is a partial solution to the social problems created by urbanization and leisure time. It is a solution, at least in part, to the fact that man is not wholly suited physiologically to meet the technological demands placed upon him. Most of the hospitalizations in the country today are emotionally based. In this vein I like to think of the new organization as the Bureau of Re-Creation. We have heard much of ORRRC. Now I like to think in terms of BORC for the Bureau of Outdoor Re-Creation.*
> **Edward C. Crafts Director, Bureau of Outdoor Recreation, June 21, 1962**

At the core of this Outdoor Recreation Resources Review Council (ORRRC) recreation campaign, exposure to nature became a way of insuring the reproduction of labor power through people's positive interactions with the environment, thus engendering social resilience (Keck and Sakdapolrak, 2013) in those taking time from their busy urban lifestyles to recreate. Here we see the government take to enhancing the welfare, health, wealth, and longevity of the population to ensure vital labor power (Agrawal, 2005). For Foucault (2003) labor power is biopower. The state depends on biopower. Building on this conception, the following paragraphs develop the unlikely but real influence of biopower and eco-governmentality on the protection of rivers.

First, the tourism industry sought to capitalize on the production of recreation spaces made legible through ORRRC resource surveys conducted in the early 1960s. Conservation policy recommendations resulted from these surveys (Olson, 2010), leading to the territorialization of recreation spaces. In this case, Bureau of Recreation documents at the National Archives at Denver reveal the ORRRC suggested that certain rivers with unusual values remain free-flowing for recreation purposes. An official interagency study for identifying such rivers through a national survey was later announced in 1963. Concurrently, neoMalthusian discourse of out-of-control population growth, or the "Population Bomb" (Robbins, 2012), also factored squarely into demarcating the public estate for conservation spaces. Namely, notions of scarcity were produced by the growing population's new demands on recreational lands. As evidenced by a June 1964 Population Bulletin featured in Fig. 8.3, land and water resources now valued for recreation potential by a burgeoning tourism industry were considered at risk due to "excess procreation." For Malthus, land was the primary factor of capital production (Brown et al., 2006). In line with Scott (1998) and Parenti (2015), arguably, the perceived threats to recreation spaces that could reproduce the conditions of capital must be addressed by the state to protect the growing tourism sector of the economy. Ultimately, discourse on the scarcity of spaces for recreation opportunities coupled with invocations of recreation's broad benefits to society were able to transform state practices of resource governance, leading to the territorialization of river resources. Thus, in 1964, the Wild Rivers Bill, the first iteration of what would 4 years later become the WSRA, was proposed.

Biermann and Mansfield's (2014) analysis of conservation policy being an eco-governmentality exercise of the state for nonhuman nature aligns with Agrawal's (2005) notions of eco-governmentality through legibility practices to include human nature. That

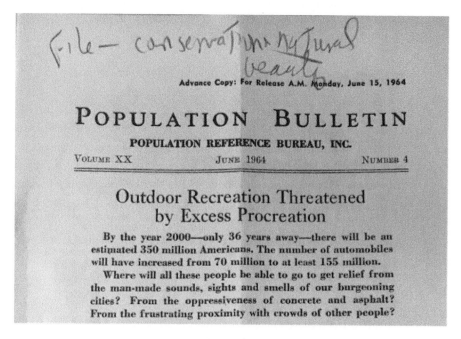

Fig. 8.3
Population growth spurred notions of scarce recreation lands.*From Population Reference Bureau. Population bulletin, vol. XX, number 4. Austin, TX: LBJ Presidential Library; 1964.*

is to say, in this case, the proposed protection of rivers not only assured the conditions for the reproduction of economic opportunities for capital in an expanding tourism industry but also suggested a relief on public health services would ensue through urbanite access to nature. This state eco-governmentality initiative took shape by incorporating recreation into resource planning through Multiple Use mandates. Now federal land management agencies would balance out public land uses from one solely focused historically on conservation for exploitative and extractive resource uses to one of environmental conservation for biopower.

Water: The wellspring of ecological planning concerns based on biopower, legitimacy, and rationality

As tourism and population growth spurred conversations centered on economic growth and human well-being, emergent ecological science and nongovernmental organizations were articulating concerns over societal impacts on riverine ecosystems and the necessity for biodiversity conservation. With the launching of the environmental movement of the 1960s by Rachel Carson's (1962) *Silent Spring*, which linked pesticide use to species decline, interest in ecology swelled in the United States and abroad. Concerns mounted over "the impact of human progress and development on the natural world [which] had produced what amounted to a state of emergency for wildlife"—as stated in the World Wildlife Fund's first report (Scott, 1965). Local, regional, and international evidence of a biodiversity "crisis" mounted, indicating rapid species extinctions were eminent. In response, the state (at the behest of special interest groups) pursued what Dempsey (2013) describes as the mobilization of policy to protect the integrated role water and biodiversity play in providing potential future value to capital.

This mobilization is captured in archived White House files. For instance, nearly 200 letter exchanges were made between concerned citizens and First Lady Johnson, who was leading the Beautification movement to clean up the nation's dirty water, highways, and cities. These letters detailed growing pollution problems in waterways from the Willamette River on the West Coast to the Potomac on the East Coast, while concern for posterity heightened. A typical excerpt from Lady Bird Johnson's responses follows: "… Every week I read letters from children –and their parents -- in Florida, or Ohio, or New York, or Arizona, and they, too, have polluted rivers, and they, too, want to do something about cleansing them" (Johnson, 1965). These letters described how youths were unable to enjoy local swimming holes or eat fish caught on family outings in places where their elders had long enjoyed the sport. Such accounts calling on the government to act on behalf of its citizenry were instrumental in driving federal policy advances to secure water quality. Notably, the 1966 Clean Rivers Restoration Act was passed to fund sewage treatment facilities and amendments to the Water Pollution Control Act took up estuary and transboundary river pollution—this policy eventually became the Clean Water Act (Bureau of Budget, 1966).

Correspondence from families and businesses alike also highlighted looming impacts of dam development on beloved rivers, such as the Lochsa, Selway, and Clearwater Rivers of Idaho, and the human and nonhuman communities they support (Hagen, 1967). Perhaps the most compelling evidence of how concerns for the biodiversity crisis drove the eventual creation of river conservation policy in this era came from the Pacific States Marine Fisheries Commission. This interstate compact agency is tasked with sustaining the fishing industry across five Western states (PSMFC, 2012). In a 1965 report, the agency detailed salmon and steelhead losses on the Columbia River due to increased dam building at such a high rate that "the future of the anadromous fishery resources may be endangered" (James, 1966, p. 6) on account of dams in that watershed blocking passage between the Pacific Ocean and spawning beds upstream (Vogel, 2012). Thus, recognizing needs to preserve the remaining free-flowing tributaries for reproductive habitat connectivity, the agency recommended permanent protection of Idaho's Salmon and Clearwater River systems and California's Klamath River in the form of the proposed Wild Rivers Bill (James, 1966). Explicitly protecting certain stretches of river from dam development was ostensibly a rational resource use approach to maintaining resilient fisheries that were facing the degradation of their vital spawning habitats. Mobilizing conservation policy in this manner exemplifies how the state and special interests can work together to secure ecosystem services critical for the reproduction of capital; in this case the conservation of habitat for biodiversity, hence, also the provision of fish stocks (Perry, 2017). In so doing, the state was able to maintain legitimacy as the purveyor of a healthy environment and protector of ecosystem services (Dryzek et al., 2002).

Meanwhile on a global scale, the International Union for the Conservation of Nature (IUCN) beseeched the United States to assume a leading role in promoting the rational use of resources and to promote preservation of wild nature, particularly "rare and vanishing species" (Cater, 1965). Calling on the United States' "record of accomplishment in conservation," the IUCN extended an invitation of membership in hopes that the US presence would catalyze other countries (i.e., the United Kingdom) to join (Cater, 1965). That same year, The Nature Conservancy, a member of the IUCN, directly entreated the United States to protect "biotic communities," suggesting "the earth's most valuable natural resources is its stock of different species, races and strains of living organisms, each of which has unique attributes and potentialities" (Cater, 1965). Notwithstanding, Cold War geopolitics ultimately limited direct US engagement in the IUCN due to conflict with states represented by the organization, namely China and Vietnam. However, Washington, DC recommended for "an internal subdivision of the" US Government to participate in IUCN activities. The Department of the Interior, specifically the National Park Service, was the candidate. However, it was not until 1981, according to Farnham (2007), that the US government per se found its way into a serious engagement with biological diversity. Nonetheless, this archival evidence suggests the IUCN and its many members (largely US-based interest groups) influenced discussions over biodiversity conservation in US natural resource policy. Notably, in 1967 the Fish and Wildlife Conservation and Protection Act, precursor to the present-day Endangered Species Act of 1973, was passed to address mounting concerns over biodiversity loss.

As biodiversity protections took shape, another paradigm of water resource management based in rational use was emerging. In line with Powell's century-old recommendation, legibility exercises now centered on the river basin for what Biswas (2009) describes as the organizing principle for integrating sustainable water management strategies. This model influenced the application of IWRM principles. IWRM was conceived as a process promoting coordinated efforts for the development and management of water and land resources for the equitable maximization of social and economic well-being while ensuring sustainable ecosystems (Gilman et al., 2004; Hering and Ingold, 2012; Poff et al., 2015; Vörösmarty et al., 2018). Ideally, successful IWRM strategies are meant to strike a balance between human resource use and ecosystem protection (Vörösmarty et al., 2010). These underlying principles influenced the creation and design of several new river governance strategies, including the Water Resources Planning Act.

The Water Resources Planning Act established a Water Resources Council and the River Basin Commissions "to plan for the best use and development of the resources of the river and adjoining land" (Johnson, 1969). As basin planning commenced, "procedures to bring about a flexible but systematic meshing of the [proposed] scenic river and comprehensive basin planning efforts" were deemed necessary "to provide best use of the nation's rivers" (Macy, n.d.). No longer would federally funded development be left to construct projects without taking into consideration the negative externalities on nonhuman nature, the environment, and society. This objective was achieved through formulating wild and scenic river eligibility and suitability studies to be conducted by federal agencies. These legibility studies set out to investigate: (a) the extent of water development projects that impaired flow, quality, and/or values of the river; (b) the degree to which plans existed for new development projects; and (c) future water development needs of watershed communities. In conducting these studies stakeholders within study basins could voice their concerns over what parts of the riverine ecosystem got developed and what parts get protected. To accommodate varied levels of current or future development plans while preserving outstanding terrestrial and aquatic values and the free-flowing nature of the river, a three-tiered classification system of wild, scenic, and recreational river segments was designed. Planned to make both development and other important river values legible, this conservation policy model is grounded in IWRM principles for social, economic, and ecological equity thereby engendering the potential for sustainable river governance.

After 10 years of Congressional negotiations and designing policy components, the WSRA was signed into law on October 2, 1968. Taking into consideration the state's needs to protect water quality, biodiversity, and other ecosystem services while accounting for population growth, and integrative planning, the WSRA includes fundamental elements of the Water Pollution Control Act, the Fish and Wildlife Conservation and Protection Act, and the Water Resources Planning Act as a unique policy designed for flexibility with generalizable parameters and intended for broad utilization. To expand policy application, federal land management agencies were mandated to survey rivers in their jurisdiction to identify segments that met two minimum requirements: possession of free-flowing waters of a high quality or with the potential to restore the water quality, and at least one ORV. ORVs as depicted in Fig. 8.4 include scenery, recreation, history, culture, fish, wildlife, geology, and other similar values to "fulfill other vital national conservation purposes" (WSR, n.d.-a). Included in this project is the establishment of protected riparian zones of up to a quarter-mile wide (0.4 km) on either side of the designated river segment (in Alaska this area reaches to a half-mile (0.8 km)) to capture values that are not in the immediate river channel and banks yet are river dependent and/or connected (i.e., waterfalls, fossils, historic and cultural sites, and amphibious species' habitat) (Diedrich and Thomas, 1999).

Once designated, a Comprehensive River Management Plan (CRMP) must be created within 3 years for the protection and enhancement of ORVs (USFS, 2015). These management plans allow for adaptive management of riverine resources to meet the protection and enhancement mandates for ORVs. Federally funded infrastructure development with the potential to diminish the river's free-flowing nature, water quality, and/or the integrity of ORVs, is subject to the National Environmental Policy Act of 1969 (NEPA) and WSRA Section 7 review. NEPA requires all federal agencies to examine the environmental impacts of their real or proposed actions, generally through an Environmental Impact Assessment (Harm Benson and Garmestani, 2011). Section 7 of the WSRA, one of the most powerful regulatory tools

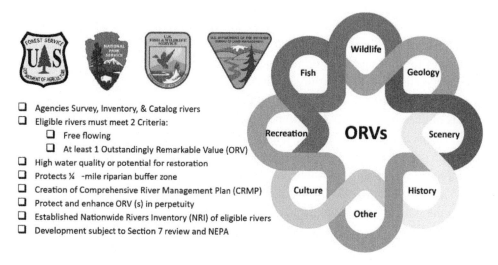

- ☐ Agencies Survey, Inventory, & Catalog rivers
- ☐ Eligible rivers must meet 2 Criteria:
  - ☐ Free flowing
  - ☐ At least 1 Outstandingly Remarkable Value (ORV)
- ☐ High water quality or potential for restoration
- ☐ Protects ¼ -mile riparian buffer zone
- ☐ Creation of Comprehensive River Management Plan (CRMP)
- ☐ Protect and enhance ORV (s) in perpetuity
- ☐ Established Nationwide Rivers Inventory (NRI) of eligible rivers
- ☐ Development subject to Section 7 review and NEPA

Fig. 8.4
Major components of the National Wild and Scenic Rivers Act.

built into the policy, provides a process for river-administering agencies to evaluate and prevent certain federally assisted projects from proceeding as proposed if they fail to meet the standards for river resource protection in the WSRA (Personal Communication, 2017).

Adopting Gretchen Daily's (1997) definition of ecosystem services as the conditions and processes through which natural ecosystems, and the species that make them up, sustain and fulfill human life, ORVs can be viewed as the quintessence of complex ecosystem services deemed important for the production and reproduction of capital, culture, human development, and well-being. Fig. 8.5, adapted from Perry (2017, p. 46), situates ORVs within the four broad categories of ecosystem services (i.e., provisioning, regulating, supporting, and cultural services) revealing their benefits to society. Moreover, the state designates river segments and ORVs through legibility exercises rationalized on notions of scarcity and efficiency to be worthy of protection and enhancement in perpetuity. The crown jewel of this legibility exercise, the Nationwide Rivers Inventory (NRI), catalogs eligible rivers identified for potential permanent protection in the National Wild and Scenic River System (WSR, n.d.-b). In line with Scott's (1998) definition of legibility being a state

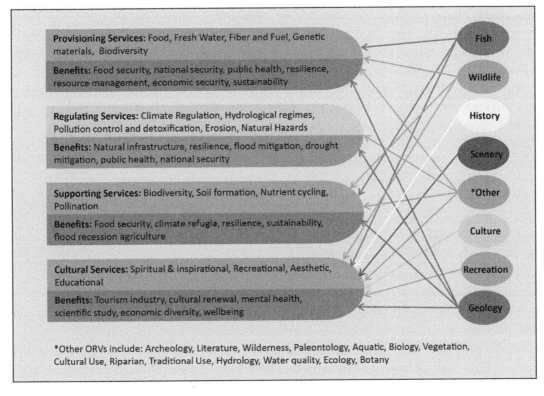

Fig. 8.5
Outstandingly remarkable values in an ecosystem service framework.

exercise of abstracting resources from nature through surveying, cataloging, and management practices, this policy is clearly a state exercise in territorializing regionally and nationally significant river resources in a federal system where water governance is devolved to the states (Amos, 2006; Davis, 2001; Doyle et al., 2013).

As of March 12, 2019, when over 600 miles were designated on the heels of the 50th anniversary, the system includes 214 named designations consisting of 579 rivers, forks, and identified tributaries totaling 14,037 miles of protected rivers (rivers.gov). In addition, there are more than 3200 rivers officially listed on the NRI and countless others deemed potentially eligible by land management agencies since the last official inventory update in 1993 (Perry, 2017). NRI rivers are afforded baseline protections of their free-flow and ORV integrity until suitability is determined. As managers and conservation advocates look to nature-based solutions for climate adaptation, this inventory of eligible rivers provides a springboard for expanding the application of wild and scenic designations to advance freshwater biodiversity conservation, natural infrastructure protection and enhancement, and riverine ecosystem resilience. Considering the small fraction of rivers protected under the WSRA (less than 0.25% of all the rivers in the United States), arguably there is room for the system to expand.

Adopting legibility language toward climate resilience and adaptation policy

Fifty years on, the underlying themes in today's environmental discourse are much the same as those of LBJ's era. Debates on population, water scarcity, and economic expansion, as well as the US role in geopolitics, are still paramount in the public arena. However, instead of expanding policies to protect water quality and biodiversity, under the Trump administration, rollbacks occurred on the Clean Water Act, the Endangered Species Act, and NEPA. Meanwhile, renewed interests in hydropower were ostensibly aimed at generating renewable energy, even though climate change was handily denied by the Trump administration (New York Times, n.d.). Yet, despite a climate of denialism (Kenrick, 2013; McCright and Dunlap, 2011; Nuccitelli, 2017; Wainwright and Mann, 2012), anthropogenic climate change is listed alongside terrorism as a threat to national security, a narrative that surfaced in the early 2000s (Baldwin, 2013; Campbell et al., 2007; CNA, 2007; Schwartz and Randall, 2003) and was most recently reified in the fourth National Climate Assessment (2018). Resilience to climate change "attacks" on the built environment depends on securing ecological infrastructure to ensure state vitality (Baldwin, 2013). Though Baldwin sets ecosystem security apart from conservation or ecosystem protection practices, the notions he borrows from Walker et al. (2004) ring true for rivers: there should be a focus on cultivating socio-ecological systems resilient to turbulence that support life in all forms through ecosystem services. Thus using natural or green infrastructure restoration and protection as proactive climate adaptation strategies can offer large benefit to cost ratios for riverine ecosystems and surrounding communities (USGCRP, 2018).

For rivers, floodplains capture and store water during floods, later releasing it in dry periods, providing a vital water security function (Arthington et al., 2006; Grill et al., 2019). Within the National Wild and Scenic River System, reserve areas, such as the quarter-mile protected buffer zones contiguous to designated rivers, provide a flexible adaptation buffering function in flood-prone areas allowing for water to move, spread out, and infiltrate (Adger et al., 2001; Knieling and Fellmer, 2013). Furthermore, protected free-flowing rivers stand to be the most resistant and resilient to climate change, buffering against temperature and flow variations, unlike clear-cut or urbanized watersheds (Palmer et al., 2009). These buffers in turn provide climate refugia for biodiversity (Isaak et al., 2016; Mawdsley et al., 2009).

Despite undeniable evidence, climate change is impacting socio-ecological systems already and despite the prevalence of possible adaptation strategies, Henstra (2015) time and again found this knowledge was not incorporated into action by specialists. "How the state responds to the climate crisis is a different question: sometimes it fails, but always it is called" (Parenti, 2015, p. 829). Henstra's findings are reflected in the responses to one interview question posed in the study at hand, which simply asked, "Do you feel climate change is making an impact on decision-making for ORV identification or management within the WSR System?" Respondents signaled that while climate change was being discussed at the upper levels of agency management, it had yet to trickle down to field offices and was far from factoring into CRMPs. The question then remains, how can the WSRA be employed as a climate adaptation policy if climate change is still not a driving factor in river management?

Parenti (2015, p. 829) explains that for conservationists to be effective at implementing climate adaptation policy, they must "create strategies that engage and transform the state." Transforming the bureaucratic engines of the federal government, however, does not have to be a top-down maneuver. Instead it can work through the web of social production, employing power in diffused ways through discourse mobilization. Sharp and Richardson (2001) found that reforming institutional structures can produce linguistic changes that in turn influence social change through new discourse (Hastings, 1999). In that vein, Henstra (2015) explains that an important method for bridging the gap between adaptation knowledge and implementation is mobilizing climate knowledge and language. This mobilization can be achieved through relationships and a meaningful dialogue between scientists doing freshwater research and the stakeholders using it. As noted by Harrison et al. (2018), "scientists must translate [their] science into recommendations for action for practitioners and policy-makers." In other words, such mobilizations can occur through social production networks comprising of river advocacy groups and technical experts (Munang et al., 2013; Stern, 2000).

In the Wild and Scenic Rivers System, personnel working on questions of river management within federal agencies and advocacy groups are trained in scientific fields well positioned to conduct research on river processes and apply this knowledge to river management. For example, Scott Bosse, Director of the Northern Rockies region of American Rivers (https://www.americanrivers.org), not only advocates for rivers by working directly with local stakeholders to build support for river protection but also produces resilience knowledge relevant to public policy. Employing his training as an environmental scientist and fisheries biologist, Bosse's (2010) piece, "Conserving Native Trout at the Landscape Scale using the Wild and Scenic Rivers Act," directly informed the one reported case of climate refugia used in an eligibility study for "Other" ORV identification. The eligibility study focused on a 600-mile network of headwaters streams in Bosse's advocacy region, Montana. As a result, a legislative proposal in favor of designating 50 rivers in this region is being considered in the 2021 Congressional session (Scott Bosse, 2020, personal communication).

This example encapsulates an arena ripe for consideration: the expediency of using the "Other" ORV category to render the WSRA and CRMPs suitable for employment as climate adaptation policy. The WSRA's flexible design enables the incorporation of emerging environmental priorities deemed important to society, or as the WSRA states, "to fulfill other vital national conservation purposes" (WSR, n.d.). For example, as climate change impacts the temperature and flow regimes of rivers, it becomes a vital national conservation need to protect natural habitats to which species have adapted that offer physical features, such as cool water pools, well-connected tributaries, and riparian shading, and that help the species survive periodic temperature changes and other disturbances (Glick et al., 2009). It is also fundamentally important to protect the natural infrastructure provided without charge for flood and drought mitigation by free-flowing rivers that are able to move in their floodplains as an alternative to costly gray infrastructure projects. Moreover, incorporating ecosystem service conservation priorities for riverine resilience into future designations through the Other category allows for adaptive management through the attendant Comprehensive River Management Plan mandates for ORV protection and enhancement in perpetuity.

The work of the River Management Society provides another example of a social production network mobilizing climate knowledge. Through biennial symposia, this organization provides technical training and expertise to river managers from federal and state agencies, conservation advocates, and scholars. These symposia facilitate knowledge exchange around timely themes such as Rivers and Recreation in a Changing Climate in 2016 and the 2018 Wild, Scenic, and Beyond! Advocates, equipped with climate knowledge, can employ the language of resilience and adaptation via ecosystem service protection in their interactions with agency personnel, politicians, and constituents. For instance, Thomas O'Keefe of American Whitewater frequents Washington, DC to advocate for WSRA designations and access to clean rivers. Reflecting on these experiences during an interview, he stated "the adjectives 'Wild and Scenic Outstandingly Remarkable Values' just don't get politicians on your side" (Personal communication, 2016).

O'Keefe highlights a general sentiment held by many conservation advocates and river managers: grounding calls for environmental protection at what is perceived to be the expense of economic development provokes a general lack of political will in many Congressional delegates. At the suggestion of framing ORVs as ecosystem services, he exclaimed, "That's the missing link!" Reflecting on his training in limnology and ecological economics, O'Keefe's enthusiasm comes with a recognition that such a framing situates biodiversity and river conservation within the dominant political-economic field by placing value on the ecosystem processes on which humanity depends (e.g., species' habitat, fresh water, food, flood mitigation, cultural values, and recreation) (MEA, 2005). This approach, in turn, provides policymakers a tool to evaluate tradeoffs between development and conservation (Liu et al., 2010).

Daniel Henstra's (2015) work analyzing adaptation policy options included Gifford and Comeau's (2011) findings that messages emphasizing potential benefits to individuals and communities motivated behavior changes geared toward climate action and adaptation. Negative messages centered on consequences, according to Henstra's review of studies, were less effective at inciting adaptation actions. Thus, for the WSRA to serve as climate adaptation policy, supplementing "Outstandingly Remarkable Values" with positive descriptors of ecosystem services (e.g., fresh water, biodiversity) and their benefits (e.g., water security, food security) may help institutionalize climate knowledge and reform linguistic structures that currently prevent the broad application of the policy (Perry, 2017).

As Albrechts (2001, p. 738) notes, "institutionalization is a process by which ideas and practices become durable reference points for social action. Institutionalization requires a certain degree of consensus about underlying values and a commitment to administrative and financial agreements between different levels of government, sectors and private institutions" (Küle et al., 2013, p. 71). The Interagency Wild and Scenic Rivers Coordinating Council has the power to institutionalize such ecosystem service language in eligibility and suitability studies and CRMP guidelines. Just as recreation transformed state natural resource governance practices with the ORRRC eco-governmentality discourse centered on public health benefits, institutionalized discourse on broad ecosystem-based adaptation benefits such as flood control and refugia can be employed in negotiating new river designations today.

More broadly, IWRM is encouraged across the developing world for countries seeking to establish human water security for the first time while preserving biodiversity. Nonetheless, compromise "between social equity, ecological integrity, and economic growth" is seen as a problematic reality in IWRM decision-making as water intensive economic development initiatives threaten to subjugate water needs to land-use decisions (Bakker and Morinville, 2013, p. 4). According to Lovejoy (2019), "Freshwater biodiversity has been particularly neglected because freshwater is widely understood and managed more as a physical resource vital to survival rather than as the special and delicate habitat that it provides for an extraordinary array of organisms." Consequently, while IWRM is seemingly based on parity, today there is a perceived lack of commitment to addressing the socio-ecological impacts of large dam development, canal and irrigation diversions, and other water development projects with negative repercussions on watershed integrity and local populations (Conca, 2006; Richter, 2010). This fact complicates the application of IWRM principles as emerging economies are faced with balancing social, economic, and environmental water needs in the face of climate change (Tickner and Acreman, 2013, p. 137). What is more, panacea policies designed for universal implementation often exclude consideration of place-specific variables found from one watershed, state, or country to the next (Meinzen-Dick, 2007). Thus, in places where IWRM practices are desired for socio-ecological resilience across sectors, the WSRA provides a durable yet flexible conservation policy framework to address these concerns.

This legibility exercise can inform IWRM strategies in the service of policy creation for riverine ecosystem service protection around the globe. To start, a careful inventory of rivers with free-flowing conditions, such as that conducted by Grill et al. (2019), coupled with an investigation of water quality and riverine ecosystem services, could provide a catalog of eligible rivers for conservation decision-making. The three-tiered classification system designed to distinguish between river corridors that have minimal to moderate development can further inform protection priorities in emergent conservation systems.

The Balkan states provide a timely case in point. In this region once riddled by civil war, rampant hydropower development is now proliferating during peace times. Extant and proposed dams touted as green energy infrastructure total over 2800 projects in the region that contains the largest number of remaining free-flowing rivers in Europe. Despite being a key freshwater biodiversity hotspot and one with strong cultural roots to river systems, no comprehensive river conservation policy yet exists (Neslen, 2017a,b, 2018; RiverWatch, 2018; Zarfl et al., 2014). Interest groups across Europe (e.g., Save the Blue Heart of Europe) and the United States (e.g., Wild and Scenic Rivers experts) are working to change that reality. On September 27–29, 2018, the First European Rivers Summit was held in Sarajevo, Bosnia Herzegovina. Here over 250 people joined to discuss strategies for sustainable integration of water management practices in this region and across Europe. Recognizing the limitation of the European Union's Water Framework Directive to explicitly protect free-flowing rivers, attendees drafted a declaration aimed at the European Union asking the regional governing body to conduct a legibility exercise to "establish and publish a pan-European inventory of ecologically outstanding, free-flowing rivers" and for the establishment of "a European-wide system for permanent protection for particularly outstanding and free-flowing rivers, modelled on the 'Wild & Scenic Rivers Act' of the USA" (Gruber, 2018; RiverWatch, 2018). Returning to Parenti (2015), we see here the European states being called upon to consider legibility acts for ecosystem protections—how they respond remains to be seen.

Conclusions

Increasing population demands, widespread water pollution, and habitat degradation leading to species decline, prolonged drought, and the awareness that water resources are finite all collectively produced notions of scarcity in the mid-20th century United States. As Parenti (2015) notes, if the functions of nonhuman nature are key sources of wealth, the state with its territorial imperative delivers those ecosystem services to capital. In this vein, the state set out to protect these commonly held ecosystem services and goods in a compendium of unprecedented water-centered environmental laws, policies, and programs. The Johnson administration of the 1960s set a policy course that focused on studying, cleaning, restoring, protecting, and enhancing water resources for human development, recreation, and biodiversity protection. This included the establishment of research partnerships at land-grant universities, water quality standards on interstate water bodies, and legislation protecting endangered species and their habitat. This policy course culminated in the 1968 passing of the WSRA. The overall design of the National Wild and Scenic Rivers System and policy authority reflects an environmental era under the Johnson administration centered on governing water resources through acts of legibility and rational planning for the purpose of protecting riverine ecosystem services.

Today as some interests lean toward big infrastructure for climate adaptation, others view ecosystem-based adaptation measures as complementary to, or substitutes for, more costly infrastructure investments, thus providing win-win solutions for resilient rivers. Just as the federal government mobilized recreation as an adaptive mental-health solution in an act of eco-governmentality to secure its labor biopower, the state can now mobilize legibility for ecosystem-based adaptation to simultaneously protect reproductive conditions for capital accumulation, freshwater, biodiversity, ecosystem services, and the intrinsic value of nature.

The state responded to the emerging biodiversity and freshwater crisis in the 1960s by mobilizing land management agencies to protect important river ecosystems. Today those agencies can play a pivotal role in climate adaptation by identifying rivers and ORVs with resilience characteristics. Legislators and the Secretary of the Interior can designate more rivers to capitalize on low-cost, high-benefit ecosystem-based adaptation measures that promote resilience and contend with the wide-ranging socio-ecological impacts of changing climate.

Lastly, as when agencies and advocacy groups stepped in to answer IUCN calls for assistance in addressing global issues of biodiversity during the LBJ administration, today these interest groups and government agencies can help other regions of the world shape policies for resilient rivers taking lessons from the WSRA. Such collaborative efforts are already underway between US scholars and conservation groups and European special interest groups. The legibility exercise that is the WSRA, with its flexible design and adaptive management capacity, could serve around the globe at once as an IWRM policy model for sustainable water development, ecosystem service protection, and climate adaptation.

Acknowledgment

The author wishes to thank the land management agency and advocacy organization personnel who took part in the interviews for their time and contributions. I also thank Alec Murphy, Diana Stuart, Ben Ruddell, Kate Berry, Ian Harrison, and two anonymous reviewers for their feedback on an earlier draft.

This work was supported by the Lyndon Baines Johnson Moody Research Grant.

References

Abell R, Harrison IJ. A boost for freshwater conservation integrating freshwater and terrestrial conservation planning has high returns. Science 2020;370(6512):38–9.

Adger WN, Kelly PM, Ninh NH, editors. Living with environmental change. Social vulnerability, adaptation and resilience in Vietnam. London: Routledge; 2001.

Agrawal A. In: Escobar A, Rocheleau D, editors. Environmentality: Technologies of government and the making of subjects. Durham, NC: Duke University Press; 2005.

Albrechts L. How to proceed from image and discourse to action: as applied to the Flemish diamond. Urban Stud 2001;38(4):733–45. https://doi.org/10.1080/00420980120035312.

Amos AL. The use of state instream flow laws for federal lands: respecting state control while meeting federal purposes. Environ Law 2006;35(4):1237–44.

Arthington AH, Bunn SE, Poff NL, Naiman RJ. The challenge of providing environmental flow rules to sustain river ecosystems. Ecol Appl: A Publication of the Ecological Society of America 2006;16(4):1311–8. Retrieved from http://www.ncbi.nlm.nih.gov/pubmed/16937799.

Arthington AH, Naiman RJ, McClain ME, Nilsson C. Preserving the biodiversity and ecological services of rivers: new challenges and research opportunities. Freshw Biol 2010;55(1):1–16. https://doi.org/10.1111/j.1365-2427.2009.02340.x.

ASCE American Society of Civil Engineers. Conditions & capacity. Available at: http://www.infrastructurereportcard.org/wp-content/uploads/2017/01/Dams-Final.pdf; 2017. [Accessed 12 April 2017].

Auerbach DA, Deisenroth DB, McShane RR, McCluney KE, LeRoy Poff N. Beyond the concrete: accounting for ecosystem services from free-flowing rivers. Ecosyst Serv 2014. https://doi.org/10.1016/j.ecoser.2014.07.005. Elsevier.

Bakker K, Morinville C. The governance dimensions of water security: a review. Philos Trans R Soc A 2013;371(2002). https://doi.org/10.1098/rsta.2013.0116.

Baldwin A. Vital ecosystem security: emergence, circulation, and the biopolitical environmental citizen. Geoforum 2013;45:52–61. https://doi.org/10.1016/j.geoforum.2012.01.002.

Benson MH. Intelligent tinkering: the endangered species act and resilience. Ecol Soc 2012;17(4):28. https://doi.org/10.5751/ES-05116-170428.

Bernhardt JR, Leslie HM. Resilience to climate change in coastal marine ecosystems. Ann Rev Mar Sci 2013;5(1):371–92.

Bernhardt ES, Palmer MA. River restoration: the fuzzy logic of repairing reaches to reverse catchment scale degradation. Ecol Appl 2011;21(6):1926–31. https://doi.org/10.1890/10-1574.1.

Biermann C, Mansfield B. Biodiversity, purity, and death: conservation biology as biopolitics. Environ Plan D Soc Space 2014;32:257–73. https://doi.org/10.1068/d13047p.

Biswas AK. Integrated water resources management: a reassessment. Water Int 2009;29(2):248–56. https://doi.org/10.1080/02508060408691775.

Bosse S. Conserving native trout at the landscape scale using the Wild and Scenic Rivers Act; 2010, September. Paper presented at the "Conserving Wild Trout" symposium, West Yellowstone, MT.

Brown D. Feed or feedback. Utrecht: International Books; 2005.

Brown TC, Bergstrom JC, Loomis JB. Ecosystem goods and services: Definition, valuation and provision. Retrieved from https://www.fs.fed.us/rm/value/docs/ecosystem_goods_services.pdf; 2006, May 31.

Burce SB. Wild rivers and the boundaries of cooperative federalism: the wild and scenic rivers act and the Allagash wilderness waterway. Environ Aff Law Rev 2008;35(77):77–110. Retrieved from http://lawdigitalcommons. bc.edu/ealr/vol35/iss1/4.

Bureau of Budget. Letter to Mr. Califano White House aide. White House Aide Files Joseph Califano, Box. 64, Austin, TX: LBJ Presidential Library; 1966.

Campbell KM, et al. The age of consequences: The foreign policy and national security implications of global climate change. Washington, DC: Center for Strategic and International Studies and Center for a New American Security; 2007.

Carson R. Silent spring. Boston, MA: Houghton Mifflin; 1962.

Cater D. Memorandum on the International Union for the Conservation of Nature and Natural Resources with attachments. White House Aide Files Douglas Cater—Conservation, Box. 96, Austin, TX: LBJ Presidential Library; 1965.

Churchill DJ, Larson AJ, Dahlgreen MC, Franklin JF, Hessburg PF, Lutz JA. Restoring forest resilience: from reference spatial patterns to silvicultural prescriptions and monitoring. For Ecol Manage 2013;291:442–57.

CNA. National security and the threat of climate change. Alexandria: CNA Corporation; 2007.

Collard R-C, Dempsey J. Capitalist natures in five orientations. Capital Nat Social 2017;28(1):78–97. https://doi.or g/10.1080/10455752.2016.1202294.

Conca K. Governing water: Contentious transnational politics and global institution building. Cambridge, MA: MIT Press; 2006.

Cosens B, Gunderson L, Allen C, Benson MH. Identifying legal, ecological and governance obstacles, and opportunities for adapting to climate change. Sustainability (Switzerland) 2014;6(4):2338–56. https://doi. org/10.3390/su6042338.

Costanza R, de Groot R, Braat L, Kubiszewski I, Fioramonti L, Sutton P, et al. Twenty years of ecosystem services: how far have we come and how far do we still need to go? Ecosyst Serv 2017. https://doi. org/10.1016/j.ecoser.2017.09.008.

Daily GC, editor. Nature's services: Societal dependence on natural ecosystems. Washington, DC: Island Press; 1997.

Daniels TL. A trail across time: American environmental planning from city beautiful to sustainability. J Am Plann Assoc 2009;75(2):178–92. https://doi.org/10.1080/01944360902748206.

Darwall W, Bremerich V, De Wever A, Dell AI, Freyhof J, Gessner MO, et al. The Alliance for Freshwater Life: a global call to unite efforts for freshwater biodiversity science and conservation. Aquat Conserv Mar Freshwat Ecosyst 2018;(June):1–8. https://doi.org/10.1002/aqc.2958.

Davis K S. The politics of water scarcity in the western states. The Social Science Journal 2001;38(4):527–42. https://doi.org/10.1016/S0362-3319(01)00148-3.

Dempsey J. Biodiversity loss as material risk: tracking the changing meanings and materialities of biodiversity conservation. Geoforum 2013;45:41–51. https://doi.org/10.1016/j.geoforum.2012.04.002.

Diedrich J, Thomas C. The wild & scenic river study process. Retrieved from https://www.rivers.gov/documents/ study-process.pdf; 1999.

Doyle MW, Lave R, Robertson MM, Ferguson J. River federalism. Ann Assoc Am Geogr 2013;103(2):290–8. https://doi.org/10.1080/00045608.2013.754686.

Dryzek JS, Hunold C, Schlosberg D, et al. Environmental transformation of the state: the USA, Norway, Germany and the UK. Polit Stud 2002;50(4):659–82. https://doi.org/10.1111/1467-9248.00001. SAGE Publications SAGE UK: London, England.

Edward C. Bureau of Outdoor Recreation—Declaration of Intent. Office Files Bill Moyers—Recreation Advisory Council, Box 75, Austin, TX: LBJ Presidential Library; 1962.

Evenden M. Mobilizing rivers: hydro-electricity, the state, and World War II in Canada. Ann Assoc Am Geogr 2009;99(5):845–55. https://doi.org/10.1080/00045600903245847.

Farnham TJ. Saving nature's legacy: Origins of the idea of biological diversity. London: Yale University Press; 2007.

Finlayson MC, D'Cruz R, Davidson N, et al. Ecosystems and human well-being: Wetlands and water synthesis. Washington, DC: Millennium Ecosystem Assessment; 2005. Available at: http://www.millenniumassessment. org/documents/document.358.aspx.pdf. [Accessed 29 March 2017].

Finley B. Colorado Springs ready to turn on $825 million river water siphon. Denver Post 2016, April.

Foucault M. In: Bertani M, Fontana A, Ewald F, editors. "Society must be defended": Lectures at the Collège de France, 1975–1976. New York, NY: Picador; 2003.

Gates SF, Getches DH, MacDonnell LJ, Wilkinson CF. Searching out the headwaters: Change and rediscovery in western water policy. Washington, DC: Island Press; 1993.

Gerisch M, Dziock F, Schanowski A, Ilg C, Henle K. Community resilience following extreme disturbances: the response of ground beetles to a severe summer flood in a Central European lowland stream. River Res Appl 2012;28(1):81–92.

Gifford R, Comeau L. Message framing influences perceived climate change competence, engagement, and behavioral intentions. Glob Environ Chang 2011;21(4):1301–7. https://doi.org/10.1016/j.gloenvcha.2011.06.004.

Gilman RT, Abell RA, Williams CE. How can conservation biology inform the practice of integrated river basin management? Int J River Basin Manag 2004;2(2):135–48. https://doi.org/10.1080/15715124.2004.9635228.

Glick P, Staudt A, Stein B. A new era for conservation : Review of climate change adaptation literature. National Wildlife Federation; 2009. Retrieved from https://www.nwf.org/pdf/Reports/NWFClimateChangeAdaptationLiteratureReview.pdf.

Graf WL. Dam nation: a geographic census of American dams and their large-scale hydrologic impacts. Water Resour Res 1999;35(4):1305–11. https://doi.org/10.1029/1999WR900016.

Grantham TE, Matthews JH, Bledsoe BP. Shifting currents: managing freshwater systems for ecological resilience in a changing climate. Water Secur 2019;8:100049. https://doi.org/10.1016/j.wasec.2019.100049.

Grill G, Lehner B, Thieme M, Geenen B, Tickner D, Antonelli F, et al. Mapping the world's free-flowing rivers. Nature 2019;569(7755):215–21. https://doi.org/10.1038/s41586-019-1111-9.

Gruber V. First European river summit to save Europe's last wild rivers. Available at: https://wilderness-society.org/first-european-river-summit-to-save-europes-last-wild-rivers/; 2018. [Accessed 25 November 2018].

Hagen G. Postcards about Salmon River in Idaho. White House Social Files, Alpha Box 131—Beautification—Howard, Austin, TX: LBJ Presidential Library; 1967.

Harden CP, Chin A, English MR, et al. Understanding human-landscape interactions in the 'anthropocene'. Environ Manag 2014;53(1):4–13. https://doi.org/10.1007/s00267-013-0082-0.

Harm Benson M, Garmestani AS. Embracing panarchy, building resilience and integrating adaptive management through a rebirth of the National Environmental Policy Act. J Environ Manag 2011;92(5):1420–7. https://doi.org/10.1016/j.jenvman.2010.10.011. Elsevier Ltd.

Harrison I, Abell R, Darwall W, Thieme ML, Tickner D, Timboe I. The freshwater biodiversity crisis. Science 2018;362:1369.

Harvey D. Spaces of capital: Towards a critical geography. New York, NY: Routledge; 2001.

Hastings A. Analysing power relations in partnerships: is there a role for discourse analysis? Urban Stud 1999;36(1):91–106. https://doi.org/10.1080/0042098993754.

Henstra D. The tools of climate adaptation policy: analysing instruments and instrument selection. Clim Pol 2015;1–26. https://doi.org/10.1080/14693062.2015.1015946.

Hering JG, Ingold KM. Water resources management: what should be integrated? Science 2012;336(6086):1234–5. https://doi.org/10.1126/science.1218230.

Hermoso V, Abell R, Linke S, Boon P. The role of protected areas for freshwater biodiversity conservation: challenges and opportunities in a rapidly changing world. Aquat Conserv Mar Freshwat Ecosyst 2016;26(April):3–11. https://doi.org/10.1002/aqc.2681.

Hobbs F, Stoops N. Demographic trends in the 20th century. Available at: https://www.census.gov/prod/2002pubs/censr-4.pdf; 2002. [Accessed 22 April 2017].

IPCC. Climate change and water: IPCC technical paper VI. In: Climate change and water, vol. 403; 2008. https://doi.org/10.1016/j.jmb.2010.08.039.

Isaak DJ, Young MK, Luce CH, et al. Slow climate velocities of mountain streams portend their role as refugia for cold-water biodiversity. Proc Natl Acad Sci U S A 2016;113(16):4374–9. https://doi.org/10.1073/pnas.1522429113. National Academy of Sciences.

James MC, editor. 18th annual report of the Pacific Marine Fisheries Commission for the year 1965. Austin, TX: LBJ Presidential Library; 1966. White House Central Files, Subject File FG 711. Pacific Marine Fisheries Commission.

Mrs. Johnson LB. Response letter to Mr. Rickie Livingston of Beaverton, Oregon. White House Social Files, Alpha Box 144—Beautification—Lykes, Austin, TX: LBJ Presidential Library; 1965.

President Johnson LB. Speech on report of River Basin Commissions. White House Aide Files, James Gaither—First Annual Report of Four River Basin Commissions, Box 239, Austin, TX: LBJ Presidential Library; 1969.

Keck M, Sakdapolrak P. What is social resilience? Lessons learned and ways forward. Erdkunde 2013. https://doi.org/10.2307/23595352.

Kendrick J. Foucault, Biopower & IR: A methodological discussion. Available at: https://ecpr.eu/Filestore/PaperProposal/03e082fa-7372-4728-9a24-2b4c1a9e2ed0.pdf; 2012.

Kenrick J. Emerging from the shadow of climate change denial. ACME: An International E-Journal for Critical Geographies 2013;12(1):102–30. Available at: http://www.acme-journal.org/vol12/Kenrick2013.pdf. [Accessed 3 June 2013].

Kirsch S. John Wesley Powell and the mapping of the Colorado plateau, 1869-1879: survey science, geographical solutions, and the economy of environmental values. Ann Assoc Am Geogr 2002;92(3):548–72. Available at: https://www.jstor.org/stable/pdf/1515476.pdf. [Accessed 22 April 2017].

Knieling J, Fellmer M. Climate adaptation in metropolis Hamburg: paradigm shift in urban planning and water management towards 'living with water'? In: Climate change adaptation in practice: From strategy development to implementation; 2013. p. 83–93. https://doi.org/10.1002/9781118548165.ch7.

Küle L, Briede A, Kļaviņš M, Eberhards G, Locmanis A. Adaptation to floods in Riga, Latvia: historical experience and change of approaches. In: Schmidt-Thomé P, Klein J, editors. Climate change adaptation in practice: from strategy development to implementation. Oxford: John Wiley and Sons Ltd.; 2013. p. 51–64. 65–81 https://doi.org/10.1002/9781118548165.ch6.

Lake PS. Resistance, resilience and restoration. Ecol Manag Restor 2013. https://doi.org/10.1111/emr.12016.

Lawrence MP. Damming rivers, damning cultures. Am Indian Law Rev 2011;30(1):247–89.

Leal CG, Lennox GD, Ferraz SFB, Ferreira J. Integrated terrestrial-freshwater planning doubles conservation of tropical aquatic species. Science 2020;121(October):117–21.

Linton J. Modern water and its discontents: a history of hydrosocial renewal. Wiley Interdiscip Rev Water 2014;1(1):111–20. https://doi.org/10.1002/wat2.1009.

Liu S, Costanza R, Farber S, Troy A. Valuing ecosystem services: theory, practice, and the need for a transdisciplinary synthesis. Ann N Y Acad Sci 2010;1185:54–78. https://doi.org/10.1111/j.1749-6632.2009.05167.x.

Lovejoy TE. Eden no more. Sci Adv 2019;5(5):eaax7492. https://doi.org/10.1126/sciadv.aax7492.

Macy J. Scenic rivers and river basin planning miscellaneous document. Office Files John Macy—River Basin Commissions, Box 816. Austin, TX: LBJ Presidential Library; n.d.

Malhi Y, Roberts JT, Betts RA, Killeen TJ, Li W, Nobre CA. Climate change, deforestation, and the fate of the Amazon. Science 2008;319(5860):169–72.

Mawdsley JR, O'Malley R, Ojima DS. A review of climate-change adaptation strategies for wildlife management and biodiversity conservation. Conserv Biol: The Journal of the Society for Conservation Biology 2009;23(5):1080–9. https://doi.org/10.1111/j.1523-1739.2009.01264.x.

McCool D. Command of the waters: Iron triangles, federal water development, and Indian water. Tucson: University of Arizona Press; 1987.

McCright AM, Dunlap RE. Cool dudes: the denial of climate change among conservative white males in the United States. Glob Environ Chang 2011;21(4):1163–72. https://doi.org/10.1016/j.gloenvcha.2011.06.003.

MEA. Summary for decision makers. In: Ecosystems and human well-being: Synthesis. Washington, DC: Island Press; 2005. p. 1–24.

Meehan K. Water rights and wrongs: illegality and informal use in Mexico and the U.S. In: Sultana F, Loftus A, editors. The right to water: Politics, governance and social struggles. New York: Earthscan/Routledge; 2012. p. 159–73.

Meinzen-Dick R. Beyond panaceas in water institutions. Proc Natl Acad Sci U S A 2007;104(39):15200–5. https://doi.org/10.1073/pnas.0702296104.

Munang R, Thiaw I, Alverson K, et al. Climate change and ecosystem-based adaptation: a new pragmatic approach to buffering climate change impacts this review comes from a themed issue on terrestrial systems. Curr Opin Environ Sustain 2013;5:67–71. https://doi.org/10.1016/j.cosust.2012.12.001.

Naiman RJ, Bilby RE, Bisson PA. Riparian ecology and management in the Pacific coastal rain forest. AIBS Bull 2000;50(11):996–1011.

Neslen A. Balkan hydropower projects soar by 300% putting wildlife at risk, research shows. Available at: https://www.theguardian.com/environment/2017/nov/27/balkan-hydropower-projects-soar-by-300-putting-wildlife-at-risk-research-shows; 2017a. [Accessed 30 November 2017].

Neslen A. Green groups condemn UN plan to use $136m from climate fund for large dams. The Guardian; 2017b. 4 April. Available at: https://www.theguardian.com/environment/2017/apr/04/green-groups-condemn-un-plan-to-use-136m-from-climate-fund-for-large-dams?CMP=twt_a-environment_b-gdneco. [Accessed 12 April 2017].

Neslen A. 'Horror list' of dams threaten some of the world's last wild rivers: The clean energy that may not be so clean. Huff Post; 2018. Available at: https://www.huffingtonpost.com/entry/horror-list-dam-projects-wild-rivers-balkans_us_5b9fbf2ae4b04d32ebfad081?guccounter=1. [Accessed 9 October 2018].

New York Times. Environmentalists and dam operators, at war for years, start making peace; October 14, 2020. Available from: https://www.nytimes.com/2020/10/13/climate/environmentalists-hydropower-dams.html [Retrieved 15 April 2021].

Norden N, Chazdon RL, Chao A, Jiang YH, Vílchez-Alvarado B. Resilience of tropical rain forests: tree community reassembly in secondary forests. Ecol Lett 2009;12(5):385–94.

Nuccitelli D. Trump has launched a blitzkrieg in the wars on science and Earth's climate. The Guardian; 2017, March 28. Retrieved from https://www.theguardian.com/environment/climate-consensus-97-per-cent/2017/mar/28/trump-has-launched-a-blitzkrieg-in-the-wars-on-science-and-earths-climate.

Olson BA. Paper trails: the Outdoor Recreation Resource Review Commission and the rationalization of recreational resources. Geoforum 2010;41(3):447–56. https://doi.org/10.1016/j.geoforum.2009.11.014.

Palmer T. Wild and scenic rivers of America. Washington, DC: Island Press; 1993.

Palmer T. Wild and scenic rivers: An American legacy. Corvallis, OR: Oregon State University Press; 2017.

Palmer MA, Reidy Liermann C, Nilsoon C, Florke M, Alcamo J, Lakw P, et al. Climate change and the world's river basins: anticipating management options. Front Ecol Environ 2008;6:1–18. https://doi.org/10.1890/060148.

Palmer MA, Lettenmaier DP, Poff NL, Postel SL, Richter B, Warner R. Climate change and river ecosystems: protection and adaptation options. Environ Manag 2009;44(6):1053–68. https://doi.org/10.1007/s00267-009-9329-1.

Palmer MA, Filosa S, Fanelli RM. From ecosystems to ecosystem services: stream restoration as ecological engineering. Ecol Eng 2014;65:62–70. https://doi.org/10.1016/j.ecoleng.2013.07.059.

Parenti C. The 2013 ANTIPODE AAG lecture the environment making state: territory, nature, and value. Antipode 2015;47(4):829–48. https://doi.org/10.1111/anti.12134.

Parry M, Canziani O, Palutikof J, Linden P, Hanson C, editors. Climate change 2007: Impacts, adaptation and vulnerability. New York, NY: Cambridge University Press; 2007.

Perramond EP. Water governance in New Mexico: adjudication, law, and geography. Geoforum 2013;45:83–93. https://doi.org/10.1016/j.geoforum.2012.10.004.

Perry D. Reframing the wild and scenic Rivers act: ecosystem-based resilience and adaptation. Int J Wilderness 2017;23(2):41–8.

Perry DM, Praskievicz SJ. A new era of big infrastructure ? (re) developing water storage in the U.S. West in the context of climate change and environmental regulation. Water Altern 2017;10(2):437–54.

Platt L, Bunten BD, Hearey O, et al. Urbanization in the United States, 1800-2000. 19041. Cambridge. Available at: http://www.nber.org/papers/w19041; 2013. [accessed 22 April 2017].

Poff NL. Beyond the natural flow regime? Broadening the hydro-ecological foundation to meet environmental flows challenges in a non-stationary world. Freshw Biol 2018;63(8):1011–21. https://doi.org/10.1111/fwb.13038.

Poff L, Brown CM, Grantham TE, Matthews JH, Palmer MA, Spence CM, et al. Sustainable water management under future uncertainty with eco-engineering decision scaling. Nat Clim Chang 2015. https://doi.org/10.1038/NCLIMATE2765. Nature Publishing Group.

Polanyi K. The great transformation. New York, NY: Farrar and Rinehart; 1944.

PSMFC. Pacific States Marine Fisheries Commission: Overview. Retrieved April 26, 2017 from http://www.psmfc.org/psmfc-info/overview; 2012.

Richter B. Re-thinking environmental flows: from allocations and reserves to sustainability boundaries. River Res Appl 2010;26:1052–63. https://doi.org/10.1002/rra.

Rittenhouse RC, editor. Letter to President Johnson and clippings from marine products magazine. Austin, TX: LBJ Presidential Library; 1965. White House Aide Files Douglas Cater—Conservation, Box. 96.

RiverWatch. River summit. Available at: https://riverwatch.eu/en/content/river-summit; 2018. [Accessed 26 November 2018].

Robbins P. Political ecology: A critical introduction; 2012. https://doi.org/10.1017/CBO9781107415324.004.

Rose N, O'Malley P, Valverde M. Governmentality. Annu Rev Law Soc Sci 2006;2(1):83–104. https://doi.org/10.1146/annurev.lawsocsci.2.081805.105900.

Schwartz P, Randall D. An abrupt climate change scenario and its implications for United States National Security; 2003.

Scott P. The launching of a new ark: First report of the President and Trustees of the World Wildlife Fund; 1965. an international foundation for saving the world's wildlife and wild places; 1962-1965 (Collins).

Scott J. Seeing like a state: How certain schemes to improve the human condition have failed. New Haven, CT: Yale Press; 1998.

Sharp L, Richardson T. Reflections on foucauldian discourse analysis in planning and environmental policy research. J Environ Policy Plan 2001;3(3):193–209. https://doi.org/10.1002/jepp.88.

Stachowicz JJ, Graham M, Bracken ME, Szoboszlai AI. Diversity enhances cover and stability of seaweed assemblages: the role of heterogeneity and time. Ecology 2008;89(11):3008–19.

Stern PC. New environmental theories: toward a coherent theory of environmentally significant behavior. J Soc Issues 2000;56(3):407–24. https://doi.org/10.1111/0022-4537.00175.

Strayer DL, Dudgeon D. Freshwater biodiversity conservation: recent progress and future challenges. J N Am Benthol Soc 2010;29(1):344–58. https://doi.org/10.1899/08-171.1.

Sundstrom SM, Allen CR, Barichievy C. Species, functional groups, and thresholds in ecological resilience. Conserv Biol 2012;26(2):305–14.

Thompson ID. An overview of the science–policy interface among climate change, biodiversity, and terrestrial land use for production landscapes. J For Res 2015;20(5):423–9.

Thompson PL, Rayfield B, Gonzalez A. Loss of habitat and connectivity erodes species diversity, ecosystem functioning, and stability in metacommunity networks. Ecography 2017;40(1):98–108.

Thrush SF, Hewitt JE, Lohrer AM, Chiaroni LD. When small changes matter: the role of cross-scale interactions between habitat and ecological connectivity in recovery. Ecol Appl 2013;23(1):226–38.

Tickner D, Acreman M. Water security for ecosystems, ecosystems for water security. In: Lankford B, Bakker K, Zeitoun M, Conway D, editors. Water security: Principles, perspectives, and practices. New York, NY: Routledge; 2013. p. 130–47.

Tickner D, Opperman JJ, Abell R, Acreman M, Arthington AH, Bunn SE, et al. Bending the curve of global freshwater biodiversity loss: an emergency recovery plan forum. Bioscience 2020;70(4):330–42. https://doi.org/10.1093/biosci/biaa002.

Tuusa R, Kankaanpää S, Viinanen J, et al. Preparing for climate change: planning adaptation to climate change in the Helsinki metropolitan area, Finland. In: Climate change adaptation in practice: From strategy development to implementation; 2013. p. 51–64. https://doi.org/10.1002/9781118548165.ch5.

UNGCF. GCF approves eight projects at its first board meeting in 2017—News—Green Climate Fund. Retrieved April 13, 2017, from http://www.greenclimate.fund/-/gcf-approves-eight-projects-at-its-first-board-meeting-in-20-1?inheritRedirect=true&redirect=%2Fhome; 2017.

Urwin K, Jordan A. Does public policy support or undermine climate change adaptation? Exploring policy interplay across different scales of governance. Glob Environ Chang 2008;18(1):180–91. https://doi.org/10.1016/j.gloenvcha.2007.08.002.

USFS. Wild and scenic rivers. In: Fsh 1909.12—Land management planning handbook; 2015. [chapter 80]. Retrieved from https://www.fs.fed.us/im/directives/fsh/1909.12/wo_1909.12_80.docx.

USGCRP. Fourth national climate assessment. Available at: https://nca2018.globalchange.gov/; 2018. [Accessed 25 November 2018].

Van Looy K, Tonkin JD, Floury M, Leigh C, Soininen J, Larsen S, …Wolter, C. The three Rs of river ecosystem resilience: resources, recruitment, and refugia. River Res Appl 2019;35(2):107–20. https://doi.org/10.1002/rra.3396.

Vogel E. Parcelling out the watershed: the recurring consequences of organising Columbia river management within a basin-based territory. Water Altern 2012;5(1):161–90. Available at: http://www.water-alternatives.org/index.php/alldoc/articles/vol5/v5issue1/163-a5-1-10/file. [Accessed 23 April 2017].

Vörösmarty CJ, McIntyre PB, Gessner MO, Dudgeon D, Prusevich A, Green P, et al. Global threats to human water security and river biodiversity. Nature 2010;467(7315):555–61. https://doi.org/10.1038/nature09549.

Vörösmarty CJ, Osuna VR, Cak AD, Bhaduri A, Bunn SE, Corsi F, et al. Ecosystem-based water security and the sustainable development goals. Ecohydrol Hydrobiol 2018;18:317–33. https://doi.org/10.1016/j.ecohyd.2018.07.004.

Wainwright J, Mann G. Climate leviathan. Antipode 2012;45(1):1–22.

Walker B, et al. Resilience, adaptability and transformability in social-ecological systems. Ecol Soc 2004;9(2):5.

WCD. Dams and development: A new framework for decision-making: Overview of the report by the World Commission on Dams; 2001. Issue paper (Drylands Programme); no. 108 (December): 17 p.

Weiser M. Bear River: The biggest dam project you've never heard of. News Deeply; 2016, August 29. Retrieved from https://www.newsdeeply.com/water/articles/2016/08/29/bear-river-the-biggest-dam-project-youve-never-heard-of.

Wilkinson CF. Crossing the next Meridian: Land, water, and the future of the American west. Washington, DC: Island Press; 1992.

Wilson E. Adapting to climate change at the local level: the spatial planning response. Local Environ 2006;11(6):609–25. https://doi.org/10.1080/13549830600853635.

Wohl E. Compromised rivers: understanding historical human impacts on rivers in the context of restoration. Ecol Soc 2005;10(2). Retrieved from https://www.ecologyandsociety.org/vol10/iss2/art2/.

Worster D. Rivers of empire. New York: Oxford University Press; 1985.

WSR. About the WSR Act; n.d. Available from https://www.rivers.gov/wsr-act.php.WSR [Retrieved 13 April 2017].

WSR. About the WSR Act. n.d.-a. Retrieved April 13, 2017 from https://www.rivers.gov/wsr-act.php.

WSR. Nationwide Rivers Inventory. n.d.-b. Retrieved May 6, 2017 from https://www.nps.gov/ncrc/programs/rtca/nri/index.html.

WWAP (World Water Assessment Programme). The United Nations World Water Development Report 3: water in a changing world. Paris, London: UNESCO and Earthscan; 2009.

WWAP (United Nations World Water Assessment Programme)/UN-Water. The United Nations world water development report 2018: Nature-based solutions for water. Paris: UNESCO; 2018.

Zarfl C, Lumsdon AE, Berlekamp J, Tydecks L, Tockner K. A global boom in hydropower dam construction. Aquat Sci 2014;77(1):161–70. https://doi.org/10.1007/s00027-014-0377-0.

Nature-based solutions in action

CHAPTER 9

Nature-based solutions for source water protection in North America

Jan Cassin

Forest Trends, Washington, DC, United States

Key messages

- Water utilities and municipalities increasingly look to nature-based solutions as core strategies for source water protection.
- Source water protection concepts and practice are evolving, moving from an emphasis on managing sources of contamination to watershed protection to secure the valued hydrological services on which water utilities and their customers depend.
- Nature-based solutions become more central to source water protection as utilities and municipalities recognize that protecting source water quality is one component of a broader concern with water security and resilience.
- As the linkages between watershed health, water security, and source water protection become more evident, utilities and municipalities are engaging in more diverse regional collaborations to implement nature-based solutions and moving natural assets and nature-based solutions into core activities and planning processes, such as infrastructure planning and asset management.
- Source water protection not only provides the first barrier in a multibarrier approach to protecting water supplies. Nature-based source water protection also helps mitigate hazards that can reduce the functionality and viability of subsequent barriers, such as filtration and disinfection, and hazards (e.g., from flooding or wildfires) to gray water infrastructure, such as storage reservoirs and distribution systems.

Introduction

In much of North America, easy access to safe drinking water has been taken for granted for a long time, despite inequities affecting rural, Indigenous, Black, and Hispanic communities. However, recent years have seen a growing awareness that these supplies are not as secure as once thought, as land degradation and climate change increasingly threaten the health of source watersheds. In 2019, towns and cities along the Missouri River in the United States had their drinking water supplies disrupted by devastating floods that impacted water quality and damaged water treatment plants, with some residents waiting up to 6 months before water

Nature-Based Solutions and Water Security. https://doi.org/10.1016/B978-0-12-819871-1.00005-1

services could be restored (Associated Press, 2019). When wildfires destroy forest lands, as in the massive 2016 fires in Alberta's Fort McMurray and 2020 fires in the Western United States, they cost utilities and ratepayers hundreds of millions to repair damaged infrastructure, dredge reservoirs, and purchase water from alternative sources. Between 2000 and 2016, just four of many wildfires during this time in Colorado, Montana, and California resulted in over $120 million in costs borne by water utilities (Headwaters Economics, 2018). This does not include the costs of lives lost or to federal and state agencies and local communities for fire suppression and damage to property, transportation, and energy infrastructure.

Surface waters in lakes and rivers provide North American communities with between 75% and 90% of with their drinking water (Statistics Canada, 2016; Wickham et al., 2011), and forests, wetlands, and grasslands in source watersheds provide valuable ecosystem services that protect water quality and regulate supplies (Brauman et al., 2007). These ecosystems help control erosion and sedimentation, filter and retain nutrients and other pollutants, and regulate flows to surface and groundwater (Emelko et al., 2011). The loss of natural land cover through conversion to urban or agricultural land uses impairs water quality, threatening water supplies and increasing costs for water utilities (Alcott et al., 2013; Ernst, 2004).

This chapter reviews how drinking water utilities and municipalities in the region are embracing nature-based solutions (NBS) as a core strategy for source water protection, driven increasingly by emerging threats and risks posed by watershed degradation and climate change. The chapter first provides an overview of NBS and related concepts (natural assets, natural infrastructure, green infrastructure, ecosystem-based management) and how NBS benefit the drinking water sector. Then how source water protection practice is evolving in North America in response to concerns about water security and climate change is discussed. Finally, cases are reviewed that illustrate two of the recent trends in implementing NBS for source water protection in North America: (1) the development of ever more complex multistakeholder, multisectoral collaborations; and (2) practices and policy frameworks for mainstreaming NBS into water infrastructure development.

Source water protection: An evolving practice

Drinking water utilities in both the United States and Canada use a multibarrier approach (MBA) to protecting potable water supplies. The first barrier is the protection of water sources to maintain raw water quality, followed by filtration, treatment and disinfection, distribution, monitoring, and emergency response plans if other barriers fail (Murphy and Carpenter, 2020). Traditional source water protection has focused until recently primarily on assessing potential sources of contamination and risks to water quality (e.g., sources of chemical, pathogen, nutrient pollution) and developing plans to address or mitigate these risks. This is despite the fact that municipalities and drinking water utilities in North America have long recognized the importance of the natural infrastructure of source watersheds in maintaining supplies of clean water for people.

Starting in the 1860s cities in the Western United States, such as San Francisco, Portland, and Seattle began buying land in their cities' watersheds to prevent development, logging, and mining activities that would degrade water quality. In 1905 the government of British Columbia protected forest lands for drinking water in Vancouver's Capilano watershed under a provision of the Land Act that defines how Crown Lands can be set aside for particular uses (Green et al., 2016). Even earlier, in the 1850s, Salt Lake City, recognizing the importance of the Wasatch Mountains watershed to the city's water supply, began instituting a series of local ordinances that gave the city regulatory control over activities in the watershed to protect water sources from pollution (Blanchard et al., 2015). Salt Lake City and the federal government subsequently partnered to incorporate a portion of watershed lands into the National Forest system as the Salt Lake Forest Reserve. While many cities incidentally rely on watersheds on US Forest Service (USFS) land, the primary goal of these USFS forests is the protection of Salt Lake City's water sources.

For many cities, existing development and land ownership precludes setting aside large areas of source watersheds so that more complex collaborations between downstream water users and upstream land managers are required, as in the pioneering case to protect New York City's watershed (Hanlon, 2017; Pires, 2004; Vintinner, 2009). Emerging threats from climate change, such as uncertainty about future water supplies and increased frequency and severity of wildfires mean that passive protection of watersheds alone, even when possible, is no longer sufficient. As water managers assess the state of natural infrastructure and critical hydrological services, the risks of losing this infrastructure, and the actions needed to protect or restore this infrastructure, this tends to expand the traditional set of threats and risks evaluated in source water protection. This in turn brings into consideration how climate risks and loss of ecosystem functions, such as increased wildfire risks or loss of floodplain wetlands, pose threats to source waters.

As a result, over the past several decades, source water protection has evolved in new directions as the importance to water security of sustainable management of ecosystems through NBS is increasingly recognized (Abell et al., 2017; Matthews et al., 2019). This is beginning to turn traditional approaches around—the health of the natural infrastructure that is providing critical hydrological services becomes the first barrier not only for protecting raw water quality but also for the present and future security of a utility's or municipality's entire water system—from source to distribution.

The role of nature-based solutions in source water protection

"Nature-based solutions" is a relatively new term that encompasses many existing approaches and captures a number of related concepts under a convenient unifying umbrella framework (Cohen-Sacham et al., 2016; Eggermont et al., 2015). As defined by the United Nations, NBS for water are "inspired and supported by nature and use, or mimic, natural processes to contribute to the improved management of water. An NBS can involve conserving or rehabilitating natural ecosystems and/or the enhancement or creation of natural processes in modified or artificial ecosystems. They can be applied at micro- (e.g., riparian buffers) or macro- (e.g., watershed restoration) scales" (WWAP (United Nations World Water Assessment Program)/UN-Water, 2018). While the concept of NBS is not new and is closely related to concepts such as natural assets, ecosystem-based adaptation, green infrastructure, and natural infrastructure, NBS has become a useful term to encompass a wide array of conceptual frameworks for practice that share the following key elements:

- They include ecosystems themselves (natural assets, natural features, natural infrastructure) and the practices (NBS) that are based on or use natural features, processes, and functions to meet human needs.
- They recognize a continuum of practices and assets based on the degree of management or intervention in natural processes, including natural ecosystems not actively or minimally managed, enhanced or actively managed assets, and hybrid natural-engineered assets (e.g., urban green infrastructure such as green roofs).
- They aim to maximize the ability of ecosystems to provide the ecosystem services that help address societal challenges.
- While there is a continuum from minimal to intensive intervention or management, NBS are distinguished by the reliance on ecosystem structures, processes, and functions to deliver outcomes. For example, using native vegetation and soils to enhance infiltration and groundwater recharge to store water for later use would be considered an NBS; while using a storage reservoir to store water for later use or using injection wells to artificially recharge groundwater do not use ecological functions or processes and would not be considered an NBS.
- NBS typically are designed for a specific priority benefit (e.g., improving water quality) but are by nature multifunctional and can simultaneously address multiple societal goals (e.g., biodiversity conservation, climate mitigation, food security, health).
- Emerging best practice requires that the design and implementation of NBS need to consider the target benefit, mitigate trade-offs, and incorporate principles of equity, inclusion, sustainability, and resilience.

North American natural assets important for source water protection

Terrestrial and freshwater ecosystems of North America are critical watershed assets for municipalities and water utilities, contributing a variety of hydrological ecosystem services of direct relevance to water sector service providers (Table 9.1). Actual delivery of these services and the magnitude of benefits depends on context and can be variable depending on factors such as location in the watershed, size of the natural asset, location of the asset within the watershed, vegetation and soil type, underlying geology, and climate regime (see also Vigerstol et al., 2021).

Table 9.1: Major North American natural assets and their hydrological ecosystem services.

Hydrological ecosystem service	Natural assets and hydrological ecosystem services benefiting source water protection					
	Mountain snowpack/glaciers	Forest (esp. source watershed and riparian zone forests)	Wetlands and floodplains	Grasslands	Aquifers	Lakes, rivers, and streams
Regulate Water Quantity; Reliable Supplies						
Maintain dry season flows/dry season supply	By slowly releasing water stored in snowpack and glaciers during spring and summer, mountain snowpack and glaciers contribute significantly to maintaining dry season flow.	By increasing infiltration (root depth and soil depth) forested vegetation can maintain or increase dry season flows; however, in some cases forests can decrease dry season flows, if increases in ET reduce infiltration.	Evaporation from wetlands can be relatively high (compared to other land cover types) and in some cases may reduce dry season flow in rivers downstream. In other cases, especially where wetlands occur over permeable soils or rocks, dry season flows can be increased (e.g., floodplain wetlands over permeable gravels or sands).	Grasslands with deeper rooted native vegetation and soils that have not been compacted also promote infiltration and can contribute to dry season flows downstream, although the effect is generally smaller than for forests.	Aquifers that are connected to surface waters (e.g., floodplain aquifers, springs in headwaters to streams) can augment dry season flows as groundwater is discharged to streams or rivers. In these cases, protecting aquifer recharge zones can contribute to maintaining dry season flows.	
Aquifer recharge, maintain groundwater supplies	Mountain snowpack and glaciers can contribute to groundwater recharge via streamflow as well as diffuse recharge from mountainous areas to adjacent alluvial aquifers.	Forests can affect groundwater recharge in a variety of ways; if deep infiltration is greater than interception and ET, recharge can be enhanced. Removal of forest (e.g., by development or wildfire) tends to increase runoff and decrease the potential for recharge.	Wetlands with hydraulic connectivity with groundwater contribute to groundwater recharge.	Grasslands with native vegetation and deep soils over permeable substrates, can be significant sources of groundwater recharge.	Natural vegetation that maintains or enhances infiltration in critical aquifer recharge zones will contribute to maintaining groundwater recharge.	Lakes, rivers, and streams with hydraulic connectivity to groundwater can be important sources for groundwater recharge.

Continued

Table 9.1: Major North American natural assets and their hydrological ecosystem services—cont'd

Hydrological ecosystem service	Natural assets and hydrological ecosystem services benefiting source water protection					
	Mountain snowpack/glaciers	Forest (esp. source watershed and riparian zone forests)	Wetlands and floodplains	Grasslands	Aquifers	Lakes, rivers, and streams
	Maintain or Improve Water Quality					
Control erosion and reduce sediment in water supplies	Gradual melt from snowpack and glaciers can help maintain soil moisture and organic matter in soils, as well as helping maintain vegetation, contributing to reducing erosion and sedimentation.	Forest vegetation reduces erosion by protecting soils from the erosive force of rainfall, enhances infiltration which reduces runoff. Tree roots help stabilize soils and hold sediment in place.	By slowing the movement of water across the landscape and storing water for short to long periods, sediments settle out and are retained in wetlands.	Healthy grassland vegetation will protect soils from erosion and hold sediments in place, especially compared to bare soil or developed areas.		
Reduce nutrient and pollutant levels in water supplies (surface and groundwater)	Gradual melt from snowpack and glaciers can help maintain soil moisture and organic matter in soils, which supports microbial activity, nutrient uptake, and processing of pollutants and toxins. Snowmelt contribution to surface waters can help moderate water temperatures and minimize occurrence of harmful algal blooms.	Pollutants can be carried with sediments so by reducing erosion forests help keep pollutants out of water supplies. Forest vegetation and soil microbes also take up nutrients and process pollutants. Shading from forest vegetation, especially in riparian areas can reduce water temperatures, which can help minimize occurrence of harmful algal blooms.	Wetlands are particularly important for uptake of nutrients, denitrification, and bioremediation of some pollutants. Wetlands can help maintain cooler water temperatures and by removing nutrients can help minimize occurrence of harmful algal blooms. Floodplains are zones of very high microbial activity, processing of nutrients and bioremediation of pollutants, as well as contributing to moderating water temperatures.	Grassland vegetation and soil microbes also take up nutrients and process pollutants, helping to keep excess nutrients and pollutants out of water supplies.		

Reduce/avoid saltwater intrusion to coastal water supplies	Other assets are more important but where snowpack and glaciers are important sources for freshwater flow in coastal areas (e.g., coastal mountain systems) they contribute to maintaining fresh groundwater elevations.	Forests that maintain infiltration (i.e., maintain pervious land cover) and contribute to recharge of coastal aquifers help maintain groundwater (freshwater) elevations that prevent saltwater intrusion.	Wetlands in coastal areas can help maintain groundwater (freshwater) elevations and minimize saltwater intrusion; draining coastal wetlands can lower groundwater levels and allow saltwater intrusion.	Grasslands in coastal areas (e.g., salt marshes) retain sediments and build up organic matter, keeping land elevations high relative to sea level and minimizing saltwater intrusion.	Streams and rivers are important sources of freshwater flow into coastal areas, including into coastal aquifers, maintaining high groundwater (freshwater) levels that can minimize saltwater intrusion. Dams and/or consumptive uses that reduce freshwater flows can result in saltwater intrusion to aquifers.		
Mitigate Water-Related Hazards							
Buffering the effects of drought conditions (including risk of wildfire and resulting water quality impacts)	Mountain snowpack and glaciers in particular serve as storage reservoirs; providing continuous supplies downstream during warmer months, helping to mitigate small to moderate drought conditions. Elevated soil moisture In the vicinity of snowfields as well as runoff from snowmelt can help protect from drought conditions and may help reduce risks of wildfire.	Healthy forests that contribute to maintaining soil moisture, infiltration, dry season flow, and groundwater recharge can provide some buffer against seasonal and short duration drought conditions.	Wetlands with hydraulic connectivity to groundwater can buffer some drought conditions by maintaining groundwater supplies. Some wetlands with hydraulic connectivity to surface water (e.g., floodplains, lakeshore) can also provide some buffering against drought by augmenting river or lake levels during dry periods.	Healthy grasslands that contribute to maintaining soil moisture, infiltration capacity, dry season flow, and groundwater recharge can provide some buffer against seasonal and short duration drought conditions.	Water storage in lakes and rivers can provide buffering against short- to moderate-duration drought conditions.	Aquifers provide some of the most important buffering for droughts but are vulnerable to over exploitation.	

Continued

Table 9.1: Major North American natural assets and their hydrological ecosystem services—cont'd

Hydrological ecosystem service	Natural assets and hydrological ecosystem services benefiting source water protection				
	Mountain snowpack/glaciers	Forest (esp. source watershed and riparian zone forests)	Wetlands and floodplains	Grasslands	Aquifers	Lakes, rivers, and streams
Reduce flooding risk	Very rapid snowmelt contributes to flooding in spring or early summer. With climate change "rain on snow" events in some mountain areas (e.g., Pacific Northwest) may increase, contributing to increased flood event.	Forests can reduce peak flows and provide some moderation of downstream flooding. The effect on reducing peak flows and downstream flooding declines as the size of precipitation events increases so that flood mitigation may be effective at moderate-sized events but not for very large precipitation events.	Floodplain wetlands in particular can reduce and/or delay flooding by providing short-term flood storage. Wetlands can reduce flooding from small to moderate precipitation events but have limited effect on flooding from very large events; small wetlands in headwaters areas are unlikely to reduce flooding significantly and can increase runoff during larger events.	Grasslands can reduce surface runoff compared to agricultural or developed land and can provide some flood protection during moderate events.		Some protection that is dependent on maintaining adjacent floodplain or lakeshore wetlands that can absorb and store floodwaters.

Reduce landslide/catastrophic sedimentation risk	By stabilizing soils, forests can protect against landslides, but this effect is limited in the case of large or extreme events, such as long periods of intense rainfall on steep slopes. Healthy forests that reduce risk of wildfire can minimize sedimentation to storage reservoirs, dredging costs, and supply disruptions.		Grasslands may provide some limited protection by stabilizing soils, but this effect is limited in the case of large or extreme events, such as long periods of intense rainfall on steep slopes.		
References	Bates et al. (2008), Schindler and Donahue (2006), Milner et al. (2009), Luce (2018)	Filoso et al. (2017), Brauman et al. (2007), Binder et al. (2017), Deal et al. (2012), Dadson et al. (2017), Birkinshaw et al. (2011), Gartner et al. (2013), Ozment et al. (2016)	Acreman and Holden (2013), Fisher and Acreman (2004), Maltby and Acreman (2003), Bullock and Acreman (2003), Burek et al. (2012), Caldwell et al. (2014), Dadson et al. (2017)	Maczko et al. (2011)	Griebler and Avramov (2015), López-Morales and Mesa-Jurado (2017)	Grizzetti et al. (2016), Auerbach et al. (2014)

Nature-based solutions encompass a diversity of practices for source water protection

Source water protection for water security and resilience includes a wide variety of approaches and practices, not all of which are nature-based (Matthews et al., 2019). NBS are most easily visualized, and most frequently discussed, as a set of conservation, restoration, management, or hybrid green-gray practices or actions (Table 9.2). However, these practices, when most effective, are supported by a set of enabling policies, institutional frameworks, governance and funding models, and technical tools and evidence that support the scoping, planning, design, funding, implementation, maintenance, and evaluation of NBS (see for example, Coxon et al., 2021; Dominique et al., 2021; Mauroner et al., 2021; Ofosu-Amaah et al., 2021). As emerging evaluations of NBS and evolving best practices suggest, to deliver desired outcomes, these more easily visualized NBS practices must be embedded in larger policy and governance systems for sustainability, equity, and resilience (Matthews et al., 2019). While there are many ways to organize or group NBS (see for example, Cohen-Sacham et al., 2016; Eggermont et al., 2015; Gómez Martín et al., 2020), the possible solutions are diverse and there is no single classification or exhaustive list of NBS. Table 9.2 lists some of the more common NBS for source water protection, organized by the degree of human intervention, following Eggermont et al. (2015).

Benefits of NBS source water protection for water utilities

Given the central role of natural assets in delivering hydrologic services, the direct benefits to drinking water utilities include the economic benefits from reduced and/or avoided costs, maintaining access to consistent supplies, and mitigation of risks to utilities from natural or human-caused hazards.

Economic benefits

Natural assets can reduce operations and maintenance (O&M) costs for utilities and reduce and/or avoid capital costs for new and replacement gray infrastructure (see for example, Postel and Thompson, 2005).

Reduced operations and maintenance costs

There is a substantial literature on the economic benefits of high source water quality in terms of reduced treatment costs (Freeman et al., 2008; Gartner et al., 2013; Ozment et al., 2016; Warziniack et al., 2016). Natural assets such as forests and wetlands reduce treatment costs by keeping sediments, pollutants, and pathogens out of source waters and removing nutrients, which can produce toxic algal blooms that require treatment plants to shut down or increase use of chemicals and additional treatment steps. Postel and Thompson (2005) evaluated drinking water treatment costs for 27 water suppliers in the United States and found that treatment costs increased in watersheds as forest cover declined. Watersheds with 60%

Table 9.2: Common NBS practices, examples of supporting policies, and associated cobenefits (size of the dot indicates relative potential for each cobenefit).

NBS activity for source water protection (water supply, water quality)	Examples	Types of supporting policies	Cobenefits				
			Climate impacts	Biodiversity	Hazard mitigation	Livelihoods	Health\well-being
Low Human Intervention—Protection of Existing Natural Assets							
Conservation/ Protection of Targeted Terrestrial Ecosystems	Forest, Wetland, Grassland Protection	Policies Establishing National Parks or Protected Areas, Wilderness Areas, or Strategic Biological Reserves; Conservation Easements; Land Acquisition and Land Trusts	X	X	X		X
Conservation/ Protection of Targeted Aquatic Ecosystems	Protection of Rivers, Lakes, Aquifers, Springs	Riparian Buffer Regulations; Environmental Flows Standards; Wild and Scenic Rivers Designation; Designation of Critical Aquifer Recharge Areas	X	X	X		X
Moderate to High Human Intervention—Restoration of Natural Assets and Sustainable Land Management							
Restoration of Targeted Ecosystems	Forest, Wetland, Grassland, River/Floodplain Restoration	Payments for Ecosystem Services; In-Stream Buy-Backs; Integrated Water Resources Management; Watershed Management Plans, Basin Planning; Conservation Incentives (US Farm Bill); Habitat Conservation Plans,	X	X		x	X
Sustainable Land Management	Sustainable Agroforestry, Agricultural and Ranching Best Management Practices	Water Quality Trading Policies; Standards and Certification Systems (e.g., Forest Stewardship Council, Rainforest Alliance, Responsible Wool Standard); US Farm Bill Conservation Incentives Programs; Local Land Use/ Zoning Regulations; Regenerative Agriculture Policies	X	x		X	X
Riparian Restoration	Establishment or Enhancement of Riparian Vegetation	Local Land Use/Zoning Regulations; Local Critical Areas Ordinances; Payments for Ecosystem Services	x	x	X		

Continued

Table 9.2: Common NBS practices, examples of supporting policies, and associated cobenefits (size of the dot indicates relative potential for each cobenefit)—cont'd

NBS activity for source water protection (water supply, water quality)	Examples	Types of supporting policies	Cobenefits				
			Climate impacts	Biodiversity	Hazard mitigation	Livelihoods	Health\well-being
Fire Risk Management	Fuels Management, Controlled Burn/Prescribed Fire	Local Land Use/Zoning Regulations; Forest Management Plans	x		X		x
Human Intervention—Hybrid Systems, Ecological Engineering							
Wetland Creation	Construction of Treatment Wetlands to Manage Nutrient Runoff From Agricultural Fields; Created Wetlands for Floodwater Storage	Conservation Incentives (Farm Bill); Clean Water Act Water Quality Regulations; Water Quality Trading Programs; Local Land Use/Zoning Regulations; Stormwater Management Regulations		x	x		x
Managed Aquifer Recharge	Using Irrigation Canals to Divert and Infiltrate Excess River Flows to increase Groundwater Storage (e.g., Nebraska, Platte River); Restoring Ponds/Wetlands in Coastal Areas to Recharge Groundwater and Protect Wells from Saltwater Intrusion	Integrated Water Resource Management Plans; Inter-State/Transboundary Water Agreements; Water Sharing Programs; Payments for Watershed Services and Other Incentives/Market-Based Programs	X		X		
Urban Green Infrastructure	Green Streets, Green Roofs, Urban Forests, Permeable Pavements can Retain Sediments, Filter Water and Help Protect Downstream Water Sources (e.g., Receiving Rivers).	Land Use/Zoning Regulations; Stormwater Management Regulations; Water Quality Regulations; Incentives and Offsets Programs (e.g., Stormwater Retention Credits)	X	x	x		X

or more forest cover had reduced treatment costs compared to watersheds with less than 30% forest cover. An analysis by the US Environmental Protection Agency (EPA) of source water protection programs in six communities in the United States found that for each $1 that was invested in source water protection efforts (incl. reforestation), there was an average savings of $27 in water treatment costs (Winiecki, 2012).

While many studies show that forest cover is related to treatments costs, the relationship is complex and depends on factors such as loading rates, amount of forest cover above intakes, type of forest cover, soils, and underlying geology, and location of forest cover in a watershed (Freeman et al., 2008). Recent studies confirm that effects may be relatively small (1% increase in turbidity increased treatment costs by 0.19%) (Warziniack et al., 2016) and depend on the variable selected (turbidity, total organic carbon, N, P, or sediment loading) (Price and Heberling, 2018), yet still can deliver significant benefits to utilities. Warziniack et al. (2016) found a 1% decrease in forest cover resulted in a 3.9% increase in turbidity, meaning that for an average treatment plant, a 10% loss of forest cover could increase chemical treatment costs by $65,000 per year.

A detailed assessment of the relationship between forest cover and chemical treatment costs for three treatment plants in the Potomac River Basin found that treatment cost savings were smaller than the costs of protecting forest lands (WRF (Water Research Foundation), 2020). The study evaluated reduced water treatment chemical costs only and did not assess other operations and capital costs, such as energy consumption or filter replacement. There were only modest improvements in source water quality under the different forest protection scenarios evaluated; however, existing forest cover in these watersheds was about 50%, and the maximum change in forest cover in the scenarios was 2%. The marginal benefits of increased forest cover to improving water quality will depend on the total amount of forest cover, so the greater the forest cover the more cost-effective forest protection activities are likely to be. While the effects of forest cover on reducing chemical treatment costs were small, forests can reduce other costs to water utilities from risks such as wildfire, climate change, pests, and urbanization.

Evaluating the cost savings or cost-effectiveness of NBS can be challenging; it is important to take a comprehensive view of costs and benefits, for example, by assessing the market and nonmarket values of NBS cobenefits and the full life-cycle costs of NBS and gray infrastructure alternatives (Browder et al., 2019). In addition, hard-to-quantify attributes of NBS, such as flexibility and resilience, are not easily incorporated in cost–benefit or cost-effectiveness analyses but are some of the most important arguments for implementing NBS (see Mauroner et al., 2021).

Reduced or avoided capital costs

Natural assets that protect source waters can result in avoided capital costs, such as avoiding the need for new treatment plants (Gartner et al., 2013; López-Morales and Mesa-Jurado,

2017; Postel and Thompson, 2005). The case of New York City is perhaps the most well-known example of avoided costs to a utility from managing natural assets: with an estimated total investment of about $1.5 billion in maintaining the natural infrastructure in the watershed, the city was able to avoid building a new treatment plant at a cost of $8–$10 billion (Kenny, 2006). Other utilities in the United States have realized significant avoided costs from natural assets, including Portland, Maine, which has a filtration avoidance waiver and is not required to install a conventional filtration system (estimated cost $97–$155 million) due to the high quality of water from the forested watershed around Lake Sebago (Gartner et al., 2013; Ozment et al., 2016).

Natural assets mitigate risks to water utilities

Healthy forests, along with wetlands and grasslands, can also help provide resilience to a number of natural or human-caused hazards, which can pose material risks to water utilities. These risks are evident when natural assets are lost—as in the case of wildfire destroying forests and eliminating the hydrological services that regulate supplies and maintain high source water quality. In North America the major risks to utilities include wildfire (with impacts to water quality, supplies, and damage to infrastructure from resulting floods and landslides), drought, pests/diseases that affect forests, agricultural runoff, and urbanization/land use change. All of these risks are exacerbated by the impacts of climate change.

Forest fire is a major and growing factor in watershed degradation globally (e.g., recent examples in Australia, Spain, Greece), impacting water supply and quality. In the Western United States and Canada in particular, there has been a general trend to bigger, more frequent wildfires over the past three decades (Dennison et al., 2014; Westerling et al., 2006). Wildfires have tended to come hand in hand with drought and are linked to warmer temperatures and earlier spring snowmelt (Anderegg et al., 2013; Westerling et al., 2006). A history of fire suppression has resulted in increased fuel loads in forests, and growth of cities and suburbs has greatly increased the wildland-urban interface, which has been associated with an increase in human-caused fires (Gibbens et al., 2013). These increased fuel loads and the density of vegetation, along with increased frequency and severity of droughts, have been associated with larger, more intense, and rapidly moving fires, or mega-fires (Bladon et al., 2014; The Brookings Institution, 2005).

Wildfire's effects in watersheds can include increased runoff and higher peak flows in streams, more rapid snowmelt in the spring (due to reduced shading from trees) with more rapid filling of reservoirs, increased erosion and sedimentation in streams and reservoirs, hazardous debris flows, and impaired water quality (Bladon et al., 2014; Emelko and Sham, 2014). Very hot wildfires destroy all above-ground vegetation and organic materials in the soil. This can result in hydrophobic soil conditions, where soil becomes virtually impervious to water, causing rapid runoff of precipitation, resulting in increased erosion and debris flows. Sediments and ash clog streams, irrigation canals, drinking water intakes, and treatment plants, disrupting operations and increasing the cost of treatment (Emelko and Sham, 2014). In addition to increased stream turbidity and suspended sediments, increased organic carbon (TOC) and phosphorus, volatized nitrogen, and oxidation of heavy metals found in soils and plants affects water quality following heavy rainfall in burn areas (Bladon et al., 2014; Emelko et al., 2011; Gibbens et al., 2013). The large amounts of sediment released following fires can be deposited in riverbeds and floodplains, reducing floodwater conveyance and storage and increasing flood damage to water infrastructure far downstream. The impacts and costs to utilities from wildfires in the past 15–20 years in Colorado illustrate many of the risks and costs to utilities from the loss of forest assets from wildfire (see Box 9.1).

Box 9.1   Impacts and costs of Colorado wildfires.

Wildfires in Colorado provide some of the clearest evidence of the costs to utilities from losing forest assets. A series of highly destructive wildfires over the past 20 years have resulted in extensive infrastructure damage, water supply disruption, and costly rehabilitation for water suppliers, bringing wildfire risk front and center for water managers. In 1996 and 2002, Denver Water, a water provider for 1.3 million customers in the Denver metropolitan region, experienced two severe fires, one in Buffalo Creek and one in Hayman. Together, the fires burned almost 150,000 acres in one of Denver Water's most critical watersheds, the South Platte. This watershed not only supplies about half of Denver's water but is also the watershed through which 80% of Denver's water is moved to the city via diversions that move water from Dillon Reservoir (in the Colorado River Basin) across the Continental Divide to the South Platte River. Together the two fires resulted in about 765,000 m^3 (1 million cubic yards) of sediment deposition into the Strontia Springs Reservoir, the equivalent of 40 years' worth of normal accumulation. Since the Buffalo Creek and Hayman fires, Denver Water has spent over $26 million on water quality, reclamation, and restoration treatments in the South Platte Watershed and dredging sediment out of Strontia Springs Reservoir.

In 2012, Northern Water, the regional supplier to communities north of Denver, had its own "wake-up call," (Gibbens et al., 2013) when the nearby High Park fire left Northern Water participants Fort Collins and Greeley with unreliable water quality (to the degree that it was sometimes untreatable) during rainfall and spring runoff events. At times, these users have had to depend entirely on water supplies from Horsetooth Reservoir (a Colorado Big Thompson, CBT, reservoir). A year after the fire, a rainstorm resulted in the Monroe Gravity Canal diversion serving the North Poudre Irrigation Company to be so clogged with debris and sediment that service was interrupted for 10 days during which irrigators had to depend on in-reservoir water (Gibbens et al., 2013). Uncertainty around supply has also affected the availability of temporary water for rent to irrigators by municipalities, exerting negative pressure on the region's agricultural economy. Though Northern Water is a raw water provider, and thus not responsible for treatment, fires in its source areas had a material impact on its users.

Climate change will exacerbate other threats to source waters

Climate change is expected to magnify the risks and impacts from wildfires (Schoennagel et al., 2017). Warmer temperatures and reduced snowpack lead to drier conditions that favor fire initiation and spread, create more favorable conditions for pests such as the pine bark beetle, and make forests more susceptible for large, destructive fires (Adams et al., 2009; Logan and Powell, 2001; Luce, 2018). Climate change is also predicted to significantly impact water supplies via changes in precipitation patterns and increased frequency and intensity of droughts and floods, creating new risks and uncertainties for utilities (Bates et al., 2008; Luce, 2018; Muir et al., 2018).

Forests, wetlands, and other natural assets, through their influence on regulating runoff, groundwater recharge, storage of floodwaters, and maintaining soil moisture can protect physical water infrastructure from climate-related hazards such as floods (Machado et al., 2014). These natural assets also help buffer water utilities against the effects of droughts and other sources of supply disruption (Table 9.1, and see for example, Brauman et al., 2007).

Nature-based solutions for resilient source water protection

Matthews et al. (2019) distinguish between traditional source water protection and resilient source water protection. Resilient source water protection "recognizes that hydrology, ecosystems, water management, and climate are intertwined, and that the integrity of ecosystems is necessary for the integrity of communities—and *vice versa*" (Matthews et al., 2019). To be resilient in the face of future changes, water systems must balance robustness and flexibility (Matthews et al., 2019; Smith et al., 2019). NBS is important for resilient source water protection for a number of reasons. Through green-gray hybrid approaches, NBS can be used to support and reinforce more traditional infrastructure and increase overall system robustness. NBS are inherently flexible solutions; they can evolve with a changing environment and adapt to new conditions. They are typically much easier to undo and adjust than gray infrastructure if conditions and needs shift over time. Using a hybrid approach that involves both NBS and gray infrastructure allows for the distribution of risks to source waters across a variety of assets, reducing risks of overall system failure if one asset fails.

Emerging innovations for nature-based source water protection
New multistakeholder, multisectoral collective action efforts promote NBS for source water protection

Over the past several decades, a growing number of municipal drinking water utilities have explored or implemented incentive schemes—investments in watershed services (IWS)—to compensate or provide incentives to upstream land managers for implementing NBS that contribute to source water protection (Bennett and Ruef, 2016; Ozment et al., 2016). Early concerns in most of these programs focused primarily on threats from impending regulations, such as new filtration requirements, or from increased costs of treating raw water due to increased development in source watersheds, the loss of forest cover, and the increase in sediment, nutrient, chemical, and pathogen pollutants in raw water (Ozment et al., 2016; Pires, 2004). As in the New York case, many early IWS initiatives in the United States were bilateral agreements (i.e., between a single utility and a single or multiple upstream land managers) driven almost entirely by the needs of the utility to protect or improve raw water quality.

More recently the need for more robust funding, concerns about increased development and land use changes, and threats from climate change, extreme flooding, and catastrophic wildfires have driven the emergence of more complex programs, especially in the Midwestern United States, Western United States, and Canada. Water utilities and municipalities are recognizing that source water protection must include not just a concern for contaminants but also considerations of the resilience of future supplies in the face of climate impacts, managing climate-related hazards to water quantity and quality and accounting for how the health of source watersheds may mitigate or exacerbate threats to the rest of the water system (e.g., flood damages to water treatment plants and distribution systems) (Matthews et al., 2019). Protecting source watersheds provides the first barrier in a multibarrier approach to protecting drinking water quality, and watershed health (or lack thereof) can influence whether the other barriers will continue to function well in the face of changing conditions. As lessons from US watershed investment programs show, effective NBS for source water protection require going beyond bilateral agreements and building partnerships across multiple stakeholders and sectors (Ozment et al., 2016).

Water funds, such as the Rio Grande Water Fund in the United States, are examples of multistakeholder partnerships that are concerned with source water protection but also emphasize the multiple benefits of NBS for improving rural livelihoods, strengthening adaptation to climate change, and building overall watershed resilience (Matthews et al., 2019). Water funds have been extensively reviewed elsewhere (McCarthy, 2014; Brauman et al., 2019) and are an increasingly important global tool for source water protection. A more recent development in the United States are regional agriculture-utility alliances between federal, state, and municipal governments and conservation and business stakeholders to collaboratively manage watersheds for improved water quality and mitigating flood risks.

Bridging the rural-urban divide: Agriculture-utility partnerships in the Mississippi Basin

Runoff from agricultural areas, particularly from intensive row crop agriculture, is a significant contributor to impaired drinking water quality in the United States and Canada (Hanson et al., 2016; Liang et al., 2020). Nitrogen and phosphorus runoff from farmlands has contributed to toxic algal blooms that have shut down treatment plants and resulted in "no drink" advisories in several cities (Seewer, 2015). In the United States, the links between agriculture and water quality are particularly critical in the Mississippi River Basin; land areas generating over 50% of the nation's agricultural production drain into the rivers and streams in the basin. Farmlands in the Mississippi River Basin are the major contributors to diffuse nonpoint source (NPS) pollutants (e.g., nitrogen and phosphorus nutrients in runoff) that affect drinking water quality for towns and cities (David et al., 2010; Spahr et al., 2010).

Nitrate in surface water supplies has long been of concern for water utilities in the region, as elevated nitrates in drinking water can cause a fatal disease in infants "blue baby syndrome" as well as a number of chronic health issues with long-term exposure (USEPA, 1987; Van Grinsven et al., 2006). Agricultural activities are the primary contributor of nitrogen to surface waters across the Mississippi River Basin (Robertson et al., 2014). Agricultural runoff is also the major source of nutrients causing the large, low-oxygen (hypoxic) zone in the Gulf of Mexico, which significantly affects the fisheries-based economy of the region. The State of Iowa alone is responsible for 41% of the nitrate load to the Gulf of Mexico (Jones et al., 2018) and 92% of the nitrate in Iowa's waters comes from agriculture (IDALS (Iowa Department of Agriculture and Land Stewardship), IDNR (Iowa Department of Natural Resources), and Iowa State University College of Agriculture and Life Sciences, 2017).

Farms and cities: Conflicts over water quality

Under the Clean Water Act (CWA), the US EPA has limited statutory authority to regulate NPS from farms, and the United States relies instead on subsidies and voluntary programs, primarily via the US Farm Bill, to control agricultural runoff (OECD, 2015; Secchi and Mcdonald, 2019). Voluntary NBS such as no-till farming, cover crops, edge-of-field and riparian buffers, and wetland restoration implemented through incentives and conservation programs in the US Farm Bill, while important, have not been of sufficient scale to address the magnitude of impacts to water quality from soil erosion and nutrient pollution (Secchi and Mcdonald, 2019). Deteriorating water quality has impacted water utilities, increasing treatment costs and, in some cases, requiring utilities to secure alternate supplies when primary sources do not meet drinking water standards (Canning and Stillwell, 2018). Conflicts between municipalities and farmers have arisen over who is responsible and should pay the costs of poor water quality (Des Moines Water Works, 2017). Adding to these tensions is the fact that stormwater runoff—an urban NPS—is regulated under the CWA, meaning that cities must bear the cost of meeting water quality standards for both raw water from agricultural runoff and urban stormwater runoff.

Des Moines Water Works versus drainage districts

For drinking water utilities and the customers they serve in the region, the costs of nutrient pollution can be extremely high. As one example, to address elevated nitrate levels in their source water, the City of Des Moines' water utility, Des Moines Water Works, constructed one of the world's largest nitrate removal facilities in 1991, at a cost of $4.1 million (Elmer, 2017). The cost of operating the facility has averaged around $500,000 per year but went as high as $1.4 million in 2015 (Des Moines Water Works, 2017). Yet since the early 1990s, nitrate levels in two surface water sources, the Raccoon and Des Moines Rivers, have continued to increase, requiring the Water Works to double the size of this facility at an estimated cost of $15 million. These costs put significant strain on municipal budgets and may be unsustainable, especially for smaller cities and towns.

In 2015, the situation in Des Moines led the utility to file a lawsuit against several upstream agricultural drainage districts, seeking damages for the cost of nitrate removal, alleging that agricultural runoff from the districts was responsible and that the drainage districts were illegally discharging pollutants from point sources as water leaves the districts via tile drains (Canning and Stillwell, 2018; Church et al., 2020). The lawsuit was contentious and over the 2 years that the suit was considered, relationships between the utility (and the city) and agricultural communities deteriorated. The suit was decided in 2017, going against the utility. The courts did not rule on whether tile drains are point sources; they acknowledged that the utility had likely suffered an injury, did not dispute that the drainage districts cause water quality problems, but determined that the districts lack the legal ability (regulatory powers) to redress the injury (Canning and Stillwell, 2018; Des Moines Water Works, 2017). In effect, the courts recognized that while voluntary conservation efforts by farmers have been insufficient to reduce nitrogen runoff and protect water quality, given the regulatory context in the United States, voluntary measures are the only ones available.

Farms and cities: Shared vulnerabilities

A series of increasingly devastating floods in the Mississippi Basin, beginning with the record-breaking 1993 flood, have disrupted urban and rural communities alike. In 2019, farmers in the Basin suffered devastating losses from massive spring flooding as over 62 levees were breached or overtopped and hundreds of miles of levees damaged along the Missouri River, with the entire spring and early summer crops lost in some cases and access to markets restricted as roads were damaged (New York Times, 2019). Damage to water treatment plants and contamination of wells meant that many towns were left without drinking water service for weeks or even months (Associated Press, 2019). Along with climate change, the long history of land use changes in the Basin make both urban and rural communities increasingly vulnerable to flood damages. In the period between 1780 and the 1980s, from 60% to over 90% of the wetlands in the upper Mississippi Basin were lost (Dahl, 1990). Soil erosion and the loss of organic matter in soils has reduced the water holding capacity of soils in the basin by an estimated 18 million acre-feet by the mid-1990s, representing 45% of the flood volume of the 1993 flood (Hey and Philippi, 1995). The NBS that reduce runoff and help improve water quality draining from farmlands can also provide flood mitigation benefits and lead to reduced flood damages to downstream infrastructure (Antolini et al., 2019).

Shared solutions: Iowa's multisectoral, multistakeholder collaboration to scale nature-based solutions

Driven by these growing crises, stakeholders across the urban-rural divide began coming together in new and innovative partnerships to find shared solutions that address the needs of both communities through science-based and collaborative governance and decision-making.

Iowa's collaborative, multipronged approach is becoming a model for other communities in the Mississippi Basin. NBS are at the heart of this shared approach with the recognition that a strategy of watershed-wide NBS can address water quality and flooding while also providing additional social and economic benefits; give farmers flexible options and appropriate incentives; lower costs for utilities and cities; reduce vulnerabilities to hazards for all communities; and improve overall resilience in the face of current and future shocks (Fig. 9.1).

Initially, as water quality and flooding crises became more urgent, Iowa developed single-issue, sector-based approaches to address each crisis in turn. However, as declining water quality, expensive gray infrastructure solutions, flood damages, and conflict between urban and rural communities became more severe, communities, agricultural businesses, and water sector leaders began a series of innovative new approaches (Fig. 9.2). These were designed to jointly address multiple problems via a focus on NBS as preventative measures that are more cost-effective than post-ante remedies, with the added feature that NBS help build resilience to future shocks and provide a host of social, ecological, and economic benefits.

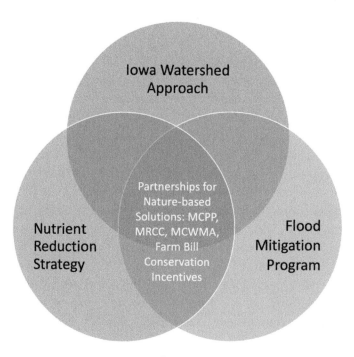

Fig. 9.1
Multisector partnerships for nature-based solutions are at the heart of Iowa's strategy to address water quality, flood hazards, and climate resilience: Middle Cedar Watershed Management Authority (MCWMA); Middle Cedar Partnership Project (MCPP); Midwest Row Crop Collaborative (MRCC).

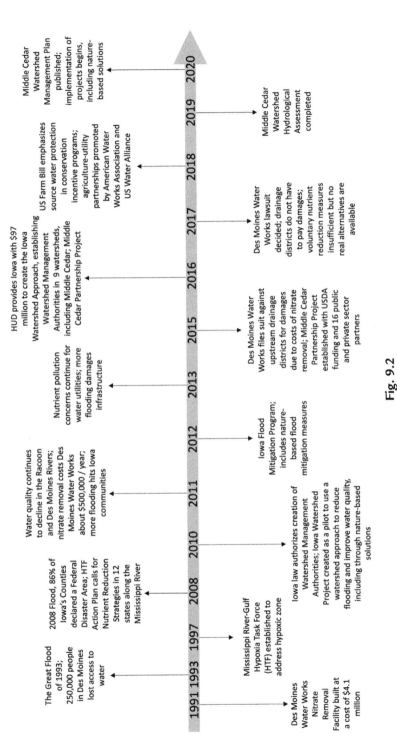

Fig. 9.2

Timeline of initiatives in Iowa, from single-issue, single-sector initiatives to multisectoral, collaborative watershed partnerships for NBS.

Iowa nutrient reduction strategy

In 1997, as the zone of low oxygen in the Gulf of Mexico grew in size, the EPA established the Mississippi River Gulf of Mexico Watershed Nutrient Task Force (the Hypoxia Task Force—HTF) to coordinate activities to reduce the severity and duration of the hypoxic zone in the Gulf. In 2008, the HTF Action Plan called for the 12 states along the river to develop Nutrient Reduction Strategies (NRS). Because most of the nutrients to the Gulf are from agricultural NPS, the EPA did not mandate federal criteria for nutrient loads but relied on each state to develop their own strategy. The EPA provides recommendations for what should be included in these strategies but does not actively monitor compliance, and it is unclear that these measures are improving water quality (Secchi and Mcdonald, 2019; Stoner, 2011). The Iowa NRS developed by the Iowa Department of Natural Resources, the Department of Agriculture and Land Stewardship, and Iowa State University's College of Agricultural and Life Sciences was designed to address both point (e.g., wastewater treatment plans) and nonpoint (mostly agricultural runoff) sources (IDALS, IDNR, and Iowa State University College of Agriculture and Life Sciences, 2017). Iowa's plan includes many NBS and has been recognized as one of the more effective plans, with science-based assessments of nutrient sources and targets for reduction, as well as taking a watershed approach (Secchi and Mcdonald, 2019).

Iowa flood mitigation plan

In response to increasingly severe flooding in the aftermath of the large 2008 flood event, Iowa created the Iowa Flood Center at the University of Iowa. The Center was tasked with creating a data system to better monitor precipitation and stream flows to improve understanding of flood risks and provide technical assistance on flood mitigation projects, including flood-related NBS. In 2012 the state developed a Flood Mitigation Program, within the state's Department of Homeland Security and Emergency Management, to address gaps in communities' preparedness and resilience, and focusing on prevention and mitigation, rather than solely on emergency response. Nature-based approaches promoted in the plan include stream and wetland restoration and green infrastructure in cities, including green streets and parks. As this plan was being developed, Iowa was also initiating a series of policy changes to promote watershed approaches, which would become the umbrella policies for flood mitigation and nutrient reduction.

Iowa watershed approach

Around 2010, Iowa took two steps to strengthen the state's ability to implement integrated watershed approaches that could be more effective in addressing water quality, flooding, and community well-being needs across both urban and rural communities. In 2010, new legislation authorized the creation of Watershed Management Authorities (WMA), which allowed for intergovernmental agreements between cities, counties, soil and water conservation districts, and other stakeholders to collaboratively engage in watershed planning and management. These WMA's had five primary goals: reduce flood risk, improve water quality, monitor federal flood risk planning and activities, educate residents about flood risks and water quality, and allocate funds for projects, including NBS projects.

Using funds from the US Department of Housing and Urban Development (HUD), Iowa then initiated a pilot project, the Iowa Watershed Project. The Watershed Project created WMAs in five priority watersheds to assess conditions and develop plans to minimize flood risks and improve water quality. Given the success of the pilot project, when it ended in 2016, Iowa expanded the program, receiving $97 million from HUD's National Disaster Resilience Competition to help fund the larger program, the Iowa Watershed Approach for Urban and Rural Resilience. Nine watersheds could now develop WMAs, and the goals were broadened to include building resilience, strengthening collaborative governance, and improving quality of life for more susceptible populations (Box 9.2).

New public-private partnerships: Middle cedar partnership project

As watershed approaches were emerging in Iowa, at a national level, organizations such as the American Water Works Association and the US Water Alliance were working to forge more partnerships between water utilities and the agricultural sector, based on their shared vulnerabilities and interests (Mehan and Carpenter, 2019; Mehan et al., 2020; Murphy and Carpenter, 2020). As a result of this advocacy on the part of water utilities, the 2018 Farm Bill in the United States mandates that 10% of all conservation funds be spent on source water protection, or about $4 billion over 10 years. These developments opened the door to new public-private partnerships that are creating a series of aligned and interrelated initiatives to implement more systemic and scaled NBS in Iowa (see Fig. 9.2).

One such regional initiative, the Middle Cedar Partnership Project (MCPP) was created in response to rising nitrate levels in source waters, a series of floods, and a severe drought in 2012, which together threatened the industrial and agricultural value chains that are the basis of the region's economy. As one of the nine priority watersheds identified under the state's NRS, controlling nutrient runoff in the Middle Cedar is extremely important for meeting

---

Box 9.2 Iowa Watershed Approach: Integrating source water protection into a broader set of environmental and social goals.

Iowa's Watershed Approach encompasses six goals:
- Reduce flood risk from upper to lower watershed.
- Improve water quality for downstream water users and the environment.
- Increase flood resilience across all communities.
- Engage stakeholders through collaboration, outreach, and education.
- Improve quality of life and health, especially for susceptible populations.
- Develop a program that is scalable in the Midwest and elsewhere in the United States.

the state's nitrate reduction targets. The City of Cedar Rapids gets its drinking water from shallow alluvial wells around and under the Middle Cedar River. More than 70% of the water produced by the city's treatment plants goes to large industrial users such as Cargill, PepsiCo, General Mills, and Archer Daniels Midlands, which require high-quality water for their production processes.

Led by the City of Cedar Rapids, 16 public and private sector partners (Box 9.3) started the MCPP to work together to improve water quality, reduce flood risk, and improve soil health. A collaboration between upstream land managers and downstream water users, the MCPP is one of the partners implementing the watershed management plan for the Middle Cedar Watershed (MCWMA, 2020). Initial seed funding of $2 million came from the Natural Resources Conservation Service (NRCS) under a relatively new program to facilitate larger and more diverse regional partnerships focused on NBS, the Regional Conservation Partnership Program (MCWMA, 2020). Partners then leveraged about $2.3 million in

---

Box 9.3   Middle Cedar Partnership Project partners.

Farmers and Agricultural Producers
- Individual farmers and producers

Farmers' Associations:
- Iowa Soybean Association (ISA)
- Iowa Pork Producers Association (IPPA)
- Iowa Corn Growers Association (ICGA)
- Iowa Farm Bureau (IFB)

Farm-related Businesses:
- Dupont Pioneer (DP)

Municipalities:
- City of Cedar Rapids

State Entities:
- Iowa Department of Agriculture and Land Stewardship (IDALS)
- Iowa Department of Natural Resources (IDNR)
- Iowa State University Extension Service (ISUES)

Conservation Organizations:
- Benton Soil and Water Conservation District (BSWCD)
- Tama Soil and Water Conservation District (TSWCD)
- Benton/Tama Counties and Miller Creek WQI projects
- Black Hawk Soil and Water Conservation District (BHSWCD)
- Sand County Foundation (SCF)
- The Nature Conservancy (TNC)

Federal Government:
- Natural Resources Conservation Service (USDA NRCS)

additional funds for planning and implementing NBS for water quality, soil health, and flood mitigation. The first phase of the project developed watershed plans for 5 subwatersheds, while in the second phase, farmer-led implementation of NBS resulted in mobilizing $1.4 million for over 54 contracts in the first 3 years. In line with Iowa's science-based approach to watershed management, the project will evaluate results of individual NBS to improve understanding of their effectiveness.

Under the umbrella of the Middle Cedar Watershed Management Plan, the MCPP also coordinates with another innovative public-private partnership, the Midwest Row Crop Collaborative (MRCC). Started in 2016, the MRCC brings together farmers and conservation groups with businesses across the food and agriculture value chain. Companies such as Cargill, Walmart, Unilever, Kellogg's, PepsiCo, General Mills, Land O'Lakes, McDonalds, Monsanto, and Bayer are working with The Nature Conservancy, Environmental Defense Fund, and World Wildlife Fund to implement regenerative agriculture projects that deliver soil health, water quality, and climate benefits. With a focus on the scalability of nature-based watershed management tools, this program aims not only to improve nutrient management for source water protection but also to create more resilient agricultural systems in the region.

The MCPP has expanded the funding available for implementing NBS in the watershed and made available technical assistance to farmers that helps them access a wider range of federal conservation incentive programs. By evaluating outcomes of NBS for nutrient and flood risk reduction, the MCPP will contribute to targeting NBS in the watershed for greater effectiveness and can help build a stronger evidence base for the cost-effectiveness and cobenefits of NBS. Emerging multisector regional partnerships such as the MCPP are not only improving water quality but also demonstrating how rural and urban communities share interests and can work together toward common goals.

A broader conception of source water protection

As noted in Mauroner et al. (2021) the way that issues or problems are defined and scoped can constrain (or expand) the potential solutions that are considered. As in the MCPP case previously described, a broader conception of source water protection requires utilities to embrace new roles as watershed stewards, engaging in diverse partnerships with multiple stakeholders. The transformation of institutional cultures within utilities is part of this change (see also Coxon et al., 2021), along with the development of new guidance and management tools for more fully integrating natural assets and NBS as alternatives or complements to gray infrastructure (Matthews and de la Cruz, 2020; https://www.adb.org/projects/documents/reg-50159-001-tacr-3). The two cases below show how consideration and management of natural assets are moving into core utility and municipal decision-making processes.

Natural asset management
Natural asset management framework for water utilities

Water utilities in the United States and Canada are moving toward more formalized asset management (AM) practices as recognized best practice in the industry. As defined by the American Water Works Association, AM is the "set of activities within an organization to realize the overall value from all its assets through stronger governance and accountability....the combination of management, financial, economic, engineering and other practices applied to all assets (infrastructure, people, processes, and systems) with the objectives of providing the required level of service at an acceptable level of risk at an optimal life cycle cost" (AWWA, 2018). As the risks of watershed degradation or loss of hydrological ecosystem services become clear to water managers, there is growing recognition of the need to include natural assets as a class of assets that are key to service delivery. Water utilities can access guidance on developing AM plans for gray infrastructure from a variety of sources, including the US EPA, the American Water Works Association, and the Water Research Foundation, but there is no comparable guidance for natural AM. To fill this need, the Water Research Foundation in 2019–20 developed a Natural Asset Management Framework to help guide water utilities in beginning to incorporate critical natural assets in their AM plans (WRF, 2020; https://www.waterrf.org/resource/asset-management-framework-forested-and-natural-assets).

Basing the framework and guidance on established AM best practices for gray infrastructure, the guidance for natural AM walks utilities through a series of guiding questions and describes how the steps in the AM process for gray infrastructure can be adapted to address the special circumstances of natural assets. For example, a utility's first step in AM for their gray infrastructure may be to conduct an inventory—to ask which assets the utility owns—to inventory their treatment plants, storage reservoirs, and distribution system. For natural assets a utility needs to look more broadly at natural assets owned by others, in source watersheds and other lands (e.g., lands that can contribute sediment to drinking water reservoirs) that can impact (positively or negatively) the utility's ability to deliver services. The natural AM framework describes five steps along with guiding questions that utilities can use to begin to consider natural assets in their AM plans (Box 9.4) (WRF, 2020).

One of the biggest differences between gray infrastructure AM and natural AM is ownership and control of the assets. While some utilities own watershed lands (e.g., Seattle) and are already managing these natural assets, most utilities do not own or control most (or any) significant areas of their source watersheds. There is a particular need for guidance on how AM frameworks and systems can incorporate natural assets that are not owned or managed by the utility but are still critical to the utility's core mission and, if lost

---

Box 9.4   Steps and guiding questions for natural AM.

1. Identify relevant natural assets and describe the current state
   a. Which natural assets are important for utility performance?
   b. Who owns those assets?
   c. What is the current condition of those assets?
   d. Are there threats to the future conditions of those assets?
2. Determine desired level of service
   a. For core business goals, how do natural assets affect performance?
   b. What level of service from the asset is required to maintain or improve that performance?
   c. For other utility goals (ESR), what level of service will meet community expectations and regulatory requirements?
3. Assess business risks associate with natural assets
   a. How can natural assets degrade and fail to deliver desired levels of service?
   b. What is the likelihood of degradation?
   c. What are the potential consequences of degraded level of service on the utility's vulnerability to reduced performance?
4. Identify capital investment and maintenance opportunities
   a. Are there feasible acquisition, conservation easement or restoration options?
   b. Are there long-term maintenance or enhancement options?
   c. What partnerships might be required to execute these?
5. Create a long-term funding strategy
   a. Can the utility capture spending on natural assets through rates?
   b. Can external funds be leveraged through partnerships?
   c. Can investments be made through the capital budget (e.g., debt financed)?

---

or degraded, can pose significant risks to or increase costs of service delivery. This makes it necessary for utilities to enter into partnerships to effectively manage critical natural assets. Multistakeholder partnerships such as the MCPP regional partnership previously described provide models for how utilities may effectively do this. A utility's natural AM plan can be integrated within overall watershed management plans, allowing utilities to collaborate with partners and leverage the processes of watershed assessment, planning, and management to identify and collaboratively manage the watershed assets of importance to the utility.

Municipal natural asset initiative

One of the new models for natural AM is the Municipal Natural Assets Initiative in Canada. In the early 2000s, led by the Town of Gibsons in British Columbia, municipalities in the region were recognizing that their ability to serve their communities were dependent on a diverse array of natural assets. The ability to provide clean water, manage destructive river flooding, control urban stormwater flooding, and provide recreational amenities depend on natural assets that were traditionally not part of how the city planned and managed their built assets such as roads and water treatment plants. The Town of Gibsons faced two important realizations (Town of Gibsons, 2015):

1. An inventory of traditional gray infrastructure assets (roads, bridges, treatment plants) revealed how many assets the town owned, and that they needed to plan for annual replacement costs that were almost one-third of the total annual revenues available for all existing asset operations, maintenance, and repair, as well as construction of new assets. This level of spending could threaten the town's long-term financial sustainability.
2. At the same time a study showed that the aquifer underneath the town held enough high-quality water to supply the current population, as well as future anticipated growth. The only cost to the town of the aquifer was an annual monitoring effort, but if the aquifer were degraded the town would be faced with the costs and liabilities of constructing new water treatment and storage infrastructure.

Gibsons needed a more cost-effective asset strategy overall, and by treating the aquifer as an asset, it could be officially included in the town's AM plan, ensuring actions and funds were secured for maintaining the long-term health of the aquifer. Gibsons subsequently developed a comprehensive "eco-assets" strategy to assess other natural assets that provide valuable water filtration, coastal storm protection, flood mitigation, stormwater management, and recreation services to the community. In 2009, the town passed a municipal AM policy, the first of its kind in North America. This policy explicitly defines and recognizes natural assets as an asset class, and requires that the town "operate, maintain, and replace" natural assets along with traditional gray infrastructure assets and have natural AM strategies and funding in place to maintain them (Town of Gibsons, 2015).

Building on this experience, Gibsons joined the MNAI in 2016 as a founding member, along with the David Suzuki Foundation, Smart Prosperity Institute, and Brookes and Associates Consulting (https://mnai.ca/). The MNAI provides scientific, economic, and local government expertise to municipalities across Canada in identifying, valuing, and accounting for natural assets in their AM plans. To date about a dozen municipalities across Canada are developing natural AM plans in collaboration with MNAI, including in Ontario, where new provincial regulations explicitly support integrating natural assets into municipal AM plans (Box 9.5). And a new multicommunity, municipal-First Nations partnership is developing a natural AM plan focused on protecting drinking water supplies in the Comox Valley, BC (see Comox Case Study below).

Box 9.5   Municipal Natural Assets Initiative approach.

The guiding principles of MNAI are the following:

1. Municipal natural assets "are the stock of natural resources or ecosystems that are relied upon, managed, or could be managed by a municipality, regional district, or other form of local government for the sustainable provision of one or more municipal services" (https://mnai.ca/media/2018/01/FCMPrimer_Jan1_2018.pdf).
2. These assets include both ecosystems and green infrastructure such as green roofs, parks, and permeable pavement. Natural assets provide many of the same services to municipalities that engineered assets do (e.g., water filtration, flood control). Managing these assets can reduce costs to municipalities (e.g., CAPEX, OPEX, reduced water treatment cost) and/or reduce risks to municipalities (e.g., from wildfire impacts, flood damages).
3. Protecting and proactively managing and maintaining natural assets is a no-regrets, flexible strategy that can help communities strengthen resilience in the face of uncertain future changes.
4. Natural assets are multifunctional so that a single investment in managing a natural asset can deliver multiple benefits to the community (e.g., wetland protection can reduce flooding, improve water quality, provide recreational benefits). Source water protection is just one of the reasons municipalities have had for developing natural AM plans.

The MNAI Steps for Natural AM Plans:

*   Develop an AM policy, bylaw, or financial management statement directing the municipality to consider natural assets.
*   Identify key natural assets and the services they provide.
*   Determine the condition of natural assets in the community, do an initial valuation.
*   Conduct a risk analysis to determine which assets are highest priority based on risk (hazards, severity of impact, likelihood of occurrence).
*   Determine scenarios of possible future conditions to understand how natural assets will perform over time.
*   Begin managing natural assets (e.g., NBS to restore or enhance degraded assets; NBS management plans; funding and financing plans).

Ontario's Regulation for Asset Management Planning for Municipal Infrastructure (O. Reg. 588/17)

Ontario's Regulation 588/17 requires that all municipalities in Ontario prepare an AM plan for core infrastructure assets by July 1, 2021, and in respect of all other municipal infrastructure assets by July 1, 2023. In accordance with the regulation, "*core municipal infrastructure assets* include water, wastewater, and stormwater management assets, *including green infrastructure assets,* as well as roads, bridges, and culverts. Core infrastructure could include *green infrastructure, which the regulation defines as infrastructure assets consisting of natural or human-made elements that provide ecological and hydrological functions and processes, including natural heritage features and systems, parklands, stormwater management systems, street trees, urban forests, natural channels, permeable surfaces, and green roofs*" (MNAI, 2019).

Comox Watershed Initiative: Strengthening the Role of Nature in Protecting Drinking Water

The Comox Valley Watershed is located on the traditional territory of the K'ómoks First Nation on Vancouver Island, British Columbia. The watershed's glaciers, forests, lakes, and rivers provide drinking water for about 49,000 people in a number of communities in the Comox Valley, including the Village of Cumberland, the City of Courtenay, and the Town of Comox. The Comox Valley Regional District (CVRD) is a federation of municipalities and rural areas, providing a local government for unincorporated areas, a forum for regional cooperation, and serves as the provider of essential services (e.g., water system, fire response). Watershed lands, lakes and rivers are the source of drinking water for the region, provide recreational opportunities, support a logging economy, provide critical fish and wildlife habitat, and are important for the cultural, spiritual, social, economic, and environmental well-being of the K'ómoks First Nation.

To improve management of the natural assets that provide beneficial services, local communities, K'ómoks First Nation, the CVRD, and the MNAI signed a memorandum of understanding in 2019. By working with the MNAI, regional stakeholders will integrate natural assets such as forests, wetlands, and riparian areas into core AM and financial processes. By understanding, managing, and valuing these assets for the services they provide, the coalition can plan for actions that can proactively secure these services into the future. Protecting drinking water supplies, controlling erosion, and mitigating flood hazards are the focal watershed services of concern. The collaboration with the K'ómoks First Nation to manage natural assets for source water protection is particularly important, as there has been an ongoing challenge in providing Indigenous communities in Canada with access to affordable, clean drinking water and ensuring full Indigenous participation in source water protection efforts (Patrick, 2011; Marshall et al., 2018; Sarkar et al., 2015).

The MNAI brings to the partnership tools, methodologies, and evidence gathered from partnering with other local governments across Canada. Guidance materials, workshops, technical support, economic analysis, modeling, and the development of strategies and methods from MNAI will help inform the long-term management of natural assets in the watershed. The overall goals of the project are to: (1) understand the roles (current and future) of natural assets in the watershed in providing safe, reliable drinking water supplies and reducing flood risks; (2) develop effective management strategies for these assets; (3) quantify the value of natural assets; and (4) compare costs and benefits of maintaining natural assets with constructing new gray infrastructure alternatives (e.g., new filtration plant) or the long-term O&M of existing gray infrastructure.

Source: Comox Valley Regional District and MNAI.

Conclusions

Natural assets and NBS are increasingly embedded as core elements in source water protection. New multisectoral, multistakeholder partnerships are putting NBS at the heart of strategies for improving watershed health, supporting rural and urban communities, and improving the flexibility and resilience of water systems throughout North America. New approaches are making it possible for NBS to be considered within the same planning and project development processes as gray infrastructure, enabling nature-based options to be fully considered from initial decision-making steps. However, sustaining these initiatives and replicating them in more communities still face a number of challenges.

Funding for water infrastructure in general (gray and green) in North America has been inadequate compared to the need, and NBS projects still do not fit easily into existing procurement and project finance models. New funding and financing instruments for NBS are in development, and existing funding models for gray infrastructure are being applied to NBS but not yet at a scale that is needed. While new tools such as natural AM plans can begin to bring NBS into the same project development and procurement system as gray infrastructure, risk averse institutional cultures can still create roadblocks to implementation. To fully integrate NBS with gray options in planning and procurement decisions, approaches for evaluating the costs and benefits of NBS need to be strengthened to evaluate the full suite of benefits, including difficult-to-value aspects such as flexibility and resilience.

References

Abell R, Asquith N, Boccaletti G, Bremer L, Chapin E, Erickson-Quiroz A, Higgins J, Johnson J, Kang S, Karres N, Lehner B, McDonald R, Raepple J, Shemie D, Simmons E, Sridhar A, Vigerstøl K, Vogl A, Wood S. Beyond the source: The environmental, economic, and community benefits of source water protection. Arlington, VA: The Nature Conservancy (TNC); 2017. www.nature.org/beyondthesource.

Acreman M, Holden J. How wetlands affect floods. Wetlands 2013;33(5):773–86. https://doi.org/10.1007/s13157-013-0473-2.

Adams HD, Guardiola-Claramonte M, Barron-Gafford GA, Villegas JC, Breshears DD, Zou CB, et al. Temperature sensitivity of drought-induced tree mortality portends increased regional die-off under global-change-type drought. Proc Natl Acad Sci 2009;106(17):7063–6. https://doi.org/10.1073/pnas.0901438106.

Alcott E, Ashton MS, Gentry BS. Natural and engineered solutions for drinking water supplies. Boca Raton, FL: CRC Press, Taylor & Francis Group; 2013.

Anderegg WRL, Kane JM, Anderegg LDL. Consequences of widespread tree mortality triggered by drought and temperature stress. Nat Clim Chang 2013;3(1):30–6. https://doi.org/10.1038/nclimate1635.

Antolini F, Tate E, Dalzell B, Young N, Johnson K, Hawthorne P. Flood risk reduction from agricultural best management practices. J Am Water Resour Assoc 2019;1–19.

Associated Press. Flooded midwest towns may not have water for weeks, https://weather.com/news/news/2019-04-05-midwest-drinking-water-problems-weeks-after-flooding; 2019. [Accessed 5 June 2020].

Auerbach DA, Deisenroth DB, McShane RR, McCluney KE, LeRoy Poff N. Beyond the concrete: accounting for ecosystem services from free-flowing rivers. Ecosyst Serv 2014;10:1–5. https://doi.org/10.1016/j.ecoser.2014.07.005.

AWWA (American Water Works Association). AWWA asset management definitions guidebook. Version 1.0. American Water Works Association Asset Management Committee; 2018. January 2018. [Accessed 10 December 2020] https://www.awwa.org/Portals/0/AWWA/ETS/Resources/AMGuidebook.pdf?ver=2018-12-13-100101-887.

Bates BC, Kundzewicz ZW, Wu S, Palutikof JP. Climate change and water. Climate change and water. Intergovernmental Panel on Climate Change; 2008. https://doi.org/10.1016/j.jmb.2010.08.039.

Bennett G, Ruef F. Alliances for green infrastructure: State of watershed investments 2016. Forest Trends Ecosystem Marketplace; 2016. https://www.forest-trends.org/publications/alliances-for-green-infrastructure/.

Binder S, Haight RG, Polasky S, Warziniack T, Mockrin MH, Deal RL, Arthaud G. Assessment and valuation of forest ecosystem services: State of the science review. Forest Service General Technical Report NRS-170 Northern Research Station, (May). Retrieved from https://www.fs.fed.us/nrs/pubs/gtr/gtr_nrs170.pdf; 2017.

Birkinshaw SJ, Bathurst JC, Iroumé A, Palacios H. The effect of forest cover on peak flow and sediment discharge-an integrated field and modelling study in central-southern Chile. Hydrol Process 2011;25(8):1284–97. https://doi.org/10.1002/hyp.7900.

Bladon KD, Emelko MB, Silins U, Stone M. Wildfire and the future of water supply. Environ Sci Technol 2014;48(16):8936–43. https://doi.org/10.1021/es500130g.

Blanchard L, Vira B, Briefer L. The lost narrative: ecosystem service narratives and the missing Wasatch watershed conservation story. Ecosyst Serv 2015;16:105–11.

Brauman KA, Daily GC, Duarte TK, Mooney HA. The nature and value of ecosystem services: an overview highlighting hydrologic services. Annu Rev Env Resour 2007;32(1):67–98. https://doi.org/10.1146/annurev.energy.32.031306.102758.

Brauman KA, Benner R, Benitez S, Bremer L, Vigerstol K. Water funds. In: Mandle L, Ouyang Z, Salzman J, Daily GC, editors. Green growth that works. Washington, DC: Island Press; 2019.

Browder G, Ozment S, Bescos I, Gartner T, Lange G-M. Integrating green and gray: Creating next generation infrastructure. Washington, DC: World Bank and World Resources Institute; 2019. https://files.wri.org/s3fs-public/integrating-green-gray-executive-summary.pdf.

Bullock A, Acreman M. The role of wetlands in the hydrological cycle. Hydrol Earth Syst Sci 2003;7(3):358–89. https://doi.org/10.5194/hess-7-358-2003.

Burek P, Mubareka S, Rojas R, Roo D, Bianchi A, Baranzelli C, et al. Evaluation of the effectiveness of natural water retention measures support to the EU blueprint to safeguard Europe's; 2012. https://doi.org/10.2788/5528.

Caldwell P, Muldoon C, Miniat CF, Cohen E, Krieger S, Sun G, et al. Quantifying the role of national forest system lands in providing surface drinking water supply for the southern United States. Asheville, NC: US Forest Service; 2014. p. 135.

Canning JE, Stillwell AS. Nutrient reduction in agricultural green infrastructure: an analysis of the Racoon River watershed. Water 2018;10:749. https://doi.org/10.3390/w10060749.

Church SP, et al. How water quality improvement efforts influence urban-agricultural relationships. Agric Hum Values 2020. https://doi.org/10.1007/s10460-020-10177-8.

Cohen-Sacham E, Walters G, Janzen C, Maginnis S, (eds.). Nature-based solutions to address global societal challenges. Gland: IUCN; 2016. https://doi.org/10.2305/IUCN.CH.2016.13.en.

Coxon C, Gammie G, Cassin J. Mobilizing funding for nature-based solutions: Peru's drinking water tariff. In: Cassin J, Matthews JH, Lopez-Gunn E, editors. Nature-based solutions and water security: An action agenda for the 21st century. Elsevier; 2021.

Dadson SJ, Hall JW, Murgatroyd A, Acreman M, Bates P, Beven K, et al. A restatement of the natural science evidence concerning catchment-based 'natural' flood management in the UK. Proc R Soc A Math Phys Eng Sci 2017;473(2199):20160706. https://doi.org/10.1098/rspa.2016.0706.

Dahl TE. Wetlands losses in the United States 1780's to 1980's. Washington, DC: US Department of the Interior, Fish and Wildlife Service; 1990.

David MB, Drinkwater LE, McIsaac GF. Sources of nitrate yields in the Mississippi river basin. J Environ Qual 2010;39:1657–67. https://doi.org/10.2134/jeq2010.0115.

Deal RL, Cochran B, LaRocco G. Bundling of ecosystem services to increase forestland value and enhance sustainable forest management. Forest Policy Econ 2012;17:69–76. https://doi.org/10.1016/j.forpol.2011.12.007.

Dennison PE, Brewer SC, Arnold JD, Moritz MA. Large wildfire trends in the western United States, 1984–2011. Geophys Res Lett 2014;41:2928–33. https://doi.org/10.1002/2014GL059980.

Des Moines Water Works. Des Moines Water Works remains focused on source water protection. May 8, 2017. [Accessed 12 October 2020] https://www.dsmh2o.com/des%e2%80%88moines-water-works-remains-focused-on-source-water-protection/; 2017.

Dominique K, Matthews N, Danielson L, Matthews JH. Why governments embrace nature-based solutions: the policy rationale. In: Cassin J, Matthews JH, Lopez-Gunn E, editors. Nature-based solutions and water security: An action agenda for the 21st century. Elsevier; 2021.

Eggermont H, Balian E, Azevedo JMN, Buemer V, Brodin T, Claudet J, Fady B, Grube M, Keune H, Lamarque P, Reuter K, Smith M, van Ham C, Weisser WW, Le Roux X. Nature-based Solutions: new influence for environmental management and research in Europe. GAIA 2015;24(4):243–8.

Elmer M. Water works plans $15 million for expanded nitrate facility. Des Moines Register; 2017. May 24. [Accessed 10 December 2020] https://www.desmoinesregister.com/story/news/2017/05/25/water-works-plans-15-millionexpanded-nitrate-facility/336648001/.

Emelko M, Sham CH. Wildfire impacts on water supplies and the potential for mitigation: Workshop report; 2014.

Emelko MB, Silins U, Bladon KD, Stone M. Implications of land disturbance on drinking water treatability in a changing climate: demonstrating the need for "source water supply and protection" strategies. Water Res 2011;45:461–72. https://doi.org/10.1016/j.watres.2010.08.051.

Ernst C. Protecting the source. The Trust for Public Land and American Water Works Association; 2004. Retrieved from http://www.tpl.org/publications/books-reports/reportprotecting-the-source.html.

Filoso S, Bezerra MO, Weiss KCB, Palmer MA. Impacts of forest restoration on water yield: a systematic review. PLoS One 2017;12(8):e0183210. https://doi.org/10.1371/journal.pone.0183210.

Fisher J, Acreman MC. Wetland nutrient removal : a review of the Wetland nutrient removal: a review of the evidence. Hydrol Earch Syst Sci 2004;8(4):673–85. https://doi.org/10.5194/hess-8-673-2004.

Freeman J, Madsen R, Hart K. Statistical analysis of drinking water treatment plant costs, source water quality, and land cover characteristics white paper. Retrieved from http://wren.palwv.org/library/documents/landnwater_9_2008_whitepaper.pdf; 2008.

Gartner T, Mulligan J, Schmidt R, Gunn J, editors. Natural infrastructure: Investing in forested landscapes for source water protection in the United States. Washington, DC: World Resources Institute; 2013.

Gibbens PG, Johnson PA, Piehl B. Wildfires and forest health-Colorado-big Thompson project. Northernwater. Org; 2013. (970), 18. Retrieved from http://www.northernwater.org/docs/WaterQuality/WG_studies_new/WildfiresForestHlthWhtPaper2013.pdf.

Gómez Martín E, Máñez Costa M, Schwerdtner Máñez K. An operationalized classification of nature-based solutions for water-related hazards: From theory to practice. Ecol Econ 2020;167.

Green TL, Kronenberg J, Andersson E, Elmqvist T, Gomez-Baggethun E. Insurance value of green infrastructure around cities. Ecosystems 2016;2:23–6.

Griebler C, Avramov M. Groundwater ecosystem services: a review Christian. Freshw Sci 2015;34(1):355–67. https://doi.org/10.1086/679903.

Grizzetti B, Lanzanova D, Liquete C, Reynaud A, Cardoso AC. Assessing water ecosystem services for water resource management. Environ Sci Policy 2016;61:194–203. https://doi.org/10.1016/j.envsci.2016.04.008.

Hanlon JW. Complementary safeguards for robust regional watershed governance in a federation: New York city and its municipal water supply. Environ Sci Policy 2017;75:47–55.

Hanson MJ, Keller A, Boland MA, Lazarus WF. The debate about farm nitrates and drinking water. Choices 2016;31(1):1–7. Agricultural and Applied Economics Association.

Headwaters Economics. The full community costs of wildfires. Research Report. May 2018, Bozeman, MT: Headwaters Economics; 2018.

Hey DL, Philippi NS. Flood reduction through wetland restoration: the upper Mississippi River Basin as a case history. Restor Ecol 1995;3:4–17.

IDALS (Iowa Department of Agriculture and Land Stewardship), IDNR (Iowa Department of Natural Resources), Iowa State University College of Agriculture and Life Sciences. Iowa nutrient reduction strategy: A science and technology-based framework to assess and reduce nutrients to Iowa waters and the Gulf of Mexico. Des Moines, IA: Iowa Department of Natural Resources; 2017. http://www.nutrientstrategy.iastate.edu/. [Accessed 10 September 2020].

Jones CS, Nielsen JK, Schilling KE, Weber LJ. Iowa stream nitrate and the Gulf of Mexico. PLoS One 2018;13:e0195930. https://doi.org/10.1371/journal.pone.0195930.

Kenny A. Ecosystem services in the New York City Watershed. Forest Trends Ecosystem Marketplace; 2006. February 10, 2006 https://www.forest-trends.org/ecosystem_marketplace/ecosystem-services-in-the-new-york-city-watershed-1969-12-31-2/.

Liang K, Jiang Y, Qi J, Fuller K, Nyiraneza J, Meng FR. Characterizing the impacts of land use on nitrate load and water yield in an agricultural watershed in Atlantic Canada. Sci Total Environ 2020;729:138793.

Logan JA, Powell JA. Ghost forests, global warming, and the mountain pine beetle. Am Entomol 2001;47(3):160–73.

López-Morales CA, Mesa-Jurado MA. Valuation of hidden water ecosystem services: the replacement cost of the aquifer system in central Mexico. Water (Switzerland) 2017;9(8). https://doi.org/10.3390/w9080571.

Luce CH. Effects of climate change on snowpack, glaciers, and water resources in the northern rockies. In: Halofsky JE, Peterson DL, Dante-Wood SK, Hoang L, Ho J, Joyce LA, editors. Climate change vulnerability and adaptation in the Northern Rocky Mountains (Part I). Gen. Tech. Fort Collins, CO: U.S. Department of Agriculture, Forest Service, Rocky Mountain Research Station; 2018. p. 48–86. https://doi.org/10.1007/978-3-319-56928-4_3.

Machado E, Newman D, Coughlin L, Boel A, Fischer L, Brooke R. Towards an eco-asset strategy in the town of Gibsons. Retrieved from http://waterbucket.ca/gi/files/2015/09/Town-of-Gibsons_Eco-Asset-Strategy.pdf; 2014.

Maczko K, Tanaka JA, Breckenridge R, Hidinger L, Heintz HT, Fox WE, et al. Rangeland ecosystem goods and services: values and evaluation of opportunities for ranchers and land managers. Rangelands 2011;33(5):30–6. https://doi.org/10.2111/1551-501X-33.5.30.

Maltby E, Acreman MC. Ecosystem services of wetlands: pathfinder for a new paradigm. Hydrol Sci J 2015;6667(December):37–41. https://doi.org/10.1080/02626667.2011.631014.

Marshall RE, Levison JK, McBean EA, Brown E, Harper SL. Source water protection programs and indigenous communities in Canada and the United States: a scoping review. J Hydrol 2018;562:358–70.

Matthews J, de la Cruz EC. A Practitioner's guide to nature-based solutions. Protecting and investing in natural capital in Asia and the Pacific. ADB Technical Assistant Consultant's Report, Project Number: 50159-001. October 2020, Asian Development Bank; 2020.

Matthews J, Matthews N, Simmons E, Vigerstol K. Wellspring: Source water resilience and climate adaptation. Arlington, VA: The Nature Conservancy; 2019. https://www.nature.org/content/dam/tnc/nature/en/documents/Wellspring_FULL_Report_2019.pdf.

Mauroner A, Anisha NF, dela Cruz E, Barrios E, Guntoju SS. Operationalizing NBS in low and middle income countries: redefining and 'greening' project development. In: Cassin J, Matthews JH, Lopez-Gunn E, editors. Nature-based solutions and water security: An action agenda for the 21st century. Elsevier; 2021.

McCarthy LF. Water source protection funds as a tool to address climate adaptation and resiliency of southwestern forests. USDA Forest Service, RMRS-P-71; 2014. p. 329–43.

MCWMA (Middle Cedar Watershed Management Authority). Middle cedar watershed management plan. Des Moines, IA: Middle Cedar Watershed Management Authority; 2020. http://www.middlecedarwma.com/the-watershed-plan.html. [Accessed 5 January 2021].

Mehan III GT, Carpenter AT. Bringing agriculture and drinking water utilities together for source water protection. J Am Water Works Assoc 2019;111(8):35–9.

Mehan III GT, Carpenter AT, Sullivan N. The farm bill and source water protection: one year later. J Am Water Works Assoc 2020;112(1):7.

Milner AM, Brown LE, Hannah DM. Hydroecological response of river systems to shrinking glaciers. Hydrol Process 2009;23(1):62–77. https://doi.org/10.1002/hyp.7197.

Muir MJ, Luce CH, Gurrieri JT, Matyjasik M, Bruggink JL, Weems SL, et al. Effects of climate change on hydrology, water resources, and soil; 2018 [chapter 4].

Municipal Natural Assets Initiative (MNAI). Advancing and integrating municipal natural asset management through asset management planning in Ontario. December 2019. [Accessed 22 December 2020] https://mnai.ca/media/2020/01/MNAI_MNAPOntario.pdf; 2019.

Murphy JD, Carpenter AT. USDA source water protection funding: Successes and opportunities. J Am Water Works Assoc 2020;112(4):50–9.

New York Times. 'Breaches everywhere': Flooding bursts midwest levees, and tough questions follow. New York Times; 2019. March 31, 2019. [Accessed 25 November 2020] https://www.nytimes.com/2019/03/31/us/midwest-floods-levees.html.

OECD. Public goods and externalities: Agri-environmental policy measures in selected OECD countries. Paris, France: OECD Publishing; 2015.

Ofosu-Amaah N, Abell R, Fabre J, Fleming P, Matosich M, Morrison J, Varghese T. Nature-based solutions and corporate water stewardship. In: Cassin J, Matthews JH, Lopez-Gunn E, editors. Nature-based solutions and water security: An action agenda for the 21st century. Elsevier; 2021.

Ozment S, Gartner T, Huber-Stearns H, Difrancesco K, Lichten N, Tognetti S. Protecting drinking water at the source: Lessons from watershed investment programs in the United States. Washington, DC: World Resources Institute; 2016.

Patrick R. Uneven access to safe drinking water for first nations in Canada: connecting health and place through source water protection. Health Place 2011;17:386–9.

Pires M. Watershed protection for a world city: the case of New York. Land Use Policy 2004;21:161–75.

Postel S, Thompson B. Watershed protection: Capturing the benefits of nature's water supply services. Nat Resour Forum 2005;29:98–108.

Price JI, Heberling MT. The effects of source water quality on drinking water treatment costs: a review and synthesis of empirical literature. Ecol Econ 2018;151:7–8. https://doi.org/10.1016/j.ecolecon.2018.04.014.

Robertson DM, Saad DA, Schwarz GE. Spatial variability in nutrient transport by HUC8, state, and subbasin based on Mississippi/Atchafalaya River Basin SPARROW models. JAWRA J Am Water Resour Assoc 2014;50(4):988–1009.

Sarkar A, Hanrahan M, Hudson A. Water insecurity in Canadian indigenous communities: some inconvenient truths. Rural Remote Health 2015;15:3354.

Schindler DW, Donahue WF. An impending water crisis in Canada's western prairie provinces. Proc Natl Acad Sci 2006;103(19):7210–6. https://doi.org/10.1073/pnas.0601568103.

Schoennagel T, Balch JK, Brenkert-Smith H, Dennison PE, Harvey BJ, Krawchuk MA, et al. Adapt to more wildfire in western North American forests as climate changes. Proc Natl Acad Sci 2017;114(18):4582–90. https://doi.org/10.1073/pnas.1617464114.

Secchi S, Mcdonald M. The state of water quality strategies in the Mississippi River Basin: Is cooperative federalism working? Sci Total Environ 2019;677:241–9.

Seewer J. Ohio river's huge algae bloom a warning for water suppliers. Phys.org; 2015. Oct. 19 https://phys.org/news/2015-10-ohio-river-huge-algae-bloom.html. [Accessed April 2019].

Smith DM, Matthews JH, Bharati L, Borgomeo E, McCartney M, Mauroner A, Nicol A, Rodriguez D, Sadoff C, Suhardiman D, Timboe I, Amarnath G, Anish N. Adaptation's thirst: accelerating the convergence of water and climate action. In: Background paper prepared for the 2019 report of the global commission on adaptation; 2019. Rotterdam and Washington, DC www.gca.org.

Spahr N, Dubrovsky N, Gronberg J, Franke O, Wolock D. Nitrate loads and concentrations in surface-water base flow and shallow groundwater for selected basins in the United States, water years 1990–2006. U.S. Geological Survey Scientific Investigations Report 2010–5098, Reston, VA: U.S. Geological Survey; 2010.

Statistics Canada, https://www150.statcan.gc.ca/n1/pub/16-201-x/2017000/sec-2-eng.htm; 2016. [Accessed 10 October 2020].

Stoner N. Memorandum to regional administrators, regions 1–10. Washington, DC: US Environmental Protection Agency; 2011. https://www.epa.gov/sites/production/files/documents/memo_nitrogen_framework.pdf. [Accessed 10 September 2020].

The Brookings Institution. The mega-fire phenomenon: Toward a more effective management model (a concept paper). Washington, DC: The Brookings Institution Center for Public Policy Education; 2005.

Town of Gibsons. Towards and eco-asset strategy in the town of Gibsons [Accessed 22 December 2020] https://gibsons.ca/wp-content/uploads/2017/12/Eco-Asset-Strategy.pdf; 2015.

USEPA. Integrated risk information system (IRIS)—Chemical assessment summary (nitrate). U.S. Environmental Protection Agency; 1987. https://cfpub.epa.gov/ncea/iris2/chemicalLanding.cfm?substance_nmbr=76.

Van Grinsven HJ, Ward MH, Benjamin N, De Kok TM. Does the evidence about health risks associated with nitrate ingestion warrant an increase of the nitrate standard for drinking water? Environ Health 2006;5(1):26.

Vigerstol K, Abell R, Brauman K, Buytaert W, Vogl A. Addressing water security through nature-based solutions. In: Cassin J, Matthews JH, Lopez-Gunn E, editors. Nature-based solutions and water security: An action agenda for the 21st century. Amsterdam: Elsevier; 2021.

Vintinner E. Thirsty metropolis: a case study of New York City's drinking water. Lessons Conserv 2009;2:110–32.

Warziniack T, Sham CH, Morgan R, Feferholtz Y. Effects of forest cover on drinking water treatment costs. Retrieved from https://www.fs.fed.us/rm/pubs_journals/2016/rmrs_2016_warziniack_t001.pdf; 2016.

Westerling AL, Hidalgo HG, Cayan DR, Swetnam TW. Warming and earlier spring increase western U.S. forest wildfire activity. Science 2006;313(August):940–3.

Wickham JD, Wade TG, Riitters KA. An environmental assessment of United States drinking water watersheds. Landsc Ecol 2011;7(1):605–16. Springer, New York, NY.

Winiecki E. Economics and source water protection, http://yosemite.epa.gov/r10/water.nsf/c6e3c862e806dd688825688200708c97/04a73c144395fda18825702e00650eb2/$FILE/Economics_of_SWP_E_Winiecki_EPA.ppt#8; 2012. [Accessed May 2013].

WRF (Water Research Foundation). Asset management framework for forests and natural assets; 2020. Project No. 4727.

WWAP (United Nations World Water Assessment Programme)/UN-Water. The United Nations World Water Development Report 2018: Nature-based solutoins for water. Paris: UNESCO; 2018.

CHAPTER 10

Protecting source waters in Latin America

Marta Echavarria[a], Jan Cassin[b], and Jose Bento da Rocha[c]
[a]*Ecodecision, Quito, Ecuador,* [b]*Forest Trends, Washington, DC, United States,* [c]*Secretary of State of Environment and Sustainable Development, Goias, Brazil*

Key messages

- Locally led nature-based solutions (NBS) initiatives in Latin America have innovated a diversity of flexible approaches that helped inspire national-level programs and/or replication in many countries across the region, including water funds, investments in watershed services, and reciprocal agreements.
- Local efforts also helped enable and inform regional and national-level policies, while national policies provided frameworks that supported local programs; local and national efforts are mutually reinforcing.
- Regional networks for sharing information, learning, and capacity-building, combined with a strong push toward results-based programs and monitoring is helping build confidence that NBS can deliver desired water quality, quantity, or hazard reduction outcomes as well as climate and food security benefits.

Introduction

Investments in natural infrastructure and in the nature-based solutions (NBS) that sustain natural infrastructure have followed an interesting trajectory over the past few decades in Latin America (see also Coxon et al., 2021; Box 10.1). In many respects, this path has been similar to that seen in the "adoption curve" for innovations and new technologies in other fields (e.g., medicine) (Lorenzi et al., 2008). New innovations are implemented through a few pioneering cases that generate sufficient enthusiasm to mobilize a group of early adopters that begin to expand implementation. With sufficient evidence of success from early adopters, along with the development of supporting information, tools, and policies (e.g., social marketing, design principles), it may be possible to make the jump from the small set of early adopters to uptake more broadly, becoming a mainstream practice among a majority of users (Rogers, 2003).

Nature-Based Solutions and Water Security. https://doi.org/10.1016/B978-0-12-819871-1.00001-4

> **Box 10.1   Nature-based solutions for water.**
>
> NBS are inspired and supported by nature and use, or mimic, natural processes **to contribute to the improved management of water**. NBS can involve conserving or rehabilitating natural ecosystems and/or the enhancement or creation of natural processes in modified or artificial ecosystems. They can be applied at micro- (e.g., a dry toilet) or macro- (e.g., watershed restoration) scales.
>
> From UN-Water (United Nations World Water Assessment Programme/UN-Water). The United Nations world water development report 2018: Nature-based solutions for water. Paris: UNESCO; 2018. https://unesdoc.unesco.org/ark:/48223/pf0000261424.

In the late 1990s, conservationists began emphasizing the benefits that nature provides for people through recognizing the value of ecosystem services, particularly services related to water security, as central to social and economic well-being (Daily and Ellison, 2002). Many governmental and nongovernmental organizations (NGOs) committed to conservation realized that protecting these services, particularly those related to critical water supplies, could be a stronger motivation for investing in nature than biodiversity or species protection alone. As the growing water crisis and climate change impacts have become more evident, issues of water scarcity and water security are more to the forefront (Cassin and Matthews, 2021a,b). Most recently the global COVID-19 pandemic has put a new spotlight on the importance of natural infrastructure and water security (Cooper, 2020). Countries in Latin America have evolved a variety of NBS policies and initiatives at national, regional, and local levels to meet these challenges, and these represent a rich array of solutions for lessons and replication elsewhere.

This chapter presents an overview of the major initiatives on source water protection that prioritize NBS and describe the evolution of projects and programs that followed from early voluntary projects based on "classic" payments for ecosystem services (PES) models to a diversity of PES-type programs, national to local initiatives, and emerging multisector, collective action efforts (e.g., national government sanctioned programs, water funds). The diversity of partnerships, funding sources, and financing mechanisms are highlighted, including those associated with water funds (Quito, Tungurahua, FORAGUA, Ecuador; Northern Santander, Colombia; Fondo Quiroz, Peru; and Watershared, Bolivia). Brief case studies provide specific examples of how governments have approached NBS for source water protection in Costa Rica, Ecuador, Peru, and Brazil.

Latin America: A water superpower still facing water insecurity

Latin Americans live on a water-rich continent. As home to the Amazon and the Andes, the region is a water superpower. But this wealth in water sources is unequally distributed and highly threatened. With some exceptions, there is a gradient of relatively more to less water from north to south. Water is not always available all year round, and some areas face marked dry periods of seasonal scarcity. Human populations are not always settled where water is available; for example, most Peruvians live on the arid Pacific side of the Andes, where water is scarce.

The region's ecosystems underpin water security but are under threat

The region's iconic landscapes of tropical rainforests, montane cloud forests, high-altitude grasslands, and undisturbed wetlands play a crucial role in delivering the hydrological services that underpin water security but also contribute to the spatial variation in water supply and quality. Forests in the region help maintain water quality by stabilizing soils, reducing erosion, and filtering pollutants. Mountain forests in particular play an important role in regulating runoff and reducing erosion, helping maintain the clean, sediment-free water that is important for drinking water supplies and for hydroelectricity generation (Birkinshaw et al., 2011; FAO, 2013; Warziniack et al., 2016).

Forests also influence both local and regional water quantity. Locally, forests can affect infiltration, runoff rates, peak flows, and dry season flows (Brauman et al., 2012). Although the effects of forest cover on water yield are complex, cloud forests play a unique role by collecting moisture from clouds and fog, and they have a positive effect on augmenting water supply, as cloud water interception can account for 10%–35% of total precipitation (Bruijnzeel, 2004, 2005). Tropical montane cloud forests, or mid-elevation forests between 700 and 3500 m elevation, are literally clouded in humidity, providing a rich habitat for mosses and other water-loving plants, like orchids and bromeliads, filtering clean and bountiful water. The Amazon basin extending across portions of Venezuela, Colombia, Ecuador, Peru, Brazil, and Bolivia is the largest rainforest on Earth and affects regional precipitation patterns. Large areas of forest recycle precipitation through evapotranspiration, which recharges atmospheric moisture that then contributes to regional rainfall (Ellison et al., 2017). The contribution to regional precipitation can be significant in some regions, for example the Río de Plata Basin is estimated to receive more than 70% of its rainfall from evapotranspiration from the Amazon forest (Van der Ent et al., 2010).

Montane grasslands, such as the high-altitude Andean *paramos*, which run from Costa Rica to Northern Peru, can be considered veritable water factories, since the soil has a high organic matter content, characterized as a sponge that infiltrates and slowly releases precipitation, contributing to maintaining stream flows (Buytaert et al., 2006; Echavarria and Stern, 2019). Paramos, especially the wet paramos of the northern Andes, are characterized by a highly buffered flow regime, a baseflow-dominated response, and high water yield (Ochoa-Tocachi et al., 2016). The region's wetlands are also important for source water protection and flood control. Floodplain wetlands moderate flood flows, reducing the magnitude and duration of flooding (Acreman and Holden, 2013), and wetlands help maintain water quality by retaining sediments and absorbing and processing nutrients and pollutants (Fisher and Acreman, 2004).

This wealth in water resources is, however, highly threatened. Urbanization, deforestation, agricultural expansion, and pollution are all threatening these ecosystems that provide valuable hydrological services. Rates of deforestation across most forest types in the region continue to accelerate (Armenteras et al., 2017). The *Global Wetland Outlook* (Ramsar Convention on Wetlands, 2018) estimates that in Latin American, about 60% of historical wetlands have been lost and 70% of the region's remaining wetlands are threatened. Water pollution in almost all rivers across the region has worsened since 1990, and severe levels of pathogen pollution affects one-third of the region's rivers (Ramsar Convention on Wetlands, 2018). An assessment of land use change in the Colombian Andes found that paramos had one of the highest rates of conversion, primarily from natural vegetation to pasture (Rodriguez et al., 2012).

The continued expansion of agricultural areas and urban sprawl are causing a loss of vegetative cover with resulting impacts to reliable water flows and water quality. The climate crisis is also impacting the region, which has already experienced increasing droughts in the Amazon, flooding and landslides in the Andean highlands, and unusually strong El Nino events that have caused extensive damage. The impacts of climate change are predicted to vary greatly across this diverse region, and there is uncertainty about the direction and magnitude of impacts on water resources, especially at the local level. However, there is consensus that warmer temperatures, increased glacial melt, decreased snowpack with reduced runoff and groundwater recharge, increased water scarcity in some regions with longer dry periods, and increased variability in precipitation, including more intense rainfall in some regions, will characterize climate impacts in Latin America (Magrin et al., 2014). There is growing concern across the region about the need to gain resilience and be able to respond to the changing climate.

Unequal access and infrastructure gaps drive interest in nature-based solutions

Beyond the uneven geographical distribution of water, the region has to confront issues of equity in terms of access. It is estimated that 77 million people still do not have safe and secure drinking water (OECD/CAF/ECLAC, 2018). Water infrastructure is generally well-developed in urban areas, providing these populations with relatively good access to water and sanitation services. In contrast, many rural areas lack infrastructure and access to safe drinking water or sanitation services. Where water infrastructure is inadequate, poorer urban and rural populations are not able to pay for expensive alternatives. Even where infrastructure exists, poorer populations may not be able to pay for reliable services. Unchecked pollution from domestic sewage, solid waste, and even overuse of agricultural products pollutes streams and rivers throughout the region.

As in other regions (see also Cassin and Matthews, 2021a), water utilities and water authorities in Latin America have traditionally relied exclusively on gray infrastructure, such as desalination plants, interbasin transfers, and large storage reservoirs, to improve water security. New and improved gray infrastructure will continue to be needed to secure equitable access to clean drinking water and sanitation services, ensure supplies for irrigation and energy generation, and reduce flood risks. However, this infrastructure comes with high capital and operating costs (including social and environmental costs), is long-lasting and inflexible, making it less resilient and more vulnerable to climate impacts and other shocks (Poff et al., 2016). To address threats to water security and recognizing the limitations of traditional gray infrastructure, over the past two decades, the region has increasingly embraced NBS as a complement and, in some cases, an alternative to gray infrastructure.

Evolution of payments for ecosystem services programs

Latin American countries were early innovators in PES schemes as an alternative to approaches that relied only on setting aside protected areas or command-and-control approaches to environmental policy. As global studies estimated the economic value of nature to the global economy, greater interest was directed toward recognizing direct payments and in-kind compensation to landowners for their effort to protect and restore the region's forests (Landell-Mills and Porras, 2002). In particular, rather than focusing on making polluters pay for impairments to water resources, the agricultural and drinking water sectors in the region became interested in using incentives to pay for NBS to improve water security (Echavarria et al., 2015). Although an early adopter of PES approaches, the region has continually adapted these schemes, evolving "classic PES" into a variety of approaches for investing in NBS. Importantly, most of these regional innovations originated in efforts to protect source watersheds—primarily to protect drinking water supplies and hydroelectric generation capacity—and these source water protection efforts opened the door to the variety of NBS approaches we see today.

From "classic" PES models for NBS to a broader conception and practice

While the classical model[a] for PES (Engel et al., 2008; Wunder, 2005) may apply in a few actual cases, more than 20 years of experience with PES programs around the world demonstrates that PES in practice rarely conform to the theoretical definition, with a variety of local, regional, and national schemes developed under the general PES policy instrument (Salzman et al., 2018; Shapiro-Garza et al., 2020). To better capture the broad set of practices, Salzman et al. (2018) suggest defining PES as "the exchange of value for land management practices (NBS) intended to provide or ensure ecosystem services."

---

[a]   PES have been defined as a voluntary transaction where ecosystem managers (e.g., land owners), are compensated through conditional payments by ecosystem beneficiaries (often governments, with the public being the general beneficiary) for the additional cost of maintaining ecosystem services above legally required levels.

The values that incentivize NBS are diverse and can be cash payments, technical assistance, or materials (e.g., fending, seedlings, beehives). The variety of PES mechanisms includes government subsidies, collective action (e.g., water funds), bilateral agreements (e.g., water users: drinking water utilities, hydropower companies, upstream communities), and compliance-based programs such as water quality trading (Bennett and Carroll, 2014; Bennett and Ruef, 2016). In many regions, including Latin America, benefit-sharing or reciprocal agreements are the prevalent mechanisms for incentives rather than quid pro quo payments. An assessment of water funds in Latin America shows that even within a single PES mechanism in one region, there are significant differences across programs in terms of the biophysical and socio-economic objectives, the quantitative targets used to assess effectiveness, funding mechanisms, models of decision-making and implementation, activities, compensation strategies, and who receives compensation (Bremer et al., 2016).

Given this complexity, it is not surprising that early assessments of PES show that very few if any programs fit the assumptions of the "classical" PES model. These assumptions include, for example, that voluntary participants will engage in conservation practices or change land management practices if payments or incentives cover opportunity costs or that individual decisions to conserve or degrade ecosystems are primarily economic decisions and can be changed through financial incentives (Gomez-Baggethun et al., 2010). More recent assessments have focused on how, in practice, this classical model has been adapted to fit local contexts and needs and to ask a different set of questions about effectiveness, equity, and sustainability (Hendrickson and Corbera, 2015; Osborne and Shapiro-Garza, 2018; Shapiro-Garza et al., 2020).

As a way to understand why certain PES permutations are more or less successful in a given context, recent studies focus less on market principles and more on the specific histories of existing PES programs that reflect a multitude of geographical, institutional, political and cultural contexts (Nelson et al., 2020; Shapiro-Garza et al., 2020). Many existing PES programs (e.g., water funds and Peru's compensation for ecosystem services) developed from the recognition that water infrastructure needs to include natural infrastructure as well as gray infrastructure, and payments for watershed services or investments in watershed services addressed the lack of existing funding mechanisms for watershed conservation (Bennett and Carroll, 2014; Nelson et al., 2020). Over time, there has been a tendency for these programs to move away from a central focus on the financing mechanism per se (e.g., payments for services). For example, many programs in Latin America have evolved toward multistakeholder, participatory watershed management institutions that emphasize natural infrastructure investments for hydrological outcomes, but also for improving local livelihoods, supporting equitable benefit-sharing between upstream and downstream communities, and engaging multiple investors to broaden the funding base (Bremer et al., 2016; Kauffman, 2014). These adaptations that go beyond the economic efficiency emphasis of the classical PES model better ensure community and stakeholder buy-in, effectiveness, and long-term sustainability of the program (Adhikari and Agrawal, 2013; Mahanty et al., 2013; Wegner, 2016).

Multiple dimensions of effectiveness

During the early development of PES, only a very few programs were subject to rigorous expost analysis in either high- or low-income countries (Engel et al., 2008; Pattanayak et al., 2010). Despite this, Pattanayak et al. (2010), using an available sample of such studies, found generally positive but limited effectiveness in producing additional ecosystem services. While assessing effectiveness in terms of NBS outcomes is still challenging (e.g., determining counterfactual conditions, quantifying changes in ES provision), over the past 5–10 years, a number of rigorous studies have begun to assess effectiveness. Many of these document that PES programs are effective in delivering NBS outcomes, even though the magnitude of changes may be small (generally related to the small areas impacted by PES programs). For example, recent studies in Brazil and Bolivia document significant changes in the targeted design indicators, including increased forest cover, reduced deforestation, and increased baseflow attributed to practices implemented through PES programs (Ruggiero et al., 2019; Sone et al., 2018; Wiik et al., 2018).

In addition, given the multiple objectives of most PES programs and the linkages between equity (benefit-sharing, equity of access, and participation in decision-making; see also Bremer et al., 2021), assessments increasingly take a broad view of effectiveness. Numerous recent studies document the degree to which, and how, PES programs are effective not only in the delivery of services but also in terms of inclusion and improvement in social welfare (e.g., poverty alleviation, gender equity, security of land tenure) (see, for example, Schreckenberg et al., 2018). Similarly, much recent attention has been paid to the ways in which poorly designed PES can result in negative impacts to the well-being of local communities, reinforce existing inequities, and exclude traditionally marginalized groups, such as women and Indigenous communities (Richards, 2012; Wegner, 2016).

Finally, while challenges remain in quantifying effectiveness and cost-effectiveness, there are now many examples of the relative cost-effectiveness of investments in NBS. The bulk of studies come from urban NBS cases; in comparisons to gray infrastructure, NBS have lower capital costs and may have lower operations and maintenance costs, can result in significant avoided cost savings, and provide many cobenefits (e.g., lowered energy costs) (Gartner et al., 2014; Odefey et al., 2012). For example, in the United States, for every $1 spent on watershed investments, an average of $27 was saved in water treatment costs (Winiecki, 2012); while 11 of 12 urban NBS projects had lower total costs when compared to equivalent gray infrastructure projects (EPA, 2014).

This evolution of practice is reflected in the array of NBS approaches—institutions, policies, and financing mechanisms—that have developed in Latin America.

From pilots to national, government-led PES-type programs

In several cases, early nongovernmental pilot projects at the local level helped inspire and/or refine national efforts. For example, in Costa Rica in the late 1990s, the water company in Heredia (ESPH) began to pay land managers in their source watershed for NBS practices (forest protection) that would help maintain water quality. An additional fee was included on each water user's monthly water bill to fund NBS, and the fee was approved by the water authority (ARESEP) in 2000 (Redondo-Brenes and Welsh, 2010). The success of this model inspired the government to scale this to the national level to create an additional source of funding for the national PES scheme.

Similarly, in Ecuador, early source water NBS pilots that were partnerships between foundations, NGOs, and international development institutions became models for the later national-level program. To protect the water supply for the town of Pimampiro, the United Nations Food and Agriculture Organization (FAO) along with the private Inter-American Foundation provided initial seed funding for compensating land managers for forest and paramo protection. These funds were then pooled with money collected as part a 20% increase in charges on water users' monthly water bill in the municipality (Echavarría et al., 2004). Part of a larger sustainable development initiative, the incentives not only include compensation for forest and paramo protection but also technical assistance and capacity-building for improved organic farming and agroforestry practices (Echavarría et al., 2004; Wunder, 2011). Rather than being a simple PES between water users and NBS providers, this multipartner project required coordination and collaboration among development institutions, government agencies (Ministry of Environment), municipalities, water utilities, and land managers. Along with combining multiple sources of funding from private, public, and development entities, this type of local project reflected an approach that has been replicated at the national level as government-led programs evolved and provided models for the emerging regional programs in Latin America.

While many of the NBS programs for source water protection in the region remain local efforts, the early national-level programs, building on local pilots, helped pave the way for broad interest in and implementation of NBS. Beginning in the 1990s, governments in Mexico, Costa Rica, and Ecuador initiated national-level PES schemes to compensate land managers for implementing NBS for water security, as well as biodiversity, carbon sequestration, and other societal values. Costa Rica started its PES Program (PSA) scheme in 1997, coordinated by the National Forestry Financing Fund (FONAFIFO) administers and funds incentive payments to landholders for: greenhouse gas (GHG) mitigation, hydrological services, biodiversity conservation, and provision of scenic beauty (Arriagada and Perrings, 2009; Miranda et al., 2007; Pagiola, 2007; Zbinden and Lee, 2005). Mexico started its Hydrological Environmental Services Program (PSAH) in 2003 with earmarked funds from national water fees. The PSAH involves payments to landowning "ejido" and "agrarian communities," as well as individual landowners, for maintaining forest in hydrologically important areas. In 2004, the Payments for Carbon and Biodiversity Services Program (PSA-CABSA), which includes agroforestry systems, was added. These programs, managed by the National Forest Commission (CONAFOR), have now been integrated into the Program of Payments for Environmental Services (PSAB).

More recently, Ecuador created the Socio Bosque program of conservation incentives in 2008, and in 2009 the Ministry of Environment established the "Páramo Chapter" of Socio Bosque resulting in the additional conservation of Andean paramo ecosystems for protecting and regulating water resources (De Koning et al., 2011; Fehse, 2012). The goals for the program as a whole identify the bundle of valued ecosystem attributes and/or services that are the focus of conservation: (a) conserving globally important biodiversity; (b) reducing GHG emissions from deforestation; (c) protecting soils and water; (d) controlling natural disasters and climate adaptation; and (e) increasing incomes and protecting human capital in the poorest rural communities (De Koning et al., 2011). Public funding from annual budget allocations pays the costs of the program; annual funding levels are set by the National Secretary for Planning and Development (SENPLADES).

These programs all for the most part share common elements:

- Programs were initiated with public funding for public goods (e.g., gasoline tax, water tariff, general funds).
- They have evolved to include some mix of public and private funding—for example, engaging hydroelectric and beverage companies. Ecuador is the exception, although they are exploring bringing in additional funding sources to complement government funding.
- New national funds and institutions were established to manage and allocate funds to land managers.
- Early drivers were protecting water sources that were becoming impaired.
- While couched as PES, most are really subsidies to land managers for incorporating NBS and not payments for services.
- The NBS that are supported by these programs are predominantly forest protection and forest restoration, with the exception of Socio Paramo.
- Evaluation of effectiveness in terms of hydrological outcomes is still rare.
- Over time, programs have incorporated some goals beyond the targeted ecosystem services (e.g., poverty alleviation, equity).

National policies promoting nature-based solutions

Latin America has also pioneered several national policy frameworks for implementing NBS for source water protection. One of the most interesting is Peru's regulatory reforms in the drinking water sector that now require water utilities to allocate a portion of the revenue from water user fees to nature-based watershed investments. Three new laws have been instrumental in widespread adoption of NBS in Peru (Coxon et al., 2021):

- The 2013 Sanitation Services Modernization law requires the national water regulator (SUNASS), together with each water utility, to establish environmental compensation mechanisms and watershed management in their master plans. Based on these plans, SUNASS approves the new five-year tariff rates. The law also allows a portion of these tariff revenues to be invested in the natural infrastructure of source watersheds via NBS.
- The 2014 Mechanisms for Ecosystem Services Compensation (MRSE by its acronym in Spanish) law allows for these public financial resources to be invested in the conservation, recovery, and sustainable use of sources of ecosystem services through agreements between parties and stipulates that the actions of those who retain these services can be paid.
- The 2015 modification of the Sanitation Services Modernization law mandates that SUNASS and water utilities must incorporate compensation for ecosystem services (MRSE) into water tariffs, providing new funds for implementing and executing NBS projects.

Colombia since the 1980s has mobilized funding for environmental management from the sale of electricity, which is dependent on hydropower. Utilities are required to pay 3% of their sales to their local environmental agency and another 3% to the municipalities where their source watershed is located. This tax was envisioned to fund investment to protect the watersheds where the electricity was generated, however, to date, the tax has become a leading source of funding for the national environment agency (Rodriguez, 2019). Other examples of enabling policies in the region include the national Water Reserves program in Mexico and the Water Producers Program in Brazil.

Environmental flows/water reserves in Mexico: Natural flow regimes for water security

The natural flow regime in rivers plays a critical role in sustaining the array of ecosystem services that aquatic ecosystems provide to people, including freshwater provision, flood management, fisheries, energy generation, and recreation (Bunn and Arthington, 2002; Poff et al., 1997, 2017: Richter et al., 1997). Environmental flows is an approach to basing river management on maintaining the quantity, quality, and timing of river flows sufficient to maintain ecosystem functions and ecosystem services that address societal needs, simultaneously providing human well-being and biodiversity benefits (Poff et al., 2017). Translated into policies, this entails establishing flow volumes based on allocation mechanisms that allow rivers to be managed more sustainably, providing water to both people and ecosystems. Environmental flow requirements have been established for many individual rivers (e.g., Murray-Darling in Australia, Zambezi in Africa), and South Africa has established an Ecological Reserve to guide allocation on its rivers (Horne et al., 2017). Most recently, Mexico has initiated a national program to establish environmental flow allocations on 189 river basins (Salinas-Rodriguez et al., 2018). This ambitious program is significant because it puts nature-based approaches at the heart of water allocation and management in the country.

The national water law (LAN), published in 1992, required an official assessment to be made for all river basins of the water availability balance, and in the late 1990s, it became evident that overexploitation and overallocation were risks to the country's water security (Barrios, 2014). In 2004, in response to increasing water conflicts, overallocation in river basins, and overexploitation of aquifers, the World Wide Fund for Nature (WWF) and the NGO Fundación Gonzalo Río Arronte (FGRA) initiated a project to develop new models for water management in Mexico (Barrios, 2014). The project sought to determine how to ensure water for the environment for each river basin by establishing an environmental flow (e-flow) that was ecologically meaningful, while also providing sustainable allocations for agricultural, domestic, industrial, or energy uses in the basin (WWF, 2011). Focusing initially on three very different pilot river basins, the project evaluated the technical, social, and political feasibility of developing an e-flow that could be the basis for allocation among basin stakeholders and the environment (Salinas-Rodriguez et al., 2018).

Based on this pilot experience, the national water authority, CONAGUA, approved a proposal to create a Mexican technical standard for developing e-flows. This would be applied nationally to guide an allocation to the environment (a water reserve) in the official water availability balances for river basins. The water reserve is a legal right to water for the environment, so that this volume of water is excluded from the amount that is available for allocation to other uses. The establishment of water reserves marks the first time that the environment was included in the water availability balances under the LAN.

Subsequently, CONAGUA partnered with WWF and FGRA to conduct a scoping study to identify potential water reserves, with an initial emphasis on maintaining e-flows in basins that were in good shape in terms of low extraction rates and high conservation value. A total of 189 basins were identified for this first phase of implementation of the National Water Reserves for the Environment Program (CONAGUA, 2011). This first phase will build capacity to apply the Mexican e-flow standard, help protect conditions in these basins, and demonstrate how water reserves contribute to healthy river systems and ecosystem services (Barrios, 2014). The second phase of the program will include basins that already have significant pressure on their water resources. A recent assessment of the program concludes that in more than 90% of the basins, the water reserve allocations are feasible under the current water availability in the basin and, in particular, consistent with securing water provision for small rural communities (Salinas-Rodriguez et al., 2018).

While implementing water reserves where water resources are already under severe pressure will be more challenging, Barrios (2014) has stressed that the power of the nature-based water reserves is in providing a buffering capacity in the face of variability and uncertainty, reducing risks around water scarcity and conflict, and contributing to the nation's water security.

Water Producers Program, Brazil: Engaging the agricultural sector for water security

Brazil's first source water protection program began in the mid-1800s in response to deforestation in the headwaters of numerous streams and springs that supplied the city of Rio de Janeiro (Drummond, 1996). The government began acquiring land in the 1850s in the area that today is the Tijuca National Park, and in 1861 initiated an ambitious reforestation project to plant (mostly) native species in areas that had been deforested. Although many water challenges remain in today's city, the water originating in the forested slopes of the national park is of good to excellent quality (Moulton et al., 2007). The National Forest Code (Decree #23,793/1934) enacted in 1934 was designed in part for the protection and sustainable use of water resources. In part, the Forest Code requires landowners to protect or restore natural features that are important hydrologically, such as riparian buffers, springs, and steep slopes. Enforcement has been an issue, and changes to the Forest Code have weakened some of these requirements (Guidotti et al., 2020; Richards et al., 2015). For example, some agricultural activities are now allowed in riparian buffer zones instead of forest restoration, which reduces buffer effectiveness in removing sediment and nutrients, impairing water quality downstream (Guidotti et al., 2020).

Watershed management in Brazil is decentralized, although water use licenses and water policy is overseen at the federal level by the National Water Agency (Agencia Nacional de Aguas, ANA). Brazil's 1997 law, the National Policy on Water Resources (Federal Law n. 9433/97), established both the National Water Resources Policy and the National Water Resource Management System (Veiga and Magrini, 2013). The law created a new institutional framework for managing water resources with the establishment of local watershed committees and local water agencies. Also, through this law, the ANA created the Water Producer Program (Produtor de Agua), which established the legal basis for permitting monetary compensation to individual land managers for implementing conservation practices, essentially by authorizing the creation of local PES programs (Bento et al., 2017). These two pieces of the law established a financing mechanism and an implementation mechanism for NBS. Funds for implementing NBS are generated through water use fees in each basin, which support local governments and institutions in implementing PES projects in watersheds that are important for drinking water provision (Richards et al., 2015; Veiga and Magrini, 2013).

Initiated in the early 2000s, the Water Producers Program has implemented at least 40 PES projects, with most being in the Atlantic Forest region in southeastern Brazil (Viani et al., 2019). Every state in the country has adapted the program with important progress in many states, including São Paolo, Para, Mina Gerais, and more recently Goias. The focus is on rural communities and providing financial incentives (including via PES) to ranchers and farmers who voluntarily adopt soil and water conservation practices (ANA; https://www.gov.br/ana/pt-br/assuntos/noticias-e-eventos/noticias/produtor-de-agua-abre-programacao-da-ana-no-espaco-brasil-do-8o-forum). As such, the program not only recognizes the heavy reliance of the agricultural sector on sustainable water management (agriculture uses about 72% of total water use in Brazil) but also the role that farmers and ranchers can play in implementing NBS that contribute to water security for others.

Not all of the practices included in the program are nature-based, as it includes practices such as road improvements, small check dams, and detention basins to control erosion. However, the majority of practices are nature-based, such as avoiding deforestation by protecting existing forests, particularly around springs, restoring forested riparian buffers, planting vegetated buffers around farm fields, and replanting native forest species in deforested areas (ANA; https://www.gov.br/ana/pt-br/todos-os-documentos-do-portal/documentos-sip/produtor-de-agua/documentos-relacionados/manual-operativo-do-programa-produtor-de-agua/view).

Proliferation of regional and local programs

The national government-led programs to protect water sources through implementing and refining approaches for investments in NBS have helped inspire and provide models for similar efforts at the regional and sector level. In addition, Latin America is home to more local, user-driven PES programs that are implementing NBS for source water protection than any other region. Recent studies have reviewed these programs to assess their effectiveness in delivering hydrological outcomes and aspects of equity and sustainability (Grima et al., 2016; Martin-Ortega et al., 2013; Tellman et al., 2018). In this section we focus on highlighting the variety of financing and institutional modalities that are being used to enable NBS for source water protection. In particular these cases highlight how additional sectors, beyond the drinking water sector, are engaging in NBS for water security.

Local implementation of the water Producers Program in Brasilia

One example of the local PES implemented under the Water Producers Program is that in the Pipiripau River, part of the Atlantic Forest ecosystem, which provides drinking water for Brasilia, the capital of the country. The Pipiripau program includes more than 16 partners, including international NGOs, local municipalities, sanitation companies, the national water authority, the national water and sanitation services regulator ADASA, and rural producers cooperative associations (Lima and Ramos, 2018). Because the program is designed to improve both soil and water conservation, which benefit both the agricultural sector and urban water users, the program includes both NBS and gray infrastructure. As of 2018, these include planting of terraces for erosion control (1400 ha), restoring vegetation (250 ha, planted with 420,000 native seedlings), road improvements (135 km), and small water reservoirs (1316) (Lima and Ramos, 2018). Between 2009 and 2019, in order to promote the conservation of native vegetation, more than 200 farmers have received $1.25 million. The program has planted over 400,000 native plants in 2166 ha in protected buffers, or what in Brazil is defined as the *legal reserve* required by the Forest Code (Box 10.2).

The combination of NBS and gray infrastructure can strengthen results and help build political will as both urban and rural communities see benefits. Support to rural communities was also an invitation for leveraging other resources, or cofinancing, from other private and public entities. The program initially also included an education component to build awareness, support, and capacity to engage in the program. Changes in social values, knowledge, and skills among farmers has strengthened cooperative associations as the benefits of working together became evident. The improved soil and water management practices also provided direct benefits to farmers, keeping soil moisture high for longer periods, making crops more resilient to seasonal droughts, and increasing on-farm productivity (GWP, 2021). Water availability in the dry season is also anticipated from the increased infiltration, reduced runoff, and increased groundwater recharged (GWP, 2021). Finally, ADASA is evaluating whether funds for the watershed investments can be incorporated into the water utility tariff to assure the sustainability of the project and providing a model for replication in other watersheds in the region (GWP, 2021).

Collective action programs for NBS

Many of the regional programs illustrate one of the leading innovations that arose in the region for implementing NBS, the diversity of collective action initiatives. Two prominent examples of these multipartner, public-private-civil society partnerships are water funds and reciprocal agreements. Evolving out of conservation efforts in the region to manage watersheds for both biodiversity and water security (Asquith et al., 2008), environmental

---

Box 10.2   Investments in the Pipiripau Water Producers Program—2009–19.

Areas	US$
Soil conservation	1,125,000
Reforestation	1,450,000
Payments for environmental services	1,250,000
Santos Dumont Channel	1,900,000
Monitoring	In-kind contributions
Environmental education	In-kind contributions
Total	5,725,000

From *Jose Bento*, personal communication.

NGOs and their local partners had to address two challenges. First, funding for environmental initiatives by itself was insufficient for the scale of conservation that was needed. There was a need to tap multiple sources of funding; public, private, philanthropic, and international development funds. New financial structures needed to be created that could provide transparency and accountability for these pooled funds. Second, moving beyond the model of setting aside protected areas, watershed management for both people and nature requires the participation of a broad set of stakeholders. In particular, managing for water security requires engaging all water users across multiple sectors (drinking water, agriculture, energy) that do not typically work together.

The water trust fund model

Water funds are institutional platforms that bring together water users across multiple sectors (e.g., domestic, agriculture, industry, and energy) with conservation organizations, cities, and philanthropic donors to invest in source water protection (Abell et al., 2017). They are a way to pool funds from multiple investors, leveraging philanthropic and conservation funds with investments from other stakeholders to implement NBS, often through PES-like incentives for upstream land managers (Cooper, 2020). As collective action funds, water funds are characterized by pooled resources and coordinated action; they both mobilize funding and coordinate action for more effective source water protection (Brauman et al., 2019). Water funds have been particularly attractive to businesses with interests in source water protection. In 2015, a global survey found that 84% of private sector investments in NBS to protect source watersheds were made through collective action mechanisms, mostly water funds, rather than through the business acting as sole investor (Bennett and Ruef, 2016).

The Nature Conservancy has been instrumental in replicating successful water fund models globally, based on three organizing components: (1) a funding mechanism to collect and allocate resources for watershed management; (2) a governance mechanism for joint planning and decision-making; and (3) a watershed management mechanism for carrying out conservation and management activities (Brauman et al., 2019). Led by The Nature Conservancy, a growing network and partnerships are creating tools and resources for water fund developers that have supported a rapid expansion in the number of water funds over the past 10 years (e.g., The Latin American Water Funds Partnership, https://www.fondosdeagua.org/en/; Water Funds Toolbox, https://waterfundstoolbox.org/). From the first water fund in Quito, Ecuador, in 2000, the water fund model has been embraced widely, particularly in Latin America, with expansion into North America, Africa, and Asia in recent years (Box 10.3).

As water funds have been established across a diversity of geographic, political, and cultural contexts, they have evolved along a similar path as PES initiatives in the region.

Box 10.3   Evolution of water funds in Latin America.

In 2000, the first water fund was established through an agreement between the drinking water utility of Quito (EPMAPS) and The Nature Conservancy (Echavarria, 2002). Motivated as a way to fund the protection of three national parks surrounding the city that provided drinking water, an endowment was created with an independent private fund manager. The fund's creation was in response to deteriorating water quality from agricultural expansion into the parks, resulting in overgrazing, erosion, and nutrient and sediment runoff. In addition, the water company has been facing lower flows, having to go farther and farther from the city for new drinking water sources. A major objective of the fund was to improve the management of the parks to control detrimental land use practices.

FONAG (Fund for the Protection of Water) soon attracted other partners and investors. Beginning with the local electrical utility, the national beer company and a signature water bottling company eventually joined creating a multistakeholder platform for source water protection. The example of FONAG soon inspired other cities in Ecuador to create water funds, for example, in Latacunga province where work was underway to improve the livelihoods of the highland communities, predominantly Indigenous, receiving international support for poverty alleviation. A water trust fund was established in 2008 to develop master paramo protection plans. Also, ETAPA, the water utility of Cuenca, promoted the creation of a fund, which allowed it to leverage resources from other stakeholders in this mid-sized city to support the utility's widely recognized environmental protection programs. ETAPA was supporting the national park above the city and developing conservation programs with farmers and other landowners to improve land use practices. Creating a fund allowed ETAPA to bring in additional resources from other water users, such as the leading hydroelectric plant in the country. The Tungurahua Water Fund supports the conservation of watersheds in the Chimborazo National Park and surrounding buffer zones through activities such as reforestation and alternative income generation projects that provide a sustainable source of income for local communities and ensure protection of natural habitats. By protecting 1.3 million acres of Andean forests and grasslands, the Tungurahua Water Fund contributes to the conservation of forests that provide water for 350,000 local residents.

The replications inspired efforts in other cities in Latin America to develop water funds with The Nature Conservancy, including in Argentina, Brazil, Chile, Colombia, Costa Rica, Guatemala, Mexico, Panama, and Peru (The Nature Conservancy, https://s3.amazonaws.com/tnc-craft/library/2018-Water-Funds-Map-TNC_180911_132459.pdf?mtime=20180911132458). With support from the Global Environmental Facility, The Nature Conservancy joined with a leading regional bottling company FEMSA and the Inter-American Development Bank to create the Latin American Water Funds Partnership to support the establishment of water funds throughout the region.

For example, while conservation and source water protection remain priorities, additional goals have become important, such as improving equity, poverty alleviation, sustainable development, and climate mitigation (Abell et al., 2017; Bremer et al., 2016; Kauffman, 2014). Interestingly, in some funds (e.g., Colombia and Peru) the framing perspective has been much less like that of a PES scheme and more in line with concerns about maintaining a sustainable water infrastructure system, which combines both NBS and gray infrastructure (Nelson et al., 2020).

Reciprocal agreements

Another approach that has proven to be effective is the Watershared Program, which uses the "reciprocal watershed agreements" (ARA in Spanish) model to bring together farmers, municipalities, and conservation NGOs in collective action. Fundación Natura Bolivia helped initiate a noncash incentive-based program in Los Negros, Bolivia, which used negotiated reciprocal agreements between municipal water users and upstream farmers for watershed protection. Although, at first glance, similar to a classic PES model, while in-kind incentives are important, the approach emphasizes strengthening rural institutions, sustainable commons management, communities' livelihoods, and behavior change, building on and aligning with the strong traditions of reciprocity in these communities (Asquith, 2016, 2020). The approach also focuses on nonmonetary motivations for participation. For example, rather than monetary or even economic motivations, Bétrisey et al. (2016) found that social recognition was a strong reason for participation among farmers in communities in Bolivia's Santa Cruz department. From the initial ARA in Los Negros in 2003, the approach has been replicated in 58 municipalities in Bolivia, with 8000 upstream farmers conserving forests on about 350,000 ha, and similar agreements have been initiated in Colombia, Ecuador, Peru, and Mexico (Asquith, 2020). The ARA approach has been particularly successful in smaller watersheds where upper and lower watershed communities are located close together, the consequences of upstream land management on downstream water quality are clear, and programs can be locally managed (Asquith et al., 2008).

What has enabled this proliferation of NBS approaches in Latin America?

Our review of the evolution of NBS in the region suggests several enabling factors and conditions in the region that has spurred strong and rapid growth in NBS for water security.

Local initiatives innovated a diversity of flexible approaches

In Costa Rica, Peru, Colombia, and elsewhere in the region, we see numerous examples of locally led projects that have demonstrated the case for NBS and spurred larger regional and national programs and policies. The local bilateral PES in Heredia, Costa Rica, between the water company and upstream land managers helped provide a model for other local PES in the region and for the larger national program to conserve forests for water, biodiversity, climate mitigation, and scenic beauty. Water funds originated as a bilateral, water user-driven local arrangement between the water utility and rural communities in Quito's watershed, evolving into a replicable but flexible approach for collective action that has been adapted to different geographical, political, and regulatory contexts.

Local to national scales working together

Locally led programs often provided models for national programs (e.g., Heredia for Costa Rica's national PES), prompted the development of enabling policies, or worked in concert with national policies. For example, in Brazil, national policy created the framework for the Water Producers Program, but local flexibility in the design of incentive programs and collective action funds helped multiply projects more rapidly. As Coxon et al. (2021) show, the iteration and mutual influence between local initiatives and national enabling policies in Peru has transformed NBS approaches in that country from isolated, individual watershed projects to national, cross-sectoral policies and programs for mainstreaming NBS. Local initiatives often provided the demonstrations—the "business case"—and the social capital for making larger policy changes.

Individual relationships among innovative leaders across local to national scales also helped ensure that local pilots or initiatives would become models or inspiration for national efforts. For example, in Peru, strong leaders in local water utilities (e.g., Cuzco, Lima), the national water regulator (SUNASS), and Ministry of Environment all worked together to leverage successful local experiences to inform enabling policies at the national level (Coxon et al., 2021). Local efforts transforming national policies are seen in the number of countries that are implementing (Peru), or moving toward (Colombia, Costa Rica), requiring or allowing water tariff allocations to fund NBS. In a number of countries, leaders from civil society organizations were appointed to government positions, with more influence in forming public policy. Several leading environmentalists became ministers of environment, for example in Peru, Colombia, and Ecuador, promoting national agendas with more integrative and innovative approaches, including NBS. While many of these leaders came from the conservation or environmental sector, several were from the water sector, which was critical for making a strong case for NBS beyond conservationists.

Building evidence and capacity from the ground up

The region's strong embrace of NBS has been helped by a rich array of locally codeveloped and user-driven technical tools and information, evidence, knowledge sharing, and capacity-building. In Peru, early cost-effectiveness assessments of NBS compared to traditional gray infrastructure options helped decision makers commit to including watershed investments in water utility master plans (Coxon et al., 2021; Gammie and De Bievre, 2014). The Nature Conservancy and WRI have conducted a number of return-on-investment (ROI) studies of NBS that have helped inform or justify investments (Kroeger et al., 2019; Ozment et al., 2018). The ROI assessment for Quito's water fund was refined in collaboration with the utility's operations staff and resulted in an ROI estimate of 1:3 making NBS a competitive investment for the water utility (De Bievre, 2019).

Programs in the region are actively addressing one of the challenges to scaling NBS—evidence for how effectively NBS interventions deliver targeted outcomes, in terms of water quantity, water quality, or hazard reduction. The long-standing ARA program in Bolivia has been the subject of one of the few randomized control trials to assess the impacts of NBS on water or conservation outcomes (Asquith, 2020; Wiik et al., 2019). While this type of study can provide more robust evidence for NBS performance, very few programs have been designed or monitored to allow for this type of assessment. The programs in Peru have been active in developing methods to estimate, ex-ante, the performance of a number of NBS interventions, helping inform project designs as well as being a first step in establishing monitoring, evaluation, and adaptive management of NBS (Forest Trends, 2020).

Finally, regional efforts are building technical capacity and strengthening monitoring and evaluation, which is essential for scaling NBS and building better understanding of NBS performance. The Nature Conservancy's Water Funds Toolbox (https://waterfundstoolbox.org/) is a platform for resources in water funds—from planning through monitoring and evaluation. Resources include guidance on monitoring methods, evidence base for water-related outcomes of NBS, and training and networking for practitioners. Through this platform and the Latin American Water Funds Partnership, water funds in the region are building and sharing a growing body of information on the effectiveness of NBS. The regional association of Nation Water Regulators, ADERASA, has established a Green Infrastructure Working Group, which supports individual national members in efforts to incorporate NBS in their utility infrastructure portfolios. This group shares information and learning on policies, governance, technical issues of planning and implementation, and funding strategies for NBS. Over the past few years, ADERASA has collaborated with Forest Trends and Ecodecision to deliver a training course for water professionals on NBS design, policy, funding, implementation, and monitoring.

The Regional Initiative for Hydrological Monitoring of Andean Ecosystems (iMHEA) is a network of organizations (local and international NGOs, public institutions, and academic partners) interested in increasing and strengthening knowledge about the hydrology of Andean ecosystems to improve decision-making regarding the integral management of water resources at the Andean region level (http://imhea.org/). Through establishing hydrological monitoring systems in more than a dozen watersheds across the region, using paired watershed studies to study relationships between land cover/land uses and hydrological impacts, supporting community monitoring efforts, supporting local students pursuing advanced degrees in hydrology, and sharing information on results across the region, iMHEA is building both a stronger evidence base and more robust technical capacity for evaluating NBS (Ochoa-Tocachi et al., 2020).

Conclusions

Challenges in scaling NBS in the region remain, including uncertainties and knowledge gaps about the performance of NBS in delivering specific water benefits, limited private sector investments in NBS, and regulatory barriers that are a legacy of policies and processes designed for gray infrastructure. Despite these challenges, other regions can learn much from Latin America in terms of successful NBS programs and viable funding models. The region has been a leader in NBS for water security, driven particularly by early pilots for investments in watershed services led by water utilities. Local innovations that provided a diversity of flexible models and inspired larger regional and national programs, iteration between local initiatives and national policy that created strong frameworks for scaling in some countries, and regional networks for sharing knowledge, lessons, and capacity-building have all contributed to the wide uptake of NBS across the region. As benefits of NBS are increasingly embraced by the water sector, other sectors, such as agriculture and energy, are beginning to explore NBS as well. Cross-sectoral initiatives, such as Peru's, are beginning to take the more systemic approach to NBS that will be needed in the future, linking water, climate, energy, and food security.

References

Abell R, et al. Beyond the source: The environmental, economic and community benefits of source water protection. Arlington, VA: The Nature Conservancy; 2017.

Acreman M, Holden J. How wetlands affect floods. Wetlands 2013;33(5):773–86. https://doi.org/10.1007/s13157-013-0473-2.

Adhikari B, Agrawal A. Understanding the social and ecological outcomes of PES projects: a review and an analysis. Conserv Soc 2013;11(4):359–74.

Armenteras D, Espelta JM, Rodriguez N, Retana J. Deforestation dynamics and drivers in different forest types in Latin America: three decades of studies (1980 – 2010). Glob Environ Chang 2017;46:139–47.

Arriagada RA, Perrings C. Making payments for ecosystem services work. Nairobi: UNEP Ecosystem Services Economic Unit, Division of Environmental Policy Implementation, UNEP; 2009.

Asquith NM. Watershared: Adaptation, mitigation, watershed protection and economic development in Latin America. London: Climate & Development Knowledge Network; 2016.

Asquith NM. Large-scale randomized control trials of incentive-based conservation: what have we learned. World Dev 2020;127:1–3.

Asquith NM, Vargas Rios MT, Wunder S. Bundling environmental services: decentralized in-kind payments for bird habitat and watershed protection in Los Negros. Ecol Econ 2008;65:675–84.

Barrios E. Water management and ecosystems: A new framework in Mexico. In: Garrick D, Anderson GRM, Connell D, Pittock J, editors. Federal Rivers: Managing water in multi-layered political systems. London: Edward Elgar with IWA Publishing; 2014. p. 128–30.

Bennett B, Carroll N. Gaining depth: State of watershed investment 2014. Forest Trends Ecosystem Marketplace; 2014. https://www.forest-trends.org/publications/gaining-depth-2/.

Bennett G, Ruef F. Alliances for green infrastructure: State of watershed investments 2016. Forest Trends Ecosystem Marketplace; 2016. https://www.forest-trends.org/publications/alliances-for-green-infrastructure/.

Bento J, Vianna JS, de Oliveira HR, Rocha AJA. Water producer project in the Pipiripau river basin, Federal District—DF, Brazil. International Water Resources Association; 2017. https://iwra.org/member/congress/

resource/ABSID304_ABSID304_Water_Producer_Project_in_the_Pipiripau_River_Basin_Federal_District__DF_Brazil___Full_version.pdf.

Bétrisey F, Mager C, Rist S. Local views and structural determinants of poverty alleviation through payments for environmental services: Bolivian insights. World Dev Perspect 2016;1:6–11.

Birkinshaw SJ, Bathurst JC, Iroumé A, Palacios H. The effect of forest cover on peak flow and sediment discharge-an integrated field and modelling study in central-southern Chile. Hydrol Process 2011;25(8):1284–97. https://doi.org/10.1002/hyp.7900.

Brauman KA, Freyberg DL, Daily GC. Land cover effects on groundwater recharge in the tropics: ecohydrologic mechanisms. Ecohydrology 2012;5(4):435–44. https://doi.org/10.1002/eco.236.

Brauman K, Benner R, Benitez S, Bremer L, Vigerstol K. Water funds. In: Mandle L, Ouyang Z, Salzman J, Dailly GC, editors. Green growth that works. Island Press; 2019. p. 119–39. Kindle Edition.

Bremer L, Auerbach D, Goldstein J, Vogl A, Shemie D, Kroeger T, Nelson J, Benitez S, Cavache A, Guimaraes J, Herron C, Higgins J, Klemz C, Leon J, Lozano J, Moreno P, Nunez F, Veiga F, Tiepolo G. One size does not fit all: natural infrastructure investments within the Latin American Water Funds Partnership. Ecosyst Serv 2016;17:217–36.

Bremer L, Keeler B, Pascua P, Walker R, Sterling E. Nature-based solutions, sustainable development, and equity. In: Cassin J, Matthews JH, Lopez-Gunn E, editors. Nature-based solutions and water security: An action agenda for the 21st century. Elsevier; 2021.

Bruijnzeel LA. Hydrological functions of tropical forests: not seeing the soil for the trees? Agric Ecosyst Environ 2004;104:185–228. https://doi.org/10.1016/j.agee.2004.01.015.

Bruijnzeel LA. Tropical montane cloud forest: a unique hydrological case. In: Bonell M, Bruijnzeel LA, editors. Forests, water and people in the humid tropics: Past, present and future hydrological research for integrated land and water management. Cambridge University Press; 2005. p. 462–84.

Bunn SE, Arthington AH. Basic principles and ecological consequences of altered flow regimes for aquatic biodiversity. Environ Manag 2002;30:492–507. https://doi.org/10.1007/s00267-002-2737-0.

Buytaert W, Iñiguez V, Celleri R, De Bièvre B, Wyseure G, Deckers J. Analysis of the water balance of small Páramo catchments in south Ecuador. In: Krecek J, Haigh M, editors. Environmental role of wetlands in headwaters. NATO science series: IV: Earth and environmental sciences, vol. 63. Dordrecht: Springer; 2006. https://doi.org/10.1007/1-4020-4228-0_24.

Cassin J, Matthews JH. Setting the scene: nature-based solutions and water security. In: Cassin J, Matthews JH, Lopez-Gunn E, editors. Nature-based solutions and water security: An action agenda for the 21st century. Elsevier; 2021a.

Cassin J, Matthews JH. Nature-based solutions and climate change. In: Cassin J, Matthews JH, Lopez-Gunn E, editors. Nature-based solutions and water security: An action agenda for the 21st century. Elsevier; 2021b.

CONAGUA. Identificación de Reservas Potenciales de agua para el Medio Ambiente in México. Alianza WWF-FGRA; 2011. p. 85.

Cooper R. Water security beyond COVID-19. K4D Helpdesk Report 803, Brighton: Institute of Development Studies; 2020.

Coxon C, Gammie G, Cassin J. Mobilizing funding for nature-based solutions: Peru's drinking water tariff. In: Cassin J, Matthews JH, Lopez-Gunn E, editors. Nature-based solutions and water security: An action agenda for the 21st century. Elsevier; 2021.

Daily GC, Ellison K. The new economy of nature: The quest to make conservation profitable. Washington, DC: Island Press; 2002.

De Bievre B. Symposium presentation; 2019. https://inaigem.gob.pe/simposio/wp-content/uploads/2019/12/Bert-De-Biere.pdf.

De Koning F, Aquinaga M, Bravo M, Chiu M, Lascano M, Lozada T, Suarez L. Bridging the gap between forest conservation and poverty alleviation: the Ecuadorian Socio Bosque program. Environ Sci Pol 2011;14:531–42.

Drummond J. The garden in the machine: an environmental history of Brazil's Tijuca forest. Environ Hist 1996;1(1):83–104.

Echavarria M. Financing watershed conservation: the FONAG water fund in Quito, Ecuador. In: Pagiola S, Bishop J, Landell-Mills N, editors. Selling forest environmental services: Market-based mechanisms for conservation and development. London: Earthscan Publications Limited; 2002 [chapter 6].

Echavarria M, Stern M. Situational analysis of Paramo ecosystem conservation. Draft Report, Quito: EcoDecision; 2019.

Echavarría M, Vogel J, Albán M, Meneses F. The impacts of payments for watershed services in Ecuador. Emerging lessons from Pimampiro and Cuenca. Markets for environmental services series, no. 4, London: Environmental Economics Programme, IIED; 2004. Available at http://pubs.iied.org/pdfs/9285IIED.pdf.

Echavarria M, Zavala P, Coronel L, Montalvo T, Aguirre LM. Infraestructura Verde en el sector de agua potable en America Latina y el Caribe: Tendencias, retos y oportunidades. Washington, DC: Forest Trends Association; 2015.

Ellison D, Morris CE, Locatelli B, Sheil D, Cohen J, Murdiyarso D, Gutierrez V, van Noordwijk M, Creed IF, Pokorny J, Gaveau D, Spracklen DV, Tobella AB, Ilstedt U, Jeuling AJ, Gebrehiwot SG, Sands DC, Muys B, Verbist B, Springgay E, Sigandi Y, Sullivan CA. Trees, forests and water: cool insights for a hot world. Glob Environ Chang 2017;43:51–61.

Engel S, Pagiola S, Wunder S. Designing payments for environmental services in theory and practice: an overview of the issues. Ecol Econ 2008;65:663–74.

EPA. The economic benefits of green infrastructure: A case study of Lancaster, PA; 2014. February 2014. EPA 800-R-14-007.

FAO. (Food and Agriculture Organization). Forests and water: International momentum and action. Retrieved from http://www.fao.org/docrep/017/i3129e/i3129e.pdf; 2013.

Fehse J. Private conservation agreements support climate action: Ecuador's Socio Bosque program. CDKN, Inside Stories, September 2012 http://cdkn.org/resource/private-conservation-agreements-support-climate-action-ecuadors-socio-bosque-programme/; 2012. [Accessed October 2013].

Fisher J, Acreman MC. Wetland nutrient removal: a review of the wetland nutrient removal : a review of the evidence. Hydrol Earch Syst Sci 2004;8(4):673–85. https://doi.org/10.5194/hess-8-673-2004.

Forest Trends. CUBHIC tools support rapid assessment of water quantity and quality benefits of nature-based solutions, https://www.forest-trends.org/blog/launch-cubhic-tools-support-rapid-assessment-of-water-quantity-and-quality-benefits-of-nature-based-solutions/; 2020.

Gammie G, De Bievre B. Assessing green interventions for the water supply of Lima. Washington, DC: Forest Trends Association; 2014.

Gartner T, Mehan GT, Mulligan J, Roberson JA, Stangel P, Qin Y. Protecting forested watersheds is smart economics for water utilities. J Am Water Works Assoc 2014;106(9):54–64.

Gomez-Baggethun E, deGroot R, Lomas PL, Montes C. The history of ecosystem services in economic theory and practice: from early notions to markets and payment schemes. Ecol Econ 2010;69(6):1209–18.

Grima N, Singh SJ, Smetschka B, Ringhofer L. Payment for ecosystem services (PES) in Latin America: analyzing the performance for 40 case studies. Ecosyst Serv 2016;17:24–32.

Guidotti V, Ferraz SFB, Pinto LFG, Sparovek G, Taniwaki RH, Garcia LG, Brancalion PHS. Changes in Brazil's Forest code can erode the potential of riparian buffers to supply watershed services. Land Use Policy 2020;94:104511.

GWP (Global Water Partnership). Water Producer Project in the Pipiripau watershed: Building resilience in a water-conflict area in Brazil [Accessed 1 February 2021] https://www.gwp.org/en/waterchangemakers/change-stories/563815/; 2021.

Hendrickson CY, Corbera E. Participation dynamics and institutional change in the Scolel T'e carbon forestry project, Chiapas, Mexico. Geoforum 2015;59:63–72.

Horne AC, O'Donnell EL, Tharme RE. Mechanisms to allocate environmental water. In: Horne AC, Webb JA, Stewardson MJ, Richter B, Acreman M, editors. Water for the environment: From policy and science to implementation and management. Academic Press; 2017. p. 361–98.

Kauffman CM. Financing watershed conservation: lessons from Ecuador's evolving water trust funds. Agric Water Manag 2014;145:39–49.

Kroeger T, et al. Returns on investment in watershed conservation: application of a best practices analytical framework to the Rio Camboriú Water Producer program, Santa Catarina, Brazil. Sci Total Environ 2019;0048-9697. 657:1368–81. https://doi.org/10.1016/j.scitotenv.2018.12.116.

Landell-Mills N, Porras TI. Silver bullet or fool's gold? A global review of markets for forest environmental services and their impact on the poor. In: Instruments for sustainable private sector forestry series. London: International Institute for Environment and Development; 2002.

Lima JEFW, Ramos AE, editors. A Experiência do Projeto Produtor de Água na Bacia Hidrográfica do Ribeirão Pipiripau [The experience of the water producer project in the Pipiripau basin]. ADASA, ANA, EMATER-DF and WWF; 2018.

Lorenzi NM, Novak LL, Weiss JB, Gadd CS, Unertl KM. Crossing the implementation chasm: a proposal for bold action. J Am Med Inform Assoc 2008;15:290–6.

Magrin GO, Marengo JA, Boulanger J-P, Buckeridge MS, Castellanos E, Poveda G, Scarano FR, Vicuña S. Central and South America. In: Barros VR, Field CB, Dokken DJ, Mastrandrea MD, Mach KJ, Bilir TE, White LL, editors. Climate change 2014: Impacts, adaptation, and vulnerability. Part B: Regional aspects. Contribution of working group II to the fifth assessment report of the Intergovernmental Panel on Climate Change. Cambridge and New York, NY: Cambridge University Press; 2014. p. 1499–566.

Mahanty S, Suich H, Tacconi L. Access and benefits in payments for environmental services and implications for REDD +: lessons from seven PES schemes. Land Use Policy 2013;31:38–47.

Martin-Ortega J, Ojea E, Roux C. Payments for water ecosystem services in Latin America: a literature review and conceptual model. Ecosyst Serv 2013;6:122–32.

Miranda M, Dieperink C, Glasbergen P. Voluntary agreements in watershed protection experiences from Costa Rica. Environ Dev Sustain 2007;9:1–19.

Moulton TP, de Souza ML, de Oliveira AF. Conservation of catchments: some theoretical considerations and case histories from Rio de Janeiro. Neotrop Biol Conserv 2007;2(1):28–35.

Nelson SH, Bremer LL, Prado KM, Brauman KA. The political life of infrastructure: water funds and alternative histories of payments for ecosystem services in the Valle del Cauca, Colombia. Dev Change 2020;51(1):26–50.

Ochoa-Tocachi B, Butaert W, De Biebre B, Celleri R, Crespo P, Villacis M, LLerena CA, Acosta L, Villazon M, Guallpa M, Gil-Rios J, Fuentes P, Olaya D, Vinas P, Rojas G, Arias S. Impacts of land use on the hydrological response of tropical Andean catchments. Hydrol Process 2016;30(22):4074–89.

Ochoa-Tocachi B, Buytaert W, de Bievre B. Participatory water resources monitoring as a science-policy tool: a decade of experience from the Andes. EGU General Assembly 2020, Online, 4–8 May 2020, EGU2020-17960 https://meetingorganizer.copernicus.org/EGU2020/EGU2020-17960.html; 2020.

Odefey J, Detwiler S, Rousseau K, Trice A, Blackwell R, O'Hara K, Buckley M, Souhlas T, Brown S, Raviprakash P. Banking on green: How green infrastructure saves municipalities money and provides economic benefits community-wide, https://www.asla.org/ContentDetail.aspx?id=31301; 2012.

OECD/CAF/ECLAC. Latin American economic outlook 2018: Rethinking institutions for development. Paris: OECD Publishing; 2018. https://doi.org/10.1787/leo-2018-en.

Osborne T, Shapiro-Garza E. Embedding carbon markets: complicating commodification of ecosystem services in Mexico's forests. Ann Assoc Am Geogr 2018;108:88–105.

Ozment S, Feltran-Barbieri R, Hamel P, Gray E, Ribeiro JB, Barreto SR, Padovezi A, Valente TP. Natural infrastructure in São Paulo's water system. World Resources Institute; 2018.

Pagiola S. Payments for environmental services in Costa Rica. Ecol Econ 2007;65(4):712–24.

Pattanayak SK, Wunder S, Ferraro P. Show me the money: do payments supply environmental services in developing countries? Rev Environ Econ Policy 2010;4(2). 254-21.

Poff NL, Allan JD, Bain MB, Karr JR, Prestegaard KL, Richter B, et al. The natural flow regime: a new paradigm for riverine conservation and restoration. Bioscience 1997;47(11):769–84.

Poff NL, Brown CM, Grantham TE, Matthews JH, Palmer MA, Spence CM, Wilby RL, Haasnoot M, Mendoza GF, Dominique KC, Baeza A. Sustainable water management under future uncertainty with eco-engineering decision scaling. Nat Clim Change 2016;6:25–34.

Poff NL, Tharme RE, Arthington AH. Evolution of environmental flow assessments science, principles, and methodologies. In: Horne AC, Webb JA, Stewardson MJ, Richter B, Acreman M, editors. Water for the environment. From policy and science to implementation and management. San Diego, CA: Academic Press; 2017.

Ramsar Convention on Wetlands. Global wetland outlook: State of the world's wetlands and their services to people. Gland: Ramsar Convention Secretariat; 2018.

Redondo-Brenes A, Welsh K. Procuencas project, Costa Rica. TEEB case. Available at: TEEBweb.org; 2010.

Richards M. What do we know about gender and other social impacts of IWS projects? A literature review. Forest Trends; 2012. https://www.forest-trends.org/wp-content/uploads/imported/literature-review_gender-and-social-impacts_pws-projects_9-19-13-pdf.pdf.

Richards RC, Rerolle J, Aronson J, Pereira PH, Goncalves H, Brancalion PHS. Governing a pioneer program on payments for watershed services: stakeholder involvement, legal frameworks and early lessons from the Atlantic forest of Brazil. Ecosyst Serv 2015;16:23–32.

Richter BD, Baumgartner JV, Wigington R, Braun DP. How much water does a river need? Freshw Biol 1997;37:231–49.

Rodriguez M. Nuestro planeta, nuestro futuro. Bogota: Penguin Random House; 2019.

Rodriguez N, Armenteras D, Retana J. Land use and land cover change in the Colombian Andes: dynamics and future scenarios. J Land Use Sci 2012. https://doi.org/10.1080/1747423X.2011.650228.

Rogers EM. Diffusion of innovations. 5th ed. New York: The Free Press; 2003.

Ruggiero GC, Metzger JP, Tambosi LR, Nichols E. Payment for ecosystem services programs in the Brazilian Atlantic Forest: effective but not enough. Land Use Policy 2019;82:283–91.

Salinas-Rodriguez SA, Barrios-Ordonez JE, Sanchez-Navarro R, Wickel AJ. Environmental flows and water reserves: principles, strategies, and contributions to water and conservation policies in Mexico. River Res Appl 2018;34:1057–84.

Salzman J, Bennett G, Carroll N, Goldstein A, Jenkins M. The global status and trends of payments for ecosystem services. Nat Sustain 2018;1:136–44.

Schreckenberg K, Mace G, Poudyal M, editors. Ecosystem services and poverty alleviation: trade-offs and governance. New York, NY: Routledge; 2018.

Shapiro-Garza E, McElwee P, Van Hecken G, Corbera E. Beyond market logics: payments for ecosystem services as alternative development practices in the global south. Dev Change 2020;51(1):3–25.

Sone JS, Gesualdo GC, Zamboni PAP, Vieira NOM, Mattos TS, Carvalho GA, Rodrigues DBB, Sobrinho TA, Oliveira PTS. Effectiveness water provisioning improvement through payment for ecosystem services. Sci Total Environ 2018;655:1197–206.

Tellman B, McDonald RI, Goldstein JH, Vogl AL, Flörke M, Shemie D, Dudley R, Dryden R, Petry P, Karres N, Vigerstol K, Lehner B, Veiga F. Opportunities for natural infrastructure to improve urban water security in Latin America. PLoS One 2018;13(12):e0209470.

Van der Ent RJ, Savenije HH, Schaefli B, Steele-Dunne SC. Origin and fate of atmospheric moisture over continents. Water Resour Res 2010;46.

Veiga LBE, Magrini A. The Brazilian water resources management policy: fifteen years of success and challenges. Water Resour Manag 2013;27:2287–302.

Viani RAG, Bracale H, Taffarello D. Lessons learned from the Water Producer Project in the Atlantic Forest, Brazil. Forests 2019;10:1031.

Warziniack T, Sham CH, Morgan R, Feferholtz Y. Effects of forest cover on drinking water treatment costs. Retrieved from https://www.fs.fed.us/rm/pubs_journals/2016/rmrs_2016_warziniack_t001.pdf; 2016.

Wegner GI. Payments for ecosystem services (PES): a flexible, participatory, and integrated approach for improved conservation and equity outcomes. Environ Dev Sustain 2016;18:617–44.

Wiik E, Asquith N, Bottazzi P, Crespo Rocha D, DAnnunzio R, Pynegar E, Jones JPG. Can payments for ecosystem services schemes reduce deforestation? A robust evaluation example from the Bolivian Andes. In: 5th European congress of conservation biology; 2018. https://doi.org/10.17011/conference/eccb2018/107826.

Wiik E, d'Annunzio R, Pynegar E, Crespo D, Asquith N, Jones JPG. Experimental evaluation of the impact of a payment for environmental services program on deforestation. Conserv Sci Pract 2019;1(2):e8.

Winiecki E. Economics and source water protection, http://yosemite.epa.gov/r10/water.nsf/c6e3c862e806dd688825688200708c97/04a73c144395fda18825702e00650eb2/$FILE/Economics of_SWP_E_Winiecki_EPA.ppt#8; 2012.

Wunder S. Payments for environmental services: some nuts and bolts. Occasional Paper No. 42, Bogor: CIFOR; 2005.

Wunder S. PES for improved ecosystem water services in the town of Pimampiro, Ecuador. In: Payments for ecosystem services and food security. Rome: Food and Agriculture Organization; 2011.

WWF. Manejo del Agua en Cuencas Hidrográficas: Desarrollo de Nuevos Modelos en México, Informe Final [Water management in river basins: Development of new models in Mexico, final report]. Mexico: WWF; 2011. Internal Report.

Zbinden S, Lee D. Paying for environmental services: an analysis of participation in Costa Rica's PSA Program. World Dev 2005;33:255–72.

Mobilizing funding for nature-based solutions: Peru's drinking water tariff

Cheyenne Coxon, Gena Gammie, and Jan Cassin
Forest Trends, Washington, DC, United States

Key messages

- This chapter examines the drivers of the rapid transformation of Peru's water sector and highlights potential lessons for other sectors and geographies.
- **In recent years, Peru has led one of the most impressive shifts toward nature-based solutions (NBS) for water management in the world. To date, Peru is the first and only country to coordinate and promote locally driven NBS watershed investments through national policy.** Peru's urban water and sanitation sector has increased commitments for NBS investment from zero to over $33 million in a short period of time. As of August 2020, 37 water utilities, including the one that serves Peru's megacity capital, Lima, had incorporated NBS investments in their master plans as percentages of revenues generated by the water tariff. That number is expected to include nearly all 50 water utilities in the country in the coming years.
- **Increased financial commitments for NBS by water utilities countrywide reflect Peru's unique approach: national policy requires and supports local water users to fund source water conservation.** In addition to reforms to Peru's public investment system, water utilities are now empowered to invest in nature as an asset, in partnership with local stewards. While the drinking water sector passed their reforms, Peru's Ministry of Environment (MINAM) and the Ministry of Economy and Finance (MEF) developed a series of new laws and regulations that enable, support, and help streamline public investments in nature. The new legal framework considers nature as an asset, much like gray infrastructure, particularly the 2014 Compensation Mechanisms for Ecosystem Services Law, which promotes partnerships with local stewards to restore and maintain natural infrastructure.
- **Our analysis finds that three key factors enabled this rapid, drastic shift toward NBS: (1) early, bottom-up initiatives, (2) visionary champions leading from multiple sectors, and (3) cross-sectoral partnerships.** Together, these factors enabled the Peruvian water sector to overcome barriers, ambiguities, and institutional inertia

Nature-Based Solutions and Water Security. https://doi.org/10.1016/B978-0-12-819871-1.00008-7

and generated the momentum and experience necessary to launch national reforms. Civil society and international cooperation played supportive roles at multiple stages. Collaborators were further enabled by the structure of institutions like SUNASS, which permitted leadership the flexibility to act on windows of opportunity, pursue new ideas, and instate widespread changes during their tenures.

- **Despite these impressive advances, there is still work to do:** implementing actions that match the ambition of these policy reforms has been a challenge. Only 7 of the 37 utilities with tariffs assigned for NBS had begun implementation at the time of this writing. Not everyone in the sector understands or buys into NBS, and the legal framework and policies are only the bare minimum necessary for investments in NBS to flow. It has been challenging for water utilities and engineers to acquire the new capacities needed to efficiently and effectively develop, implement, and manage a new type of investment.
- Peru is now in the critical phase of implementing NBS on the scale envisioned by the drinking water sector's ambitious policies and plans. The water sector is in process of creating a community of NBS practice among water utilities, practitioners, and decision-makers across sectors; strategic monitoring and evaluation, openly sharing information, and evidence-based decision-making on NBS for water are evolving, and increased participation of beneficiaries continues to be of interest as advocates work toward implementing NBS at scale. However it evolves moving forward, **the Peruvian water sector's experience in NBS offers important lessons for its consideration in other sectors and countries.**

Introduction

Despite being home to world famous glaciers and the headwaters of the Amazon River, Peru's population of 33 million people faces substantial water stress and insecurity. Peru's capital city, Lima, has a population of 11 million, and is the second-largest desert city in the world, after Cairo, Egypt. Overall, the country is rated as being at medium to high risk from the water crisis (Bloomberg, 2019). But watersheds on the dry Pacific slope of the Andes, where most Peruvians live, are rated as high or extremely high for baseline water stress (World Resources Institute, 2020). Water insecurity in Peru is largely due to both natural climate variability and recent shifts in population, as Peruvians migrate toward larger cities. These factors make Peru particularly vulnerable to climate change: longer, more frequent droughts and more intense flash flooding are expected to increase, along with the full retreat of Andean glaciers, which have long served as buffers to natural hydrological extremes.

In response to Peru's long-standing water risks, the country's water utilities, local and regional governments, and national authorities have invested in increasingly complex and expensive gray water infrastructure, including more recent investments in desalination technology and massive systems to transport water from the headwaters of the Amazon to the thirsty Pacific coast. This conventional approach to addressing water risks has not historically considered the role of natural infrastructure—forests, grasslands, and wetlands—in protecting and sustaining clean, reliable water supplies (see Box 11.1).

Box 11.1   Natural infrastructure and nature-based solutions.

Natural infrastructure is the landscape, vegetation, soils, and organisms that affect water quality and flow. In Peru, ecosystems that provide hydrologic benefits include forests, wetlands, riparian systems, native grasslands such as *punas*, *jalcas*, and *páramos*, and wetlands and peatlands (*bofedales*). NBS are activities or interventions designed to maintain or enhance the ability of natural infrastructure to provide beneficial services. NBS include a wide variety of activities, from protecting intact ecosystems and avoiding their degradation or loss to active management, such as restoration or sustainable agroforestry. Examples of NBS include:

Conservation: Investment to avoid expected loss of vegetation and soils important to modifying water flows. NBS in conservation include avoided deforestation, avoided wetland drainage, and avoided grassland degradation.

Restoration: Active interventions to restore vegetation and soil health, usually by planting and cultivating native landscapes. Ecological restoration is the process of assisting the recovery of an ecosystem that has been degraded, damaged, or destroyed (Society for Ecological Restoration, https://www.ser-rrc.org/what-is-ecological-restoration/). NBS are used in forest, wetland, and grassland restoration.

Enhancement through sustainable management: This might include rotational or low intensity grazing to maintain or enhance infiltration of water and soil conservation, linear elements such as dikes, gully control, infiltration trenches, agricultural terraces, riverbed works, rotational cultivation, and sustainable agricultural practices, such as tillage.

Managed aquifer recharge: This group of NBS includes permeable structures: artificial lagoons, qochas, terraces, and other ancestral practices for water harvesting. On the arid Pacific coast of Peru, for example, pre-Inca communities developed water harvesting practices known as *amunas* to divert water from streams during the rainy season and store it in the soils to increase dry season water availability (Ochoa-Tocachi et al., 2019).

Modified from Brauman et al. (2021).

A movement formed in the 2000s by local communities, civil society, national policymakers, and international cooperation, however, led to a remarkable shift in Peru's drinking water sector: the systematic incorporation of NBS into the sector's strategy for addressing water risks. In less than a decade, the Peruvian drinking water sector has transformed from one in which investments in watershed health were seen as outside the scope of water utility responsibility to one in which three-quarters of water utilities across the country allocate a portion of water tariffs to nature-based solutions (NBS) to protect the natural infrastructure of their watersheds. Through the tariffs, or user fees, utilities have already allocated over $33 million to compensation mechanisms for ecosystem services (MRSE[a]). The principles of NBS are just beginning to gain ground at the national level in Peru, but the MRSE tariff, as a renewable, revenue-generating mechanism, is a strong start to scaling its use to address water risks.

---

[a]   MRSE: Mecanismo de Retribución por Servicios Ecosistémicos (Payment Mechanism for Ecosystem Services)—a financial structure used to finance NBS.

This chapter explores the drivers and dynamics of this rapid, sector-wide shift toward NBS. Our analysis indicates that it was enabled by a combination of: (1) early, bottom-up initiatives; (2) visionary champions leading from multiple sectors; and (3) cross-sectoral partnerships. Studying these drivers and remaining challenges in scaling institutional shifts to their full potential offers useful lessons for advocates and policymakers seeking to increase the use of NBS in other sectors.

Local initiatives in Moyobamba and Cusco lay the foundation for nature-based solutions

In the early 2000s, water utilities and communities in some of Peru's Andean cities began noticing drastic changes in water quality and quantity. In some contexts, where urban water users were acutely aware of source watersheds and the pressures they faced, local leaders began to see possible connections between land use change and water availability. Supported by civil society and international partners familiar with payment for ecosystem services programs, local initiatives began to address source water degradation using new financing mechanisms, particularly in the cities of Moyobamba and Cusco. These efforts were led by local water companies, local communities, and local governments—requiring new kinds of partnerships and coordination among the entities involved in land management and drinking water services in Peru (see Box 11.2). They were often supported by international cooperation agencies and involved Peru's national water regulator SUNASS (*Superintendencia Nacional de Servicios y Saneamiento*). The involvement of national authorities in these cases,

---

Box 11.2   Peruvian watershed land management and drinking water services at a glance.

Land tenure in Peru's source watersheds varies. In some watersheds, like Lima's, it is largely dominated by campesino communities (communal land tenure). Others, such as Arequipa's source watersheds, are dominated by large nationally protected areas. Sometimes large tracts of land are owned privately. In many areas, especially in the Amazonian part of the country, large areas lack clear land tenure.

Water and sanitation services for urban populations are provided by *Empresas Prestadoras de Servicios de Saneamiento* (EPS): public water utilities that provide both drinking water and sanitation services and are regulated by a centralized, national agency, SUNASS. There are 50 EPS in the country, 49 of which are incorporated in their respective municipalities. The exception is SEDAPAL, the utility that provides water to Lima. The largest utility in Peru by far, SEDAPAL is incorporated in the national Ministry of Housing, Sanitation, and Construction.

Rural water and sanitation is managed by local water user boards. Recently, SUNASS was assigned responsibility for supervising water and sanitation services in rural areas, in addition to urban areas.

particularly SUNASS, helped consolidate local initiatives and build the foundation that later influenced national policy change.

Local action in Moyobamba creates Peru's first MRSE water tariff

The first successful negotiation for the inclusion of a MRSE tariff in a water utility budget began in 2007 in the city of Moyobamba, located in the Andean Amazon in the northern province of San Martín. Moyobamba's 50,000 inhabitants[b] depend primarily on the nearby Rumiyacu, Mishquiyacu, and Almendra micro watersheds—a modest 924 ha that provides 80% of the city's total water supply (SUNASS, 2014).

In the early 2000s, the montane rainforest of Moyobamba's watersheds was steadily being converted into farmland, largely driven by the arrival of low-income migrants and limited governmental capacity to control land conversion (León and Renner, 2010). As a result, Moyobamba's water utility noted a decline in water quality and quantity available for drinking, which resulted in rising production costs and the prospect of water restrictions (León and Renner, 2010). Negotiations between upstream land managers and downstream water users began in 2003, which allowed them to begin to understand each other's interests and identify solutions that could be mutually beneficial. International organizations also began supporting efforts to improve watershed management. The German Agency for Technical Cooperation (GTZ) and RARE Conservation (a US-based environmental NGO), together with local and national actors, began raising awareness and creating social marketing campaigns targeted to the general population of Moyobamba about the need for watershed protection. These campaigns significantly increased public awareness of the city's dependence on its watersheds by 2006, and in 2007, a study found that Moyobamba water users were willing to pay for watershed conservation (interviews with Fernando León and Echavarria, inter alia, 2015; Nowack, 2005; Renner, 2010).

At the same time, a management committee was established to promote and coordinate investments in watershed conservation and restoration to protect Moyobamba's water supply. The Committee included representatives from the municipal and regional governments, journalists, agricultural and environmental groups, university and community groups, and the local water utility. In 2007, SUNASS convened a public hearing to review the proposed increase of 1 PEN ($0.33) in each household's monthly water bill to go toward a watershed conservation fund, representing a 3.3% increase in overall rates. After receiving public support for the proposal, SUNASS formally approved the new tariff that year, and it went into effect in August 2009 (SUNASS, 2007; León and Renner, 2010). After the approval of the tariff, the management committee negotiated agreements with land users to support forest conservation in the watershed in exchange for technical

---

[b] All populations in this chapter are drawn from Peru's 2017 census.

assistance and materials to help implement sustainable agroforestry systems (León and Renner, 2010; Momiy Hada, 2014).

Cusqueñans learn from and expand on the Moyobamba experience

For centuries, residents of the Andean city of Cusco have depended on Lake Piuray as their primary freshwater source. Located about 20 miles from Cusco, Lake Piuray is within the Chincheros municipality and surrounded by a rural catchment area home to 16 communities. Today, Lake Piuray still constitutes approximately half of the city's drinking water supply, although the water utility, SEDACUSCO, has tapped into additional groundwater sources to serve the city's nearly half a million residents. In 2012, Cusco and Lake Piuray became the site of the second MRSE tariff approved in Peru; in contrast to Moyobamba, however, this agreement followed decades of social conflict between the downstream city and the stewards of its source water landscapes.

At least as early as 1988, SEDACUSCO, the Chincheros local government, and Piuray communities began negotiations for a potential compensatory agreement between the utility and local communities. In the late 1990s, these negotiations intensified when SEDACUSCO increased the volume of water it extracted from Lake Piuray by approximately 35% in response to an extended drought and rising population in Cusco. Lake Piuray communities began to attribute decreased agricultural output to this increased diversion. Shortly afterward, the opposite occurred: the 1998 El Niño brought a severe wet season and flooded agricultural lands in the floodplains of the Piuray catchment. The local communities argued that increased water extraction by SEDACUSCO had contributed to unstable and sinking land in the area, which, in combination with the flooding, caused landslides that severely affected crops. Upstream communities still attributed growing agricultural losses to the downstream activities of SEDACUSCO and organized a local movement to secure compensation from the utility for damage to their livelihoods. Like the negotiations that began in the 1980s, these also failed to result in a compensatory agreement.

Conversations resumed in 2011 when more flooding in 2010 and 2011 renewed the communities' calls for compensation and alerted SEDACUSCO to the decline in the quality of water extracted from Lake Piuray. In 2012, SUNASS became involved in the negotiations to help prepare SEDACUSCO'S 2012–16 Master Plan and helped the utility examine the value of working with Piuray communities to: (1) improve watershed conditions and (2) manage social risks to their primary water source. Shortly afterward, in 2012, representatives from Lake Piuray communities were invited to attend a watershed conservation event in Moyobamba, organized with the support of international cooperation agencies. Attendees say the experience inspired them to replicate that model in Cusco. As a result of their efforts, SUNASS approved an increase in SEDACUSCO's tariff to pay for watershed services in the Lake Piuray catchment in 2012. SEDACUSCO increased the tariff by 4.8% in the first year and 4.2% in the third year of the 2012–16 Master Plan (Momiy Hada, 2014). With over 70,000 contributing households, the total amount of the fund was almost 8 million soles (~\$2.6 million), almost seven times larger than Moyobamba's watershed conservation fund (Momiy Hada, 2014).

The local experiences in Moyobamba and Cusco proved to be catalysts for learning and political change throughout the sector. Implementing the first MRSE agreements between water users and upstream communities would demonstrate how such mechanisms could work, what the process to develop them could look like, and what the benefits of NBS could be. At times, as in Cusco, they served as the platform for improved dialogue between communities, water users, and water providers—the ecosystem services and other water benefits were not necessarily the primary frame or focus of negotiations, even if they became more central to later discussions and projects. These cases also clarified the need for national policy change to facilitate and officially support the use of public funds for NBS, specifically through water tariff revenues. Spurred in part by the local programs in Moyobamba and Cusco, national policies and regulations began evolving to better enable and support action at the local level, ultimately leading to more widespread change.

Legal frameworks and technical advances build momentum for nature-based solutions

As local efforts in Moyobamba and Cusco matured, champions of NBS within SUNASS, the Ministry of Environment (MINAM), and civil society began leveraging these experiences to build the evidence base, technical tools, and policies that would eventually support national level action.

National laws and regulations evolve to enable and promote nature-based solutions

When Moyobamba and Cusco began allocating tariff revenues to NBS with the approval of SUNASS, the Peruvian government did not yet have an explicit legal framework to support the participation of the drinking water sector in watershed protection. In Peru, public officials generally follow specific, positive guidance on how to invest and spend public funds. Not following such guidance can expose officials to reprimand, even legal consequences. Prior to these first cases, investing in environmental conservation above a utility's intake was broadly considered to be outside the water and sanitation sector's responsibility. NBS financed by the water sector existed in a legal gray area—they were neither prohibited nor explicitly sanctioned.

Moreover, Peru's public investment system, overseen by the MEF, did not have a method for supporting public investments in NBS; investments in nature were guided by traditional conservationist principles, rather than a desire to maintain and enhance ecosystem services. Compensating local land users for acting as stewards of the environment was also significantly limited. The development of the flagship cases in Moyobamba and Cusco revealed a clear need for legislative action to address these barriers.

An opportunity to resolve this policy gap presented itself in 2012 during the comment period for the Modernization of Sanitation Services Law. SUNASS suggested a provision for legally recognizing and promoting the role of water utilities in financing environmental conservation. The provision, adopted in the final Sanitation Services Law approved in June 2013, specifically states that water utilities should invest in watershed conservation, and that SUNASS should work with water utilities to ensure due consideration of such investments in their master plans.

While the Sanitation Services Law required utilities to start raising funds for investments in ecosystem services, spending water tariff revenues—or any public funds, for that matter—on these investments was still not explicitly authorized. This issue was resolved in 2014, with the Law on Compensation for Ecosystem Services Mechanisms (*Mecanismos de Retribucion por Servicios Ecosistémicos*—MRSE). The law formally recognizes the importance of compensation for ecosystem services projects across the forestry, agriculture, and water sectors. It also explicitly permits the use of public funds for this type of compensation, allowing for a range of voluntary agreements where those benefiting from ecosystem services pay for the work to conserve, recover, or sustainably use them (MINAM, 2015).

After the passage of these two laws, MINAM and SUNASS were tasked with developing the regulations that would support their implementation. Here, technical collaboration supported by civil society and international cooperation helped these agencies develop the necessary regulations while simultaneously promoting MRSE tariffs in the water sector. A range of actors and initiatives, including USAID-funded Initiative for Conservation of the Andean Amazon and GIZ-funded ProAmbiente, supported development of the MRSE Law with civil society leaders, including the Peruvian Society of Environmental Law (SPDA). The Ecosystem Services Compensation Mechanisms Incubator (the Incubator) stands out among the efforts focused on connecting NBS policy developments between the environmental and water sectors. Between 2012 and 2016, the Incubator, coordinated by Forest Trends, CONDESAN, and EcoDecisión with funding from the Swiss Agency for Development and Cooperation (SDC), worked closely with MINAM, SUNASS, local policymakers, and project developers to bridge the gaps between policy and implementation (see Box 11.3).

Incubator partners supported MINAM's initial efforts to design, socialize, and pass the MRSE Law in 2014; they coordinated stakeholder meetings to review the law and its regulation and organized a visit to Moyobamba for Congressional leaders to see its tariff program in action.

Box 11.3   The incubator for ecosystem services compensation mechanisms—Building technical capacity.

In 2012, Forest Trends and MINAM established a MOU to support a national MRSE Incubator. Members also included Andean watershed investment and hydrological experts at EcoDecisión and the Consortium for the Sustainable Development of the Andean Ecoregion (CONDESAN). With the support of the SDC, the MacArthur Foundation, and the D.N. Batten Foundation, the Incubator initially emphasized protecting and investing in watershed services for water security. A new type of collaborative entity, the Incubator was instrumental in creating the enabling conditions for Peru's groundbreaking advances in watershed investments—flexible and inclusive governance, overcoming institutional inertia, fostering cross-sectoral coordination, providing regulatory support, and developing technical tools.

Collaborative development of tools and analysis: In addition to policy support, the Incubator provided key technical support for evaluating and prioritizing watershed investments. This includes guidance and designs for hydrological monitoring, led by the project's key technical partner in Peru, CONDESAN, which continues to improve knowledge and help refine project design. The role of the Incubator in analyzing the cost-effectiveness of natural infrastructure has been particularly critical. Forest Trends, CONDESAN, Lima's water fund Aquafondo, SUNASS, and SEDAPAL collaborated to develop a cost curve analysis and made it open access. The positive impact of these cost curves on decision-making is partly attributed to the collaborative effort required to develop them in the first place. The cost curves provided SUNASS and SEDAPAL with an economic case for NBS and strong rationale for investment, strengthened ties between the organizations, and helped them make the case to their leadership and constituents. The methodologies and data for the cost curves are still freely available, and SUNASS continues to develop their own to support future investment decisions.

Providing targeted technical support and implementing watershed investments: The Incubator helped new watershed projects by providing technical support for key aspects of planning and design and facilitating more rapid replication of watershed investments. Strengthened institutions in Moyobamba and new finance strategy development in Jequetepeque and San Martín helped create replicable models for additional watersheds. Learning exchanges and clinics have also resulted in the establishment of local knowledge centers, which greatly expands the base of experts needed to build future capacity and incubate projects.

Such activities helped galvanize support for the MRSE Law and demonstrated its connection to the water sector. Finally, a memorandum of understanding (MOU) between MINAM and SUNASS in 2015 facilitated more formal water sector reform by ensuring that NBS investments by water utilities could function within national regulation with clear processes and public financing mechanisms. This partnership and other nongovernmental alliances ultimately strengthened governmental coordination and encouraged the water sector's shift toward investments in NBS.

Collaborative development of technical tools helps build capacity and make the case for nature-based solutions

Beyond the legal barriers, early experiences developing MRSE tariffs in the Peruvian water sector exposed the need for practical tools and new capacities to support water utilities as they planned, designed, and implemented NBS. This is an area where the Incubator also played a key role. The development of the rapid hydrological diagnostic (*diagnostico hídrologico rápido*—DHR) methodology, spearheaded by CONSESAN and SUNASS, is a strong example. The DHR tool allows water utilities to practically assess which NBS could alleviate priority water risks in their watersheds (SUNASS, 2017). Representatives from water utilities and regional governments participated in DHR courses supported by the Incubator, and the drinking water sector began applying the tool in numerous watersheds in 2015 (CONDESAN, 2014, 2015).

In the same time period, Incubator partners Forest Trends and CONDESAN worked with US-based consulting firm Kieser & Associates and Lima NGO Aquafondo to develop hydro-economic assessments of NBS with the goal of addressing water security concerns for Lima (Gammie and De Bievre, 2014). This was the first time the potential contributions of wetland and grassland restoration and restoration of pre-Incan infiltration systems were quantified in the context of alleviating Lima's growing dry season water deficit. Assessing the effectiveness of NBS in terms of hydrological outcomes and costs allowed decision-makers to appreciate the importance of natural infrastructure as a water risk management strategy and to compare it to gray infrastructure alternatives, like desalination plants and interbasin transfers. The analysis found that the NBS evaluated were cost-competitive with gray options and that implementing the most cost-effective NBS at scale could reduce the dry season deficit by approximately two-thirds (Gammie and De Bievre, 2014). This assessment helped SUNASS and SEDAPAL justify the inclusion of a 1% MRSE allocation in the budget and 2015–20 Master Plan, a commitment of approximately $25 million for NBS for water—the largest of any Latin American city at the time.

Investments in nature-based solutions are adopted across Peru's drinking water sector

With the Sanitation Services and MRSE Laws in place, the water and environmental sectors began to implement tariff-funded MRSE throughout the country in 2015 (Fig. 11.1). The Sanitation Services Law and subsequent regulations provided a clear mandate for SUNASS to work with water utilities countrywide to incorporate watershed investments into their master plans and water tariffs. The MRSE Law and regulation began to clear a path for executing those funds and empowered MINAM to promote these financing mechanisms.

Fig. 11.1
Key milestones in mainstreaming nature-based solutions in Peru's water sector.

The work that followed was a systematic push to incorporate NBS into the master plans, tariffs, and operations of dozens of water utilities across Peru. The process followed the calendar for master plan updates, which each water utility completes every 5 years, under SUNASS supervision. Each year, SUNASS's specialized team working on MRSE tariffs would support a new group of water utilities to perform the rapid hydrological diagnostic, identify and cost out investments in natural infrastructure, and include the investments in a new MRSE tariff. Since 2015, SUNASS has passed and implemented a series of resolutions and amendments to streamline the Sanitation Services Law and MRSE mechanisms, and MINAM has issued a number of new guidelines and rules that clarify and streamline investments in natural infrastructure (see Fig. 11.1).

A major milestone in this process was the approval of SEDAPAL's 2015–20 Master Plan for Lima, which was formally approved in June 2015 and included a 1% tariff increase to be used for NBS. An additional 3.8% was allocated for climate change adaptation and disaster risk reduction, and this set of funds can be used for NBS (nature-based climate solutions, disaster risk reduction). These were by far, and still are, the greatest commitments for NBS by any water utility in Latin America—the 1% fund for ecosystem services alone totaled over $25 million over the master plan's 5-year period. Between 2015 and 2018, 43 of Peru's 50 water utilities had their tariffs reviewed for the incorporation of a MRSE tariff, advancing the systematic inclusion of NBS investments in the water sector (personal communication, Oscar Angulo, 2020).

Since 2018, with the introduction of Peru's new National System of Multi-Year Programming and Investment Management, MINAM and MEF have developed a number of initiatives to accelerate and promote investments in natural infrastructure, efforts which also support the water sector's push for NBS. In contrast to Peru's previous public investment system, the new system explicitly recognizes natural infrastructure in guidelines for public investment planning wherever references to "infrastructure" are made.[c]

Public efforts to restore natural infrastructure and nature-based Indigenous water management systems could be dramatically accelerated under new guidelines approved by MINAM in late 2019 (Gammie et al., 2020). Under these new guidelines, the implementation mechanism for public funds—Investments from Optimization, Marginal Expansion, Replacement, and Rehabilitation (IOARR)—can now be applied to natural infrastructure. The IOARR implementation mechanism holds enormous potential to accelerate investment by simplifying the project design and justification process.

---

[c] El Decreto Supremo No. 284-2018-EF: Approves the regulation of legislative decree No. 1252, which creates a national system of multiyear programming and investment management. It defines the public investment gap as the difference between the optimized available supply of infrastructure (which includes natural infrastructure) and/or access to services and the demand at a specific date and geographic area. It also defines Natural Infrastructure: the network of natural spaces that preserve the values and functions of ecosystems, providing ecosystem services.

In addition to accelerating restoration of the *amunas* (see Box 11.1) that hold so much potential for Lima's water security, a regional government could utilize IOARR to respond more rapidly to forest fires and restore areas affected by such fires. Using IOARR will allow these public agencies to avoid the extended bureaucratic process associated with the alternative pathway for public investment projects. By defining strategic assets associated with natural and Indigenous infrastructure in the new IOARR guidelines, the state recognizes that natural assets, such as *amunas*, forests, and wetlands, already exist. As such, they do not need to be constructed, they need to be protected, restored, and maintained.

With the major legal barriers to scaling NBS in the drinking water sector addressed, efforts by NBS advocates, governmental and nongovernmental alike, have shifted toward improving technical tools, governance, and implementation. Significant investments by international cooperation, such as the Natural Infrastructure for Water Security project (NIWS), are helping address remaining gaps in NBS implementation (see Box 11.4).

Box 11.4   Natural Infrastructure for Water Security Project and Peru's drinking water sector.

The NIWS is a $27.5 million effort funded by the US Agency for International Development (USAID) and the Government of Canada that aims to develop natural infrastructure as a central piece of Peru's water and climate risk management strategies while reducing gender gaps. The NIWS project began in December 2017 and is implemented by a consortium led by Forest Trends, with partners CONDESAN, SPDA, EcoDecisión, and researchers from Imperial College London. Though NIWS is not focused exclusively on the drinking water sector, it has supported MINAM, SUNASS, and numerous water utilities as they follow through on the potential to scale up NBS in the sector. NIWS has continued to support institutional shifts by developing technical tools, generating information for decision-making, and compiling project portfolios that can be implemented by using MRSE tariffs. The project's end is slated for June 2023.

Institutional and political support for implementing NBS: NIWS works with MINAM, SUNASS, the National Water Authority, the Ministry of Agriculture and Irrigation, the Ministry of Housing, Construction and Sanitation, and the Ministry of Women and Vulnerable Populations to address remaining institutional and political barriers to scaled investments in natural infrastructure for water security. Among the high-level fora held to advance this objective was the first National Water Summit in 2018, organized in partnership with the National Association of Water Utilities (ANEPSSA). During the summit, presidents and general managers of 26 water utilities signed the Piuray Declaration, which recognized the dedication of these institutions to follow through on utility financial commitments to natural infrastructure. NIWS has also supported the development of new regulations by SUNASS and new guidelines by MINAM to further clarify and streamline MRSE development and execution.

Information for decision-making on NBS: NIWS has also generated greater clarity and credibility for the hydrological benefits of natural infrastructure. The project has supported decision-makers by producing a series of summaries on the impacts of NBS on water and

---

Continued

Box 11.4   Natural Infrastructure for Water Security Project and Peru's drinking water sector—cont'd

soils in Andean ecosystems, based on available scientific literature. One of the NIWS project's greatest contributions has been developing and sharing technical tools for NBS implementation and training professionals how to apply them to water security and disaster risk reduction—all particularly relevant concepts and skills for the drinking water sector. Among these tools are HIRO (*Herramienta de Identificación Rápida de Oportunidades para la Infraestructura Natural en la Gestión del Riesgo de Desastres*), a spatial analysis tool that combines existing resources on floodplains, populated centers, infrastructure, degraded areas, forests, water supply, and soil permeability to inform basic initial infrastructure planning. One of its most powerful applications is to identify areas where natural infrastructure might contribute to reducing a range of water risks, from floods and landslides to water scarcity. NIWS has also launched a series of CUBHIC methods (*Cuantificación de Beneficios Hidrológicos de Intervenciones en Cuencas*), which quantify potential water benefits of natural infrastructure. Designed for use at the project site level, each methodology is focused on an Andean ecosystem type and is a practical option for estimating the water benefits of ecosystem services. NIWS also supports the Initiative for Hydrological Monitoring of Andean Ecosystems (iMHEA), including the development of new monitoring sites and protocols and trainings. These tools have filled an important technical gap, especially because many existing hydrological quantification tools have been developed in (and for) the Global North.

Nature-based Solutions project portfolios and implementation mechanisms: In addition to strengthening enabling conditions for NBS and developing tools for decision-making, NIWS works directly with water utilities and project developers to create portfolios of NBS investments and facilitates their implementation. To date, NIWS has advanced a national portfolio of over $63 million in NBS investments in development, including over $10 million USD for financing and implementation by the drinking water sector. NIWS works especially closely with Lima's utility, SEDAPAL, to address bottlenecks in getting funds to the ground, and supported the utility to approve its first MRSE funded project, a $0.9 million USD investment in wetland restoration. NIWS complements this facilitation with capacity building—the project has already trained over 1200 professionals from 15 regional governments, 72 local and provincial governments, 34 water utilities, and 25 engineering consulting firms.

What drove the rapid uptake of nature-based solutions in Peru's drinking water sector?

In just under 15 years, Peru's drinking water sector went from a model that discouraged water utilities from investing upstream of their intakes to having some of the most progressive NBS policies in the world. While early NBS projects in Peru were local, ad-hoc efforts, a combination of leadership by advocates of NBS and cross-sectoral partnerships leveraged these early initiatives to create wider institutional change. The institutional shifts were at first discursive: traditionally conservative institutions and communities adopted a new language and approach to managing risk that included NBS. Those shifts were then formalized in policy—first opportunistically, then systematically. The structure of Peru's drinking water sector, which has one central regulator, allowed this institutional shift to have national impact. This section explores how and why these factors led to successful incorporation of NBS into national water management in Peru.

Champions drove informal and formal institutional shifts at key junctures

Champions within the water sector played a critical role in the institutional shifts that enabled Peruvian water utilities to recognize watershed conservation as part of their responsibility and that NBS would be central to those efforts. They introduced new discourse on NBS and strategically directed decisions in its favor at key junctures. At times, external pressure helped strengthen their cases—environmental disasters and threats to water security created the extra pressure and urgency Peruvian decision-makers needed to advocate for NBS. Their actions helped lead to wider acknowledgment of the water security problems Peru faces, of its connection to climate security, and that "traditional" methods alone are not enough to meet these challenges. This realization became a uniting cultural shift in the water sector and allowed NBS advocates to act on opportunities to demonstrate how it could benefit the sector.

José Salazar is widely recognized as one of Peru's first NBS champions, perhaps the first from within the sector to present a compelling concept for using payment for ecosystem services. Elected president of SUNASS in 2007, Salazar had a background in environmental economics and worked at the National Institute of Natural Resources (INRENA) for years, the institution that preceded today's Ministry of Environment in Peru (MINAM). Reflecting on his time at SUNASS, Salazar has said that his greatest contribution was the "hybridization" of environmental economics and the economics used in the institution's day-to-day evaluations. This approach included economic analyses to justify pursuing NBS, such as cost-effectiveness estimates that compared NBS to gray options.

However, the most critical changes Salazar introduced were far more subtle—his tenure began to broaden the scope of what water utilities considered their responsibility, especially in terms of watershed interventions upstream of intakes and pursuing new types of water management. Skepticism of NBS still pervaded most of the sector outside of SUNASS. This led Salazar to make two strategic choices that proved critical to wider acceptance of NBS: (1) suggest Moyobamba as the first city to incorporate an NBS fund into their water tariff—there was a solid foundation of analytical and community support, based on local studies; and (2) seek international allies, including GTZ, The Nature Conservancy, and Forest Trends to build legitimacy and credibility (personal communication, José Salazar, 2015).

Fernando Momiy[d] was appointed Salazar's successor as president of the board of SUNASS in 2012 and is considered another leading champion of NBS. Momiy was general manager of SUNASS during the development of Moyobamba's water tariff increase, where he worked on the incorporation of NBS into the drinking water sector. In 2011, he led SUNASS and others in responding to the crisis of floods and landslides in Cusco and used this as an opening to begin negotiations between upstream communities and the water utility, SEDACUSCO. Momiy took on considerable personal risk when he approved Cusco's tariff fund for NBS—use of public funds for NBS was not explicitly legal at that time. His bold choices and strong leadership at SUNASS broke a cycle of risk-averse leadership in the water sector and demonstrated what successful projects could look like on the ground.

One of Momiy's other most important contributions was acting on the policy level. This ensured that future decision-makers in the water sector would have legal, established procedures to follow for NBS investment. In 2013, he acted on an opportunity to integrate NBS into the new Sanitation Services Law. During the law's comment period, Momiy effectively explained to the state how formal recognition of the role of water utilities in watershed management would advance important sectoral interests, such as more cost-effective service delivery. He was also able to explain the benefits to the public of raising tariffs for watershed investments, a concept that encountered little active resistance (at least at the national level). Finally, he convinced Congressional representatives to include formal recognition of water utilities' roles in watershed management in the Sanitation Services Law, ensuring a full legal framework to support future projects modeled after Moyobamba and Cusco.

Much like policy changes elsewhere in the world, the passage of the Sanitation Services Law in 2014 did not guarantee successful adoption of NBS. Momiy's leadership at SUNASS during that time was therefore critical; he prioritized formal implementation of NBS investment throughout the country, hired new staff, and directed existing staff to focus on strategic execution. He also helped support a growing regional movement among utilities and water regulators through the development of a "Green Infrastructure Working Group" within the Latin American Association of National Water Regulators (ADERASA) to share information on NBS and encourage its adoption throughout ADERASA's peer network. Momiy's successor, Ivan Lucich, continued to champion the incorporation of NBS into the sector, further strengthening and expanding the team at SUNASS dedicated to the adoption of MRSE tariffs throughout the country.

Momiy's tenure at SUNASS coincided with the leadership of an NBS champion within MINAM, Minister Manuel Pulgar-Vidal.[e] As minister, Pulgar-Vidal was instrumental in developing and passing the MRSE Law, which opened the door to using public

---

[d] Fernando Momiy is now employed by Forest Trends and leads the NIWS Project from their office in Lima.

[e] Manuel Pulgar-Vidal is currently climate chief at World Wildlife Fund and serves on the Forest Trends Board.

funds for NBS, including from the water sector. His support for the Incubator helped ongoing efforts to build a stronger evidence base for NBS through targeted technical support for pilot projects in several watersheds, including Lima's. The overlapping leadership of Momiy at SUNASS and Pulgar-Vidal at MINAM proved to be a powerful coincidence—it helped foster a stronger relationship between the two institutions and brought more attention to the potential of NBS and the connections between water and climate.

Individual champions were enabled by new, cross-sectoral alliances

While individual leadership was critical to this transition, champions alone could not have achieved the scale of change that occurred in Peru's water sector. Recognizing this, actors like Salazar and Momiy sought allies in other sectors who supported their leadership with additional resources, reach, and technical capacity. Such partners often had the flexibility to experiment with new approaches, such as training farmers in sustainable agricultural practices. Civil society organizations often acted as important liaisons with communities, using existing networks and relationships to publicly promote NBS strategies and fund meetings, including the one that inspired stakeholders in Cusco to replicate Moyobamba's model.

Alliances with international cooperation agencies and civil society allowed champions to attract new resources and introduce new ideas that helped legitimize and build momentum for the adoption of NBS. National workshops from 2012 through 2015, led by the Incubator, connected project developers and national policy leaders, raised awareness, and built relationships. SUNASS and MINAM were among the participants, which advanced both formal and cultural changes in the sector. Overall, main actors at this time found ground up efforts, such as supporting local NBS pilot projects and advocates, to be the best ways to launch NBS (as opposed to top-down regulations and mandates).

Other partnerships have been critical to identifying and addressing bottlenecks to NBS implementation. The International Center for Tropical Agriculture led a study in 2015 to characterize the main obstacles faced by project developers, which was key to prioritizing and justifying policies and technical capacity at a national level (Quintero and Pareja, 2015). In the 37 cities whose water utilities have committed to funds for NBS, alliances between local and regional governments and civil society have been critical to designing, implementing, and monitoring watershed investments. More recently, the NIWS project has supported cross-sectoral alliances to strengthen the uptake of NBS, including new sectors, such as agriculture. The NIWS project is also supporting MINAM and SUNASS to systematically address remaining regulatory and management obstacles.

Institutional context facilitated leadership

The final factor that enabled such rapid formalization of NBS in Peru is the water sector's regulatory structure. Even before investments in NBS were fully sanctioned by law or considered a norm, the institutions within which champions and their collaborators worked afforded them flexibility to act quickly on windows of opportunity, introduce new ideas, and institute widespread changes. This enabled leaders like Pulgar-Vidal and Momiy to introduce ideas and methods that ultimately caused cultural shifts within MINAM, SUNASS, water utilities, and other partners in government and civil society.

Because SUNASS is an autonomous institution whose leadership is appointed for a 5-year term, and therefore somewhat insulated from politics, it can afford to assume a less risk-averse culture than many other governmental agencies in Peru (personal communication, De Bievre,[f] ten Brink,[g] 2015). Because staff at these institutions are not political appointees, personnel turnover is slower than other government agencies, and they are less beholden to shifts in national politics. This gives leadership time to explore new ideas and more leeway to develop and strategically implement their own visions.

Another important factor is that the inclusion of funding for NBS was generally not perceived as a conflict of interest within water utilities or SUNASS. Water tariffs were increased to accommodate new NBS budgets, meaning that NBS funding was not perceived to be redirecting resources from other, perhaps more conventional, projects. The water utility cost-effectiveness analyses used in the early 2010s also suggested that in some cases, NBS investments can be more cost-effective than gray infrastructure and can allow utilities to provide desired levels of service at lower cost.

The future of nature-based solutions in Peru

After significant shifts in water sector culture and policy over the last decade, Peru now boasts NBS policy that is seen as a model around the world. The Sanitation Services and MRSE Laws and their regulations created a unique and powerful legal framework for scaled implementation of NBS. Under this model, investment in NBS is: (1) driven by local water users; (2) promoted and required by national regulations; and (3) funded via a continuous, revenue-generating mechanism (the water tariff). In contrast to other examples of national ecosystem service investment programs—such as those in Costa Rica or China— the nationwide push for ecosystem services in Peru has depended on local leadership and coordination. This suggests that NBS have the potential to be more relevant in addressing local needs than previously assumed—they can help local actors make connections and

---

[f]   Bert de Bievre, CONDESAN and FONAG.
[g]   Dirk ten Brink, USAID.

governance platforms that support water and natural resource management that is sustainable at the landscape scale.

Unlike other places where local users fund and drive investments in NBS—such as successful cases in Quito, Ecuador, or New York, United States—Peru's Sanitation Services Law requires all water utilities to at least consider funding NBS when they develop their master plans and budgets every 5 years. This, and the requirement for SUNASS to support utilities in doing so, creates demand for developing capacities, assessing the potential for NBS and beginning to generate revenue that can only be used for NBS. In some cases, local water managers may have otherwise never deviated from the norm of conventional, gray infrastructure without this incentive. Finally, and critically, the tariff structure creates a stable, renewable foundation for the maintenance of NBS over time. Rather than depending on political decisions for each new investment, water managers can plan for a continued stream of funds to support NBS in the long term.

With the continued support of SUNASS, MINAM, and international cooperation, these policies have steadily translated to increased funding for NBS in the drinking water sector. As of August 2020, over $33 million has been committed for investments in ecosystem services by 37 water utilities, in cities such as Andahuaylas, Chachapoyas, Chiclayo, Cusco, Huamanga, Huancayo, Lima, Moyobamba, and Tarapoto. An additional $112 million has been allocated from water tariffs for climate change adaptation and disaster risk reduction in those same cities, much of which will be invested in NBS for adaptation and risk reduction.

All of these shifts also supported increased investments in NBS by other sectors. Public investment in NBS as a strategy to manage water risks emerged as a new segment of public spending on the environment in the last decade, although it is still relatively small at approximately $7 million in 2018 (Benites and Gammie, 2020). Regional governments lead in these types of investments, far outpacing those by the drinking water sector. Although the water sector has been successful in incorporating MRSE tariffs in the great majority of water utilities across Peru, only 7 of the 37 utilities that currently have the MRSE tariff have begun implementing those investments on the ground. If the water sector implemented investments on the same scale that it is currently generating revenue, it could surpass other sources of investment in Peru.

The implementation gap is especially apparent in the drinking water sector but also exists in other public entities seeking to scale up investments in NBS; for example, the public sector only spent 55% of what it had programmed and budgeted for natural infrastructure for water security in 2018 (Benites and Gammie, 2020). Much of this gap can be attributed to the complexity of the public investment system, which requires specialized skills for project development and evaluation to catch up with new NBS-related guidelines issued in recent years. To embrace NBS, water utilities accustomed to engineered solutions suddenly need to be well versed in fields like ecology, agricultural practices, community organizing, fundraising, and mediation. These capacities can be built in the water utilities themselves, or they can be supplemented through new partnerships—both strategies can be seen across Peru.

Peruvian water sector leaders and NBS advocates alike point to the need for more evidence on the impact of NBS on water quality and quantity. Continuing to collect data and to fine tune existing analyses will improve decision-making, enhance the credibility of these measures, and support continued adoption of NBS at scale. Given the nature of NBS, there is still uncertainty related to its performance; this not only limits water managers' abilities to best use the resources available to them, it sustains skepticism of NBS in the water sector and limits its potential uptake in new sectors, such as agriculture or hydropower. Ultimately, this uncertainty needs to be addressed by monitoring local hydrological outcomes. Most public investments in natural infrastructure for water security in Peru to date have not included a rigorous monitoring program. The notable exception is the drinking water sector—SUNASS has required all NBS investments to include a component of hydrological monitoring. Increasingly, policymakers, scientists, and practitioners are working to address this gap through multisectoral networks, like the Initiative for Hydrological Monitoring of Andean Ecosystems (iMHEA).

What are the most valuable lessons from the Peruvian water sector's experience?

As NBS are increasingly recognized as a critical part of managing today's water risks, policymakers across the globe may look to Peru for ideas, lessons, and tools for how traditionally conservative sectors, like drinking water, can rapidly shift to incorporate NBS. While this chapter has evaluated just one rather unique case there are a number of lessons that may prove valuable for those seeking to replicate this success in other sectors or countries:

- **Early, locally driven NBS projects built the initial experience, confidence, relationships, and momentum necessary to enable more comprehensive change in the water sector**. These early initiatives tended to be successful where local populations easily made the connection between their water supply and source watersheds, due to a combination of geographic proximity and cultural ties. Even where early initiatives lacked rigorous hydrological and cost-benefit analyses, the early case studies were critical references as policymakers, technical specialists, and project developers sought to understand and incorporate new concepts related to NBS. A "learning while doing" approach in the early initiatives—committing to learning over time rather than requiring all the answers up front—helped inspire an iterative and adaptive approach to national policy change as well.

- **Long-standing institutional practice and culture within regulatory and engineering communities can create barriers to adopting NBS**. Experiences in Peru highlight some of these barriers, including a risk-averse culture in the water sector, existing laws and regulations that tended to limit NBS options, and financial and institutional bias toward large capital projects. Nevertheless, relatively rapid institutional change toward NBS is possible. In Peru, the independence and stability of its national water utility regulator was a critical factor that allowed champions to enact change in the water sector.
- **The institutional shifts in Peru happened both formally, through rules and regulations, and informally, through shifts in culture and discourse.** The main cultural shifts within Peru's water sector included the recognition of the benefits of NBS and acknowledgment of the important role water utilities can play in investing in their watersheds. Advocates, like Momiy and Pulgar-Vidal, clearly communicated that if Peru could enact preventative policies and management, they would be spared the stress, expense, and losses that are associated with reacting to events such as natural disasters and water stress. This discursive change in Peru's water sector led to the cultural shifts that ultimately supported formal institutional changes. These changes were more opportunistic at first, like the inclusion of a single article in the Sanitation Services Modernization Law, and were then developed more systematically to address remaining process bottlenecks and policy gaps.
- **Early advocates of NBS incurred personal risk in approving early pilots, due to some legal gray areas. Partnerships with international cooperation and civil society helped support this leadership** by providing technical analysis, generating public awareness, building new tools and capacities, disseminating new policies, and reinforcing new discourse on the importance of incorporating NBS into water risk management strategies.
- Of course, the lessons to date are still incomplete. Peru is now in the critical phase of scaling up NBS implementation. Future assessments may be able to show whether the factors that made the water sector's uptake of NBS a success will also allow them to be effectively implemented and mainstreamed. In the meantime, drinking water sector leaders and their allies in the environmental sector, civil society, and international cooperation continue to work on filling information and capacity gaps, streamlining implementation, instilling confidence in the new system, and demonstrating how NBS can contribute to reducing the water and climate risks Peruvians are facing.

Acknowledgments

We thank Renan Claudio Valdiviezo and Mia Smith (Forest Trends) for content of the timeline graphic, Abby Lindsay Ostovar (Montgomery & Associates, Monterey, CA) and Tania Lucía Ramírez Farías (Pontifical Catholic University of Peru, Lima, Peru) for original research and draft of an earlier version, as well as Genevieve Bennett, Abel Aucasime, Oscar Angulo, Lucas Benites, and Fermando Momiy Hada (Forest Trends) for valuable input and review.

References

Benites L, Gammie G. Opening the tap: state of investments in natural infrastructure for water security in Peru. Lima: Forest Trends; 2020.

Bloomberg. These countries are the most at risk from a water crisis, www.bloomberg.com/graphics/2019-countries-facing-water-crisis/; August 6, 2019. [Accessed 24 August 2020].

Brauman K, Gammie G, Bonnesoeur V, Ochoa-Tocachi B, Arapa E, Roman F, et al. Informing natural infrastructure investment decisions with hydrologic modelling: lessons learned from a case study in Chancay-Lambayeque Watershed, Peru. Washington, DC: Forest Trends; 2021. In preparation.

CONDESAN. Memoria Curso: Hidrología De Los Ecosistemas Andinos Y La Aplicación Del Diagnóstico Hidrológico Rápido; 2014. December 2014. Report to MINAM and SUNASS.

CONDESAN. Hidrología De Los Ecosistemas Andinos Y La Aplicación Del Diagnóstico Hidrológico Rápido: Informe DHR de la Subcuenca del Rio Shullcas; 2015. January 2015.

Gammie G, De Bievre B. Assessing green interventions for the water supply of Lima, Peru. Washington, DC: Forest Trends; 2014. www.forest-trends.org/publications/assessing-green-interventions-for-the-water-supply-of-lima-peru/.

Gammie G, Benites L, Armas Y. Peru approves innovative regulation to accelerate investment in natural infrastructure. Washington, DC: Forest Trends; February 18, 2020. https://www.forest-trends.org/blog/peru-approves-innovative-regulation-to-accelerate-investment-in-natural-infrastructure/.

León F, Renner I. Conservation of water sources in Moyobamba: a brief review of the first experience in payments for environmental services in Peru. Mt Forum Bull 2010;X(1). January 2010.

MINAM. Plataforma Nacional de Servicios Ecosistemas. Ministerio de Ambiente; 2015. http://serviciosecosistemicos.minam.gob.pe/.

Momiy Hada F. "El Rol de Regulador." II Foro Internacional "Retribución por Servicios Eco-Sistémicos y Regulación de Servicios de Saneamiento en el Perú, un aporte a la conservación del agua en el país", http://www.sunass.gob.pe/doc/foro/FERNANDO_MOMIYExperiencia_de_la_SUNASS.pdf; 2014. December 4.

Nowack M. Implementación de un esquema de pago por servicios ambientales: un estudio de la voluntad a pagar. Peru: GTZ (German Technical Cooperation Agency); 2005.

Ochoa-Tocachi BF, Bardales JD, Antiporta J, et al. Potential contributions of pre-Inca infiltration infrastructure to Andean water security. Nat Sustain 2019;2:584–93. https://doi.org/10.1038/s41893-019-0307-1.

Quintero M, Pareja P. Estado de Avance y Cuellos de Botella de los Mecanismos de Retribución por Servicios Ecosistémicos Hidrológicos en Perú. Cali, CO: Centro Internacional de Agricultura Tropical (CIAT); 2015. 40 p.

Renner I. TEEBcase Compensation scheme for upstream farmers in municipal protected area, Peru, http://www.teebweb.org/wp-content/uploads/2013/01/Compensation-scheme-for-upstream-farmers-in-municipal-protected-area-Peru.pdf; 2010.

SUNASS. Aprueban fórmulas tarifarias, estructuras tarifarias y metas de gestión de la EPS Moyobamba S.R.Ltda. para el próximo quinquenio. Resolucion de Consejo Directico No. 080-2007-SUNASS-CD, Lima: SUNASS; 2007. http://www.sunass.gob.pe/doc/normas%20legales/2007/re80_07cd.pdf.

SUNASS. Proyecto Moyobamba (San Martin). Youtube video. Available at: https://www.youtube.com/watch?v=HlSFn_bCH3o; 2014.

SUNASS. Annex 1 of "Aprueban Directiva de Mecanismos de Retribución por Servicios Ecosistémicos Hídricos—MRSE Hídricos y modifican disposiciones aprobadas mediante las RR. N°s 009, 003 y 011-2007-SUNASS-CD". Resolucion de Consejo Directico. N° 045-2017-SUNASS-CD https://busquedas.elperuano.pe/normaslegales/aprueban-directiva-de-mecanismos-de-retribucion-por-servicio-resolucion-no-045-2017-sunass-cd-1577184-1/; 2017, https://www.sunass.gob.pe/wp-content/uploads/2020/09/re28cd_2017_anexo1.pd.

World Resources Institute. Aqueduct water risk atlas, aqueduct country rankings. Washington, DC: World Resources Institute; 2020. www.wri.org/aqueduct/. [Accessed 24 August 2020].

CHAPTER 12

Urban blue spaces, health, and well-being

Paula Vandergert, Pauline Georgiou, Lisa Peachey, and Sam Jelliman
Sustainability Research Institute, University of East London, London, United Kingdom

Introduction

Water is essential to life on earth, therefore its role in the urban ecosystem, human health, and ill-health is a fundamental consideration in achieving the UN Sustainable Development Goals, including climate resilience. Water bodies in urban areas, often referred to as blue spaces to distinguish them from the green and gray, are made up of multiple typologies—rivers, canals, lakes, ponds, creeks, drains—spanning natural, man-made seminatural environments (Fig. 12.1). In many urban areas, these blue spaces are often neglected, degraded or polluted, whereas there are multifunctional benefits to be gained from well-managed urban blue spaces: protecting water supplies; increasing protection from riverine and coastal flooding; urban stormwater management; improving water quality; increasing sustainable water provision for enhanced food security; and providing urban cooling. As climate change brings increasingly extreme weather events, including heatwaves and frequent rainfall leading to flooding, for urban managers water becomes at any one time an ever more precious resource and/or an acute problem to be managed. Other chapters in this book give clear evidence on the systemic and strategic thinking required to develop nature-based solutions (NBS) to increase resilience and water security, including in urban areas.

However, the health and well-being aspects of water bodies, particularly in urban environments, remain an underconsidered aspect of the overall picture. In this chapter we consider these broader health and well-being aspects of urban blue spaces in relation to emerging research and reflections on practice that often require engaging with a level of governance and design innovation that recognizes the complexities of spatial interconnectedness, broader ecosystem thinking, and essential human practices that promote well-being. By understanding this complexity, NBS that protect, restore, or enhance blue and green spaces within the urban fabric can also address a wide range of societal issues, including directly improving the health and well-being of urban residents. This "new science has real-world implications for education, public policy, health care, coastal planning, travel, real estate and business – not to mention our happiness and general wellbeing" (Nichols, 2014, p. 21). We also explore some new ways of embedding water into inclusive

Nature-Based Solutions and Water Security. https://doi.org/10.1016/B978-0-12-819871-1.00013-0

(A)

(B)

Fig. 12.1 *See the figure legend on opposite page.*

(C)

Fig. 12.1

Urban blue spaces can be large or small, neglected liabilities or valued assets, whether (A) a large river, (B) a small polluted creek, or (C) an architectural feature, urban blue spaces exist in all our cities. *Copyright: ©Paula Vandergert.*

neighborhood and community developments, highlighting cases of blue spaces and their contribution to health and well-being in London. Finally, we propose a suite of approaches that could offer holistic tools for engaging with blue spaces in urban areas as socio-ecological systems that, if enhanced through the application of NBS, could make the most of water in our cities to the benefit of health and well-being of people and planet.

What do we know about urban blue spaces, health, and well-being?

Water, as the source of life and human necessity has been studied anthropologically in relation to wellness (Ember and Ember, 2004), in symbolic observations of human/nature relations and ritual uses of water for spiritual and metaphysical connections and environmental conservation (Runk, 2009), Indigenous wisdom and society (Lansing, 1987; McAllester, 1941), and social meaning (Ballestero, 2019; Wagner, 2013).

The link between urban green spaces and health and well-being has increasingly become an accepted fact, with academics from various disciplines building the evidence base. The press is inundated with journalism to support this notion, be it exercising in a park, a new interest in a Japanese tradition—"forest bathing" (Shinrin-yoku) or planting to improve air quality in a street near a school. In 2020, parks and open spaces took on unimagined significance in the global COVID-19 pandemic. However, the scientific research is often disparate because of the academic silos embedded in the different disciplines. Variously, the relationship between health and nature has been discussed under the premise of green care sector, evidence-based design in landscape architecture, NBS, environmental psychology, "Vitamin G," and health design, being considered by landscape architects, urban planners, ecologists, sociologists, the health sector, educational, occupational, and development psychologists, and environmentalists. "Health" can include mental health, stress, heart disease, diabetes, exercise, air pollution levels, mortality, or general well-being. This is sometimes measured biometrically and at other times by self-reporting methods. Thematically, there is also a broad spectrum of approaches: passive and active space, sedentary and ambulatory use of space, and therapeutic programs such as horticultural therapy mixed with prescribed mindfulness or Cognitive Behavioral Therapy (Peachey, 2020).

More specifically related to blue spaces in cities, researchers are also increasingly studying health and well-being aspects of water bodies in cities—with the phrase urban blue spaces being coined to distinguish them from green spaces and representing natural and man-made water features, large and small, such as rivers, canals, lakes, reservoirs, ponds, and streams. The mental and physical benefits of urban blue spaces is an emerging research field with far fewer research studies than for green areas. Research has tended to focus on coastal areas (Wheeler et al., 2012) or on photographs of natural and built scenes with varying amounts of water (Karmanov and Hamel, 2008; White et al., 2010). These studies suggest positive associations between health and well-being and water, although their limitations are clear in terms of the focus on the specific characteristics of the coast and the limited nature of relying on images (and visual sense only), rather than other sensory aspects of being in nature, such as touch, sound, and smell.

Blue Health is a European-funded research program that aims to further explore the positive relationships between urban blue spaces and public health impacts (Grellier et al., 2017). As part of this program, Gascon et al. (2017) have undertaken a systematic review of quantitative studies of outdoor blue spaces, human health, and well-being. The researchers found that the literature is currently far less developed than for green spaces and is of a heterogenous nature, so comparability and synthesis is challenging. Their review of 35 studies found an association between greater exposure to outdoor blue spaces, mental health and well-being, and levels of physical activity. Less consistent was an association between outdoor blue space and general health, obesity, and cardiovascular outcomes. Gascon et al. (2017) refer to the significance of the existing qualitative (as opposed to quantitative) evidence that "already suggests a range of benefits of being exposed to blue spaces" and how valuable this is to policymakers (p. 1220).

Since Gascon et al.'s review (2017), further research is being published on health, well-being, and access to water bodies in the urban environment. Völker et al. (2018) undertook a cross-sectional analysis of over 1000 residents in two German cities, one with "poor" urban blue space supply and one with "good" urban blue space supply. The researchers have recognized a limitation in their research in terms of having a focus on measuring ill-health rather than positive health and well-being in that they used the Medical Outcomes Study Short Form (SF-12v2) for measuring self-reported mental and physical health, with an ill-health focus, rather than, for example, the Warwick-Edinburgh Mental Wellbeing Scale or the WHO Quality of Life (WHOQOL) scale, both of which focus on factors that support health and well-being (Völker et al., 2018). This distinction is important for research because a focus on ill-health may miss positive aspects of blue spaces on health and well-being because they were not measured.

Developments in psychology as a discipline recognize this distinction as significant in regard to human well-being and reflect a wider health perspective that is broadening health policy from the purely medical to preventive measures and building good health (Seligman, 2011). Völker et al. (2018) also measured perceived walking distance to blue space and type of blue space. Results showed a significant association between use frequency and perceived walking distance, but that it was the blue space use, rather than the perceived walking distance, which was a better indicator for health outcomes. The researchers suggest that the active use of large blue spaces (lakes, rivers, canals), implying a direct engagement with the water, was the most beneficial for mental health, rather than smaller water bodies embedded in residential areas: "When blue space is used, its specific health enhancing potential can then best evoke mental health benefits" (Völker et al., 2018, p. 8). However, they suggest that research is needed on the health promoting capacities of smaller, more integrated blue spaces (such as creeks and ponds) that are embedded in residential neighborhoods or green spaces networks. They point to other qualitative research that links blue space to environmental and contemplative experiences, a feeling of identify and sense of place (Völker and Kistemann, 2013), positive mood and self-esteem (Barton and Pretty, 2010), and high ratings for the restorative effects of water, even in urban parks (Nordh and Østby, 2013). Our own research has started to look at these aspects (see the case study in this chapter) and the role of blue-green spaces as restorative spaces in times of grief (Peachey, 2020).

Researchers are starting to consider factors such as environmental justice, inclusive accessibility, who benefits from access to blue space, and the quality and the type of blue space, echoing research associated with the importance of the quality and accessibility of green space (CABE Space, 2010). Haeffner et al. (2017) used an environmental justice lens to consider different demographic and socio-economic groups' access and proximity to urban water ways in Utah. Their findings supported other research (Wendel et al., 2011) that showed that the type of water appears to make a considerable difference to people's experience, with poorer groups often living closer to poorer quality water such as stormwater ponds. de Bell et al. (2017) found that social interaction and psychological benefits were the most important reported benefits from visiting blue space, with psychological benefits being more prevalent for women and older people, associated with perceptions of the presence of nature and environmental quality. They found that social interaction benefit was of greatest significance to those who were most socially disadvantaged. However, the researchers found that these groups were less likely to report visiting blue spaces frequently (de Bell et al., 2017).

Urban freshwater bathing and immersion in water

Bathing and immersion in water has a long cultural tradition in many cities. Social relations to water and immersion vary cross-culturally—though a common thread of social bathing can be found in historical and contemporary practices across the world (Good, 2010). The ancient Greeks saw public bathhouses as spaces of art and creativity (Fournet, 2013) while the ancient Romans used them as intellectual platforms (Smith and Gadeyne, 2013). In Finland, Russia, Turkey, and Japan, for example, hygiene, wellness, and socialization in water centers remain highlights of the daily experience.

Swimming is one of the most widespread, accessible forms of physical activity suitable for all ages, and indoor swimming pools are a standard feature of many urban neighborhoods. However, unless city dwellers are fortunate enough to live by or near the coast, opportunities for open-water swimming are much rarer. However, the particular health benefits of open-water swimming are increasingly being reported and supported by research, especially for mental health (Foley, 2017; Van Tulleken et al., 2018) and physical and mental health in Japan (Goto et al., 2018).

More cities are likely to have accessible rivers than coastline, although pollution and public safety considerations often mean city authorities do not permit urban river swimming (Fig. 12.2). In Basel, Switzerland, the river Rhine is so popular for swimming that the Wickelfisch[a] was invented and has become iconic of the city. After a century of swimming restrictions, parts of the river Seine, in Paris, France, have in recent years reopened to the public for swimming. As part of a €1 billion clean-up project, which aims at preparing the river for the 2024 Olympics, the popular Bassin de la Villette canal was reopened in 2018. In London, United Kingdom, the River Thames used to have stretches to the west of the city where public bathing was permitted (Davies, 2015), but now it is not. However, the growing popularity of open-water swimming in the United Kingdom has led to the creation of an open-water swimming hub in East London's Royal Docks under controlled access and the longer established outdoor swimming venues such as the city's lidos (often architecturally beautiful buildings set in Parkland) and the more natural swimming ponds of Hampstead Heath. As part of London's National Park City launch celebrations in July 2019, the first natural swimming pond to be created in decades opened in South East London's Beckenham

---

[a] A swim bag in the shape of a fish, which was invented in Basel, to keep clothes and personal artifacts dry by floating beside a fully immersed swimmer.

Fig. 12.2
People find many informal, and sometimes illegal, bathing opportunities in cities; (A) in the 2018 and 2019 heatwaves, impromptu (and often not permitted) bathing locations were spread via social media, including this little-known corner of the river Lee in East London; (B) signs had to be put up to warn of the dangers from pollution, underwater hazards, and currents. *Copyright: ©Paula Vandergert.*

Place Park (the grounds of a Georgian mansion)—and had to be temporarily closed within 3 days because of the sheer volume of people flocking to bathe there.

Blue spaces and heatwaves

Water is a challenge in extended heatwaves—reservoirs and rivers run low, drinking water has to be carefully managed, often resulting in measures to reduce unnecessary consumption of a precious resource. Water is also a core part of health advice for coping with extreme heat, including having cool baths and showers and drinking water. Panno et al. (2017) summarize the negative effects of heatwaves on physical and psychological health, such as respiratory and cardiovascular problems, distress, fatigue, and low energy. Public bathing, and collective immersion in cool water, is often an impromptu response of people in the face of heatwaves. In the summers of 2018 and 2019, Europe saw prolonged heatwave episodes, with highest recorded temperatures in many European cities (BBC News, 2019). In these circumstances, water can provide a joyful cooling element (Fig. 12.3). News reports often showed people enjoying formal and informal opportunities to cool down by and in the water, such as fountains, and impromptu bathing spots on rivers and canals being enjoyed, often with little thought to things like water quality or personal safety.[b] In London, blue spaces such as outdoor natural and seminatural pools and ponds, lidos, lakes, and water fountains are particularly attractive in the summer months. In the London heatwaves of 2018 and 2019, police had to be called to outdoor swimming spots—both formal and informal (and often illegal) because these places became overwhelmed by numbers way beyond their capacity to be safely managed.

Blue spaces as surface water management features

Whether large or small, natural or constructed, they can play an important role in a city's resilience and transport systems. Often, they are of poor ecological state, or neglected spaces. They are also increasing in number in many cities—as climate change brings increased rainfall episodes leading to flash flooding in urban areas, an increasingly widespread approach to surface water management of excess rainwater is the inclusion of sustainable urban drainage systems (SUDS) in new urban development and (albeit much less common) their retrofit into existing neighborhoods. SUDS that are designed on the basis of delivering multiple environmental, social, cultural, and health benefits are increasingly being researched and innovated through global NBS programs such as those supported by the European

Fig. 12.3
People find many ways to enjoy and play in water in the city, particularly in hot summer months, where blue spaces provide respite from the urban heat island even in the dense urban cores.
Copyright ©Paula Vandergert.

---

[b]   https://search.creativecommons.org/photos/58bbf92b-e9ba-44d8-b5b8-c6b51a4d0b8a.

Commission (Fig. 12.4). Thus, there are increasing opportunities for providing good quality new urban blue spaces that provide multiple benefits.

The health benefits of these new urban blue spaces is little researched to date (Völker et al., 2018), and indeed, SUDS can sometimes provide a challenge to neighborhoods where people are not used to such features. A report by Donald (2018) focuses on the relationship between professional surface water management practices and public perceptions of water in three areas of Glasgow where SUDs are incorporated in new development. The researchers used creative practice, observation, and interviews to understand resident attitudes to water and the design, construction, integration, and maintenance of SUDs among professionals. Their findings echo other research that indicates the quality of water is particularly

(A)

(B)

Fig. 12.4 See the figure legend on next page.

(Continued)

(C)

Fig. 12.4, cont'd
Sustainable urban drainage systems and wetlands in (A, C) London and (B) Malmö. Could an urban acupuncture approach mean more of these multifunctional sustainable urban drainage schemes and urban wetlands that bring delight and havens for wildlife to local neighborhoods?
Copyright ©Paula Vandergert.

important to residents' perceived benefits and that associated "naturalness" of the water and its environment contributes to this, whether natural or constructed water bodies. Donald (2018) found that creative practice (play and sensory experiences) helped professionals and residents communicate with each other about water and the function of SUDS, enhancing understanding on both sides. This codesign and comanagement approach to SUDs was successfully implemented in the neighborhood of Augustenborg, Malmö, where wider benefits include improved social cohesion, the start-up of local nature-based businesses, a fall in unemployment rates and void properties, and a reduction in the previous high turnover of tenancies (Kazmierczak and Carter, 2010).

A case study of blue spaces action research and practice: Thames Ward water wellness project

Between 2018 and 2019, the chapter authors were engaged in the United Kingdom's National Health Service England Healthy New Towns program. This program designated 10 places in England as Healthy New Towns and provided funding to trial different approaches to achieve three goals: planning and designing a healthier built environment; enabling strong, connected communities; and creating new ways of providing integrated health and care services. As part of London's only Healthy New Town—Thames Ward, Barking in East London—we were commissioned by the London Borough of Barking and Dagenham and Barking Riverside Ltd. to investigate blue spaces as a focus for these overall goals.

Thames Ward is—as the name suggests—a neighborhood that sits on the northeast bank of the Thames as it heads out of London and into neighboring Essex and then to the North Sea and English Channel. Thames Ward is on the tidal Thames. Thames Ward is home to not only one of the largest regeneration zones in Europe—a 10,800 home "new town" of Barking Riverside—but also to existing neighborhoods built in the 1950s and to industrial and postindustrial land. This is an area of former marshland and therefore is a "watery" environment in its essence. Barking is also one of the most deprived areas in London, with lower health, well-being, and socio-economic indicators than most other parts of the city. The stated aim of the Barking local authority, shared by the development agency (a joint venture between the Greater London Authority and housing association London and Quadrant), is that growth should be inclusive and that all residents should benefit from the new development.

Within this context, we identified a suite of related research activities at the outset and adapted and added to them as we learned. We started with a literature review, finding many more references to the health and well-being benefits of green spaces rather than blue spaces. We also undertook a survey of Thames Ward, mapping and visually assessing the location, typology, potential accessibility as amenity space, and visual state of local blue spaces in terms of whether they contained rubbish or debris, waterfowl, how overgrown they were. While we did not test water quality as part of this project, as this was not part of the research brief and therefore the funding for the chemical analysis was not available, colleagues undertake this analysis as part of specific research projects from time to time in different locations across the area. To our knowledge, a systemic review of pollution and ecological status across the area has not been undertaken, although all water bodies should be functioning within the Water Framework Directive and various statutory bodies are responsible in the area, adding to complexity of governance and data availability.

Our literature review and blue spaces survey led to the development of a "water wellness" map for Thames Ward. We also undertook a series of action research activities to engage with residents. We coorganized (with a walking charity, Living Streets) several guided walks around the neighborhood, taking in selected blue spaces and, in particular, showing residents how to walk to the River Thames. One of the walks was a themed "ecology" walk. The walks were advertised through local schools and community centers and took place in the summer of 2018. We also shared the draft map with residents and local stakeholders at community events, such as coffee mornings, community picnics, and resident groups—we chose to attend existing events and only held one specific event to share the map—and a "drop-in" cup of coffee in the local café. From these events we learned several key aspects: the large majority of residents we engaged with did not know that they lived so near the River Thames or how to reach it, regardless of how long they have lived in the area; the different communities (those in the older estates and those in the newer ones) have little interaction in each other's neighborhoods (nor consider themselves part of one single area, despite the fact they are in the same ward); most people know nothing about their local environment, in particular the blue spaces (even the River Thames), but are very interested; the SUDS in the new development are not understood and generate mixed feelings among local people in relation to nature, mosquitoes, and "untidiness."

As a result of our analysis of the mixed methods approach previously outlined, we developed an additional piece of work within the scope of this Healthy New Town project. We learned from the literature review that engaging people proactively in their local blue spaces was more likely to lead to positive well-being outcomes than just showing people where these spaces were. In Thames Ward, there are no activities related to the blue spaces that people can take part in and most people do not know about the location, function, or history of local blue spaces. Many of the blue spaces are also of poor ecological quality. We therefore decided to develop and pilot a local "nature and wildlife" course for residents (Fig. 12.5). We delivered this over two and a half days for a local resident group based at a local community center. The course consisted of an introduction to the ecology and history of the area, a local walk (based on the blue spaces map) and a visit to another part of East London—Walthamstow Wetlands, managed by a local wildlife charity, the London Wildlife Trust. The process was woven into a film made by New View Arts project (a project of the Thames Residents and Tenants Association) funded by the Arts Council, which was a public art engagement project with local children and a local group remembering the former Creekmouth Village, which was wiped out in the floods of 1953.[c]

What we learned from the cumulative work in the area is the value of this integrating research within local community and environmental groups. This not only ensures that research is responsive but also creates opportunities for developing ongoing momentum and building capacity for local people to influence design and management of their local surroundings. This was evidenced by some of these same residents actively contributing to a new project—Ripple Greenway—which started as the development of a new cycleway but expanded to focus on opportunities for ecological enhancement, including of a neglected watercourse. People who participated in the course also participated in the participatory design process, sharing stories of how, 50 years previously, they had taught their children to canoe on this apparently unprepossessing space. This enabled the charity (Sustrans) to increase their ambition for the watercourse—and we, as the Sustainability Research Institute, were brought in to advise on the ecological condition and opportunities for the site. This began building some of the well-being benefits for local residents from their local blue and green spaces.

---

[c] https://www.youtube.com/watch?v=Vl1WdstgHFE.

(A)

(B)

Fig. 12.5 See the figure legend on next page.

(Continued)

(C)

Fig. 12.5, cont'd
Evidence shows that proactively engaging local people in valuing and caring for their local blue spaces increases their health and well-being benefits; (A) community meeting Healthy New Town project; (B) informational poster; and (C) students in the "nature for wildlife" course for residents. *Copyright ©Paula Vandergert.*

Envisioning and delivering more beautiful, health-giving places with water at their heart

So, a question for us all: if heatwaves and flooding are going to be more frequent and water is therefore at any one time a scarce resource to be conserved, flash flooding to be minimized, and an essential ingredient in dealing with extreme heat, are we doing enough to put water center stage of urban design, regeneration, and management?

Many towns and cities have different types of water bodies, from large rivers to small canals, creeks, ponds, and ditches. Many of these water bodies are neglected and unloved, others are hotspots for regeneration, with developers often featuring pictures of water in their marketing materials. The opportunities for making the most of the water, through increased and equitable access, improved quality, and enhanced biodiversity are immense. This is especially important in those inner-city areas that have more blue space (water) than green space. Rather than either ignoring hidden urban water bodies or just seeing them as profit-optimizing development opportunities with no focus on their ecological functioning, we should invest in them for their health, biodiversity, recreation, cooling, and social benefits. In Malmö, Sweden, the new neighborhood in the docks area explicitly includes water management at the heart of its design, and this has included water cleaning and creating access for open-water swimming. The potential gains of taking a water-centric approach have been exemplified in an extraordinary and unexpected way in Venice, as a result of the COVID-19 pandemic of 2020. Due to the lockdown on people's movements, and thus a lack of pollution, waterways were running clearer than they have ever done, with wildlife thriving.

There is the opportunity to channel the emerging evidence across scientific disciplines, historical practices, and creative and innovative design to transform how we engage with water. For example, immersion in water as a social and cultural practice linked to urban sustainability and climate change poses a particularly interesting field for research and innovative design practices. Another innovative lens through which to take a more holistic approach to blue spaces, health, and well-being is through adapting a one-health approach, combining human, companion animals, and ecological health. This systemic approach at the human-ecological-animal interface of health problems (Binot et al., 2015; Lebov et al., 2017). How could an adapted one-health approach provide opportunities to help identify how water could be center stage of a healthy and multifunctional urban social-ecological system? Could the concept of urban acupuncture, whereby the nodes and networks of natural assets in an urban area are enhanced through multiple small-scale nature-based interventions, be the way forward for urban planning? Former mayor of Curitiba, who put into practice this approach as he led the successful transformation of Curitiba to become a globally lauded sustainable city, describes urban acupuncture as: "small interventions that can provide new energy to the city, and provide assistance during the process of long-term planning, which has to take time."[d]

The scope to test these innovative approaches in an integrative, future-proofed way while also ensuring robust and just institutions will require multifaceted ways of working (Frantzeskaki et al., 2020). If we can achieve that, our urban blue spaces have the capacity to deliver multiple health and well-being benefits for humans and the environment while also creating beautiful, adaptable places to live (Fig. 12.6). At the time of writing, it feels that the world has

---

[d]　https://www.asla.org/ContentDetail.aspx?id=30875. [Accessed 20 August 2020].

Fig. 12.6 See the figure legend on opposite page

(D)

Fig. 12.6, cont'd

Water-focused urban design is a centuries-old practice worth reviving; (A) and (B) older traditions like those found in Moorish Seville (A) and Córdoba (B) are echoed in 21st century urban design in prestigious city center developments (C) and (D). *Copyright ©Paula Vandergert.*

had a sharp shock in relation to values and meaning of nature due to COVID-19—it would be wonderful to seize this impetus to put ecologically rich, accessible blue and blue-green spaces at the heart of our decisions and practices in relation to urban management at strategic and neighborhood scales, for the health and well-being of people and places.

References

Ballestero A. The anthropology of water. Ann Rev Anthropol 2019. https://doi.org/10.1146/annurev-anthro-102218-011428.

Barton JO, Pretty J. What is the best dose of nature and green exercise for improving mental health? A multi-study analysis. Environ Sci Technol 2010;44(10):3947–55. https://doi.org/10.1021/es903183r.

BBC News. Europe heatwave: Records tumble in Belgium, Germany and the Netherlands. 24 July 2019. Available at: https://www.bbc.co.uk/news/world-europe-49100271; 2019. [Accessed: 16 April 2020].

Binot A, et al. A framework to promote collective action within the One Health community of practice: using participatory modelling to enable interdisciplinary, cross-sectoral and multi-level integration. One Health 2015;1:44–8. https://doi.org/10.1016/j.onehlt.2015.09.001. The Authors.

CABE Space. Community green: Using local spaces to tackle inequality and improve health; 2010. p. 1–61. Available at: http://www.cabe.org.uk/publications/community-green, http://scholar.google.com/scholar?hl=en&btnG=Search&q=intitle:Community+green+:+using+local+spaces+to+tackle+inequality+and+improve+health#0%5Cnhttp://scholar.google.com/scholar?hl=en&btnG=Search&q=intitle:Community+green:+Using+local+spaces+to+tackle+ine.

Davies C. Downstream: A history and celebration of swimming the river Thames. London: Aurum Press; 2015.

de Bell S, Graham H, Jarvis S, White P. The importance of nature in mediating social and psychological benefits associated with visits to freshwater blue space. Landsc Urban Plan 2017;167:118–27.

Donald M. Living, working, playing with water: Exploring perceptions of water in the urban environment through creative practice. Project Report, The Metropolitan Glasgow Strategic Drainage Partnership; 2018. https://www.mgsdp.org/CHttpHandler.ashx?id=40594&p=0.

Ember CR, Ember M. Encyclopedia of medical anthropology: health and illness in the world's cultures. Hum Relat 2004. https://doi.org/10.1016/S0014-4886(08)00242-2.

Foley R. Swimming as an accretive practice in healthy blue space. Emot Space Soc 2017;22:43–51.

Fournet T, et al. Catalog of Greek baths. In: Lucore S, Thrumper M, editors. Greek baths and bathing culture. New discoveries and approaches. Peeters; 2013.

Frantzeskaki N, et al. Land Use Policy Examining the policy needs for implementing nature-based solutions in cities: findings from city-wide transdisciplinary experiences in Glasgow (UK), Genk (Belgium) and Poznań (Poland). Land Use Policy 2020;96:104688. https://doi.org/10.1016/j.landusepol.2020.104688. Elsevier. (April 2019).

Gascon M, Zijlema W, Vert C, White MP, Nieuwenhuijsen MJ. Outdoor blue spaces, human health and well-being: a systematic review of quantitative studies. Int J Hyg Environ Health 2017;220:1207–21.

Good BJ, et al. A reader in medical anthropology: Theoretical trajectories, emergent realities. Blackwell anthologies in social and cultural anthropology, Wiley-Blackwell; 2010.

Goto Y, et al. Physical and mental effects of bathing: a randomized intervention study. Evid Based Complement Alternat Med 2018. https://doi.org/10.1155/2018/9521086.

Grellier J, et al. BlueHealth: a study programme protocol for mapping and quantifying the potential benefits to public health and well-being from Europe's blue spaces. BMJ Open 2017;7(6):1–10. https://doi.org/10.1136/bmjopen-2017-016188.

Haeffner M, Jackson-Smith D, Buchert M, Risley J. Accessing blue spaces: social and geographic factors structuring familiarity with, use of, and appreciation of urban waterways. Landsc Urban Plan 2017;167:136–46.

Karmanov D, Hamel R. Assessing the restorative potential of contemporary urban environment(s): beyond the nature versus urban dichotomy. Landsc Urban Plan 2008;86(2):115–25. https://doi.org/10.1016/j.landurbplan.2008.01.004.

Kazmierczak A, Carter J. Adaptation to climate change using green and blue infrastructure. A database of case studies. University of Manchester; 2010. p. 182.

Lansing JS. Balinese "water temples" and the management of irrigation. Am Anthropol 1987. https://doi.org/10.1525/aa.1987.89.2.02a00030.

Lebov J, et al. A framework for One Health research. One Health 2017;3:44–50. https://doi.org/10.1016/j.onehlt.2017.03.004. The Authors.

McAllester D. Water as a disciplinary agent among the Crow and Blackfoot. Am Anthropol 1941. https://doi.org/10.1525/aa.1941.43.4.02a00070.

Nichols W. Blue Mind: The surprising science that shows how being near, in, on, or under water can make you happier, healthier, more connected and better at what you do. Abacus; 2014.

Nordh H, Østby K. Pocket parks for people—a study of park design and use. Urban For Urban Green 2013;12(1):12–7. https://doi.org/10.1016/J.UFUG.2012.11.003. Urban & Fischer.

Panno A, Carrus G, Lafortezza R, Mariani L, Sanesi G. Nature-based solutions to promote human resilience and wellbeing in cities during increasingly hot summers. Environ Res 2017;159:249–56.

Peachey L. Designing for mental health: Bloedel Reserve, a case study [Masters Thesis]. University of East London; 2020.

Runk JV. Social and river networks for the trees. Am Anthropol 2009;111(4):456–67.

Seligman MEP. Flourish: A visionary new understanding of happiness and wellbeing. Nicholas Brealey Publishing; 2011.

Smith G, Gadeyne J. Perspectives on public space in Rome, from antiquity to the present day; 2013. https://doi.org/10.4324/9781315600215.

Van Tulleken C, Tipton MJ, Massey H, Harper M. Open water swimming as a treatment for major depressive disorder. BMJ Case Rep 2018. 2018:bcr-2018-225007.

Völker S, Kistemann T. I'm always entirely happy when I'm here! In: Urban blue enhancing human health and well-being in Cologne and Düsseldorf, Germany. Social science & medicine, vol. 78. Pergamon; 2013. p. 113–24.

Völker S, Heiler A, Pollmann T, Clasen T, Hornberg C, Kisteman T. Do perceived walking distance to and use of urban blue spaces affect self-reported physical and mental health? Urban For Urban Green 2018;29:109.

Wagner JR. The social life of water; 2013. https://doi.org/10.1080/00207233.2013.862465.

Wendel HEW, Downs JA, Mihelcic JR. Assessing equitable access to urban green space: the role of engineered water infrastructure. Environ Sci Tech 2011;45(16):6728–34. https://doi.org/10.1021/es103949f.

Wheeler BW, et al. Does living by the coast improve health and wellbeing? Health & Place 2012;18(5):1198–201. https://doi.org/10.1016/J.HEALTHPLACE.2012.06.015. Pergamon.

White M, et al. Blue space: the importance of water for preference, affect, and restorativeness ratings of natural and built scenes. J Environ Psychol 2010;30(4):482–93. https://doi.org/10.1016/j.jenvp.2010.04.004. Elsevier Ltd.

CHAPTER 13

Learning from indigenous and local knowledge: The deep history of nature-based solutions

Jan Cassin[a] and Boris F. Ochoa-Tocachi[b,c,d]
[a]*Forest Trends, Washington, DC, United States,* [b]*Imperial College London, London, United Kingdom,*
[c]*Regional Initiative for Hydrological Monitoring of Andean Ecosystems (iMHEA), Lima, Peru,* [d]*ATUK Consultoría Estratégica, Cuenca, Ecuador*

Key messages

- Indigenous peoples and local communities around the world have developed nature-based, resilient, and sustainable solutions for water security based on traditional knowledge-practice-value systems.
- Water sowing and harvesting technologies such as *albarradas* in Ecuador, sand dams in Africa, and wetland management and infiltration systems in Spain and the Andes work with nature to "sow" water for storage in soils, river sands, and groundwater, making water available for harvesting during times of water scarcity.
- Integrated land and water management systems rely on managing ecosystem processes in combination with irrigation and terracing infrastructure to control the timing and supplies of water to complex agroforestry, rice or wet-pond taro cultivation, home gardens, and in some cases coastal fishponds and fisheries, as well as providing clean water for households. Predating contemporary integrated watershed and ecosystem-based management, these systems that emphasize protecting source watersheds are models of sustainably working and building with nature to manage water resources.
- Continued viability of many of these practices is under threat, as Indigenous peoples and local communities remain marginalized in many countries, are under pressure from the forces of globalization, with lives and territories threatened in several places, and with traditional and Indigenous knowledge lost with the passing of older generations and forced migration.
- However, these practices and the knowledge they are based on are being revitalized as Indigenous and local peoples lead movements to protect and revitalize their heritage, while scientists, conservationists, water managers, and governments increasingly recognize the value of these traditional nature-based solutions.

Nature-Based Solutions and Water Security. https://doi.org/10.1016/B978-0-12-819871-1.00012-9

Introduction

The engineering structures that ancient societies created to capture, store, and control water have long fascinated archeologists and historians. The famous qanats, or sloping underground aqueducts in Iran and Iraq, tap into groundwater that accumulates under upland areas, using gravity to convey subterranean streams down to farms and villages in adjacent lowlands—creating oases in the desert (Fig. 13.1) (Beckers et al., 2013). In the 14th and 15th centuries in the Valley of Mexico, the Mexica built massive hydraulic structures—dikes, causeways, aqueducts, diversion dams, and canals—to create dryland, manage flooding, maintain freshwater supplies, and expand agricultural lands around their capital of Tenochtitlan (Fig. 13.2) (Candiani, 2014). On Easter Island, when Polynesian voyagers arrived, they found an island prone to droughts, with few freshwater lakes or streams. On the porous volcanic basalts of the island, rainfall rapidly vanishes from the surface, percolating into a large but deep underground reservoir. Otherwise too deep to be accessible, this freshwater emerges at the coast, in many springs and seeps. The people of Rapa Nui built numerous stone wells (*puna*) along the coast to capture and store freshwater from groundwater seeps and rainfall, protecting it from saltwater intrusion, and creating small storage reservoirs for drinking water (DiNapoli et al., 2019; Hixon et al., 2019; Rull, 2020). While the two large crater lakes on the island likely supported extensive agriculture and large populations, in times of drought the coastal seeps may have been critical, more reliable water sources (Hixon et al., 2019; Rull, 2019, 2020).

Fig. 13.1
Airshafts from a network of ancient underground aqueducts, Mehriz, Iran. © *George Steinmetz (STNMTZ_20031101_17). Reproduced with permission from George Steinmetz.*

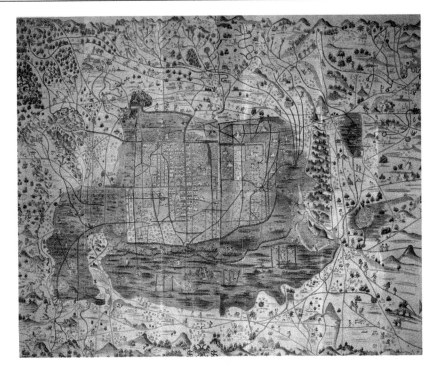

Fig. 13.2

Map of the Valley of Mexico showing Mexico City and the surrounding lakes, dikes, and water system drawn in 1560 by Santa Cruz Alonso. *Reproduced with permission from Alamy Stock Photo, Biblioteca Nacional-Coleccion, Madrid, Spain.*

Perhaps less well known but just as impressive are technologies that Indigenous peoples have created through the centuries that work with nature to control floods, store water for seasons of scarcity, and protect the quality of water supplies. No less sophisticated than ancient hydraulic structures, these nature-based technologies rely on a rich heritage of traditional knowledge that has allowed people to thrive where water is scarce, or highly variable and unpredictable from season to season and year to year. Many of these practices, which are still in use today, have a deep history, demonstrating thousands of years of sustainable living on these landscapes. Although current generations are faced with new challenges such as those from climate change, many of the current water issues have been solved before by our ancestors, and we should learn from them.

In this chapter we explore the contribution of traditional knowledge to nature-based solutions (NBS) for water, present several families of related traditional practices, describe what is known about the effectiveness of these technologies in delivering water security outcomes, and provide examples of how these practices are being revitalized, used, and adapted today.

Traditional, local, ancestral, and indigenous knowledge

Traditional ecological knowledge, Indigenous knowledge, and Indigenous and local knowledge comprise a related set of knowledge-practice-value systems that are widely recognized as important to creating resilient, sustainable responses to current water and climate crises (Berkes, 2004, 2012; McMillen et al., 2014; Thaman, 2013). What these systems have in common are customary knowledge-practice-value systems that are culturally transmitted from generation to generation, emerge from place-based understanding of the relationships between living beings, including humans, with the environment and each other, and have evolved through adaptive learning processes in specific places over time (Berkes, 2012; McMillen et al., 2014). In this chapter we use the terms indigenous and local knowledge (ILK) or traditional knowledge to encompass aspects of all three systems. Some of the cases we discuss are examples of IK; the local knowledge of Indigenous peoples, unique to a particular cultural group (Dudgeon and Berkes, 2003). Other cases are examples of local knowledge held by groups that may not be Indigenous but have a long history of adaptive learning in place over multiple generations.

Much of the literature on traditional knowledge has emphasized knowledge of and conservation of biodiversity, customary resource management in marine systems, and adaptation and resilience—especially within agricultural or pastoral systems—but very few studies have focused on traditional knowledge and nature-based water resources management (Thaman, 2013). While the components of knowledge-practice-value systems are interrelated and act together, we focus here primarily on the technologies for resource management that are the expression of knowledge and value systems. We describe traditional practices for water security that are: guided by a particular cultural worldview; adaptive, dynamic, and empirically based; embedded in multigenerational belonging to a particular place; and passed from one generation to another.

Traditional technologies for water security

There are many Indigenous technologies that incorporate water as a temporary component but in which the management of water flows is not the primary function or purpose. Many of these are agriculturally focused, incorporating soil and water conservation practices such as traditional agricultural terracing or raised fields that create suitable surfaces for agriculture. For example, in pre-Columbian Andean communities, Lane (2009) has distinguished two kinds of hydraulic architectures, dry and wet. Dry hydraulic structures are those in which water is an episodic component, such as in terraces and fields that are periodically irrigated. Wet hydraulic features have water as a constant element, via storage in soils or groundwater, such as the Tiwanaku raised fields near Lake Titicaca (Erickson, 2000) or the *qocha* pond fields in Peru (Flores Ochoa et al., 1996). In this discussion, we include technologies that have a primary purpose of maintaining or improving water security by enhancing storage and regulating flows to manage the amount and timing of water availability to avoid too much or too little water. In some cases, these technologies are embedded within a larger integrated landscape and watershed management system.

When are traditional technologies nature-based?

We include those technologies that work with nature rather than those that rely on hydraulic or engineered structures alone. Ancient and traditional water harvesting technologies, many of which are still in use today, have been central to maintaining water supply systems for thousands of years in arid and semiarid regions (Beckers et al., 2013; Bhattacharya, 2015). These technologies use a wide variety of structures and techniques to collect and store water, but many of these practices use only built structures, such as dams, storage reservoirs, diversion canals, or cisterns for capturing, routing, or storing runoff from rainfall or floodwaters (Beckers et al., 2013; Mekdaschi-Studer and Liniger, 2013; Schwilch et al., 2012).

A subset of these traditional practices rely on natural processes or components of natural infrastructure—vegetation, soils, sands, shallow aquifers—to slow the flow of water, collect, filter, and store runoff, and infiltrate and route water to desired areas. The term "water sowing and harvesting," adopted from the Spanish *siembra y cosecha del agua*, captures these approaches that use natural elements to "sow" the water for harvesting later in areas and at times where it can be used by people. These technologies are sometimes referred to as managed aquifer recharge, but this term is often used for artificial aquifer recharge, such as the use of injection wells. Water sowing and harvesting emphasizes the use of practices based on nature, relying on natural processes of infiltration and storage in soils and aquifers to provide water supplies that can be harvested when needed.

Many of the practices discussed here also fall under the umbrella of sustainable land management (SLM) (Schwilch et al., 2012). SLM is defined as: "measures and practices adapted to biophysical and socio-economic conditions aimed at the protection, conservation and sustainable use of resources (soil, water and biodiversity) and the restoration of degraded natural resources and their ecosystem functions" (Food and Agriculture Organization, http://www.fao.org/land-water/land/sustainable-land-management/slm-practices/en). We focus here on those practices that have water management as the central component. While not intended to be an exhaustive survey, we have searched for examples of traditional technologies that:

- are intended to manage or manipulate natural elements to maintain and manage catchment hydrology in a way that benefits people;
- are designed to augment water supplies, particularly by changing when water is available at key locations for use, while conserving soils, managing erosion, and minimizing flooding; and

- may be combined with built or engineered structures but rely on natural elements (vegetation, soils and streambed or alluvial sediments, aquifers) and processes (infiltration, groundwater recharge, groundwater discharge) to capture, infiltrate, store, and filter water that can then be tapped for use.

We discuss here two general groups of nature-based traditional technologies: water sowing and harvesting; and integrated land and water management systems.

Water sowing and harvesting

Indigenous peoples and local communities around the world have independently developed a rich variety of these nature-based water sowing and harvesting technologies (Table 13.1). These often (or almost always) combine some kind of structure of earth, stone, masonry, or wood in combination with vegetation, soils, sands or gravels, shallow aquifers, and infiltration zones, which convey water to shallow or deep aquifers. While each practice is somewhat unique and closely adapted to local conditions, there are recognizable categories that are found in widely divergent regions (Akpinar Ferrand and Cecunjanin, 2014; Beckers et al., 2013). The variety of nature-based practices can be grouped into some general types: percolation or infiltration ponds, hillslope infiltration basins or tanks, wadi (stream channel) terracing for floodwater harvesting, perched freshwater basins, silt dams and wetland enhancement, sand dams, and diversion and infiltration systems.

Infiltration basins

Found on almost every continent, traditional infiltration or percolation basins, created in natural depressions or on gently sloping surfaces, have been used for thousands of years for collecting and infiltrating runoff, storing water in soils and shallow aquifers for later harvesting in downstream springs or nearby wells (see Table 13.1). The *cuchacuchas* and *qochas* in the highlands of Peru, *albarradas* in Ecuador, *johads* in Rajasthan, and *eris* in southern India are all infiltration basins that share a set of common design elements (see Table 13.1). These elements are well-illustrated by the *albarradas* of coastal Ecuador.

Albarradas: Coastal Ecuador

Albarradas are infiltration basins or lagoons found on the coastal plain in portions of Santa Elena, Manabi and Guayas provinces in Ecuador and are estimated to be about 3800 years old (Marcos and Bazurco, 2006). Also known as *jagueyes*, these are small, often horseshoe- or U-shaped embankments constructed on hillslopes, across small drainages, or in natural depressions on gently sloping land (Denevan, 2001; Marcos and Álvarez, 2016; Marcos and Bazurco, 2006; Yapa, 2017). On the dry coast of northern Peru there are historical reports of similar structures for capturing and infiltrating runoff (Antunez de Mayolo, 1986; Spruce, 1864). By slowing the flow of runoff, retaining water and sediments, and infiltrating water

Table 13.1: Global examples of water sowing and harvesting practices.

Ancestral practice	Region	Description
Silt Dams/Wetland Enhancement		
Represas de limo, silt dams, artificial *bofedales*	Andean highlands, Peru	Small dams or embankments constructed across small streams and drainages to accumulate silt and water. Slowing runoff and spreading water over a larger area raises water levels to sustain a greater area of wetland plants and saturated soils, which enhances forage for herders. Also reduces erosion, increases soil moisture, increases infiltration, and recharges aquifers (Lane, 2014).
Diques de champas	Northern Highlands, Chile	Embankments, dams, and dikes constructed from transplanting sections of wetland soils, plants (seeds, rhizomes), and associated microbes to channel water to new areas and support growth of new wetland vegetation. Slowing runoff and spreading water over a larger area reduces erosion, increases soil moisture, helps dissolve salts, increases infiltration and recharges aquifers, raising water levels to sustain greater area of wetland plants and saturated soils. Improves forage biomass as well as buffering soils and vegetation during dry periods (Gonnet et al., 2016).
Borrequiles	Sierra Nevada, Spain	In mountain wetlands and wet meadows, wetlands are managed and enhanced by pastoralists through diverting water from streams and springs with ditches or ridges to spread the water out, infiltrate water, saturate the soils, and increase the water level in the wetland. This prolonged storage then allows the wetlands to discharge water downstream in the dry season, helping to feed rivers and springs lower down and improving pastures and providing forage in the dry summer (Martin-Civantos, 2014).
Perched Freshwater		
Virdas	Gujarat	The groundwater in this area is mostly brackish; by enhancing infiltration of rainwater, the *virdas* establish a lens of freshwater perched on top of the more brackish or saline and deeper groundwater. *Virdas* are shallow depressions, which are made in the sandy soils or the sands of dry riverbeds and lakes for collecting drinking water. The harvesting system depends on the grass cover of the adjacent areas, which is essential for free infiltration of rainwater. In *virdas*, the sweet freshwater remains in the upper layer from which the water is collected, and the saline water remains below the freshwater zone because of its higher density (Bhattacharya, 2015).
Puna	Rapa Nui	Stone dams that allow freshwater from coastal seeps to discharge and perch on top of more saline or brackish groundwater found at the coastal edge (DiNapoli et al., 2019; Hixon et al., 2019).

Continued

Table 13.1: Global examples of water sowing and harvesting practices—cont'd

Ancestral practice	Region	Description
Sand Dams/Streambed Dams		
Sand dams	Kenya	Stone, masonry, or brush structures built across a stream and can extend several feet into the streambed in arid or semiarid regions with intermittent streamflow. Monsoon rains carry water and sediment downstream, which are trapped behind the dam, building up a deep layer of saturated sands that serve as storage reservoirs that also recharge adjacent aquifers. Water is extracted through wells and used for domestic purposes, livestock, and crops (Nilsson, 1988).
Dohs (streambed dams)	India	Sunken streambed structures that are rectangular excavations in seasonal streambeds, which are intended to capture and hold runoff to enhance groundwater recharge, thus increasing water for irrigation from nearby shallow wells. *Dohs* are built in semiarid areas where rainfall is low and seasonal. *Dohs* are generally built in-sequence. The technology is used in conjunction with shallow wells (*odees*), which enable farmers to harvest the increased groundwater for supplementary irrigation of annual crops (Bhattacharya, 2015).
Infiltration Basins		
Cuchacuchas	Central and Southern Peruvian Highlands	Small, circular, permeable ponds formed in natural depressions and/or excavated and deepened to enhance storage and infiltration. Maintained by pastoralists; there may be hundreds per hectare in some areas. In the *altiplano, puna* grasslands trap and infiltrate runoff from snowmelt and rain. The elevated soil moisture and shallow groundwater contributes to lush vegetation and improves forage for livestock (alpaca, llamas, sheep) and recharges springs lower down (Denevan, 2001).
Qochas	Peru Highlands	Larger depressions than the *cuchacuchas* that are deepened to increase water storage, they provide surface water storage and recharge groundwater to feed springs at lower elevations. Important sources of water for people and livestock and for improving forage for livestock. Often found in large numbers or pond fields that maximize storage and recharge (Denevan, 2001).
Talabs	Rajasthan	In natural depressions, structures increase the depth and volume of the depression, trapping runoff from slopes and temporarily storing water, infiltrating to groundwater, which supplies nearby wells and step wells. The earliest known *talab* example is from the late 13th century (Bhattacharya, 2015).
Paar system	Rajasthan	*Paar* is a common water harvesting practice in the western Rajasthan region, where a relatively large depression surrounded by vegetation is used to capture rainwater runoff, which percolates into the sandy soil/shallow aquifer for storage. Stone wells within the depression then harvest the stored water for use (Bhattacharya, 2015).

Table 13.1: Global examples of water sowing and harvesting practices—cont'd

Ancestral practice	Region	Description
Albarradas, also known as *jagueyes*, *atajados*, *pataquis*.	Coastal Ecuador	Infiltration lagoons that are thousands of years old are found in the coastal provinces of Santa Elena and Manabi in Ecuador. These circular or U-shaped lagoons slow and capture runoff behind earthen embankments following rain events in this semiarid, dry forest region of coastal Ecuador. They are found on gently sloping terrain and on porous/sandy soils and sandstones. Water trapped by the embankment sinks into the soils and permeable substrate, recharging the local aquifer. Water is then harvested from wells excavated to one side of the embankment or springs downstream. Many *albarradas* were used together in a complex system (especially in Muey) that cumulatively provided reliable groundwater resources in this area with short intense rains and prolonged dry periods (Marcos, 2006).
Johads	Rajasthan	Very much like *albarradas*, *Johads* are small earthen check dams that capture and conserve rainwater, improving percolation and groundwater recharge. Simple stone/mud barriers built across the contours of a slope slow and collect runoff from rain events. U-shaped, embankments on three sides with the upstream side open for water to enter. The revival of some 3000 *johads* spread across more than 650 villages in Alwar district, Rajasthan has resulted in a general rise of the groundwater level; rivers that used to go dry immediately following the monsoon season have now become perennial. *Johads* in western Rajasthan are associated with protected sacred groves (Bhattacharya, 2015).
Ahar pynes	South Bihar, India	Similar to *johads* but in hilly areas with steeper slopes and sandy soils; during monsoons, river water is diverted to the U-shaped, three-sided embankments to capture water for use in adjacent agricultural fields and recharge groundwater (Bhattacharya, 2015).
Eri and Kumans (tanks)	Tamil Nadu, India	Very similar to the individual tanks in the tank cascades of Sri Lanka, except that the *eris* lack the integrated protected forest-agroforest-agriculture system. Serving as flood control, erosion prevention, groundwater recharge, and drinking and irrigation water. The tanks fill with the heavy monsoon rains, store water in soils and shallow aquifers. Many tanks in the region have allowed for rice cultivation, which would not be possible without the water from the tanks. The earliest tanks are thought to have been constructed between the 6th and 10th centuries (Bhattacharya, 2015).
Khadin system	Jaisalmer, Jodhpur, Bikner and Barmer Districts, Rajasthan	The *khadin* system is a runoff agricultural system in which the runoff water from the catchment area is stored with the help of a *khadin* bund or earthen embankment where it is impounded during the monsoon season. Water is stored in the *khadin* soils and shallow groundwater; soils remain moist for a long period buffering crops from seasonal drought (Bhattacharya, 2015).

Continued

Table 13.1: Global examples of water sowing and harvesting practices—cont'd

Ancestral practice	Region	Description
Wadi Terracing/Floodwater Harvesting and Sowing		
Wadi Terrace Systems (*jessour, cultivo de sanada, tabia*)	Tunisia, Spain, Rajasthan	Multiple examples in arid and semiarid regions of the Middle East, North Africa, and South Asia. Stone check dam constructed across a ravine or ephemeral stream to capture floodwater runoff. Silt deposits store water in the soil, retain moisture and nutrients to support crops, and recharge connected aquifers (Beckers et al., 2013).
Tapes	Ecuador	Small walls built in the main channel of intermittent streams or rivers, collecting water and sediment during rain events, infiltrating water to recharge springs or wells (Alvarez, 2006).
Naada/Bandha	Rajasthan	*Naada/bandha* are found in the Mewar region of the Thar desert. A stone check dam constructed across a stream or gully captures monsoon runoff. Silt deposits store water in the soil, retain moisture and nutrients to support crops, and recharge connected aquifers (Bhattacharya, 2015).
Diversion and Infiltration Systems		
Amunas, mamanteo	Peruvian Andes	System of diversion canals, infiltration canals, infiltration zones, ponds, and springs that sow water from streams by recharging groundwater, which is then harvested in downstream springs or streams (Apaza et al., 2006).
Acequias de careo	Andalusia, Spain	System of diversion canals, infiltration canals, infiltration zones, ponds, and springs that sow water from snowmelt by recharging groundwater, which is then harvested in downstream springs or streams (Pulido-Bosch and Sbih, 1995).

into underlying shallow aquifers, *albarradas* help control flooding and erosion during the rainy season, recharge aquifers, and allow for the harvesting of this water in wells or springs through the subsequent dry season (Fig. 13.3).

Structure and function

Coastal Ecuador has a semiarid to arid climate characterized by dry tropical forest or savannah vegetation and a single, but unpredictable, rainy season from late December to May (Recalde-Coronel et al., 2014). Runoff is rapid, and flooding can be significant, especially during strong El Niño storms, but most rivers are dry or carry very little water much of the year. Prolonged droughts can occur, associated with ENSO cycles (La Niña) (Recalde-Coronel et al., 2014). Freshwater supply is a critical problem, noted by one of the first chroniclers of the Spanish conquest, Augustin Zarate, in 1555: "this land is very dry, although it rains often; there runs little fresh water....everyone drinks from wells or detention ponds which the natives call *jagueyes*" (cited in Marcos, 2006, p. 48).

Fig. 13.3

One of the oldest still active *albarradas* near El Morro, Santa Elena Province, Ecuador; Albarrada de los Tamarindos. *Data from Google Earth. V 7.3.3.7721. (July 2012). El Morro, Ecuador. 2°37′57.37″ S, 80°19′40.56″ W, Eye alt 3646 ft. Image © Maxar Technologies, Landsat Copernicus. http://www.earth.google. com [November 17, 2020].*

In the early 2000s, Jorge Marcos and colleagues undertook extensive studies to understand the construction and management of *albarradas* as part of a Global Environment Facility project, the Albarradas Project (Marcos, 2006; Marcos and Álvarez, 2016; Marcos and Bazurco, 2006). *Albarradas* consist of an earthen wall or embankment constructed from excavated soils forming a shallow depression surrounded by the circular, semicircular or U-shaped wall, with the open-end facing upslope (see Fig. 13.3). Size can vary but typical dimensions are 10–30 m in length, 5–20 m wide, and 5–6 m deep (Denevan, 2001). The majority of *albarradas* in Ecuador are constructed in areas that are underlain by alluvial aquifers or the permeable sandstones (Tablazo, Pleistocene-age marine terraces) that form a large shallow coastal aquifer (Marcos and Bazurco, 2006).

When filled with water during the rainy season *albarradas* form shallow lagoons that support a diverse wetland biota atypical in the rest of the region. The Albarradas Project excavations found evidence that *albarradas* likely supported important medicinal plants and possibly aquaculture (freshwater prawns) (Marcos and Bazurco, 2006). The increased groundwater levels around *albarradas* not only supported crops but also helped sustain native forest (Marcos, 2006). *Albarradas* may have been particularly important in mitigating flooding during El Niño events and storing a significant portion of this runoff, allowing dense populations to subsist in an otherwise dry region (Aristizábal, 2019; Marcos and Álvarez, 2016).

Threats and revitalization

One of the most extensive areas of *albarradas* in Ecuador was on the coast near the Village of Muey and La Libertad, where about 250 *albarradas* were mapped in the 1960s (McDougle, 1967), but this area is now almost completely covered by urban settlements and shrimp farms. In the region as a whole, deforestation, oil exploration and exploitation, urbanization along the coast, and shrimp farming has resulted in the loss of *albarradas* (Alvarez, 2006; Marcos and Álvarez, 2016). Large floods can damage the earthen embankments, and sedimentation can reduce the ability to infiltrate water and recharge groundwater (Aristizábal, 2019). Loss of forests may have contributed to the decline of *albarradas* by increasing sedimentation and peak flows during storms. Local commune members interviewed by the Albarradas Project reported that the very large El Niño storms in 1982–83 and 1997–98 damaged or destroyed several of the still-functional *albarradas* (Marcos, 2006).

Loss of traditional knowledge

The impacts of colonization and, more recently, globalization have contributed to a loss of the Indigenous knowledge that is essential to the functioning of *albarradas*. This knowledge includes not only technical design and construction aspects but also the traditional participatory construction and management systems, governance for collective use and traditions of reciprocity (Alvarez, 2006; Aristizábal, 2019; Carrion et al., 2018). Researchers for the Albarradas Project found that in villages that are still organized as communes (less erosion of Indigenous rights and tenure) there is knowledge of the Indigenous origin of *albarradas*, recognition that these are part of their Indigenous heritage, understanding of their function, and collective management of *albarradas*, with all members of the community having access to the water (Alvarez, 2006). Where management is by parish government, many *albarradas* are privately owned, there is much less awareness of their Indigenous origin, and access is more restricted, excluding the broader community.

The consequences of losing Indigenous knowledge and the importance of supporting Indigenous and local communities to recover this knowledge are illustrated in efforts over the past several decades to renovate and revitalize *albarradas* in Ecuador. Starting in the 1950s but most actively since 2008, there have been a number of efforts to renovate existing and construct new *albarradas* to address the water needs of local communities in Santa Elena and Manabi provinces. These projects include a mix of privately owned and communal property, with some developed in a top-down manner driven by government programs or outside experts, while others have been led or codeveloped with local communities, based on local knowledge (Alvarez, 2006; Marcos and Álvarez, 2016). Some of these attempts have been considered failures, as they were not adequately informed by knowledge of local geological, environmental, or social conditions (Marcos and Álvarez, 2016). Many of the *albarradas* constructed as part of government programs were not built to promote infiltration and aquifer recharge but are basically detention ponds; they store surface water but are not able to maintain the longer-term storage of runoff in groundwater (Alvarez, 2006). A number of the new *albarradas* constructed in the 1970s did not withstand the larger El Niño floods (Marcos and Álvarez, 2016). The revitalization of *albarradas* has been more successful within Indigenous communities in Ecuador, where Indigenous councils control the management and use of the system (Marcos and Álvarez, 2016).

For example, the large pre-Hispanic *albarrada* in the community of El Morro (see Fig. 13.3) is still functional, communally managed, and regarded as part of the community's ancestral, Indigenous legacy. The community members' efforts to protect it from El Niño floods has contributed to its survival, and it plays a critical role in supplying the local population with water, especially for poorer members of the community (Alvarez, 2006). Free access to water is an important way to save money; however, use of water from *albarradas* can be considered less prestigious than buying water from tank trucks, as it may be associated with marginalized, rural populations (Alvarez, 2009). Without adequate recognition and support for the legitimacy of local, Indigenous knowledge and recognition of the importance of traditional communal resource management practices, efforts to revitalize *albarradas* and realize their potential for contributing to local water security will not be successful.

Sand dams

Sand-storage dams are a type of groundwater dam that block the flow of water and sediment in streams, impounding and storing water in the sediments that accumulate behind the dam and recharge alluvial aquifers (Lasage et al., 2015; Neal and Maddrell, 2013; Nilsson, 1988; Nissen-Petersen, 2006). These technologies based on local experience have been developed in many semiarid regions, most notably in eastern and southern Africa (Kenya, Ethiopia, Namibia) and southern India (Nilsson, 1988). Construction of sand dams has occurred most actively since the middle of the 20th century, but there is some evidence that sand dams have a much longer history. Examples may have occurred in the Libyan desert 1000–2000 years ago (Baurne, 1984; Nilsson, 1988), and from the 19th century, tappoons constructed in sandy riverbeds in the southwestern United States (Lowdermilk, 1953).

Structure and function

Sand dams are used in regions characterized by low and highly variable rainfall and flashy, intermittent, or ephemeral rivers in basins with coarse, sandy sediments (Hut et al., 2008; Maddrell and Neal, 2012; Quinn et al., 2019). In these systems, brief but intense rainstorms result in rapid runoff into rivers that then carry large volumes of water and coarse sediments, but once the rains end, surface water quickly disappears. Potential evaporation typically exceeds precipitation particularly in years of very low rainfall (Lasage et al., 2008). In these regions, dams and reservoirs designed to store surface waters are not ideal, as large amounts of water are lost to evaporation. By storing water in the sands building up behind the dams and in connected alluvial aquifers, sand dams can store large amounts of water between rain events, while reducing losses to evaporation (Quinn et al., 2018b, 2019).

In *sand rivers* with unconsolidated alluvial deposits and where the underlying geology has low permeability, alluvial river aquifers form and are recharged from surface runoff as flood waters travel down the river (Walker et al., 2018). Sand dams, in effect, serve to increase storage capacity and prolong the amount of time water is available, either by expanding an existing or creating a new and artificial alluvial aquifer (Schreiner et al., 2019). These sand dams are typically on small rivers (< 50 m wide) with a gradient of less than 5% and where bedrock is near the surface of the streambed (Hut et al., 2008; Neal and Maddrell, 2013). A masonry or concrete barrier is constructed across the river, either resting on the riverbed or extending a few feet below the riverbed, creating a barrier that impedes the flow of water both above and below the surface of the riverbed. With each flash flood during rain events, sands filled with water accumulate behind the dam while some water and finer sediments pass over the dam. Over several flood events, a large volume of saturated sands completely fill the reservoir behind the dam and future floods flow over the dam (Hut et al., 2008; Lasage et al., 2008). It may take several seasons to fill the reservoir completely, and sand dams are typically constructed in steps, gradually increasing the height of the dam to build up a large storage volume (Maddrell and Neal, 2012).

Once filled, sand dams retain large volumes of water, as sandy sediments can store 30%–40% of their total volume as water (Lasage et al., 2008). Water stored in the sands slowly seeps out into the adjacent riverbanks and into connected aquifers, not only providing storage in the sand reservoir itself but also raising groundwater levels and soil moisture in the vicinity of the dam (Ryan and Elsner, 2016). Wells within the sand dam itself or tapping into the aquifer nearby provide access to the water.

Advantages of sand dams

As a local, community-based adaptation to scarce and unpredictable water supply, sand dams have several advantages over other technologies. They are more cost-effective (cost per m^3 of water provided) than smaller household technologies, such as cisterns or rooftop harvesting, and much larger surface storage reservoirs (Lasage and Verburg, 2015). The typical locally available construction materials are low cost, and maintenance costs are also low (Lasage and Verburg, 2015). Evaporative losses are much lower than for open water in ponds or surface reservoirs. Once the water table in the sand reservoir drops a few centimeters below the surface, evaporative losses are extremely small, just a few millimeters each day (Borst and

De Haas, 2006; Quinn et al., 2018b; Liu et al., 2015 in Schreiner). The lack of surface water also means sand dams are less likely to become breeding grounds for mosquitoes, although the evidence for this is ambiguous (Cruickshank and Grover, 2012; Pauw et al., 2008). Pauw et al. (2008) found no difference in malaria incidence or stakeholder perceptions of abundance of mosquitoes in communities with and without sand dams in the Kitui region of Kenya. Mosquito breeding sites can be present if surface water ponds behind the dams, which can occur especially in the early stages of construction (Cruickshank and Grover, 2012).

Sand dams are presumed to be less likely to become contaminated and have higher quality water than water stored on the surface. Although sand dams need to be protected from external sources of pollution, for example, by keeping livestock out of riparian areas, the sands and microbial communities within the sand can help filter and purify water. While water quality can differ depending on the method of extracting water (scoop holes vs wells), several studies have documented high-quality water, meeting standards for human consumption (Avis, 2016; Quinn et al., 2018a). Sand dams can improve water quality in the same way that slow sand filters introduced in the early 1800s by sanitary engineers in Scotland helped launch the first big improvement in purifying drinking water in European and North American cities.

Finally, sand dams retain only a small portion of overall catchment runoff and therefore have minimal impact on water available to downstream areas (Aerts et al., 2007; Lasage et al., 2015; Pauw et al., 2008). Studies in Ethiopia and Kenya suggest that sand dams retain only between 2% and 4% of runoff, even when there are multiple small dams in a catchment (Aerts et al., 2007; Pauw et al., 2008). However, with a large increase in the number of dams, for example, to address projected reductions in rainfall with climate change, reduced flows downstream become more significant (Lasage et al., 2015).

Water security and climate resilience effects of sand dams

A number of studies document the increase in dry season water supply from sand dams (Borst and De Haas, 2006; Hoogmoed, 2007; Lasage et al., 2015; Pauw et al., 2008; Ryan and Elsner, 2016). For example, at a site in the Kitui region of Kenya, before dams were built, about 500 m^3 of water were available from the river sands during the dry season (Borst and De Haas, 2006). This is about 2 m^3 per day, even though about 8 m^3 were used during this time, a deficit that had to be supplied from distant water sources. The amount of water available after sand dams were built increased to 2800 m^3 or about 11 m^3 per day, sufficient to meet dry season demand (Borst and De Haas, 2006). This study estimated that the maximum amount of water that could be harvested from an average sand dam per year was about 8100 m^3. Sand dams also extend the duration during which water is available in the sand reservoir so that there is a sufficient supply throughout the year, in some cases adding 2.5 months of water reserves and providing a bridge from the dry season to the beginning of the next rainy season (Hoogmoed, 2007; Lasage et al., 2015; Pauw et al., 2008).

The increased water supply comes from the water stored in the sand reservoir itself plus the increased groundwater levels in connected aquifers. The dam not only creates a large area of water storage in the sand reservoir but also creates an extended alluvial groundwater reservoir far upstream of the dam (200–350 m) and raises the groundwater levels in the adjacent riverbanks by as much as 40% (Borst and De Haas, 2006; Hoogmoed, 2007; Hut et al., 2008; Quinn et al., 2019). As water levels in the sand reservoir begin to drop once the dry season begins, the reservoir is replenished as groundwater flows back through the riverbed and from the banks. Groundwater flows around sand dams thus reflect seasonal cycles of transfers of water to and from the surrounding aquifer. Quinn et al. (2019) showed that the groundwater dynamics around three small dams in the same region in Kenya were quite different. One dam remained completely full throughout the dry season, continually receiving water from the surrounding aquifer, the second dam began losing water to the surrounding aquifers as soon as the dry season began, while the third dam began losing water 1 month into the dry season. While early studies and design guidance suggest that sand dams should be constructed in sites with impermeable riverbeds, recent studies have documented that many sand dams have more "leaky" beds and can lose (or gain) water to connected groundwater (Quinn et al., 2019). This suggests that sand dams should not be viewed as static storage systems but must be understood (and designed) in light of the nature of the surrounding aquifers.

Climate resilience

Sand dams may be promising climate adaptation measures in drylands, improving resilience for local communities in the face of climate variability and long-term change (Aerts et al., 2007; Lasage et al., 2008). Climate change is expected to exacerbate water scarcity in drylands through a combination of higher temperatures, more intense and frequent extreme events such as droughts and heavy rainfall, and decreased overall precipitation (Huang et al., 2017; Yao et al., 2020). Using modeling to assess sand dams as adaptation strategies in Kenya, Aerts et al. (2007) estimated that although net water availability is expected to decrease by 13% under climate change, an expanded network of sand dams can capture enough water to offset this decrease in an average year. Similarly, in Ethiopia, Lasage et al. (2015) found that water storage from increasing the number of dams in the Dawa catchment could more than make up for projected increased demand and decreased precipitation due to climate change. However, the greater number of dams would capture a large proportion of total runoff, increasing impacts on downstream water supplies.

Beyond the potential to maintain adequate supplies for people and livestock under climate change, there is evidence that sand dams can provide improved overall resilience to drought. Researchers from the United Kingdom and Canada have shown that the effect of sand dams on elevating groundwater levels around dams in Kenya can extend up to 350 m upstream of the dam and 70–80 m into stream banks (Cruickshank and Grover, 2012; Ryan and Elsner, 2016). Using satellite images that captured three drought periods with major water scarcity in the Makueni District in Kenya, Ryan and Elsner (2016) compared vegetation density and response to drought between sites with sand dams and without. Across all three drought periods, including extreme drought conditions, the sites with sand dams maintain a higher biomass of vegetation and vegetation recovered more quickly from droughts compared to areas without sand dams. Following the commencement of rains after each drought period, the vegetation around dams quickly reached higher densities than the long-term average for the region (Ryan and Elsner, 2016). The area near sand dams supported healthy woodland or savannah vegetation as compared to shrubland and grasses at the sites without dams.

Other studies have also shown that sand dams increase the amount of vegetated land cover (Manzi and Kuria, 2012). Sand dams therefore can provide a buffering effect in times of drought, maintaining higher soil moisture levels, a slower drying of vegetation with the onset of drought conditions, and faster more complete recovery once rains start. Because groundwater recharge in sand dams happens very quickly following rain events, within the first few days of precipitation, sand dams can also help address the increased variability in precipitation that is anticipated with climate change (Borst and De Haas, 2006; Hut et al., 2008). Increased vegetation cover also can reduce vulnerability to the extremes of climate change by altering runoff, erosion, infiltration, flooding, and groundwater recharge. Maintaining higher levels of soil moisture can help buffer crops during times of drought, increasing yields and making local communities less vulnerable to the next drought.

Social and economic benefits of sand dams

Significant socio-economic and environmental benefits have been documented for sand dams, beyond improvements in water security. Households with access to dams had increased harvests, grew a greater diversity of crops, and realized improved incomes compared to households without access to dams (Pauw et al., 2008; Teel, 2019). In Kenya, the additional water provided by sand dams can be used for brickmaking, providing an added source of income (Cruickshank and Grover, 2012; Pauw et al., 2008). Several studies have documented improvements in women's lives with access to sand dams. The distance that women and children have to travel each day to collect water can be reduced from 4 km to less than 2 km in an average rain year, saving an average of 2.4 hours per trip (Cruickshank and Grover, 2012; Lasage et al., 2008). When distances to water sources were reduced, women in these communities became more involved in agricultural and other income-generating activities (Teel, 2019). The shorter distances to water also mean families increase their water use after sand dams are in place (Lasage et al., 2008). In addition to increased water use for crops, domestic use also increased, including for improved hygiene (washing hands and clothes) (Cruickshank and Grover, 2012; Lasage et al., 2008). There can be inequities, however, in who actually has access to more water. For example, families who live closer to the dams and those with access to donkeys or carts tend to use more water (Cruickshank and Grover, 2012).

Threats to sand dams

Increased erosion within a catchment, for example, from deforestation, overgrazing, or loss of riparian vegetation, can result in siltation, which reduces the ability of sand dams to infiltrate water (Maddrell and Neal, 2012). Aggressive groundwater pumping in the area surrounding sand dams can draw down the water stored in the sand reservoir, limiting effective storage (Hut et al., 2008; Quinn et al., 2019). Impacts to water quality from livestock, domestic waste, or chemical or mineral pollution in the catchment can make water in sand dams unusable for drinking (Avis, 2016; Quinn et al., 2018a). Finally, successful construction, maintenance, and equitable management of sand dams require coordinated collective action and governance (Teel, 2019).

Managed wetlands and infiltration enhancement systems

More than a thousand years ago and 10,000 miles apart, local peoples in the Sierra Nevada mountains of Spain and in the Andes mountains of South America independently created ingenious nature-based technologies for managing water. In both regions, local peoples created remarkably similar cultural landscapes to manage the recharge of aquifers to address seasonal water scarcity (Fernandez Escalante et al., 2019; Martos-Rosillo et al., 2019; Ochoa-Tocachi et al., 2019) and to enhance wet pasture areas for livestock—the *borreguiles* in Spain and *bofedales* in the Andes (Martin-Civantos, 2014; Verzijl and Guerrero Quispe, 2013). Long before the era of modern hydrological science, people on both continents had a deep understanding of water, soils, vegetation, and geology—knowledge of how water moves across the landscape—and they used this knowledge to ensure water supplies during times of scarcity.

In the Alpujarra region of Andalusia, a traditional infiltration enhancement system has been in use since at least the Arab settlement of the area in the 9th century and possibly earlier (Martin-Civantos, 2014). In the Sierra Nevada, spring snowmelt from the high peaks fills rushing streams and rivers, but this water is largely gone by the dry Mediterranean summer. Between March and June, the traditional *acequias de careo* (Fig. 13.4) use a system of irrigation channels to divert water from streams and rivers in the highlands to areas where the water can infiltrate the permeable soils and weathered bedrock, recharging groundwater (Martin-Civantos, 2014; Martos-Rosillo et al., 2019). As the subsurface water slowly moves downhill, springs and seeps lower down in the valleys are recharged and continue to flow during the dry season months. This ensures drinking water and water for crops during the summer, and the higher soil moisture and elevated groundwater help maintain lush and diverse vegetation on the hillslopes—supporting habitat for native animals and plants, reducing erosion, and providing some protection from wildfires (Pulido-Bosch and Sbih, 1995). In addition, diversion of water to high-altitude wet meadows, or *borreguiles*, helps maintain and expand these areas that are important summer forage for sheep (Martin-Civantos, 2014).

Fig. 13.4
Careo diversion canal in the Sierra Nevada, Spain, carrying spring runoff to infiltration galleries to store water for release during the drier summer months. *Reproduced with permission from Alamy Stock Photo.*

Traditional wetland management in the Andes

The *paramo*, *jalca*, and *puna* ecoregions in the mountain highlands of Latin America, extending from Costa Rica to Argentina, are some of the most biologically diverse and hydrologically important ecosystems in the region (Buytaert et al., 2006; Maldonado Fonkén, 2014; Oyague and Cooper, 2020). In the arid and semiarid high mountains of central Peru to Chile, *puna* grasslands, including patchy wetlands (*bofedales*) dominated by tussock grasses, sedge communities or cushion plants have been critical sources of water and pasture for highland communities since pre-Incan times (Domic et al., 2018; López-i-Gelats et al., 2015; Maldonado Fonkén, 2010; Rolando et al., 2017; Squeo et al., 2006; Yager et al., 2019). *Bofedales* are humid, saturated ecosystems in an otherwise arid to semiarid environment. Naturally occurring *bofedales* are typically found near springs fed by groundwater, on valley bottoms where runoff accumulates, and along the banks of streams and rivers (Oyague and Cooper, 2020). The system of ancestral wetland management by pastoral peoples in the Andes is an interconnected community—a community of wetlands, animals, people, and livelihoods—based on a centuries old wetland management and sustainable governance practices.

Bofedales provide critical and high-quality forage for herds of llama, alpaca, and sheep and are important sources of water for local and downstream users. These wetlands store large amounts of water from precipitation, groundwater discharge, and glacial melt in the soils and shallow groundwater and slowly release this water, providing a buffer against low stream flows during the dry season (Buytaert et al., 2006). They are examples of geologic water storage (Fairley, 2003) where the saturated soils serve to both store and filter water, retaining sediments and nutrients. For example, dry season baseflow in the upper Marañón River, which is a major tributary to the Amazon, is dependent on groundwater released from the *puna* wetlands (Hill et al., 2018). The *bofedales* in the Ramuschaka watershed near Cusco sustain dry season baseflow in the river, with baseflow positively correlated with the extent of wetlands in the upper basin (Wunderlich et al., 2019). *Bofedales* also play an important role in regulating snowmelt runoff in the steep semiarid slopes of the Andes in northern Chile. The wetlands provide a large storage reservoir that prevents spring snowmelt from immediately running off these steep slopes that otherwise have limited subsurface storage (Hevia et al., 2016). This allows a sustained and slow release of snowmelt extending through the dry season, maintaining baseflows downstream (Valois et al., 2020).

The long history of indigenous wetland management in the Andes

Multiple lines of evidence suggest that beginning about 1500–2000 years ago, Andean peoples played a significant role in creating and maintaining large areas of wetlands in the *puna*, supporting the herds of camelids (llama and alpaca) that were the basis of pastoral livelihoods. Pollen and peat core data from northern Chile show that about 2000 years ago, the area of high-altitude peatlands expanded and became dominated by a mixed grassland-sedge community (Domic et al., 2018). An expansion of peatlands and the appearance of plants that are high-quality forage and tolerant of grazing pressure also coincided with the arrival of the Inca in northern Chile and the grazing of larger herds. The pollen data also show that around this same time there may have been a shift in Indigenous management practices to managing wetlands for the higher quality forage of succulents and cushion plants rather than the less nutritious grasses typical of drier areas (López-i-Gelats et al., 2015; Villarroel et al., 2014).

A growing body of archeological studies are also providing evidence for extensive management of wetlands in the Andean highlands. Irrigation and flow management structures have been found at pre-Columbian archeological sites associated with *bofedales* in the Ancash region of Peru, near Cusco, and in Puno province (Lane, 2006, 2017; Lovera et al., 2007; Maldonado Fonkén, 2010). It is also likely that the limited extent of natural wetlands could not have supported large herds of camelids and the many areas of dried, former wetlands that exist now indicate that artificially managed *bofedales* were much larger during pre-Columbian times (Erickson, 2000; Lane, 2014). Early Colonial tax records from Puno support the idea that there were much larger areas of wetlands prior to the Spanish conquest. Tax records show that very large herds of alpaca were being raised at this time, much larger than could be supported by the current extent of wetland pastures, suggesting that artificial *bofedales* were formerly much more extensive (Erickson, 2000). Many extensive artificial *bofedales* likely became abandoned following colonization due to the decimation of the Indigenous populations.

Despite archeological, ethnographic, and contemporary evidence for the importance of Indigenous wetland management in the Andes, the true extent of these systems remains unknown. Archeological evidence of canal systems in artificial *bofedales* can be difficult to find. For example, outtake canals that feed water to new areas of wetland disappear as the wetland progressively expands outward. Although some archeological evidence has been documented for the silt dam structures in Peru (Lane, 2017), Verzijl and Guerrero Quispe (2013) argue that due to the simplicity of the technologies and the decentralized, communal control of the contemporary systems, even current practices are effectively invisible, "the system nobody sees."

Silt dams, silt reservoirs, and combined irrigation-wetland creation systems

Three different but interconnected systems of wetland management have been described from the archeological record and from contemporary Andean communities: silt dams, silt reservoirs and complex irrigation-wetland restoration practices (Erickson, 2000; Gonnet et al., 2016; Lane, 2017; Saylor et al., 2017; Verzijl and Guerrero Quispe, 2013). Silt dams have been most extensively documented in the Cordillera Negra region of Peru (Lane, 2014) where multiple silt dams and over 300,000 m^2 of artificial *bofedales* have been found. Silt dams are stone and earth structures built across small drainages that trap fine sediments, retain and store water in the resulting soil reservoir, recharge groundwater, and provide rich and saturated soils for the growth of wetland plants (Lane, 2006). These may have been constructed in steps, gradually raising the height of the dam and enlarging the area of saturated soils and wetland vegetation over time (Lane, 2017). The area of artificial wetland behind the dams can be quite large. Lane (2017) documents one site in the Chanclayo Valley of Peru that is 28.5 ha in size, with the depth of silts and peat near the dam of over 6 m, and the wetland area extends for about 3 km upstream of the dam.

Silt reservoirs are created by a series of small check dams or cross channel terraces in the highlands, in areas that are generally too high for agriculture, suggesting they were used to create additional areas of wetland pasture (Lane, 2014). The silt reservoirs in the Ancash region tend to be U-shaped or semicircular constructions between 7 and 20 m in length and 1 and 2 m high. The check dams are placed perpendicular to the flow to trap silt and also serve as an erosion barrier to higher flows of water and silt. The design allows water to spread across the widest area possible, maximizing the area of soil and water retention. Lane (2014) suggests that these may have had a dual purpose of creating additional wetland pastures while also providing some protection from the landslides (*huaycos*) that periodically discharge large amounts of rock, sediments, and water down hillslopes in this region.

Indigenous irrigation-wetland creation

Perhaps the most intriguing ancestral practice involves a combination of irrigation and water management to maintain and extend areas of saturated soils, along with sophisticated transplanting and restoration of wetland vegetation (Palacios Rios, 1977). Documented from central and southern Peru, Bolivia, and northern Chile, three different types of irrigation canals are used to maintain constant and elevated water levels in existing wetlands and to extend these wetlands by irrigating new areas (Erickson, 2000; Gonnet et al., 2016; Lane and Grant, 2016; Saylor et al., 2017). This system is actively used today in Peru, Bolivia, and Chile (Palacios Rios, 1977).

Verzijl and Guerrero Quispe (2013) described the system employed by highland communities in Huancavelica in central Peru. Wetland creation (outtake) canals are constructed parallel to and several meters from the edge of an existing wetland. Side channels then carry water into the area between the outtake canal and the existing wetland, creating saturated soils that provide suitable sites for the growth of wetland plants. Once wetland vegetation is established in the new area, the process is repeated to further extend wetland area, gradually growing the size of the wetland over many years. A second type of irrigation channel runs through existing wetlands to maintain high water levels over a larger area and keep portions of the wetland from becoming too dry, maintaining the quality of forage for longer periods during the year. Finally, some irrigation channels may function only to bring water to the wetland, for example, from nearby springs, or to convey water from the wetland to adjacent settlements. In the community studied by Verzijl and Guerrero Quispe (2013), more than 100 irrigation canals with a total length of 36 km are used to maintain and extend wetland pastures, some of which are over 100 ha in size and have been managed for hundreds of years. The authors estimate that more than 40% of the wetlands in this area would not survive if active management were to stop.

The system for transplanting and restoring wetland vegetation has been described by Squeo et al. (2006) and more recently by Gonnet et al. (2016). In northern Chile, Aymara herders have traditionally managed large areas of wetland for camelid pasture in the arid high *altiplano*. The system is referred to as *digues de champas*, as it involves using ditches and dikes created from wetland vegetation and soils to manage water. *Champas* (Quechua word for a tangle of roots and plans with soil) are small pieces of existing wetland vegetation and soil, with attached seeds, roots, and bulbs, which are transplanted into new areas to seed new wetlands. Ditches are created by excavating existing wetland, and the plants and soils are then used to build dikes and barriers that help slow the movement of water and distribute it to new areas. As water is retained for longer periods in the wetland, infiltration increases and soils remain saturated (or become saturated in newly created areas). These techniques are particularly effective in peatlands dominated by cushion plants that have a dense growth form that creates barriers to the lateral flow of water, creating many small ponds within the wetland (Fig. 13.5). In northern Chile, soils that have been periodically saturated can become saline due to high evaporation rates, and the irrigation of areas can serve to reduce salts in the soil, making them more suitable for the desirable wetland forage plants. Keeping water in place longer helps buffer against the tendency for alternating wet and dry periods that can occur in unmanaged wetlands.

Fig. 13.5

Bofedales near Volcan Parinacota, Chile; cushion plants and expanse of ponded water. *Reproduced with permission from Shutterstock Photo.*

Water security and climate resilience

Despite recent interest in traditional *bofedal* management, there have been few direct studies of the outcomes of these practices in terms of hydrological ecosystem services or climate resilience. Several studies documented an increase in the area of wetland vegetation after traditional practices have been implemented. In northern Chile, following implementation of traditional practices to restore degraded wetlands, surface water cover increased by a factor of four, the area covered by wetland vegetation increased, and this vegetation stayed green for a longer period of time throughout the year (Gonnet et al., 2016). In the four watersheds in this study, an estimated 55 ha of wetlands were restored and enlarged, while about 500 million additional liters of water were available.

In the high plans below the Nevado Sajama in Bolivia, the cover of healthy *bofedales* being used by Aymara herders declined from between 1986 and 2016 (Yager et al., 2019). However, in the communities that were actively engaged in managing wetlands using canals, dikes, and *champas*, the area of wetlands increased during this time, despite some years of below-average precipitation. Also, in the Cochabamba region of Bolivia, Aymara communities began efforts in the 1990s to rehabilitate and restore wetland areas that had been degraded by erosion, gullying, reduced baseflow, and a drop in water tables. Based on traditional practices, small check dams, terracing, and infiltration ditches were used to slow the flow of water, retain sediment, and enhance infiltration. To evaluate the effects of these activities, Hartman et al. (2016) compared paired micro-watersheds with and without restoration activities. They found increased *bofedal* vegetation and also more ponded surface water at local and landscape scales in areas with restoration activities but not in areas without restoration.

Wetland areas in the dry *puna* region have shown some resilience to drought and the maintenance and expansion of wetland areas likely provided a significant insurance and buffering to highland communities in the face of dry seasons and short-duration drought. Washington-Allen et al. (2008) evaluated change in vegetation communities in the Altiplano of Bolivia in response to the severe 1982–84 El Niño drought. Vegetation cover in the region decreased during the drought, but the *bofedales* showed the least effects and greatest resilience of any of the vegetation types, indicating an important role for wetland areas in climate resilience for the local population. Andean wetlands appear to be resistant to periodic drought, given their great age and continuing peat accumulation (Chimner and Karberg, 2008; Domic et al., 2018). Restoration and wetland management also help preserve the climate mitigation benefits of Andean peatlands, as carbon sequestration is maximized and methane emissions minimized when water tables are kept near the surface and plant diversity is higher (Villa et al., 2019). Despite dry conditions overall, peatlands in the *puna* are capable of significant amounts and rapid rates of peat formation and carbon sequestration (Hribljan et al., 2015).

Threats to indigenous wetland management

Communities in the Andean highlands face many of the same threats to maintaining their traditional water management practices as Indigenous and local communities elsewhere in the world. Ongoing land degradation in mountain watersheds can increase erosive forces, landslides, and flooding, which can cause erosion, gullying, or burial of wetlands (Salvador et al., 2014). Removal of vegetation cover on contributing hillslopes can decrease infiltration and subsurface flows to wetlands, while declining water tables dry out soils and vegetation. Earthquakes and droughts can alter water sources for wetlands, and these natural and anthropogenic factors make it much more challenging to maintain or expand healthy wetland areas (Gonnet et al., 2016).

In several basins in Peru and Chile, the construction of large storage reservoirs and water transfers from the highlands threaten to eliminate the water sources that are needed to maintain both natural and artificial *bofedales* (Gonnet et al., 2016; Verzijl and Guerrero Quispe, 2013). As water scarcity in the region becomes more severe, the pressure to move water to lowland zones dominated by export agriculture will continue to grow. Mining activities along with extensive road building can also damage to or eliminate *bofedales* and undercut traditional management practices. Climate change is also affecting the viability of *bofedal* management, for example in northern Chile, locals cite delayed rains, lack of rain, and "a strong sun that burns the bofedales" as new challenges to the effectiveness of management practices (Yager et al., 2019). Retreat of glaciers in the Andes will impact water sources to many *bofedales*, although as glaciers melt and release more water, there may be a temporary increase in wetland areas associated with peak water release (Cooper et al., 2019; Polk et al., 2017).

Perhaps the greatest threats to traditional wetland management, however, involve the loss of traditional knowledge and cultural practices that underpin the technical practices, communal governance, reciprocity, and collective action that are needed to construct, maintain, and operate water management structures. Migration from the highlands to cities for economic opportunities, especially by young people, erodes the labor force that is required and disrupts the transfer of Indigenous knowledge to the next generation. Aymara communities in Bolivia cited *individualismo* or dividing *bofedales* for individual use as creating conflicts over grazing and water allocation and disrupting the effective flow of water that is needed to maintain a healthy wetland (Yager et al., 2019). Despite these challenges, many upland communities are responding to changing conditions by adapting traditional practices with new technologies, such as capturing water from more distant springs, sprinklers for distributing water, and small tanks to add storage (Saylor et al., 2017; Verzijl and Guerrero Quispe, 2013), demonstrating the adaptive and dynamic nature of traditional knowledge (Berkes et al., 2000).

An encouraging sign is that these traditional practices are being revitalized throughout the region. In addition to the previously noted examples, Peru's Ministry of Agriculture has a national program on water sowing and harvesting for water security, including promoting traditional wetland management practices. New technical tools being developed in Peru will help practitioners estimate the hydrological benefits of wetland restoration, providing important support for encouraging these sustainable practices (Foster et al., 2020), https://www.forest-trends.org/publications/metodologias-cubhic-restauracion-y-proteccion-de-humedales/.

Pre-Incan infiltration enhancement systems in Peru: Amunas/mamanteo

Around the same time that *acequias de careo* were developed in Spain, in Peru, pre-Incan peoples created an extensive network of infiltration channels in the Andean highlands, known as *amunas/mamanteo* (Apaza et al., 2006). Water security in the tropical Andes is fraught with difficulties because of the complex topography and extreme climatic conditions. For centuries, local inhabitants of the Andes have developed water harvesting methods to cope with the strong seasonality of the region. In the present, these technologies have the potential to alleviate water scarcity and provide a relief to droughts. Pre-Inca infiltration enhancement systems can offer local and regional benefits to increase water security of a large metropolis located in arid lowland areas (Ochoa-Tocachi et al., 2019).

The capital of Lima, Peru, currently experiences a water deficit of approximately 43 million cubic meters during the dry season and depends on a total artificial storage capacity of approximately 330 million cubic meters. The water source that feeds this storage originates from the Andean rivers that provide water to Lima and the coastal region in general. However, these rivers are characterized by strongly seasonal flow regimes. This results in large water deficits during the dry season and surpluses during the rainy season that are not captured. In addition, the existing water resource is stretched because of the increasing climatic changes and the effects of human activities, in particular soil degradation and land use change.

This situation requires rethinking of the current strategies to increase water security. The limitations of conventional solutions based on "gray" infrastructure, such as artificial dams and reservoirs, are becoming more obvious: they involve long-term investments with high unrecoverable costs and require complex planning and implementation. Additionally, uncertainties in future rainfall projections and water availability complicate the design of large, fixed infrastructures with a long lifespan. As a response, there is a great interest for the implementation of NBS and "green" infrastructure, which can be gradually implemented, adjusted after implementation, and provide several benefits that make them compatible climate change adaptation strategies (UNESCO, 2018).

One of the most promoted interventions is a particular type of water harvesting method in the Andes known as *amunas* (Quechua for "retaining" water) or *mamanteo* (Spanish for "breastfeeding"). It has been developed by pre-Inca cultures in Peru (Chavín initially and Wari later) from as early as the 5th century. It consists of diverting water from natural streams during the wet season to enhance infiltration in mountain slopes. Water moves much slower inside the soils than it does over the surface, and thus the water delayed by this longer subsurface residence time increases the water yield and longevity of downslope springs during the dry season.

Structure and function

In the sierra of Lima, a local community known as Huamantanga restored one of the last remaining active pre-Inca infiltration systems still existing in Peru to cope with the extreme seasonal and interannual hydrological variability of the Andes. Huamantanga is an agropastoral village of approximately 1000 people located at an elevation of 3300 m above sea level in the central Peruvian Andes (Zulkafli et al., 2017). Local livelihoods include raising livestock for cheese production and irrigated subsistence agriculture, both of which depend on seasonal river flows. The infiltration system is designed to increase available water for irrigation during the dry season and consists of the following elements (Fig. 13.6) (Ochoa-Tocachi et al., 2019):

1. Diversion canals: There are two types: (i) long and impermeable canals that divert wet season flows from small streams to infiltration canals and hillslopes; and (ii) short and permeable canals that route excess water to ponds or from ponds to other streams downslope.

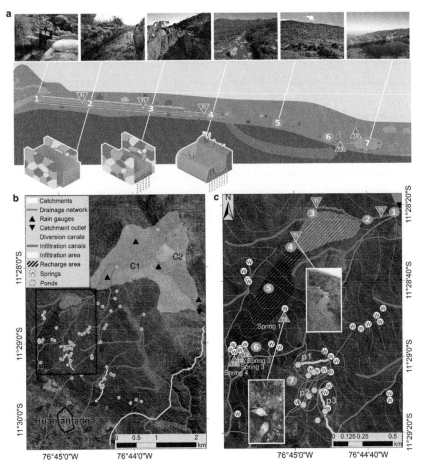

Fig. 13.6

(A)–(C) *Amuna-mamanteo* system in Huamantanga, Peru. (A) Components of the system: (1) diversion structure, (2) and (3) two types of diversion canals, (4) infiltration canal, (5) infiltration hillslope, (6) water resurfacing in springs, and (7) ponds, T1 tracer injection and T2 tracer sampling points. (B) Spatial distribution of the infiltration system and monitoring points. (C) Detail of the infiltration system. *Reproduced with permission from Ochoa-Tocachi BF, et al. Potential contributions of pre-Inca infiltration infrastructure to Andean water security. Nat Sustain 2019;2:584–93.*

2. Infiltration canals: Earthen canals and ditches that transport water further toward the infiltration hillslopes while simultaneously allowing infiltration into the subsurface.
3. Infiltration hillslopes: Rocky or stony areas that receive water from canals and ditches and spread it in the field. They use the soils as natural water reservoirs, leveraging fractured rocks that allow water to infiltrate and recharge subsurface stores while delaying water movement.
4. Springs: Typically, natural occurrences where infiltrated water resurfaces.

5. Ponds: Small water bodies (around 300 m^3 each) that are used to regulate the flow throughout the infiltration system. They are used to: (i) store water for direct human access; and (ii) enhance further subsurface water infiltration.

These systems are strongly embedded in the cultural practices of the communities that use them. Cultural practices involve rituals to clean the gates and canals that divert water and religious festivities held around them. The springs where water resurfaces and several ponds are considered sacred, and they are marked with rocks of considerable size to indicate protected zones. The community of Huamantanga is composed of approximately 120 members or *comuneros*, who are the people that have rights to water and land allocations (Grainger et al., 2020). They are in charge of cleaning and maintaining the infiltration systems every year and benefit directly from the delayed water during the dry season. Decisions are made in assemblies, where a president leads the discussion and every *comunero* member has voice and vote.

Evidence for water security/climate resilience outcomes

The study by Ochoa-Tocachi et al. (2019) documented quantitative evidence of the hydrological functioning of the infiltration system in Huamantaga. A dye tracer was injected in one of the upslope diversion and infiltration canals, and its resurgence was monitored in downslope springs using activated carbon samplers. The experiments revealed a clear hydrological connectivity between the canal and the springs, with a mean residence time of the dye tracer of 45 days, ranging from 2 weeks to 8 months. These results show that the system can store wet season flow effectively and recover it during at least a part of the dry season. Although the short residence time implies that the infiltration system is unable to replace artificial storage, it can potentially provide more storage capacity to complement the existing gray infrastructure.

To quantify potential infiltration volumes, hydrological monitoring was implemented in two headwater catchments whose outlets are located immediately upstream of the system intakes (Ochoa-Tocachi et al., 2018). It was estimated that average annual precipitation was approximately 550 mm per year with a clear monomodal seasonality. The dry season lasts for approximately 6 months (June–November) and the rainy season for another 6 months (December–May). River flows reflect this seasonality with a water yield of around 25%. This means that only approximately one fourth of the total precipitation is effectively converted into streamflow whereas the remaining is either evaporated, consumed by vegetation and transpired, or lost to other water pathways such as deep percolation. Flows during the rainy season account for most of the total annual discharge in the monitored catchments. However, the actual volume of diverted water tends to be lower than the potential volume because of the system's hydraulic capacity, operational restrictions, and minimum ecological flow for environmental purposes and other downstream water uses. It was then estimated that diverting water to the infiltration systems can increase natural dry season flows of the local stream by 3%–554%.

Lastly, the scale at which the system works in Huamantanga is too small to produce regional impacts at the basin scale. Nevertheless, it was studied how replicating such a system to the main source-water areas of Lima, in the Rimac River basin, can contribute to the city's water supply. Considering similar characteristics to Huamantanga, 1428 km² of highlands whose elevation is above 4000 m a.s.l. were identified. This area has a mean annual rainfall climatology of 505 mm of which 410 mm (81%) fall during the wet season. The Chosica station was taken as the reference point to quantify the impacts of replicating these systems, encompassing an area of 2319 km², with a mean annual rainfall climatology of 437 mm, of which 364 mm fall during the wet season (Manz et al., 2016). Using a mean runoff ratio of 22%, and assuming a diversion capacity of 50% of the wet season discharge in the highlands (in line with the actual diversion rate of 54% of the Huamantanga system, it was calculated that 34.7% of the Rimac River wet season discharge can potentially be diverted to the hypothetical systems. Assuming an effective recovery rate of 0.5, it was estimated an average increase in dry season flow for the Rimac River of 33% at the start of the dry season, reducing to less than 1% at the end of the dry. This represents approximately 100 million cubic meters per year that can be delayed from the rainy season to the dry season (Fig. 13.7).

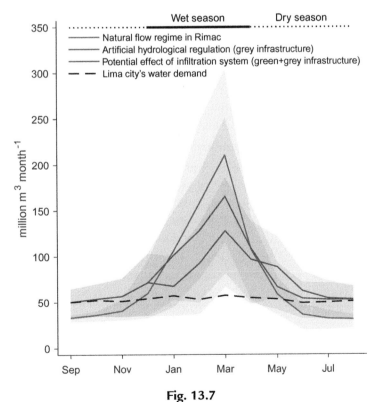

Fig. 13.7

Contribution of the *amuna-mamanteo* system to augmenting dry season flow in the Rimac River. *Reproduced with permission from Ochoa-Tocachi BF, et al. Potential contributions of pre-Inca infiltration infrastructure to Andean water security. Nat Sustain 2019;2:584–93.*

Threats and current status

Several Andean countries have become a testbed for approaches that complement traditional solutions with a wider set of catchment interventions that leverage natural processes (Somers et al., 2018). The uncertainties in future projections of precipitation and water availability complicate the design of large, fixed infrastructure with a long lifespan. The flexibility of natural infrastructure, which can be implemented and expanded more gradually and adjusted in line with future changes in climate conditions, is therefore seen as a promising way to enhance the adaptive capacity of water resource systems.

For this reason, Peru's drinking water regulator SUNASS is currently incentivizing and supporting public water utilities of Lima and other cities to complement traditional engineering infrastructure with NBS for source-water protection. Peruvian water utilities, including SEDAPAL of Lima, have responded by designing and implementing a portfolio of catchment interventions that is funded by a water tariff regulated by SUNASS. A recent law on reward mechanisms for ecosystem services (MINAM, 2014) provides the legal framework and establishes permissible interventions. It also puts strong emphasis on the integration of scientific and Indigenous knowledge, including the rehabilitation of water harvesting infrastructure. This has led to an increasing need to quantify the potential hydrological benefits of these practices and to identify explicitly the beneficiaries as a prerequisite to combine gray and green infrastructure and to maximize the cost-benefit ratio in the context of water supply and drought resilience.

Identifying which combination of interventions optimizes the return-on-investment and adaptive capacity in view of future climate uncertainty remains a challenge, especially in mountainous regions. The development of solutions that integrate elements of Indigenous practices and NBS can play a role in this. In practice, infiltration systems will have to be part of a more integrated catchment-scale water management strategy. Historically, this has included grassland conservation, headwater protection, sustainable grazing, water harvesting methods, terrace building on lower climatic zones for sustainable agriculture and erosion control, aqueducts to abstract and transport groundwater in the coastal region, and irrigation systems using earthen canals to increase aquifer recharge and promote water use efficiency. Given the current water stress pressure, the study from Ochoa-Tocachi et al. (2019) provides the scientific evidence needed to upscale Indigenous infrastructure, thus challenging the preconception that local water management traditions are outdated and supporting the uptake of NBS for water security.

Traditional integrated watershed management

Many of the individual water sowing and harvesting practices are integrated into more complex systems of water management within a basin. For example, the *amunas* in Peru are typically part of a system-wide set of practices that may include a combination of *qochas*, infiltration terraces and ditches, wetland protection or enhancement, water and soil conservation terracing, pasture management to increase infiltration, forest restoration, and protection of springs (MINAGRI, 2016). Indigenous peoples have also developed even more complex systems that integrate nature-based water sowing and harvesting technologies with SLM practices on a watershed or regional basis (Table 13.2). Based on traditional ecological knowledge, these systems anticipate and predate many of the best practices that present-day water managers seek to apply in watershed management, integrated water resources management, and SLM.

Table 13.2: Examples of nature-based Indigenous systems for sustainable land and water management.

Ancestral practice	Region	Description
Integrated Landscape and Water Management Systems		
Ifugao Muyong Forests and Rice Terraces	Philippines	Integrated protection forests, agroforestry, and terracing to protect water sources, sow water from runoff in steep catchments in soils, groundwater, and terraces for harvesting to supply paddy rice, wet-pond taro, and domestic uses (Acabado and Martin, 2015).
Village Tank Cascades	Sri Lanka; Karnataka, Tamil Nadu, Andhra Pradesh, India	Integrated protection forest, water tanks, and agricultural fields that protect water sources, sow water from seasonal monsoon rains in tanks, wetland soils, and groundwater for harvesting to supply rice fields, domestic uses, and livestock (Geekiyanage and Pushpakumara, 2013).
Moku-Ahupua'a System	Hawai'i	Mountains to sea integrated land divisions and water management system in Hawai'i consisting of protected upland forests (including cloud forests) to capture and infiltrate precipitation, diversion channels, and terracing for agricultural production but also to retain sediments (protecting coral reefs and their critical habitat for fisheries from sedimentation) and slow flow of water to enhance groundwater recharge and sustain supplies of freshwater for communities and farms on the coastal plain (Winter et al., 2018; Winter and Lucas, 2017).
Honghe Hani Rice Terraces	China	With some similarities to the *Ifugao* system, components include a four-fold integrated system of forests, water supply, terraces, and houses. The mountain top forests are the lifeblood of the terraces in capturing and sustaining the water needed for the irrigation. There are four types of forests: the ancient "water recharge" forest, sacred forest, consolidation forests, and village forests for the provision of timber for building, food, and firewood. Channels direct water to the permeable sandstone to store and then later release water as springs; a complex system of communally managed channels then spreads this water around a massive system of terraces in and between different valleys (Yongxun et al., 2016).

Continued

Table 13.2: Examples of nature-based Indigenous systems for sustainable land and water management—cont'd

Ancestral practice	Region	Description
Zabo System	Nagaland, NE Hill Ranges, India	With similarities to the tank cascades and *ahupua'a* systems, the *zabo* system has various components for water management: forest land as a catchment area, water harvesting systems like ponds with earthen embankments, with livestock and agricultural lands at lower elevations. Protected forest on the hilltop slows runoff and enhances infiltration; water flows or seeps down slopes and passes through a system of terraces, further slowing the flow, providing planting areas for crops and recharging groundwater along the slope. The water is then collected and retained in pond-like structures in the middle terraces to enhance recharge; toward the foot of the hill are paddy fields, where the runoff ultimately meanders into collection systems (Bhattacharya, 2015; Sharma and Sharma, 2004).

Three examples that are still in use today, despite being diminished by threats from colonialization and globalization, are discussed here. First, a brief discussion of the traditional moku-ahupua'a system of land divisions linked to integrated land-water management in Hawai'i, which is being revitalized by Native Hawaiian communities (Akutagawa et al., 2016; Kaneshiro et al., 2005; Winter et al., 2018; see also Bremer et al., 2021). Second, the muyong-payoh system in northern Luzon in the Philippines (Camacho et al., 2012; Soriano and Herath, 2018), and third, the Village Tank Cascades of central Sri Lanka (Bebermeier et al., 2017).

Moku-ahupua'a system in Hawai'i

Throughout Oceania, from the Solomon Islands to Hawai'i, there are remarkably similar traditional systems that integrate water and land management from "ridges to reefs" linking upland watershed health with sustaining productive nearshore marine habitats—the *puava* in the Solomon Islands, *vanua* in Fiji, and *moku-ahupua'a* in Hawai'i (Berkes et al., 2000; Hviding, 1996). In Hawai'i, this sophisticated management system seamlessly connected protected cloud forests and upper watershed forests to lower elevation mixed agroforestry plantations (breadfruit, banana, and plants for medicines, dyes, fiber, and fuel), wet-pond taro terraces and irrigation channels, coastal settlements, farm gardens, fish ponds, and productive marine fishing grounds, maintaining regular flows of freshwater from mountains to sea (Akutagawa et al., 2016; Andrade, 2008; Kaneshiro et al., 2005; Kurashima et al., 2018; Winter et al., 2018). Protecting high elevation forests, especially cloud forests, ensured continuous water flow through streams and irrigation channels to the wet terraces (Bremer et al., 2019;

Winter and Lucas, 2017) where large crops of taro, central to Native Hawaiian cultures both spiritually and nutritionally, could be grown to support historically large populations (Kirch and Kelly, 1975). The system of terracing for growing taro helped retain sediment, protect water quality, and maintain fresh groundwater discharges to the critically important coastal fish ponds and reefs that supported traditional fisheries (Koshiba et al., 2013).

Ifugao muyong-payoh system

In the central Cordilleras of Luzon Province in the Philippines, the Indigenous Ifugao people have developed an integrated water and land management system that has sustained their culture and productive rice paddy agriculture for hundreds if not thousands of years (Butic and Ngidlo, 2003; Serrano and Cadaweng, 2005). These mountains with abundant rainfall, cooler temperatures, and steep, rugged slopes are not well-suited to traditional lowland rice cultivation. The place-based landscape management system of the Ifugao people, reflecting a deep understanding of soil and water conservation, agroforestry, biodiversity, pest management, and rice cultivation, is anchored in a complex web of relationships the Ifugao have with both the human and nonhuman worlds (Acabado and Martin, 2015, 2016). The rice terraces of the Banaue are famous (Figs. 13.8 and 13.9), recognized by UNESCO in 1995 as an outstanding example of a living cultural landscape, but the system is anchored in the muyong forests that infiltrate rainwater and recharge aquifers, protecting the water that supplies the rice paddies.

Fig. 13.8

Ifugao Batad rice terraces, Banaue, showing muyong forests, swidden gardens, and settlements, central Cordillera, Philippines. *Reproduced with permission from Eric Montalban, Wikimedia Commons.*

Fig. 13.9

Flooded Ifugao terracing with surrounding muyong forest. *Reproduced with permission from Getty Images, Michael Runkel.*

Historical development of Ifugao terracing and muyong system

A clear picture of the exact timeline of the development of the Ifugao terraces, how old they are, and what drove the initial development is still emerging. While early assessments estimated the age of the Ifugao terraces at over 2000 years old, recent studies suggest that Ifugao wet-rice cultivation in the highlands was likely a response to two factors: changing climate and drier conditions associated with the Little Ice Age around AD 1300, and the expansion of Spanish colonization into the lowlands in northern Luzon in the late 16th century (Acabado and Martin, 2015; Acabado, 2018; Acabado et al., 2019; Peterson and Acabado, 2015). Peterson and Acabado (2015) suggest that development of water control features and terracing may have originally been built to support taro (*Colocasia esculenta*) wet-pond cultivation in the highlands as lowland areas became too dry. Microfossils and radiocarbon dating show that taro was grown in the highlands "as early as 1300," with wet-rice varieties introduced during the 14th century (Peterson and Acabado, 2015). Wet-pond taro cultivation in terraces has a long history and is widespread across southeast Asia and Oceania, could have been rapidly adapted for wet-rice cultivation, and taro is still grown on the rice terraces of the Ifugao today, but not as the dominant crop (Acabado, 2012).

The combination of archeological, linguistic, plant microfossil, and recent radiocarbon dating all suggest that the Ifugao moved to higher elevations and innovated wet-rice cultivation in a very short period of time, probably beginning between the 14th and 16th centuries, with intensification of the terraces largely in response and as resistance to Spanish colonization (Acabado, 2009, 2017; Horrocks et al., 2018; Keesing, 1962; Lambrecht, 1967). Suggestions of a later date for the terraces has been controversial. This is in part because suggesting an earlier date provides more validation for the Ifugao claim to the land as part of their ancestral domain (Acabado, 2012). Regardless of the exact date of the terraces, the integrated Ifugao system itself demonstrates the deep ties of the Ifugao to their mountain landscape and the dynamic nature and adaptability of Indigenous knowledge and technologies in the face of change (Acabado and Barretto-Tesoro, 2020; Berkes et al., 2000; Castonguay et al., 2016; Soriano and Herath, 2020).

Structure and function of the muyong-payoh system
The biophysical system

Although the massive terraces are the most dramatic manifestation of the Ifugao landscape, it is a landscape of complementary systems that form an integrated mixed production water-centric landscape of catchment forests, agroforestry, swidden cultivation, and irrigated rice fields (Acabado and Martin, 2015; Butic and Ngidlo, 2003). The muyong system has six major zones that all work together (Table 13.3).

Table 13.3: Complementary and integrated zones in the muyong-payoh system.

Name	Zone	Description
Inalahan or Bilid	Communal/Protected Watershed Forest	Protected communal forests are at highest elevations, difficult to access, serve as recharge zones, and help regulate the flow of water that supplies the rice terraces. Cutting down trees in the *bilid* is prohibited by customary laws.
Muyong or Pinugo	Kin-based Private Woodlot/ Agroforestry	The muyong forests are passed down within families and privately (by family groups) owned and managed. Located immediately above paddy fields, the muyong forests also serve as important recharge zones for the paddies and springs and streams originating in the muyong forests help protect the water supply to adjacent terraces.
Payoh	Rice Terraces	Just below the muyong forest, irrigated rice terraces are also private property, owned and managed by a family group, although maintenance and repairs are accomplished through a system of reciprocal labor exchanges within the community.
Wah-le/wang-wang and ala/liting	Water Management and Irrigation Infrastructure	Communally owned and maintained, these include stream and riverbeds, irrigation channels that move water to terraces, sluice gates that control water levels in the terraces, and movement of water from one terrace to another downhill.

Continued

Table 13.3: Complementary and integrated zones in the muyong-payoh system—cont'd

Name	Zone	Description
Habal or Uma	Swidden Fields	Considered common property and usually on slopes too steep for terracing or lacking a water source. They support rainfed farming in upland areas that are temporarily cleared for planting and used for crops that complement rice, mostly root crops such as sweet potatoes (*Ipomea batatas*).
Boble or labangan	Settlement Areas/ Households and Livestock	Located either above or below the terraces but generally located with easy access to swidden fields and rice terraces. Families with agricultural fields and muyongs within the same area live together in these settlements or hamlets, and these social units form the basis for the cooperative management of the agricultural-water control system.

Data compiled from Butic M, Ngidlo R. Muyong forest of Ifugao: assisted natural regeneration in traditional forest management. In: Dugan PC, et al., editors. Advancing assisted natural regeneration (ANR) in Asia and the Pacific. 2003. p. 23–8; Camacho LD, et al. Indigenous knowledge and practices for the sustainable management of Ifugao forests in Cordillera, Philippines. Int J Biodivers Sci Ecosyst Serv Manag 2016;12:5–13; Rabena MAF, Macandog DM. Contemporary knowledge of Woodlot (Muyong) resource management: a case study of key-informants' perceptions in Brgy. Kinakin, Banaue, Ifugao, Philippines. J Manag Dev Stud 2017;6:14–22; Soriano MA, Herath S. Quantifying the role of traditional rice terraces in regulating water resources: implications for management and conservation efforts. Agroecol Sustain Food Syst 2018;42:885–910.

Muyong forests

The muyong forests are located in the upper watershed immediately above the rice fields and are managed for watershed protection, fuel, construction materials, food, and medicinal plants (Butic and Ngidlo, 2003; Serrano and Cadaweng, 2005). Muyong forests are privately held by individual families or clans, and along with the associated swidden farms and rice terraces, these can be passed down through inheritance. The Ifugao people recognize the central role of the muyong in maintaining the stable water supply essential for the sustainability and function of the entire integrated farming system (Camacho et al., 2012; Soriano and Herath, 2018). Maintaining water flow from the forest helps maintain the overall physical structure and function of the terraces (Butic and Ngidlo, 2003) as well as providing a source of nutrients that fertilize the rice paddies (Rabena and Macandog, 2017; Serrano and Cadaweng, 2005). Traditional practices ensure sustainable management of these forests through a set of customary laws, norms, and ritual beliefs (Camacho et al., 2016; Serrano and Cadaweng, 2005). Ritual beliefs protect important old-growth trees from cutting (e.g., endemic dipterocarps) and traditions such as replacing any tree cut down by planting two new seedlings ensures the forests are not lost (Serrano and Cadaweng, 2005).

The Ifugaos have developed a sophisticated, sustainable, agroforestry system for managing the muyong forests, through a system of assisted, natural regeneration (Butic and Ngidlo,

2003), maintaining a high level of biodiversity and an efficient production landscape (Camacho et al., 2012). Several studies have documented the diversity of plants in muyongs, including as many as 45 species that the Ifugao use for medicines, 171 species used for fuelwood, 112 species used for construction, 70 species that provide food, and a variety of species used for woodworking and basket-making, including multiple varieties of rattan (Butic and Ngidlo, 2003). The Ifugao also use their knowledge of aromatic herbs in the muyong forest to develop plant-based pesticides to control pests that can damage rice crops. Management activities include very selective harvesting of trees, thinning and pruning to encourage growth, and salvage cutting.

Rice fields and irrigation infrastructure

The rice terraces are the most famous elements of the Ifugao landscape and have been recognized by contemporary engineering societies as Historic Civil Engineering Landmarks as they embody "engineering principles of hydrology, sustainable development, and efficient use of water resources" in their design (https://www.asce.org/project/ifugao-rice-terraces/). Occupying an area of about 4000 km^2, terracing extends up to an altitude of 5000 ft and can occur on slopes as steep as 70 degrees (Conklin, 1980). The terraces are traditionally kept flooded the entire year for cultivation of the Indigenous upland and cold-tolerant tinawon rice varieties, but they also support edible fish and snails (Nozawa et al., 2008; Soriano and Herath, 2018). The terrace wet-pond fields are used for taro cultivation during rice fallow periods (Horrocks et al., 2018). Keeping the terraces flooded throughout the year helps maintain the stability of slope and terrace walls (Conklin, 1980). The water and irrigation components of the system are intimately associated with the rice terraces and include a network of irrigation canals that convey water from the forest to the terraces, drainage canals that move water from flooded terraces or from upper terraces to lower ones. Spillways on each terrace can be opened to drain excess water during wet periods or closed to prevent water loss during the dry season (Soriano and Herath, 2018).

Social, cultural, and ritual elements

Cooperative kin-based work groups, organized by micro-catchments or "water districts," worked together to build the terraces and maintain irrigation channels, repair terrace walls, and plant and harvest rice. The systems of reciprocity and community-wide cooperation have allowed for the long-duration maintenance of this complex landscape (Acabado and Martin, 2015; Gonzalez, 2000). Separate workgroups were organized, one for agricultural activities such as planting, transplanting seedlings, weeding, and harvesting (mostly done by women), and another for maintenance and repair of the terraces and irrigation system (mostly done by men) (Araral, 2013; Gonzalez, 2000). Organization of activities by the water district help ensure equitable water distribution and the synchronization of planting activities to enhance pest control (Araral, 2013).

The Ifugao have traditionally mostly cultivated annual Indigenous varieties of japonica rice called tinawon, meaning "once a year" (Nozawa et al., 2008). Tinawon rice has larger grains than indica varieties, a long growing season (7–9 months) and can tolerate higher altitudes and cooler temperatures than other varieties (Soriano and Herath, 2018). The complex agricultural calendar based on the growing season for tinawon rice is embedded in ritual practices that also serve to help coordinate the collective water and terrace management activities of the community (Acabado and Martin, 2016; Soriano and Herath, 2018).

Muyong-payoh system, water security, and climate resilience

Traditional knowledge and recent studies attribute multiple ecosystem services to the muyong forests and the payoh rice terraces, including erosion control and soil conservation; provision of nutrients to the rice fields; maintaining supply of surface water (via cloud water interception) (Conklin, 1980); regulation of surface runoff; groundwater recharge, maintenance of shallow subsurface flow and baseflow (or dry season flow) in streams; provision of food, medicines, timber, and firewood; microclimate regulation and carbon sequestration (Acabado and Martin, 2015; Borromeo et al., 2019; Camacho et al., 2012; Soriano and Herath, 2018). The muyong-payoh system has sustained areas of high biodiversity, including agrobiodiversity, for example, in the many varieties of rice with different nutritional characteristics and adapted to different soil, climate, and cultivation conditions (Camacho et al., 2016; Nozawa et al., 2008).

Relatively few studies have directly assessed the hydrological services provided by the muyong rice terracing system, but the few studies available do document the hydrological benefits. The most detailed study of muyong-terrace hydrology is that of Soriano and Herath (2018). Combining field measurements and modeling, this study evaluated the relative contribution of forested areas and rice terraces to total streamflow, baseflow, interflow, surface runoff, and groundwater recharge (Soriano and Herath, 2018). The muyong forests reduce surface runoff by increasing infiltration, shallow subsurface flow, and groundwater recharge that feeds springs and provides water to the terraces, especially during dry periods (Soriano and Herath, 2018, 2020). Forests cover more area than terraces and therefore dominate the hydrological response, however, terraces were responsible for three times the surface runoff as forests. The terraces remain flooded throughout the year, so they are fully saturated and any rain event will generate runoff from the terraces. Forests were significant for generating more interflow per unit area than the terraces, consistent with the role traditionally ascribed to muyong forests as infiltrating water and slowly releasing it to the terraces (Acabado and Martin, 2015).

Soriano and Herath (2018) found that the rice terraces were responsible for significantly more groundwater recharge per unit area than forests, and this recharge was consistent through the entire year. Forests contribute to groundwater recharge during the wet season when precipitation is greater than canopy interception and evapotranspiration, but the permanently inundated terraces are recharging groundwater during both wet and dry seasons. This study confirmed that the terraces play a significant role in maintaining baseflow (including to streams that supply downstream terraces) by enhancing groundwater recharge. The forests and terraces in the Ifugao mountains play a significant role in maintaining flows to downstream lowland areas, including to a major dam that supplies electricity and irrigation to lowland farms and the city of Luzon (Soriano and Herath, 2018).

Threats and consequences of losing indigenous knowledge

While the muyong-payoh system has been resilient in the face of some past climate extremes (floods, droughts), current and future climate change may pose a challenge. Given the role that inundated terraces place in runoff, especially during the wet season, increased precipitation or more intense storms predicted by climate models could damage or destroy terracing if water control structures and practices are not modified (Soriano and Herath, 2020). Prolonged droughts and reduced precipitation may require more measures for storing water through the dry season, especially if terraces are abandoned due to other pressures given the important role the terraces play in groundwater recharge (Soriano and Herath, 2018).

The most significant threats to the sustainability of the Ifugao cultural landscape arise from the erosion of Indigenous knowledge, global market forces, and migration of Ifugao people. A number of these forces intersect with the introduction of new commercial rice varieties. Martin et al. (2020) describe how the replacement of tinawon rice varieties with the more economically viable commercial varieties has eroded the Indigenous knowledge that is the foundation of the muyong-payoh system. Traditional rituals and religious beliefs are central to maintaining Indigenous culture, and the rice fields are the heart of this system, with the ritual calendar aligned with the planting and harvesting phases of the annual tinawon rice. The social cohesion needed to organize the substantial amount of labor needed to build and maintain the terraces and the synchronization of water management that is needed for pest control is maintained through the observance of these annual rituals. Replacing tinawon rice with commercial varieties has changed the relevance of the ritual calendar and resulted in the loss of Indigenous knowledge as the number of ritual specialists who hold this knowledge has declined (Martin et al., 2020).

Gender roles have also changed with the introduction of new rice varieties. The availability of new varieties has helped displace the traditional role of the "elder female farmer" who were the keepers of knowledge about the traditional heritage tinawon varieties (Acabado and Martin, 2015; Martin et al., 2020). These elders had the knowledge of seed selection and how to maintain and improve the heritage varieties, but this knowledge is no longer needed when seeds of commercial varieties can be purchased (Acabado and Martin, 2015). Acabado and Martin (2015) argue that Ifugao society was traditionally gender equal, in part because of women's central role in the planting of tinawon rice. Planting and transplanting was traditionally the domain women who were the experts at handling the sensitive tinawon seedlings. Because commercial varieties are hardier, less expertise is required for transplanting these seedlings, and men are frequently now hired to do this work. As more men become wage earners, they have less time to spend maintaining the terraces, a role traditionally performed by the men of the community (Acabado and Martin, 2015). The fertilizers and pesticides used with high-yielding commercial varieties have impacted the complex agro-ecosystems of the traditional terrace fields, reducing the effectiveness of pest control measures that relied on synchronized flooding and fallow periods and natural enemies to control pests (Acabado and Martin, 2015; Camacho et al., 2012).

The pressures of a global market economy have also contributed to the abandonment of terraces and the weakening of customary laws that governed the watershed protection forests. Urbanization within Ifugao territory, migration to cities, and alternative economic opportunities have all drawn young people away from rural communities and traditional practices (Araral, 2013; Magcale-Macandog, 2018). Labor shortages have led to the abandonment or disrepair of terraces. The transmission of traditional knowledge between generations is reduced as young people leave for economic opportunities elsewhere (Araral, 2013). Since the awarding of World Heritage Status, tourism has increased. While providing some new economic opportunities, this has also contributed to increased demand for the traditional Ifugao wood carvings, increased harvesting of trees, and the loss of tree cover in the muyong and communal forests, with potential impacts on water availability to the terraces (Magcale-Macandog, 2018).

Current status and revitalization

While the extent and condition of the rice terraces and muyong forests have declined over the past several decades and there are many challenges to maintaining this complex adaptive system, the Ifugao have recently been engaged in a process of revitalization of their traditional practices. UNESCO and the Government of the Philippines have developed programs to protect the rice terraces, beginning in the late 1990s, and these have had mixed success (Acabado and Martin, 2020). More recent community-led efforts are showing promise as they emphasize working with young Ifugao members and recovering and adapting traditional knowledge as the key to sustaining traditional water and land management practices (Acabado and Martin, 2020; Martin et al., 2020).

The Save the Ifugao Terraces Movement (SITMo) is an Ifugao nongovernmental organization (NGO) that focuses on community-led heritage conservation, emphasizing a revitalization of the culture-based, sustainable livelihoods centered around the traditional muyong-payoh system (Martin et al., 2020). Working with the Ifugao Archeological Project, SITMo established the first community-led Indigenous Peoples Education (IPEd) to promote Indigenous education, local history, and heritage management. SITMo also helps strengthen Ifugao identity while developing new economic opportunities within the community, for example, by supporting traditional Ifugao weavers in reaching global markets through artisan cooperatives and communities of entrepreneurs (e.g., Cambio & Co., https://www.shopcambio.co/). The Ifugao Community Heritage Galleries, hosted by IPEd, give the community and particularly young Ifugao access to and a showcase of material culture and traditional knowledge, an especially important vehicle for increasing awareness and pride in their heritage. Through a community-led eco-tourism program, which renews the performance of the rice rituals and supports the few remaining ritual specialists (*mumbaki*) in transferring their knowledge to a new generation, SITMo is helping ensure the continuance of the dynamic traditional knowledge on which the survival of the muyong-payoh system depends. Although many challenges remain, the resilience of the Ifugao social-ecological system over many centuries is a testament to the dynamism of their culture and traditional knowledge.

The village tank cascades in Sri Lanka

Thousands of years ago communities in the dry zone of Sri Lanka began creating systems of cascading village "tanks" (Fig. 13.10) within small watersheds to collect and store water during the brief monsoon season, providing a reliable supply of surface and groundwater

Fig. 13.10

Typical Village Tank Cascade, Toruwewa, Sri Lanka. Flow of water is from south to north. *Data from Google Earth. V 7.3.3.7721. (February 17, 2018). Toruwewa, Sri Lanka. 8°06′38.82″ N, 80°35′06.93″ E, Eye alt 12544 ft. Image © 2020 CNES/Airbus.*

for drinking, livestock and irrigating crops (Abeywardana et al., 2019; Bandara, 1985; Panabokke et al., 2002). Both surface and groundwater cascades by gravity from one tank to another in the series, either through spillways and irrigation channels connecting upstream and downstream tanks or by the flow of groundwater between tanks (Fig. 13.10). Although in a region quite different from the wet mountains of the Ifugao, tank cascades utilize a similar integration of forest, hydraulic structures, ponds, tanks, rice paddies, and farm fields to manage their water resources and create productive cultural landscapes where agriculture would otherwise be challenging (Geekiyanage and Pushpakumara, 2013).

The dry zones of Sri Lanka with low undulating hills and gently sloping valleys are characterized by a short rainy season when 80% of the annual precipitation falls in intense monsoon storms that can cause significant flooding (FAO, 2017). Shallow, unconfined aquifers underlie much of the area and are recharged during the rainy season, however, groundwater levels fall rapidly once the rains have ended. During the long dry season of about 8 months, evaporation is typically greater than annual rainfall resulting in a soil moisture deficit. Prolonged droughts and desiccating winds during the dry season further aggravate water stress for people and crops. Despite these challenges, the area has historically supported large populations and sustainable paddy rice cultivation in both wet and dry seasons (Dharmasena, 2009).

Structure and function

The tank cascades are composed of a series of small- to medium-sized reservoirs (*wewas* or tanks) constructed in a series from upstream to downstream within each small watershed, with the tanks embedded within a complex agroecosystem (Fig. 13.11). The four major zones of a tank cascade are: protected watershed forest, rainfed uplands; tank bund (embankment) and tank bed; irrigation channels and paddy fields; and household areas or settlement areas (Geekiyanage and Pushpakumara, 2013; Panabokke, 2009). Within each of these four zones there are a variety of managed landscapes and structures that control erosion, manage flooding, recharge aquifers, maintain soil moisture, manage surface water for irrigation, provide nontimber forest products, and support a diversity of plants and animals that contribute to controlling pests in the rice fields (Abeywardana et al., 2019).

Watershed forests

The protected watershed forests (*mukalana*) are owned and managed communally and can be used for nontimber forest products, such as medicinal plants, but are managed primarily to protect water quality, enhance infiltration, and recharge groundwater, which can then be harvested in the dry season. Within the *mukalana* there may be small silt-trapping tanks or depressions that help to slow surface runoff, enhance infiltration, and minimize erosion during heavy rains, reducing siltation within the tanks. Rainfed forest gardens between the protected forests and the tanks provide additional sources of food and medicinal plants.

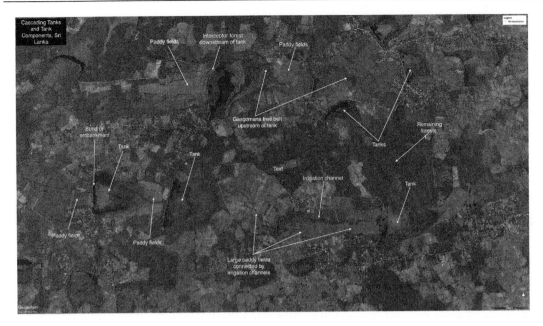

Fig. 13.11

Components of a cascading series of tanks: Protection forests, tanks, embankment and paddy fields; irrigation channels connecting tanks and irrigating paddy fields, interceptor forests, and tree belts that help protect tanks and paddy fields. *Data from Google Earth. V 7.3.3.7721. (September 12, 2018). Kahatagasdigiliya, Sri Lanka. 8°29′00.54″ N, 80°35′08.93″ E, Eye alt 23129 ft. Image © 2020 CNES/Airbus*

Tanks

The tanks (*wewas*) themselves include a combination of hydraulic structures and natural infrastructure to manage water and sediments. A bund or embankment across the downstream end of each tank creates a dam that impounds water behind the bund, creating a shallow pond as the tank area fills with water following rains. The shallow edges of the inundated area support wetland plants and animals, creating wetlands (*wew thaulla*) that are very similar to the natural wetlands along major river floodplains. The wetlands help improve water quality in the tanks by filtering or processing sediments, nutrients, and salts. Sluice gates, water level indicators, and stone lining for the embankment allow water levels within the tank to be controlled, allowing for spilling of water to the paddy fields and downstream tanks in the series, while controlling erosion.

At the upstream boundary of the tank a tree belt (*gasgommana*) acts as a wind break protecting the tank from desiccating winds, reducing wave action and evaporation from the tank surface. In the rainy season the tree belt is flooded, providing habitat for fish and other aquatic species. Finally, between the tank and the tree belt there is a water filter (*perahana*), an area of grasses, sedges, and shrubs that further serves to retain sediments and filter water coming into the tank.

Water stored in the tanks gradually recedes as the tanks begin to dry out after the rainy season ends, although all except the smallest tanks typically retain some water throughout the year (Perera, 2018; Schütt et al., 2013; Shanmuganathan et al., 2010). When surface water has disappeared from the tank beds, the moist soils may be used to cultivate a variety of vegetable crops and flowers, such as lotus and water lily for nearby temples. Water also continues to seep out of the tanks during the year, maintaining groundwater levels near the tanks and paddy fields. Within a cascading series of tanks, some may be used primarily for wildlife or as silt traps, reserved for livestock, domestic uses, and paddy irrigation. Frequently the lowest tank in a series is associated with a village and primarily supplies drinking water or larger rice fields.

Paddy fields and irrigation channels

Immediately below the tank is a belt of trees, the interceptor or *katakaduwa*. With a diversity of deep-rooted native trees, this helps stabilize soils, filter water coming from the tank, create a wind barrier to protect the paddy fields below, and provide habitat for wildlife, and it is a source of medicinal plants, firewood, and fodder for livestock. The cultivated paddies are fed by surface and groundwater from the tank, allowing for rice production in both wet and dry seasons. Drainage canals through the rice fields help control water levels for optimal rice production and also to convey excess water to the downstream tanks. The irrigated paddy system is communally managed and traditionally involved rituals and planting calendars based on traditional knowledge. During times of drought, agreed rules for sharing water typically reduce the size of the irrigated area to match the available water, while proportional allocation of fields among community members gives everyone access to farmable land.

Because much traditional knowledge of the system has been lost there are gaps in our understanding of how the entire system functioned historically. For example, small areas of rice paddy on the edges of each field were traditionally not harvested but left for foraging birds. While the exact function of this area is not well understood, it may have been to attract birds to this area to reduce damage to the rest of the rice fields or to keep insect pests down. Shrub lands located away from the rice fields may also serve as buffers where wildlife—such as elephants—can feed, keeping them away from rice and other crops (Anuradha et al., 2019).

Tank cascades and hydrological services

Surface and groundwater flow dynamics have not been studied extensively, but there are some tank-scale and basin-scale studies that document how the tank cascades function. In many of the areas with tanks, there is now active groundwater pumping for irrigation and domestic uses, and this makes assessing the influence of tanks on groundwater dynamics difficult. Nevertheless, recent studies document relatively rapid recharge of groundwater once tanks have filled from rains, with recharge rates declining as the tanks dry (Perera, 2018). Along the cascade, there can be variation in whether a tank primarily receives inflow from groundwater or provides outflow to recharge groundwater. In cases that have been studied, the tanks higher in the series provide greater recharge to the shallow aquifer, while lower tanks tend to receive more inflow from groundwater (Panabokke, 2009). Assessments of the effects of tank cascades on basin-scale water balance using modeling suggests that runoff can be reduced significantly—up to 75%, while recharge increased by about 40%. The abandonment or loss of tanks in some areas has been associated with a lowering of the water table, although this could also be due to increased groundwater pumping or density of settlements (Dharmasena, 2002).

Challenges and revitalization of traditional practices

All of the traditional communities noted here, their practices, and the traditional knowledge underpinning these practices are under significant threat. The examples discussed here, from Peru to Spain to Sri Lanka, are being impacted by a similar suite of threats. Climate change, increasing and competing demands for water from more powerful users, the introduction of commercial crop varieties and the pesticides and fertilizers they require, tourism, urbanization and forest loss, and competition from global market forces, which spur migration to cities in search of better economic opportunities, all pose threats to the viability of traditional practices. Due to these forces, many of the Ifugao terraces and Sri Lankan tank cascades have been abandoned or eliminated since the middle of the 20th century. Almost 20,000 tank cascades are estimated to have been developed in Sri Lanka and to have existed at the time of British colonization, but only a few hundred are functioning today (Panabokke, 2009).

The overarching threat is the loss of traditional knowledge and values, which are the foundation of these nature-based practices. Traditional knowledge is passed from generation to generation by elders, in many cases only through oral transmission. As elders pass away and a new generation is no longer interested or able to follow traditional ways, the knowledge can be lost. All of these technologies have common elements built around complex cultural and knowledge traditions that are easy to disrupt. These include the centrality of rituals and religious ceremony to guide practices, consensus governance and cultural norms that guide the use of common pool resources, and the oral transmission of traditional ecological knowledge. Without the vibrant Indigenous communities themselves, their traditional ecological knowledge will not be maintained, and this valuable heritage will be lost for all of us.

Despite the threats, as seen in Hawai'i, Peru, and the Philippines, Indigenous and local communities are actively revitalizing and adapting traditional practices. In Peru, a series of national Siembra y Cosecha del Agua projects under the Sierra Azul program of the Ministry of Agriculture, is rehabilitating and constructing *qochas*, *amunas*, and *zanjas de infiltración* to improve water security in rural areas. In Kenya and Ethiopia, sand dams are being constructed to improve local water supplies and provide water security under anticipated climate change impacts. Local communities in Spain and Peru are actively rehabilitating the *careos* and *amunas* systems, with the help of local NGOs, regional and national governments, and international development agencies. In many cases, such as the revitalization of the traditional Hawaiian moku-ahupua'a system and the Ifugao save the terraces movement, these efforts were begun and are led by Indigenous peoples.

As the world community searches for more sustainable and resilient technologies in the face of growing water and climate crises, it is critical that we learn from traditional peoples who have crafted sustainable NBS for centuries. To do this, the first step must be supporting Indigenous peoples and traditional communities so that their knowledge is not lost. Locally led, community initiatives that use traditional NBS to protect their own resources through collective action mechanisms (e.g., Hawai'i, Peru, Philippines) while providing multiple benefits beyond the local community are likely to be the most equitable and durable actions. There is a need to connect local actions and innovations to the growing international focus on NBS. International legitimacy and recognition can help support and value local efforts if these are focused on enabling and building local, bottom-up efforts and supporting the rich diversity of place-based and nature-based approaches that arise from local knowledge.

References

Abeywardana N, et al. Indigenous agricultural systems in the dry zone of Sri Lanka: management transformation assessment and sustainability. Sustainability 2019;11:910.

Acabado S. A Bayesian approach to dating agricultural terraces: a case from the Philippines. Antiquity 2009;83:801–14.

Acabado S. The Ifugao agricultural landscapes: agro-cultural complexes and the intensification debate. J Southeast Asian Stud 2012;43(3):500–22.

Acabado S. The archaeology of pericolonialism: responses of the "unconquered" to Spanish conquest and colonialism in Ifugao, Philippines. Int J Hist Archaeol 2017;21:1–26.

Acabado S. Zones of refuge: resisting conquest in the northern Philippine highlands through environmental practice. J Anthropol Archaeol 2018;52:180–95.

Acabado S, Barretto-Tesoro G. Places, landscapes, and identity: place making in the colonial period Philippines. In: Beaule CD, Douglass JG, editors. The global Spanish empire, five hundred years of place making and pluralism. Tuscon: The University of Arizona Press; 2020. p. 215–35.

Acabado S, Martin M. Between pragmatism and cultural context: continuity and change in Ifugao wet-rice agriculture. In: Willems WJH, van Schaik HPJ, editors. Water & heritage: Material, conceptual, and spiritual connections. Leiden: Sidestone Press; 2015. p. 273–95.

Acabado S, Martin M. The sacred and the secular: practical applications of water rituals in the Ifugao agricultural system. TRaNS Trans Regional Natl Stud Southeast Asia 2016;4:307–27.

Acabado S, Martin M. Decolonizing the past, empowering the future: community-led heritage conservation in Ifugao, Philippines. J Community Archaeol Herit 2020;7:171–86.

Acabado S, et al. The short history of the Ifugao Rice Terraces: a local response to the Spanish conquest. J Field Archaeol 2019;44:195–214.

Aerts J, et al. Robustness of sand storage dams under climate change. Vadose Zone J 2007;6:572–80.

Akpinar Ferrand E, Cecunjanin F. Potential of rainwater harvesting in a thirsty world: a survey of ancient and traditional rainwater harvesting applications. Geogr Compass 2014;8:395–413.

Akutagawa M, Williams H, Kamaka'ala S. Traditional and customary practices report for Mana'e, Moloka'i: Traditional subsistence uses, mālama practices and recommendations, and Native Hawaiian rights protections of kamaʻāina families of Manaʻe Moku, East Molokaʻi, Hawaiʻi; 2016. Prepared for Office of Hawaiian Affairs, Honolulu, HI.

Alvarez SG. The use and traditional knowledge of pre-Hispanic hydraulic systems amongst Indigenous and non-Indigenous populations on the Ecuadorian coast. In: Jacobsen FF, McNeish J, editors. From where life flows: The local knowledge and politics of water in the Andes. Trondheim, Norway: Tapir Academic Press; 2006. p. 53–63.

Alvarez SG. Representaciones, saberes y gestión de los recursos naturales y culturales, a nivel local. In: Alvarez SG, editor. Prácticas, creencias y valores que condicionan la reproducción de los sistemas de albarradas en la Península de Santa Elena, Ecuador. Abya-Yala, Quito: ESPOL; 2009. p. 89–117.

Andrade C. Ha'ena: Through the eyes of the ancestors. Honolulu, HI: University of Hawaii Press; 2008.

Antunez de Mayolo R. El riego en Aija. Allpanchis 1986;18:47–71.

Anuradha JMPN, et al. The role of agricultural land use pattern dynamics on elephant habitat depletion and human-elephant conflict in Sri Lanka. Sustainability 2019;11:2818.

Apaza D, Arroyo R, Alencastre A. Las amunas de Huarochirí: Recarga de acuíferos en los Andes. Lima: GSAAC/Embajada de Países Bajos/IICA; 2006.

Araral E. What makes socio-ecological systems robust? An institutional analysis of the 2,000 year-old Ifugao society. Hum Ecol 2013;41:859–70.

Aristizábal MFV. Albarradas: pertinencia de los saberes ancestrales frente a la colonialidad del desarrollismo. Rev Ciênc Sociais 2019;50(3):223–47.

Avis O. A microbial analysis of water in sand dams and associated abstraction methods. In: 7th RWSN forum: Water for everyone, vol. 1; 2016. p. 730–46.

Bandara CMM. Catchment ecosystems and village tank cascades in the dry zone of Sri Lanka a time-tested system of land and water resource management. In: Strategies for river basin management. Springer; 1985. p. 99–113.

Baurne G. "Trap-dams": artificial subsurface storage of water. Water Int 1984;9:2–9.

Bebermeier W, et al. Tank cascade systems as a sustainable measure of watershed management in South Asia. Water 2017;9:231.

Beckers B, Berking J, Schütt B. Ancient water harvesting methods in the drylands of the Mediterranean and Western Asia. J Anc Stud 2013;2:145–64.

Berkes F. Traditional ecological knowledge in perspective. In: Inglis JT, editor. Traditional ecological knowledge: Concepts and cases. Ottawa: International Program on Traditional Ecological Knowledge; 2004. p. 1–6.

Berkes F. Sacred ecology. London: Routledge; 2012.

Berkes F, Colding J, Folke C. Rediscovery of traditional ecological knowledge as adaptive management. Ecol Appl 2000;10:1251–62.

Bhattacharya S. Traditional water harvesting structures and sustainable water management in India: a socio-hydrological review. Int Lett Nat Sci 2015;37:30–8.

Borromeo TH, et al. Assessing resiliencies, biodiversity of global significance and environmental goods and services of GIAHS-designated Ifugao Rice Terraces in comparison to conventional rice paddies. World Agricultural Heritage Foundation; 2019. Reviewed September 10, 2020 http://34.90.113.78/~worldagr/wp-content/uploads/2019/10/GIAHS_IRT_final-Report.pdf.

Borst L, De Haas SA. Hydrology of sand storage dams: a case study in the Kiindu catchment, Kitui District, Kenya [Master of Science thesis]. Vrije University; 2006.

Bremer LL, et al. Contributions of native forest protection to local water supplies in East Maui. Sci Total Environ 2019;688:1422–32.

Butic M, Ngidlo R. Muyong forest of Ifugao: assisted natural regeneration in traditional forest management. In: Dugan PC, et al., editors. Advancing assisted natural regeneration (ANR) in Asia and the Pacific; 2003. p. 23–8.

Buytaert W, et al. Human impact on the hydrology of the Andean páramos. Earth Sci Rev 2006;79:53–72.

Camacho LD, et al. Traditional forest conservation knowledge/technologies in the Cordillera, Northern Philippines. Forest Policy Econ 2012;22:3–8.

Camacho LD, et al. Indigenous knowledge and practices for the sustainable management of Ifugao forests in Cordillera, Philippines. Int J Biodivers Sci Ecosyst Serv Manag 2016;12:5–13.

Candiani VS. Dreaming of dry land: Environmental transformation of colonial Mexico City. Stanford, CA: Stanford University Press; 2014.

Carrion P, et al. Practical adaptations of ancestral knowledge for groundwater artificial recharge management of Manglaralto coastal aquifer, Ecuador. WIT Trans Ecol Environ 2018;217:375–86.

Castonguay AC, et al. Resilience and adaptability of rice terrace social-ecological systems: a case study of a local community's perception in Banaue, Philippines. Ecol Soc 2016;21(2):15–28.

Chimner RA, Karberg JM. Long-term carbon accumulation in two tropical mountain peatlands, Andes Mountains, Ecuador. Mires Peat 2008;3:1–10.

Conklin HC. Ethnograhic atlas of Ifugao. London, New Haven: Yale University Press; 1980.

Cooper DJ, et al. Drivers of peatland water table dynamics in the central Andes, Bolivia and Peru. Hydrol Process 2019;33:1913–25.

Cruickshank A, Grover VI. These are our water pipes—sand dams, women and donkeys: dealing with water scarcity in Kenya's arid and semi-arid lands. In: Filho WL, editor. Climate change and the sustainable use of water resources: Climate change management. Berlin, Heidelberg: Springer Berlin Heidelberg; 2012. p. 701–26.

Denevan WM. Cultivated landscapes of native Amazonia and the Andes. Oxford University Press; 2001.

Dharmasena PB. Integrated management of surface and groundwater resources in tank cascade systems. In: Pathmarajah S, editor. Use of groundwater for agriculture in Sri Lanka. Peradeniya, Sri Lanka: Agricultural Engineering Society of Sri Lanka, University of Peradeniya; 2002. p. 53–65.

Dharmasena PB. Sustainability of small tank irrigation systems in Sri Lanka at the 21 st century. In: Bandara SM, editor. Traversing no man's land: Interdisciplinary essays in honour of Professor C.M. Madduma Bandara. Colombo: Godage International Publishers; 2009. p. 233–52.

DiNapoli RJ, et al. Rapa Nui (Easter Island) monument (ahu) locations explained by freshwater sources. PLoS One 2019;14(1):e0210409.

Domic AI, et al. Two thousand years of land-use and vegetation evolution in the Andean Highlands of Northern Chile inferred from pollen and charcoal analyses. Quaternary 2018;1:32.

Dudgeon RC, Berkes F. Local understandings of the land: traditional ecological knowledge and indigenous knowledge. In: Nature across cultures. Springer; 2003. p. 75–96.

Fairley JP. Geologic water storage in Precolumbian Peru. Lat Am Antiq 2003;14:193–206.

FAO, Food and Agriculture Organization. The cascaded tank-village system (CTVS) in the dry zone of Sri Lanka: A proposal for declaration of a GIAHS; 2017. p. 128.

Fernandez Escalante ÁE, et al. Ancient techniques of Managed Aquifer Recharge: Spanish careos and Peruvian amunas as an adaptive complex system, breakdown, anthropology and comparative analysis. In: ISMAR10 symposium: Managed aquifer recharge: Local solutions to the global water crisis; 2019. p. 389–96.

Flores Ochoa J, Percy Paz Flores M, Rozas W. A (re-) descubrimiento reciente: la agricultura en lagunas temporales (qocha) en el Altiplano. In: Morlon P, editor. Comprender la agricultura campesina en los Andes Centrales: Peru—Bolivia. Lima: IFEA/CBC; 1996. p. 247–55.

Foster ME, Chen D, Kieser MS. Metodologías CUBHIC. Restauración y Protección de Humedales; 2020. p. 16.

Geekiyanage N, Pushpakumara DKNG. Ecology of ancient tank cascade systems in island Sri Lanka. J Mara Island Cult 2013;2:93–101.

Gonnet JM, et al. Manual introductorio al manejo de vegas y bofedales mediante prácticas tradicionales de culturas andinas en el norte de Chile. Corporación Norte Grande; 2016.

Gonzalez RM. Platforms and terraces: Bridging participation and GIS in joint-learning for watershed management with the Ifugaos of the Philippines [PhD thesis]. Wageningen University; 2000.

Grainger S, et al. Tailoring infographics on water resources through iterative, user-centered design: a case study in the Peruvian Andes. Water Resour Res 2020;56. e2019WR026694.

Hartman BD, Bookhagen B, Chadwick OA. The effects of check dams and other erosion control structures on the restoration of Andean bofedal ecosystems. Restor Ecol 2016;24:761–72.

Hevia A, et al. The role of upland wetlands in modulating snowmelt runoff in the semi-arid Andes. Geophys Res Abstr 2016;18:10804. EGU General Assembly 2016.

Hill AF, Stallard RF, Rittger K. Clarifying regional hydrologic controls of the Marañón River, Peru through rapid assessment to inform system-wide basin planning approaches. Elem Sci Anthropocene 2018;6:37.

Hixon SW, et al. The ethnohistory of freshwater use on Rapa Nui (Easter Island, Chile). J Polynesian Soc 2019;128:163–89.

Hoogmoed M. Analyses of impacts of a sand storage dam on groundwater flow and storage: Groundwater flow modeling in Kitui District, Kenya [Master of Science thesis]. Vrije University; 2007.

Horrocks M, Acabado S, Peterson J. Plant microfossil results from Old Kiyyangan Village: looking for the introduction and expansion of wet-field rice (*Oryza sativa*) cultivation in the Ifugao Rice Terraces, Philippine Cordilleras. Asian Perspect 2018;57:159–76.

Huang J, et al. Dryland climate change: recent progress and challenges. Rev Geophys 2017;55:719–78.

Hut R, et al. Effects of sand storage dams on groundwater levels with examples from Kenya. Phys Chem Earth Parts A/B/C 2008;33:56–66.

Hviding E. Guardians of Marovo lagoon: Practice, place and politics in maritime Melanesia. Honolulu, HI: University of Hawaii; 1996.

Kaneshiro KY, et al. Hawai'i's mountain-to-sea ecosystems: social–ecological microcosms for sustainability science and practice. Ecohealth 2005;2:349–60.

Keesing F. The ethnohistory of northern Luzon. Stanford: Stanford University Press; 1962.

Koshiba S, et al. Palau's taro fields and mangroves protect the coral reefs by trapping eroded fine sediment. Wetl Ecol Manag 2013;21:157–64.

Kurashima N, et al. Āina Kaumaha: the maintenance of ancestral principles for 21st century indigenous resource management. Sustainability 2018;10:3975.

Lambrecht F. The Hudhud of Dinulawan and the Bugan at Gonhadan. Saint Louis Q 1967;5:527–71.

Lane K. Through the looking glass: re-assessing the role of agro-pastoralism in the north-central Andean highlands. World Archaeol 2006;38:493–510.

Lane K. Engineered highlands: the social organization of water in the ancient north-central Andes (AD 1000–1480). World Archaeol 2009;41:169–90.

Lane K. Water technology in the Andes. In: Encyclopaedia of the history of science, technology, and medicine in non-western cultures. Dordrecht: Springer Netherlands; 2014. p. 1–24.

Lane K. Water, silt and dams: prehispanic geologic storage in the Cordillera Negra, north-central Andes, Peru. Rev Glaciares Ecosistemas Montana 2017;2:41–50.

Lane K, Grant J. A question of altitude: exploring the limits of highland pastoralism in the prehispanic Andes. In: Capriles JM, Tripcevich N, editors. The Archaeology of Andean pastoralism. UNM Press; 2016. p. 139–57.

Lasage R, Verburg PH. Evaluation of small scale water harvesting techniques for semi-arid environments. J Arid Environ 2015;118:48–57.

Lasage R, et al. Potential for community based adaptation to droughts: sand dams in Kitui, Kenya. Phys Chem Earth Parts A/B/C 2008;33:67–73.

Lasage R, et al. The role of small scale sand dams in securing water supply under climate change in Ethiopia. Mitig Adapt Strat Glob Chang 2015;20:317–39.

Liu X, et al. The response of infiltration depth, evaporation, and soil water replenishment to rainfall in mobile dunes in the Horqin Sand Land, Northern China. Environ Earth Sci 2015;73(12):8699–708.

López-i-Gelats F, et al. Adaptation strategies of Andean pastoralist households to both climate and non-climate changes. Hum Ecol 2015;43:267–82.

Lovera D, et al. Asimilación y Transferencias de Conocimientos y Tecnologías en la Vinculación Universidad–Gobierno Local–Población: Caso Lacabamba. Rev Inst Investig Fac Ing Geol Minera Metal Geogr 2007;10:31–44.

Lowdermilk WC. Some problems of hydrology and geology in artificial recharge of underground aquifers. In: Ankara symposium on arid zone hydrology; 1953. p. 158–64.

Maddrell S, Neal I. Sand dams, a practical guide. London: Excellent Development; 2012.

Magcale-Macandog D. Understanding the dynamic interactions and environmental problems in the Muyong-Payoh system of Banaue, Ifugao, Philippines through participatory rural approaches. Philipp BIOTA 2018;48:1–18.

Maldonado Fonkén MS. Comportamiento de la vegetación de los bofedales influenciados por las actividades antrópicas [Master's thesis]. Pontificia Universidad Católica del Perú; 2010.

Maldonado Fonkén MS. An introduction to the bofedales of the Peruvian High Andes. Mires Peat 2014;15:1–13.

Manz B, et al. High-resolution satellite-gauge merged precipitation climatologies of the Tropical Andes. J Geophys Res Atmos 2016;121:1190–207.

Manzi HK, Kuria DN. The use of satellite images to monitor the effect of sand dams on stream bank land cover changes in Kitui District. J Agric Sci Technol 2012;13(2):133–50.

Marcos JG. Albarradas in coastal Ecuador: rescuing traditional knowledge on sustainable use of biodiversity. In: Jacobsen FF, McNeish J, editors. From where life flows: The local knowledge and politics of water in the Andes. Trondheim: Tapir Academic Press; 2006.

Marcos JG, Álvarez SG. Campos de camellones y jagüeyes en Ecuador: una visión integral desde la arqueología al presente socioambiental. Intersecc antropol 2016;17:19–34.

Marcos JG, Bazurco M. Albarradas y camellones en la region costera del antiguo Ecuador. In: Valdez F, editor. Agricultura ancestral camellones y albarradas: contexto social, usos y retos del passado y del presente. Editorial Abya Yala; 2006. p. 93–110.

Martin M, Acabado S, Macapagal RA. Hongan di Page: the sacredness and realism of terraced landscape in Ifugao culture, Philippines. In: Liljeblad J, Vershuuren B, editors. Indigenous perspectives on sacred natural sites: Culture, governance and conservation. Oxford: Routledge; 2020. p. 167–79.

Martin-Civantos JM. Mountainous landscape domestication: management of non-cultivated areas in Sierra Nevada (Granada-Almeria, Spain). Eur J Post-Class Archaeol 2014;4:99–130.

Martos-Rosillo S, et al. The oldest managed aquifer recharge system in Europe: new insights from the Espino recharge channel (Sierra Nevada, southern Spain). J Hydrol 2019;578:124047.

McDougle E. Water use and settlements in changing environments of the southern Ecuadorian coast [Master of Art thesis]. Columbia University; 1967.

McMillen HL, et al. Small islands, valuable insights: systems of customary resource use and resilience to climate change in the Pacific. Ecol Soc 2014;19(4):44–60.

Mekdaschi-Studer R, Liniger H. Water harvesting: Guidelines to good practices. Rome: International Fund for Agricultural Development (IFAD); 2013.

MINAGRI, Ministry of Agriculture and Irrigation, Peru. Rumbo a un Programa Nacional de Siembra y Cosecha de Agua: Aportes y reflexiones desde la practica. Lima: Ministerio de Agricultura y Riego del Perú (MINAGRI), Viceministerio de Políticas Agrarias; 2016. p. 2016.

MINAM, Ministry of Environment, Peru. Ley No. 30215 De Mecanismos De Retribución Por Servicios Ecosistémicos—Law No. 30215 on Reward Mechanisms for Ecosystem Services. Ministry of Environment of Peru; 2014.

Neal I, Maddrell S. Sand dams: harvesting rain-water from sand rivers. In: Proceedings of the 36th WEDC international conference, Nakuru, Kenya: Delivering water, sanitation and hygiene services in an uncertain environment; 2013. p. 1–5.

Nilsson A. Groundwater dams for small-scale water supply. Practical Actions Publishing, Intermediate Technology Publications; 1988. https://doi.org/10.3362/9781780442297.

Nissen-Petersen E. Water from dry riverbeds. ASAL Consultants Ltd., Danish International Development Assistance (Danida); 2006.

Nozawa C, et al. Evolving culture, evolving landscapes: the Philippine rice terraces. In: Amend T, et al., editors. Protected landscapes and agrobiodiversity values. Protected landscapes and seascapes, vol. i. Heidelberg: Kasparek Verlag; 2008. p. 71–93.

Ochoa-Tocachi BF, et al. High-resolution hydrometeorological data from a network of headwater catchments in the tropical Andes. Sci Data 2018;5:180080.

Ochoa-Tocachi BF, et al. Potential contributions of pre-Inca infiltration infrastructure to Andean water security. Nat Sustain 2019;2:584–93.

Oyague E, Cooper DJ. Peatlands of the Central Andes Puna, South America. Wetl Sci Pract 2020;37(4):255–60.

Palacios Rios F. Pastizales de regadío para Alpacas. In: Flores Ochoa J, editor. Pastores de Puna: Wywamichiq Punarunakuna. Lima: Instituto de Estudios Peruanos; 1977.

Panabokke CR. Small village tank systems of Sri Lanka. Colombo: Hector Kobbekaduwa Agrarian Research and Training Institute; 2009.

Panabokke RR, Sakthivadivel R, Weerasinghe AD. Evolution, present status and issues concerning small tank systems in Sri Lanka. Colombo: International Water Management Institute; 2002.

Pauw WP, et al. An assessment of the social and economic effects of the Kitui sand dams. IVM Report R-08/08, Instituut voor Milieuvraagstukken, Amsterdam, viewed 20 October 2020 https://research.vu.nl/ws/portalfiles/portal/2415504/R-08-08.pdf; 2008.

Perera MP. Shallow groundwater behavior of tank Cascade areas in Sri Lanka: a study based on geo-spatial technology. In: Seventh UGIT international conference on "climate change, disaster risk reduction, and sustainable development through geospatial technologies", CDSGeo-2018; 2018. p. 29–32.

Peterson JA, Acabado SB. Did the little ice age contribute to the emergence of rice terrace farming in Ifugao, Philippines. Natl Mus J Cult Herit 2015;1:1–10.

Polk MH, et al. Exploring hydrologic connections between tropical mountain wetlands and glacier recession in Peru's Cordillera Blanca. Appl Geogr 2017;78:94–103.

Pulido-Bosch A, Sbih YB. Centuries of artificial recharge on the southern edge of the Sierra Nevada (Granada, Spain). Environ Geol 1995;26:57–63.

Quinn R, et al. An assessment of the microbiological water quality of sand dams in southeastern Kenya. Water 2018a;10:708.

Quinn R, Parker A, Rushton K. Evaporation from bare soil: lysimeter experiments in sand dams interpreted using conceptual and numerical models. J Hydrol 2018b;564:909–15.

Quinn R, Rushton K, Parker A. An examination of the hydrological system of a sand dam during the dry season leading to water balances. J Hydrol X 2019;4:100035.

Rabena MAF, Macandog DM. Contemporary knowledge of Woodlot (Muyong) resource management: a case study of key-informants' perceptions in Brgy. Kinakin, Banaue, Ifugao, Philippines. J Manag Dev Stud 2017;6:14–22.

Recalde-Coronel GC, Barnston AG, Muñoz ÁG. Predictability of December–April rainfall in coastal and Andean Ecuador. J Appl Meteorol Climatol 2014;53:1471–93.

Rolando JL, et al. Key ecosystem services and ecological intensification of agriculture in the tropical high-Andean Puna as affected by land-use and climate changes. Agr Ecosyst Environ 2017;236:221–33.

Rull V. Climate change, deforestation patterns, freshwater availability and cultural shifts on prehistoric Easter Island (SE Pacific). Peer J Preprints 2019. https://doi.org/10.7287/peerj.preprints.27680v1.

Rull V. Drought, freshwater availability and cultural resilience on Easter Island (SE Pacific) during the Little Ice Age. Holocene 2020;30:774–80.

Ryan C, Elsner P. The potential for sand dams to increase the adaptive capacity of East African drylands to climate change. Reg Environ Chang 2016;16:2087–96.

Salvador F, Monerris J, Rochefort L. Peatlands of the Peruvian Puna ecoregion: types, characteristics and disturbance. Mires Peat 2014;15:1–17.

Saylor CR, Alsharif KA, Torres H. The importance of traditional ecological knowledge in agroecological systems in Peru. Int J Biodivers Sci Ecosyst Serv Manag 2017;13:150–61.

Schreiner L, Duval S, Mendez BL. Sand storage dams: A tool to cope with water scarcity in arid and semi-arid regions. Germany: RUVIVAL; 2019.

Schütt B, et al. Characterisation of the Rota Wewa tank cascade system in the vicinity of Anuradhapura, Sri Lanka. DIE ERDE J Geol Soc Berlin 2013;144:51–68.

Schwilch G, Hessel R, Verzandvoort S, editors. Desire for greener land: Options for sustainable land management in drylands. Bern, and Wageningen: University of Bern—CDE, Alterra—Wageningen UR. ISRIC—World Soil Information and CTA—Technical Centre for Agricultural and Rural Cooperation; 2012.

Serrano RC, Cadaweng EA. The Ifugao muyong: sustaining water, culture and life. In: Durst PB, et al., editors. In search of excellence: Exemplary forest management in Asia and the Pacific. Bangkok: Asia-Pacific Forestry Commission, FAO; 2005. p. 103–12.

Shanmuganathan S, Manobavan M, Fernando GWAR. The importance of the small-tank cascade system for the sustainable production of water in the dry zone of Sri Lanka. In: Proceeding of the Asia-Pacific round table for sustainable consumption and production; 2010. p. 13.

Sharma UC, Sharma V. The Zabo soil and water management and conservation system in northeast India: tribal beliefs in the development of water resources and their impact on society—an historical account of a success story. In: Proceedings of the UNESCO/IAHS/WHA symposium: The basis of civilization—Water science, vol. 286; 2004. p. 184–94.

Somers LD, et al. Does hillslope trenching enhance groundwater recharge and baseflow in the Peruvian Andes? Hydrol Process 2018;32:318–31.

Soriano MA, Herath S. Quantifying the role of traditional rice terraces in regulating water resources: implications for management and conservation efforts. Agroecol Sustain Food Syst 2018;42:885–910.

Soriano MA, Herath S. Climate change and traditional upland paddy farming: a Philippine case study. Paddy Water Environ 2020;18:317–30.

Spruce R. Notes on the valleys of Piura and Chira, in Northern Peru, and on the Cultivation of Cotton Therein. London: Eyre and Spottiswoode; 1864. London. Google Books, reviewed August 18, 2020.

Squeo F, et al. Bofedales: high altitude peatlands of the central Andes. Rev Chil Hist Nat 2006;79:245–55.

Teel WS. Catching rain: sand dams and other strategies for developing locally resilient water supplies in semiarid areas of Kenya. In: Climate change management: Agriculture and ecosystem resilience in sub Saharan Africa. Cham: Springer International Publishing; 2019. p. 327–42.

Thaman RR. The contribution of indigenous and local knowledge systems to IPBES: Building synergies with science; 2013. IPBES Expert Meeting Report.

UNESCO. Nature-based solutions for water; 2018. World Water Development Report 2018.

Valois R, et al. Characterizing the water storage capacity and hydrological role of mountain peatlands in the arid Andes of north-central Chile. Water 2020;12:1071.

Verzijl A, Guerrero Quispe S. The system nobody sees: irrigated wetland management and alpaca herding in the Peruvian Andes. Mt Res Dev 2013;33:280–93.

Villa JA, et al. Carbon sequestration and methane emissions along a microtopographic gradient in a tropical Andean peatland. Sci Total Environ 2019;654:651–61.

Villarroel EK, et al. Local management of Andean wetlands in Sajama national park, Bolivia. Mt Res Dev 2014;34:356–68.

Walker D, et al. Alluvial aquifer characterisation and resource assessment of the Molototsi sand river, Limpopo, South Africa. J Hydrol Reg Stud 2018;19:177–92.

Washington-Allen RA, et al. Quantification of the ecological resilience of drylands using digital remote sensing. Ecol Soc 2008;13(1):33.

Winter KB, Lucas M. Spatial modeling of social-ecological management zones of the Ali'i era on the island of Kaua'i with implications for large-scale biocultural conservation and Forest restoration efforts in Hawai'i. Pac Sci 2017;71:457–77.

Winter K, et al. The Moku system: managing biocultural resources for abundance within social-ecological regions in Hawai'i. Sustainability 2018;10:3554.

Wunderlich W, et al. Quantifying water storage capacity in, and dry season water yield from Bofedales, Andean Wetlands. American Geophysical Union; 2019. Fall Meeting, A21U-2668.

Yager K, et al. Socio-ecological dimensions of Andean pastoral landscape change: bridging traditional ecological knowledge and satellite image analysis in Sajama National Park, Bolivia. Reg Environ Chang 2019;19:1353–69.

Yao J, et al. Accelerated dryland expansion regulates future variability in dryland gross primary production. Nat Commun 2020;11:1665.

Yapa KAS. Nurturing water: ancestral ground water recharging in the Americas. In: Proceedings of the 7th RWSN forum: Water for everyone, 29 Nov – 02 Dec 2016, Abidjan, Côte d'Ivoire. Papers & films, vol. 1; 2017.

Yongxun Z, et al. Values and conservation of Honghe Hani rice terraces system as a GIAHS site. J Resour Ecol 2016;7:197–204.

Zulkafli Z, et al. User-driven design of decision support systems for polycentric environmental resources management. Environ Model Software 2017;88:58–73.

Bremer L, Keeler B, Pascua P, Walker R, Sterling E. Nature-based solutions, sustainable development and equity. In: Nature-based solutions and water security: An action agenda for the 21st century. Amsterdam: Elsevier; 2021 [this volume].

Hribljan JA, Cooper DJ, Sueltenfuss J, Wolf EC, Heckman KA, Lilleskov EA, et al. Carbon storage and long-term rate of accumulation in high-altitude Andean peatlands of Bolivia. Mires Peat 2015;15:1–14.

Erickson C. The Lake Titicaca Basin: A pre-Columbian built landscape. In: Lentz D, editor. Imperfect balance: Landscape transformations in the Precolumbian Americas. New York: Columbia University Press; 2000. p. 311–56.

Kirch PV, Kelly M, editors. Prehistory and ecology in a windward Hawaiian valley: Halawa Valley, Molokai, Pacific Anthropological Records No. 24. Honolulu, Hawaii: Bernice Pauahi Bishop Museum; 1975.

CHAPTER 14

Nature-based solutions and corporate water stewardship

Naabia Ofosu-Amaah[a], Robin Abell[b], Jehanne Fabre[c], Paul Fleming[d,*], Michael Matosich[a], Jason Morrison[e,f], and Tara Varghese[g]

[a]The Nature Conservancy, Arlington, VA, United States, [b]Conservation International, Arlington, VA, United States, [c]Danone, Paris, France, [d]Microsoft, Redmond, WA, United States, [e]Pacific Institute, Oakland, CA, United States, [f]UN Global Compact CEO Water Mandate, New York, NY, United States, [g]PepsiCo, Purchase, NY, United States

[*]Contributor

Key messages

- By at least one estimate, the private sector is investing on the order of USD 1–1.5 trillion annually in water-related infrastructure, with institutional investors accounting for 30%–40% and corporations accounting for the remainder (Money, 2019). While there are no firm numbers for private sector investment in nature-based solutions (NBS) for water security (i.e., green infrastructure), figures for ecosystem-based adaptation spending to address water-related climate risks suggest a very small fraction of the total (Goldstein et al., 2019).

- Reasons for this limited investment are both internal (e.g., corporate risk aversion) and external (e.g., regulatory disincentives) but are largely surmountable.

- At the same time, NBS within the context of collective action—at the more mature end of the corporate water stewardship progression—will almost certainly be more complicated than NBS within or directly adjacent to an operational site, where the company is the main or sole actor.

- Leading edge private sector entities with the resources and willingness to invest in advancing NBS for water security can contribute to building much-needed demonstrations, evidence, and businesses cases, which can encourage NBS engagement by a new wave of companies and investors. More involvement, in turn, can contribute to taking NBS to scale.

- The public sector may always be the largest investor in both conventional and green water infrastructure, but companies can make distinct contributions, including by wielding influence to create enabling conditions for yet more NBS.

Nature-Based Solutions and Water Security. https://doi.org/10.1016/B978-0-12-819871-1.00014-2

Introduction

Water is a key driver of the global economy. Without water security, businesses would not be able to access raw materials and other inputs needed to produce goods and services for the world. Industry accounts for 19% of total global water withdrawals, and an additional 70% of global withdrawals come from agricultural supply chains (FAO, 2016). Industrial waste and agricultural runoff are also major drivers of water pollution. Therefore, companies have a significant role to play in ensuring water security. They also have a significant stake, as lack of water security has an impact on businesses' bottom line; in 2018, water-related financial losses (e.g., due to disruptions in supply) totaled USD38.5 billion (CDP, 2018). While challenges and barriers exist, harnessing the potential of nature-based solutions (NBS) for water security through corporate water stewardship presents opportunities for businesses to reduce their risks, create value, and contribute to moving NBS to scale.

Drivers and models of nature-based solutions in corporate water stewardship

Corporate water stewardship, water risks, and value creation

Water stewardship is defined as "the use of water that is socially and culturally equitable, environmentally sustainable and economically beneficial, achieved through a stakeholder-inclusive process that includes both site- and catchment-based actions" (AWS, 2019). While any entity can take a stewardship approach to their use of water, business adoption of stewardship principles is termed *corporate water stewardship* . Corporate water stewardship is aimed at ensuring that companies, as water users, can manage their own risks and seize water-related opportunities.

Water stewardship actions often take place along a progression, beginning inside the company's operations and then moving outside into the catchment (Fig. 14.1) (UNGC CEOWM, n.d.-a). The majority of companies sit on the "operations" section and develop measures to improve water efficiency and manage water pollution (Newborne and Dalton, 2019). Companies in leading industry sectors, such as food and beverage, mining and metals, and clothing and apparel, sit on the "context" section and beyond . Leading companies understand that improving water efficiency within their operations (e.g., the volume of water consumed per unit of production) is insufficient to improve their resilience against water risk. They need to understand the catchment context—water challenges and water users within the catchment—and take that into account when reducing their water consumption and improving their water pollution across their operations and supply chains (Newborne and Dalton, 2019; CDP, 2019).

The Water Stewardship Journey

A typical progression from beginning to mature corporate water stewardship practice

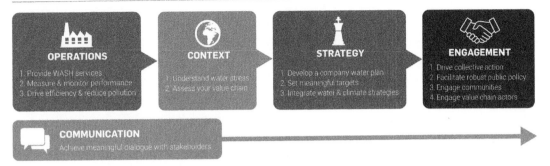

Fig. 14.1

The water stewardship journey. *Used with permission from the Pacific Institute/CEO Water Mandate.*

An increasing number of companies are implementing water stewardship activities and setting targets that can address water-related externalities and help ensure water security (UNGC CEOWM, n.d.-a). According to the CDP (2017), in 2017 companies committed US\$23.4 billion across more than 1000 projects to address water risks in 91 countries. And, as of 2021, over 190 companies had endorsed the CEO Water Mandate and in so doing committed to continuous improvement in their water stewardship practice (UNGC CEOWM, n.d.-b). At the same time, private sector water withdrawals have continued to increase and water quality has continued to decline, suggesting that corporate water stewardship has yet to fulfill its potential (CDP, 2018; Sarni and Share, 2019).

Companies are driven to take a stewardship approach due to the material nature of water to their business. This materiality varies across sectors (Table 14.1), and some sectors and subsectors (e.g., food and beverage) have invested substantially more in corporate water stewardship than others (e.g., retail) (CDP, 2018).

Cutting across materiality are water-related business risks, often broken down into physical, reputational, regulatory, and litigation (Table 14.2) (Ceres, 2012). All of these risks are exacerbated by increases in water insecurity.

Broadly speaking, mitigating reputational risks has been a prime motivator of many higher-profile corporate water stewardship commitments, investments, and activities, which have typically been undertaken as part of companies' corporate social responsibility (CSR) strategies (CI, 2018). As companies move increasingly from more siloed CSR to more integrated environmental social governance (ESG) as a vehicle for positive contribution, corporate water stewardship may become mainstreamed into company decisions (Thygesen, 2019). Shifting from mitigating water-related risks to business value

Table 14.1: Why water is material to different sectors.

Sector	Subsector	Materiality
	Biotech, Health Care, and Pharmaceuticals	Water plays a critical and varied role as a reagent, solvent and cleaning agent, and companies often require high-quality and ultra-pure water in the production and delivery of products. Monitoring water discharge quality is also a key consideration as companies must comply with regulations concerning the removal of pharmaceutical compounds from discharged water. The water-energy nexus is often explicit in this sector as manufacturing plants require cooling systems that can vary in their water and energy efficiency.
Consumer Staples	Apparel	Water is used for dyeing, bleaching, cooling, cleaning, and painting processes. Thousands of chemicals are used in these processes, and companies can face regulatory and reputational risks if wastewater is not properly treated. In addition, textile manufacturers rely heavily on water in the value chain for the irrigation of cotton and other natural fibers and for the irrigation of feedstock for cattle that produce leather. Finally, the use of products is an emerging risk for the sector, for example, as microplastics enter water bodies from the washing of synthetic materials.
	Food, Beverage, and Agriculture	The agricultural sector accounts for almost 70% of the world's water consumption for the growing of crops and animals. Water is also used in the washing, boiling, packaging, and transportation of agricultural commodities. Cleaning water can account for as much as 70% of overall water use in soft drinks manufacturing. Excessive or poor application of fertilizers and pesticides can lead to nitrate and phosphorus runoff, polluting waterways, and contaminating groundwater.
	Retail	Retail has a unique position to effect change in farming, production, manufacturing, and consumer behavior. While water use in direct operations may be limited to taps and toilets in stores, the production of food, textiles, and pharmaceuticals in the value chain is water-intensive and carries significant water pollution risks.
	Fossil Fuels	Water is involved in most parts of the value chain from construction to drilling, pumping, treatment, and cooling. For some production techniques, such as hydraulic fracturing and oil sands, large volumes of water are required to extract the resource. A major risk within the industry is water pollution, which can arise from spills, cross-aquifer mixing, drill muds and cuttings, disruption and pollution of surface water, and groundwater through water discharge, contamination by pump leakages, and oil spillages during transportation. The extraction of hydrocarbons also produces large volumes of water, which must be safely managed. Oil and gas refining is a water-intensive process, and there is also high water pollution potential if products are spilled. Retailers, midstream, and oil and gas services companies share water pollution and reputational risk with their clients.
	Hospitality	The hospitality sector has water-intensive supply chains through agricultural production of the food and beverages sold. These companies often have large real estate holdings that can be severely affected by water pollution and scarcity. For example, water consumption for hotel chains can be high. Water accounts for 10% of utility bills in many hotels.

Sector	Subsector	Materiality
Materials	Manufacturing	In the *chemicals subsector*, water is used for cooling purposes (90%) but also as a raw material, in cleaning, transport, as a solvent and as part of the final product. The sector faces the risk of spillage of hazardous chemicals, which can affect the quality of local water resources. Ultra-pure water is required to manufacture many items in the *electrical equipment subsector*, including semiconductors. Water is used primarily for cooling and cleaning in the *light manufacturing subsector*, and ultra-pure water is required for some processes. Furthermore, rinse water used in the production of aluminum, titanium and other composites produces high levels of hazardous wastes and must be precisely treated. The *transportation equipment subsector* covers auto manufacturing, which uses water for surface treatment and coating, paint spray booths, washing, rinsing, hosing, cooling, air-conditioning systems, and boilers. Wastewater then needs to be treated to high standards to meet environmental regulations. This subsector also has a water-intensive supply chain, as water is required to produce many auto components.
	Construction, Packaging, and Other Materials	For *wood products such as pulp and paper* , water is used to create the pulp and flush away unwanted impurities. Industrial effluent from mills can contain toxic and nonbiodegradable organic materials. Deforestation in service to this sector poses a substantial risk for water security worldwide. *Cement production* requires significant amounts of water for cooling heavy equipment and exhaust gases, in emission control systems such as wet scrubbers, as well as for preparing slurry in wet process kilns. Discharged water may be altered in terms of temperature, acidity or suspended solids with potential impacts for the discharge environment. Companies involved in *metal smelting, refining and forming* have a water-intensive supply chain (mining) and require large quantities of cooling water. Water is used in virtually all *ceramic* processes, and wastewater can contain minerals and some heavy metals that must be treated carefully.
	Mineral Extraction	In mining, water is used to extract the raw material from the ground, to extract the desired element from the raw material, in the transport and storage of excess slurry, and for other processes such as dust suppression, cooling, and employee requirements on-site. Water pollution is a significant risk; to access minerals below the water table, mines need to "dewater" and then safely dispose of this groundwater. Acid mine drainage and water pollution from tailings dams is a significant risk, even after the lifetime of a mine. Mining and mineral processing is also energy intensive, and water is often required to produce this energy.
Utilities	Infrastructure	The construction subsector is dependent on cement and other water-reliant construction materials. For utility networks, there is water-related risk through flooding, water access, and water pollution. Energy utility networks often have water-intensive supply chains (coal, natural gas, shale gas fracking) and a potential reputational risk through media interest.
	Power Generation	The power generation sector is heavily dependent on water for cooling, and, in the case of hydroelectric generation plants, for power generation itself. For this reason, thermal power plants are often located near water bodies, and companies rely on these resources for the success of their business. Water usage and risk exposure will depend on the power generation source, the cooling technology used by the power plant, and the location of the plant. Furthermore, power generation plants face reputational risk from the impact that their activities can have on local water resources. Reservoirs and dams can affect aquatic life and the hydrologic cycle.

Modified from CDP. Treading water: Corporate responses to rising water challenges. 2018. https://www.cdp.net/en/reports/downloads/4232. [Accessed 26 October 2019] and CI (Conservation International). Corporate water stewardship and the case for green infrastructure. 2018.

Table 14.2: Four types of corporate water-related risks.

Physical risks	Reputational risks	Regulatory risks	Litigation risks
Physical water risks are defined as current or predicted changes in water quantity (e.g., droughts or floods) or quality that may impact a company's direct operations, supply chains, and/or logistics. Physical water risks also include disruption of needed electric power due to water issues, as many electricity sources require water for cooling (e.g., nuclear or coal plants) or for generation (hydropower).	Reputational risks are defined as current or potential conflicts with the public regarding water issues that can damage a company's brand image or result in a loss of the company's license to operate in a certain community. Reputational risks are particularly common in developing countries where infrastructure and/ or regulation may not be sufficient to provide all users with access to safe and reliable drinking water supplies. The United Nations formally recognized access to safe water as a fundamental human right in 2010, and the human right to water is gaining visibility globally.	Regulatory risks are defined as the impacts of current and/or anticipated water-related regulations on a given company. As physical and reputational pressures increase, many local and national governments are responding with more stringent water policies. If unanticipated, these regulatory changes can prove costly to companies and, in some cases, limit industrial activities in particular geographies. The United States, EU, and China have all instituted stricter water-related regulations in recent years. As demand for and stress on water resources accelerate, this regulatory trend is likely to continue.	Litigation risks refer to the consequences of lawsuits or other legal actions related to the company's impacts on water levels and water quality. As water challenges continue to gain more attention and water-related physical, reputational, and regulatory risks increase, companies face increased litigation risks.

Modified from Ceres. Clearing the waters: A review of corporate water risk disclosure in SEC filings. 2012. https://www.comunicarseweb.com/ sites/default/files/biblioteca/pdf//1340395029_Clearing_the_Waters_Ceres_2012.pdf . [Accessed April 2021].

creation—as has been advocated by Sarni and Share (2019)—may leverage substantially more investment for water stewardship commitments, allowing for deeper engagement in collective action.

Risk mitigation through NBS: PepsiCo's experience in Cape Town, South Africa

Background
In 2018, all eyes were on Cape Town, South Africa, during the countdown to "Day Zero," when taps were projected to run dry in many areas across the city. PepsiCo's nearby manufacturing facility in Parow faced a real-world manifestation of the operations and supply chain risks that the company had been tracking. The top three vulnerabilities in a company-commissioned source water assessment included availability of water resources, drought, and competition over water resources. The 2018 water crisis spurred additional focus on new opportunities to employ the company's multipronged approach to mitigating water risks and improving water security, including supporting NBS.

Risk mitigation through NBS: PepsiCo's experience in Cape Town, South Africa—cont'd

Rationale

Within its water stewardship strategy, PepsiCo has a replenishment goal that aims to return an equivalent volume of water to what they use in their high water risk facilities to the same watershed from which extraction has occurred. As a primary reason to include NBS in their water risk response in Cape Town, PepsiCo determined that ecological infrastructure restoration offered a way to address a water supply issue—sustainably—at the source. In addition, ongoing studies led by The Nature Conservancy in the Greater Cape Town Region 's catchments (2023) were pointing to NBS increasing water supplies at one-tenth the cost as compared to other supply augmentation strategies. Collective action was needed, as the scale of the required landscape restoration was beyond what PepsiCo could do alone.

Details

PepsiCo joined the Greater Cape Town Water Fund as a founding member, and the company's investments are focused on recovering water losses in priority mountain catchment areas by removing water-intensive invasive species and restoring the native fynbos landscapes. Over two-thirds of the Greater Cape Town regio n' s source watersheds are covered with invasive alien plants, such as pines with deeper root systems that abstract more groundwater than the native fynbos species. From a water stewardship lens, PepsiCo's objectives were to improve the hydrologic conditions, conserve and restore native cover and associated ecosystem services, and improve water supply. The water benefits of the company-supported activities were measured by examining the evapotranspiration before and after the interventions were implemented within the Olifants and Du Toits catchments of the Wemmershoek Dam. The average evapotranspiration rate for Pinus panister (cluster pine) is 825 mm/year, as opposed to mountain fynbos' evapotranspiration rate of 627 mm/year. Streamflow monitoring stations are being set up in a control catchment and the project's targeted catchments. In addition to the water security improvements, these NBS interventions also had positive impacts in climate change resilience, fire risk reduction, job creation, and community empowerment, particularly for women who received specialized skills training.

Challenges

PepsiCo evaluated potential barriers that might impact the sustainability of the Greater Cape Town Water Fund's interventions. For example, would the NBS activities be protected under any governance mechanism, or would landscapes be at risk of being degraded once again? Complementary policy and management changes, requiring engaged watershed stakeholders and decision-makers, can go a long way in protecting the sustainability of these NBS solutions. PepsiCo found reassurance that both the City of Cape Town and the Western Cape Government sat on the Water Fund's Steering Committee and would be engaged in all planning to ensure the enduring benefits and impacts of these interventions. Gaining internal buy-in took some time but was aided by internal capability development to educate PepsiCo's cross-functional teams on NBS and how landscape restoration in upstream catchments would benefit manufacturing facilities in the long run.

Lessons learned

PepsiCo's participation in the Greater Cape Town Water Fund has demonstrated how the company's water stewardship investments can also promote resilience for ecosystems, communities, and businesses. Cape Town may have held off Day Zero in 2018, but the city—and many others around the world—continue to face the threat of their next water crisis. At PepsiCo, NBS has helped change the conversation from how to mitigate water as a risk to how to treat water as a value.

Incorporation of NBS into corporate water stewardship

Corporate water stewardship has often prioritized reducing water use via innovations and improved efficiencies—activities that tend to cluster on the operational end of the water stewardship progression — and have relied less on NBS. But, companies are increasingly considering on-site NBS for benefits that go beyond reduced water use; for instance, a company may construct wetlands to help filter pollutants from water before discharging it into downstream water bodies. There are fewer documented examples of corporate investment in NBS on the "mature" end of the progression, in part because companies are only beginning to enter that stage. The engagement stage can include direct or indirect investments in NBS. Direct investments in larger-scale NBS can be made within companies' operational boundaries (see Dow case study in this chapter), in the broader catchments from which they source (see Danone and PepsiCo case studies in this chapter), or in their value chains (see example from Heineken, 2019 in the references). Indirect investments can be made via helping to build enabling conditions for NBS, such as by advocating for sustainable NBS financing policy or contributing to building the scientific case for NBS as a substitute or complement to conventional built solutions (CI, 2018).

The relevance of NBS for a company's corporate water stewardship strategy varies for different sectors depending on their risks, dependencies, and opportunities (Table 14.3). For instance, sectors that require high-quality water inputs may invest in upstream NBS designed to reduce nutrient or sediment pollution in water sources. Sectors with infrastructure at risk from flooding may invest in upstream NBS designed to promote soil water infiltration. Sectors that produce effluents requiring treatment may invest in NBS to enhance filtration. And those whose operations create reputational risks through real or perceived impacts on water security—which could include any of the previously mentioned—may tailor an NBS investment principally to address the needs and interests of catchment stakeholders.

In addition to the potential benefits that NBS can provide to different sectors, companies can also choose to invest in NBS through provision of financing to the direct users/beneficiaries across their value chains and in other high opportunity areas in the form of debt or equity instruments that seek a risk-adjusted return (CI, 2018).

Available case studies demonstrate how NBS can be applied by companies for a range of purposes (CI, 2018; Dow et al., 2013; Heineken, 2019; TNC, 2019a; WBCSD, 2015; WWAP, 2018) (Table 14.4).

Range of implementation models

Companies pursue a range of implementation models once an NBS solution is identified and depending on where the NBS will be implemented and the stakeholders involved. Moving from operations to collective action requires engaging with increasing number of partners and stakeholders (Table 14.5).

Table 14.3: Relevance of NBS for water for select sectors.

Sector/subsector	Relevance of NBS for water
Forestry, packaging, and paper	Forestry operations in particular can require conversion of natural areas into planted systems, often using nonnative species. At the same time, the health of forest plantations depends on broader hydrological functioning of the surrounding landscape. Common forest product certification programs include requirements around maintaining areas of high conservation value, including those supplying critical ecosystem services like flood regulation and water purification.
Materials—metals and mining, construction materials	Investments in NBS can in certain contexts ameliorate water quantity and quality issues for this sector. NBS can also complement gray infrastructure solutions aimed both at improving operational water use inside the factory and at reducing downstream pollution.
Consumer staples	As a consumer-facing sector with often-broad supply chain footprints, investments in NBS can help address reputational risks through collaboration with stakeholders to improve water access and achieve broader sustainability goals.
Utilities	Protection and restoration of natural areas can help prevent soil erosion and resulting sedimentation and can better regulate nutrient loads and downstream water flows, which is critical for water utilities, hydropower companies, and thermoelectric plants. In the case of cloud forests, natural areas even capture rainfall; riparian zones can regulate water temperature; and in some regions, NBS can reduce the risk of catastrophic fires, which can lead to landslides and massive sedimentation.

Modified from CI (Conservation International). Corporate water stewardship and the case for green infrastructure. 2018.

Table 14.4: Selected examples of NBS applied for water quantity and quality purposes.

Adequate Supply	• Revegating and monitoring to improve soil water infiltration • Engaging watershed stakeholders to promote source water protection • Employing strategic planting for reduced irrigation and maintenance needs
Acceptable Quality	• Constructing wetlands and treatment channels for wastewater treatment • Leveraging multifunctional field margins • Treating produced water (water that comes out of the well with crude oil during crude oil production) using reed beds • Employing phytoremediation for groundwater decontamination • Reusing decommissioned sites through wetland restoration or other NBS • Using buffer zones to mitigate contaminants • Implementing best land management practices to reduce erosion and sediments reaching waterways • Engaging watershed stakeholders to promote source water protection
Managing Runoff	• Managing stormwater runoff with wetlands, bioswales, bioretention ponds, rain gardens, green roofs, and other natural solutions • Revegetating and monitoring to improve soil water infiltration

Table 14.5: Corporate implementation models for NBS.

Implementation model	Details	Area of influence	Example
Company-Led Implementation A round Operations	A company assigns internal experts to design and implement on-site solutions, using company resources and publicly available tools	• Immediately within or adjacent to operations	• A beverage company hires an internal manager to reduce water consumption of a facility while maintaining expected output and install a treatment wetland on-site .
External Partnerships for On-Site Solutions, Community Engagement, and Value Chain Engagement	A company fills potential information or technical gaps through developing external partnerships to provide expertise, share best practices, and streamline project design and implementation. External partnerships can also increase the transparency and impact of the work by providing third-party verification and amplifying the story within their networks.	• Immediately within or adjacent to operations • In discrete locations in catchments of interest to the company across and beyond the value chain	• A pharmaceutical company hires a third-party verification company to calculate the company's replenishment activity to ensure calculations accurately reflect replenishment standards and procedures without biases.
Collective Action	A company partners with government, businesses, communities, NGOs, and others with shared water challenges to address these challenges in a more effective, efficient, and holistic manner.	• In discrete locations in catchments of interest to the company across and beyond the value chain • At the scale of a catchment	• A food company encourages other stakeholders in the water basin (farmers, water utility, city government, etc.) to join them in participating in a water fund—a collective action mechanism to achieve the common goal of water security (TNC, 2019b).

Modified from CI (Conservation International). Corporate water stewardship and the case for green infrastructure; 2018 ; TNC (The Nature Conservancy), Business Council Natural Infrastructure Working Group. Strategies for operationalizing nature-based solutions in the private sector. 2019. https://www.nature.org/content/dam/tnc/nature/en/documents/NBSWhitePaper.pdf. [Accessed 25 October 2019] ; and UNGC CEOWM (UN Global Compact CEO Water Mandate). Water stewardship 101: The basics; n.d.-a. https://ceowatermandate.org/ university/101-the-basics/lessons/the-water-stewardship-journey/. [Accessed 26 October 2019] .

Companies may choose one or more implementation models depending on the water challenges they are facing. As critical stakeholders within watersheds, companies have a vested interest in managing and protecting water resources. However, companies must understand and recognize the leading role that governments and local authorities must play (UNGC CEOWM, n.d.-b) and also the shared nature of water challenges that require shared solutions to achieve scale. NBS to achieve certain water outcomes and cobenefits will be most effective when implemented at a large (catchment-level) scale (CI, 2018), so coordination with other actors is likely ideal, although transaction costs in terms of coordination and negotiation may be higher.

This collective action with others "allows organizations with diverse sets of expertise, knowledge, technology, and financial connections to collaborate on water issues that affect them all (such as collective impacts they might have)" (UNGC CEOWM, n.d.-a). Collective action can provide:

- an expanded pool of expertise, capacity, or financial resources focused on fostering change;
- more durable outcomes with strong support from the engaged parties;
- establishment and maintenance of credibility and legitimacy with key interested parties, resulting in improved legal and social license to operate; and
- stronger, more sustainable water governance by engaging multiple stakeholders (UNGC CEOWM, n.d.-a).

Companies can finance collective action or other implementation models using available private sector financing mechanisms for NBS, including private grants/donations, direct company spending, institutional investing, payments for watershed services and/or market mechanisms, or insurance (CI, 2018).

NBS and business decision making

Businesses can decide to use NBS for a range of reasons, but there are some elements that might make companies more likely to select an NBS option. Having a company culture that supports environmental solutions—including elements such as Board pressure, CEO commitments, leadership at multiple levels within the company, and incorporation of water issues into corporate governance and other corporate policies—can contribute to the facilitation of businesses selecting NBS (TNC, 2019a). Operationalizing NBS into company decision making can include:

- organizing like-minded individuals into resource groups where they can share knowledge and experiences and proactively educate company stakeholders through storytelling;
- securing top-level leadership buy-in on NBS;
- engaging employees across all levels of the organization;
- building NBS into standard design;
- increasing the payback period for projects to more accurately capture lifetime project costs; and
- developing NBS projects that complement gray infrastructure to move from an "either/or" to a "both" approach (TNC, 2019a).

Steps and timeframes for operationalizing NBS within a company have been recommended (Fig. 14.2) (TNC, 2019a). A key conclusion is that NBS is not a "quick fix" and can take

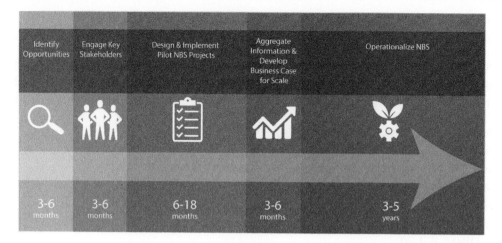

Fig. 14.2
Operationalizing NBS within a company. *From TNC (The Nature Conservancy), Business Council Natural Infrastructure Working Group. Strategies for operationalizing nature-based solutions in the private sector. 2019. https://www.nature.org/content/dam/tnc/nature/en/documents/NBSWhitePaper.pdf. [Accessed 25 October 2019].*

4–7 years in total, from conceptualization through to operationalization. However, NBS should, if designed and implemented effectively, produce enduring, resilient outcomes.

The case study from Danone's experience in Pasuruan, Indonesia (see the following case study), outlines key considerations in an external partnership that is driving for collective action.

Watershed collective action: Danone's experience in Pasuruan, Indonesia

Background
Danone's second largest bottled water facility in Indonesia is in Pasuruan, East Java. As a result of land use change and climate-related flow variations, flow from the natural spring that feeds the city had fallen by more than 20% between 2007 and 2019. The 62,773 ha Rejoso watershed, where the spring originates, has a combination of upstream horticulture, mid-catchment agroforestry, and downstream rice fields. Without intervention, studies predicted zero water discharge from the spring by 2040 (Universite De Montpellier, Universitas Gajah Mada and Danone Waters, 2019). Physical and potential reputational risks drove Danone to take action.

Rationale
Danone was facing a complex landscape, with upstream deforestation, increased industrial activities mid-catchment, and downstream "thirsty" rice production and other land uses. Collective action appeared to be the only pathway for achieving water security. Consequently, Danone aimed to develop an integrated watershed scale approach using NBS not only to improve water security outcomes (SDG 6) but also to provide additional cobenefits, such as carbon capture and sequestration (SDG 13), increased standards of living (SDG 1), improvements in gender equality (SDG 5), and social capital among local communities (SDG 16).

Watershed collective action: Danone's experience in Pasuruan, Indonesia—cont'd

Danone's 20 years of experience in Evian, Volvic (France), and Villavicencio (Argentina), where NBS generated multiple benefits, led water stewardship to become instrumental and fully integrated into Danone's approach to its business. The Danone Ecosystem Fund (DEF) provided seed funding for hydrogeological and socio-economic studies to underpin a robust, science-based business case and build-up a governance structure able to sustain the project in the long run.

Through a partnership between the University of Montpellier and the University Gadja Mada in Indonesia, PhD students developed a comprehensive water cycle study, which led to recommendations for the most impactful actions for improving water security. Socio-economic studies aimed to understand the local context of Pasuruan households' livelihoods, the value chain of the ingredients grown in the watershed, and strategies for increasing household incomes. Through the combined set of studies, Danone targeted actions to address both water insecurity and improvement of local livelihoods.

Details

Danone, DEF, and World Agroforestry Centre (ICRAF) have joined together with public authorities to drive collective action and strengthen water governance in the Rejoso watershed. This governance aims to incentivize the change of practices within the Rejoso landscape through a new decree signed by the local authorities and all the partners that could provide funding to farmers per liter of water saved.

Maintaining and rehabilitating tree-based farming systems in the upstream and mid-catchment of the Rejoso watershed is predicted to decrease runoff by 1.5%–4% and increase water infiltration by 9%–23% (Leimona et al., 2018). Tree planting in horticultural and agroforestry areas is expected to sequester about 42.8 tons of CO_2 per hectare, or about 677,777 tons of CO_2 annually. Water and soil conservation will increase soil health and smallholders' farming productivity, which is predicted to increase farmers' income by up to 40% from horticulture and 15% from agroforestry. In the lower catchment, Danone is replacing conventional paddy rice cultivation practices, with the improved System of Rice Intensification (SRI) method, which maintains soil moisture without flooding, keeping it well-drained and aerated to facilitate aerobic conditions, thereby reducing the formation of methane. A 1 ha paddy cultivation plot abates an estimated 0.9 tons of CO_2 equivalent per year and can reduce water dependency by 25%. ICRAF will be monitoring carbon and water outcomes via soil sampling and the use of water meters.

Challenges

A key challenge is wide adoption by farmers. Appropriate support is required to mitigate risk during the transition phase from old to new approaches, as farmers not only need to learn new technical skills but also gain adequate experience to master more sustainable, water-friendly practices. In picking these NBS, Danone also faced the internal challenge of managing expectations among company leadership about the time frame over which impacts would be observed.

Lessons learned

NBS allows a company like Danone to account for and achieve multiple benefits simultaneously. For companies with multiple environmental targets, NBS can avoid having to choose among, for instance, a water, carbon, or regenerative agriculture project. This integrated approach allows CAPEX and OPEX project decisions to combine the multiple environmental benefits of an investment, which de-siloes the decision-making process. Tracking these impacts will be key to ensuring that the company contributes to Sustainable Development Goals (SDGs) and to accelerating the transformation needed for the wide adoption of this approach through many projects at the landscape level.

Corporate adoption of NBS : Opportunities and challenges
Opportunities that NBS present to businesses

The business case for adopting NBS includes both monetary and nonmonetary arguments. On the monetary side, NBS can be less expensive than conventional built alternatives, with potential savings in terms of capital, operating, labor, and maintenance costs (American Rivers, 2012; Waage, 2016). The value of maintained NBS may also increase over time, as compared to depreciation of built infrastructure (Scott et al., 2018).

On the side of typically nonmonetized values, a host of cobenefits can include contributions to a company's climate change mitigation and other SDGs (e.g., biodiversity); health and wellbeing benefits to its employees, their families, and the broader community; increased lifespan and resiliency of operational infrastructure to future climate change impacts; and reduced risk of paying fees when new regulations related to private property are imposed in the future (Abell et al., 2017; Scott et al., 2018 ; Brill et al., 2021).

As previously noted, there are also reputational benefits from corporate NBS involvement, not only with local stakeholders (including government) but also with investors and consumers. The climate change mitigation benefits of some NBS are especially noteworthy, given evidence that companies investing in mitigation are outperforming those that are not (Duetsche Bank, 2019). According to a Bank of America survey, climate change risk is the top-ranked ESG concern for ESG asset managers, and there is broad agreement among those managers that ESG is the best measure at present for signaling future earnings risk (Reinicke, 2019). A Morgan Stanley Institute for Sustainable Investing report (2019) confirms that finding, noting that 86% of individual investors surveyed believed that corporate ESG practices can potentially lead to higher profitability and may be better long-term investments. NBS that are designed to generate both water and climate mitigation benefits, and are communicated as such, may therefore generate even greater business value than investments in water NBS alone.

Calculations of returns on investment (ROI) of large-scale NBS have typically focused on public sector investments (e.g., public water utilities) (Feltran-Barbieri et al., 2018; Kroeger et al., 2017), but new examples involving private sector actors are coming online (TNC, 2015). Importantly, ROI increases when the values of cobenefits (e.g., carbon) are stacked on top of water benefits (Abell et al., 2017).

The value of a wetland: The case of the Dow Chemical Company

Background
In 1995–96, the Dow Chemical Company piloted its first green infrastructure project to expand water treatment capacity at its Seadrift manufacturing site in Texas. Almost 110 acres of an existing water treatment pond was converted into constructed wetlands for naturally treating suspended solids. Investing in green infrastructure allowed Dow to ensure long-term compliance with the Clean Water Act Guidelines and to meet discharge permits.

The value of a wetland: The case of the Dow Chemical Company—cont'd

Rationale

Dow saw potential in investing in NBS projects to address long-term watershed management and water risk at their most water-stressed manufacturing sites. Further, this project fell under Dow's corporate water resiliency program and the company's 2025 Sustainability Goal "Valuing Nature," focused on incorporating nature into business decisions to improve the health of watersheds. A replacement cost methodology for financial assessment was used to determine the cost-effectiveness of this project. Successful implementation of the project facilitated buy-in from upper management and made a business case for future green versus gray infrastructure projects.

Details

Dow partnered with The Nature Conservancy to assess the constructed wetland project for its delivered ecosystem services associated with water and biodiversity values. Dow used the Ecosystem Services Identification and Inventory (ESII)[a] tool to evaluate benefits and to incorporate them into business and site decisions to contribute to business resilience.

As a result of the adoption of the green versus gray infrastructure, Dow saved $282 million in net present value over the expected lifetime of the project. Cost reduction was achieved through lower capital, operational, and maintenance costs, reduced electricity costs, and long-term compliance with regulations. There were also additional benefits such as the provision of wildlife habitat and recreational opportunities.

Challenges

The major challenge that Dow faced when implementing its first green infrastructure project was internal cultural change associated with the need to shift conventional engineers' thinking about green versus gray infrastructure. Also, when planning new green infrastructure projects, it was challenging to include the future probability of water price into proposed projects given the uncertainty of future water availability and associated water risks.

Lessons learned

Based on the success of this project, Dow now has dedicated staff resources and is actively developing tools to evaluate each project as an opportunity to deploy natural infrastructure solutions at sites around the globe. Currently, at Dow, NBS projects are screened and evaluated alongside other projects proposed by the Technology Leadership Group and the Environment Technology Center. Investment recommendations are made based on Dow's business strategy needs to address water risk and secure a social license to operate. Through peer-to-peer learning, other manufacturing sites have started implementing NBS projects.

---

[a] https://www.esiitool.com/.

Challenges and barriers preventing business adoption of NBS

Corporate decision-makers responsible for addressing water-related risks, including those related to climate change, may find it difficult to make the case for NBS investments for a number of reasons, many grounded in uncertainty (CI, 2018; Forest Trends, 2016; Goldstein et al., 2019; Sarabi et al., 2019; Soz Saiman et al., 2016). Potential barriers include:

- Lack of an empirical evidence base on the performance of NBS, specifically, but not only, related to reductions in water-related risk.

- Few business cases for investment that can serve as templates.
- Lack of demonstrations on the ground.
- Limited corporate expertise and local capacity as it relates to designing NBS and quantifying its cobenefits, especially for smaller enterprises.
- Mismatch between potentially longer time frames to achieve water-related outcomes and short-term funding.
- Larger space requirements as compared to gray infrastructure.

For these and other reasons, business managers may see NBS projects as too risky, given uncertainties associated with them as compared to better-known gray infrastructure solutions (Waage, 2016). This also speaks to a general lack of familiarity with NBS, both within the private sector and among the stakeholders whose engagement is critical, especially where collective action is key (Scott et al., 2018). Relatedly, policies and regulatory frameworks may encourage investment in gray infrastructure and hinder collective action (CI, 2018; Soz Saiman et al., 2016). Barriers to collective action, in turn, can make the monetization of cobenefits—which may be necessary to justify investment—even more challenging (CI, 2018). Even without institutionalized barriers, collective action across a watershed is challenging. Multiple geographic and sectoral jurisdictions, as well as a diversity of public and private actors, are likely to be involved in land and water management (UN Environment et al., 2018), and turnover of decision-makers and a fragmented policy landscape can make long-term planning and investment difficult (CI, 2018).

Companies may also underestimate the water-related risks that they face due to climate change (Goldstein et al., 2019), and as a result, fail to take necessary action that could include NBS. Goldstein et al. (2019), analyzing disclosures to CDP, found that only 3.3% of companies engaged in ecosystem-based adaptation (EbA) for all climate-related risks, whereas 30% invested in gray water infrastructure. The authors attributed a business-as-usual approach to addressing risk as a core barrier to greater NBS investment, which requires longer-term thinking and a level of comfort with uncertainty, though some EbA approaches will be low or no-regret. Among other reasons for low investment in EbA, the authors noted a potential disincentive among companies for investing in public goods with benefits for multiple actors.

More generally, despite the overall favorability conferred on companies engaging in ESG, companies may be inhibited from improving ESG performance (e.g., through NBS) due to investors' inability to evaluate ESG quality (Mayor, 2019). Many investors also continue to perceive trade-offs between sustainability and profitability and asset managers are working to better educate, inform , and support current and would-be sustainable investors (Morgan Stanley Institute for Sustainable Investing, 2019).

Spotlight: Benefit accounting of nature-based solutions for watersheds initiative

Benefit Accounting of Nature-Based Solutions for Watersheds is a multistakeholder initiative including the CEO Water Mandate, Pacific Institute, The Nature Conservancy, Danone, and LimnoTech. This initiative aims to develop a standardized guide and method to account for the stacked water, carbon and biodiversity benefits, and identify wider socio-economic cobenefits of NBS for watersheds. This method will be adapted into an online tool, where users will be able to select multiple entry points for inputting data (such as habitat type, interventions, activities, benefits, etc.) to receive suitable outcomes. Notably, benefits can be identified and accounted for by category (e.g., water, carbon, biodiversity, etc.) or by sector (e.g., environmental, social, or economic) based on the needs of the user. The dynamic tool will be highly intuitive and user friendly and will also inform the user of additional tools, methods, and approaches (including indicators and calculation methods) to further quantify the benefits identified. Both the method and tool are designed to align with existing tools and resources for NBS benefit accounting and related water stewardship initiatives, and will incorporate lessons learned from NBS case studies globally.

This initiative will have multiple stages, considering benefit identification, estimation, and valuation, based on the needs of the primary and secondary audiences. Ultimately, this initiative will demonstrate how identifying and measuring stacked benefits can quantitatively and qualitatively build the business case and show investment potential of NBS, as well as build awareness of the value of NBS.

For more information on this initiative, please visit www.ceowatermandate.org/nbs.

The way forward for NBS and corporate water stewardship
Making NBS more accessible to business

There are ways to lower the barrier to entry for corporate investment in NBS. Many involve creating enabling conditions through policy adjustments or financing innovations. Scott et al. (2018) list incentives and methods specific to urban environments in the United States, some of which will be transferrable to other countries and settings (Table 14.6).

Spotlight: Water Resilience Coalition

The Water Resilience Coalition (WRC) is spearhead ed by the UN Global Compact's CEO Water Mandate. It was established in 2020 with the mission to accelerate progress against the global water crisis in stressed geographies, powered by leading corporations with involvement of their CEOs. WRC will ground its work in building the resilience of water-stressed basins around the world—prioritizing basins that pose the greatest risk to local communities and economies. Through collaboration around investments in infrastructure, innovation, and policy advocacy, and by engaging supply chain partners and nongovernmental organizations (NGOs), businesses can help lead the way on water basin resilience.

Continued

Companies that join WRC sign a pledge to make the needed investments in their own operations and to work together through collective action in prioritized basins to accomplish three commitments by 2050:

(1) Achieve Net Positive Water Impact.
(2) Support the development of a Water Resilient Value Chain.
(3) Raise the ambition of water resilience through public and corporate outreach via its CEO and senior leadership.

WRC is connected to the UN Global Compact SDG Ambition for water that states that "businesses that have not established goals in line with achieving net positive water operations in water-stressed basins by 2050 as outlined above would fall below the SDG Ambition Benchmark."
For more information go to https://ceowatermandate.org/resilience/.

Table 14.6: Incentives for making NBS more accessible to business.

Development Incentives	Typically offered to developers during the process of applying for development permits, this incentive can take a number of forms, including expedited permitting, reduced permit fee or an increase in floor area ratio.
Building Codes	A number of city governments have adopted building codes to become more climate resilient and regulate their natural resources more effectively. Well-defined and universally applied codes provide clarity to developers on the types of interventions that are required and ultimately result in lower emissions, greater adaptive potential, and improved water resource management.
Property Assessed Clean Energy (PACE)	PACE is a financing mechanism used in the United States wherein a municipality issues revenue bonds to residential, commercial, or industrial property owners to finance green infrastructure and associated renewable energy installations, energy efficiency retrofits, or stormwater retrofits on their properties. PACE borrowers can benefit from new green infrastructure installations immediately and repay their debt over time through a set line item on their property tax bill.
Credit Enhancement	Credit enhancement is offered to improve the terms of private financing by reducing perceived financial risk to investors. City and national governments can extend loans for green infrastructure projects to accelerate private sector investment in green infrastructure, reducing initial costs. When designed appropriately, this mechanism is a cost-effective measure for the government, reducing public service expenditures by sharing the initial capital costs with the private sector and reducing longer-term O&M costs when compared to gray infrastructure. Similarly, private stakeholders and communities operating stormwater utilities may incentivize the installation of green infrastructure practices (e.g., reducing impervious surfaces, retaining stormwater) on private property through stormwater fee discounts and credits to property owners.
Payment for Ecosystem Services (PES)	PES is a financing mechanism that can support the conservation and expansion of ecosystems and green infrastructure implementation. PES is a transaction between economic actors who enhance and conserve ecosystem services, such as farmers or natural resource owners located in upstream areas of a watershed, and direct beneficiaries of the improvements in ecosystem services, such as a water company, government, donor agency, or NGO providing services in downstream areas of the watershed. In developing countries, PES is mostly financed by the public sector and donor agencies. Among private sector buyers of PES, the highest amounts recovered are from water utilities and food and beverage companies. The PES mechanism aims to: (1) increase efficiency of NBS and sustainable management of water and natural resources, agricultural land, biodiversity conservation, and carbon sequestration; and (2) enhance capacity building of farmers and natural resource owners (i.e., providing training and technical assistance), promote behavioral change and transformational change (i.e., positive incentives through PES rather than coercion), and increase community resilience and adaptive capacity.

Table 14.6: Incentives for making NBS more accessible to business—cont'd

Tax Increment Financing (TIF)	TIF is a method of financing green infrastructure initiatives based on the anticipated property tax increase that can be generated by the green infrastructure solutio n ' s implementation. The revenue generated by TIF is the property tax assessed on the increase in property value following the green infrastructure implementation, compared to the baseline property value prior to this green infrastructure development. The property value increases can be driven by the green infrastructure's effectiveness in mitigating flooding or stormwater runoff, or improving urban aesthetics or environmental health (US. PIRG Education Fund, 2011). TIF can be a highly valuable option for a local government since it allows for financing green infrastructure without raising property tax rates or exceeding municipal debt limits (Georgetown Climate Center, 2015).
Green B onds	Green bonds are mechanisms by which governments, corporations, state-owned utilities and multilateral development banks can fund green infrastructure projects linked to climate resilience or mitigation. This bond enables borrowers to access lower interest rates and incentives such as tax deferrals to help offset some of the uncertainty associated with quantifying and allocating monetary benefits. Green bonds have been one of the fastest growing sectors on the bond market, with nearly $161 billion in green bonds sold globally by 2017. Some limited examples of green bonds for green infrastructure implementation and climate adaptation include improvement of water-related infrastructure and transport, forest restoration in China, coral reef rehabilitation in Indonesia, and sustainable management of forests in Mexico.
Stormwater F ee Di scount	This incentive requires that a stormwater management fee is in place and is being collected. It functions by offering a discount on the fee as property owners reduce the volume of runoff. Depending on a municipality's goals, the discount can incentivize various green infrastructure interventions, including reducing impervious areas, increasing infiltration, increasing the number of buildings with green roofs, or increasing the practice of rainwater harvesting.
Grants	Municipalities can provide direct funding to property owners or community groups to stimulate green infrastructure buy-in and implementation. In the developing world, a grant program may or may not be feasible, but a donor could provide a grant pool to incentivize green infrastructure as part of a larger project.
Rebates and F inancing	This incentive requires that participants have the upfront capital necessary to fund the green infrastructure intervention directly; they then receive support from the government through low-interest loans, tax credits, or reimbursements. In developing countries, rebates would likely be best targeted at businesses, as residents may not have the capital to fund green infrastructure improvements without assistance.
Water F unds	Water funds provide an opportunity for private investors and companies to invest in green infrastructure in exchange for the product they receive: clean water. The fund, in turn, pays for watershed protection and water quality restoration activities.

Modified from Scott O, Fatoric S, Merritt R, Fulton Lee V, Miley D. Engaging the private sector in green infrastructure development and financing: A pathway toward building urban climate resilience. USAID; 2018. https://www.chemonics.com/wp-content/uploads/2019/06/ Engaging-the-Private-Sector-in-Green-Infrastructure.pdf. [Accessed 20 November 2019].

Opportunities for businesses to advance NBS for water security

There are ample opportunities for corporations to contribute to growing the science, practice, and scope of NBS for water security, be it to reduce water-related risks and associated costs, to create business value, or to make a positive contribution to society (Money, 2019). These opportunities range from creating enabling conditions to investing in on-the-ground implementation at scale (CI, 2018), and include the following:

Filling information and evidence gaps
- Investing in demonstration projects and designing them in ways that test uncertainties and collect clear metrics to build the evidence base and outline a path for other companies to follow.
- Developing and disseminating additional business cases that consider and demonstrate the full range of benefits that can accrue from NBS investment.
- Contributing technical expertise to new or existing NBS projects, especially (but not only) where they are being integrated with conventional gray infrastructure to deliver targeted water-related services.

Influencing the policy landscape
- Advocating for stronger watershed governance and policies that allow for and/or promote NBS and watershed investment.

Investing in NBS on the ground, at scale
- Expanding existing gray WASH initiatives to include watershed protection and restoration as complementary, nature-based WASH solutions for local communities.
- Making water stewardship commitments that involve collective action at the watershed scale and dedicating commensurate budgets to those commitments.
- Convening other private sector actors and key stakeholders in specific geographies to develop shared NBS investments at the watershed scale, including through the seeding of green infrastructure funds.
- Initiating or joining a regional or basin-scale water fund.
- Participating in payment for watershed services programs.

Providing sector leadership
- Pushing for sector associations to elevate the expectations for member performance, taking water stewardship beyond operational fence lines.
- Encouraging companies along their supply chains to support NBS investment.

Advancing innovative sustainable financing for green infrastructure
- Issuing or investing in a bond for NBS.
- Designing and/or purchasing insurance for NBS.

References

Abell R, et al. Beyond the source: The environmental, economic and community benefits of source water protection. Arlington, VA: The Nature Conservancy; 2017. https://www.nature.org/content/dam/tnc/nature/en/documents/Beyond_The_Source_Full_Report_FinalV4.pdf. [Accessed 20 November 2019].

American Rivers. Banking on green: A look at how green infrastructure can save municipalities money and provide economic benefits community-wide, https://www.asla.org/uploadedFiles/CMS/Government_Affairs/Federal_Government_Affairs/Banking%20on%20Green%20HighRes.pdf; 2012. [Accessed 20 November 2019].

AWS (Alliance for Water Stewardship). Definition of water stewardship, https://a4ws.org/about/; 2019. [Accessed 24 October 2019].

Brill G, Shiao T, Kammeyer C, Diringer S, Vigerstol K, Ofosu-Amaah N. Benefit accounting of nature-based solutions for watersheds. Oakland, CA: Guide United Nations CEO Water Mandate and Pacific Institute; 2021.

CDP. A turning tide: Tracking corporate action on water security, https://6fefcbb86e61af1b2fc4-c70d8ead6ced55 0b4d987d7c03fcdd1d.ssl.cf3.rackcdn.com/cms/reports/documents/000/002/824/original/CDP-Global-Water-Report-2017.pdf?1512469118; 2017. [Accessed 19 November 2019].

CDP. Treading water: Corporate responses to rising water challenges, https://www.cdp.net/en/reports/downloads/4232; 2018. [Accessed 26 October 2019].

CDP. Cleaning up their act: are companies responding to the risks and opportunities posed by water pollution?; 2019. https://www.cdp.net/en/reports/downloads/5165 [Accessed 26 April 2021].

Ceres. Clearing the waters: A review of corporate water risk disclosure in SEC filings, https://www.comunicarseweb.com/sites/default/files/biblioteca/pdf//1340395029_Clearing_the_Waters_Ceres_2012.pdf; 2012. [Accessed April 2021].

CI (Conservation International). Corporate water stewardship and the case for green infrastructure; 2018.

Dow, SwissRe, Shell, Unilever, The Nature Conservancy. The case for green Infrastructure: Joint-industry white paper, https://www.nature.org/content/dam/tnc/nature/en/documents/the-case-for-green-infrastructure.pdf; 2013. [Accessed 24 October 2019].

Duetsche Bank. Climate change and corporates: Past the tipping point with customers and stockmarkets, https://www.dbresearch.com/PROD/RPS_EN-PROD/PROD0000000000500285/Climate_change_and_corporates%3A_Past_the_tipping_po.pdf; 2019. [Accessed 20 November 2019].

FAO (Food and Agriculture Organization of the United States—AQUASTAT). Water withdrawal by sector, around; 2016. p. 2010. http://www.fao.org/nr/water/aquastat/tables/WorldData-Withdrawal_eng.pdf. [Accessed 26 October 2019].

Feltran-Barbieri R, Matsumoto M, Ozment S, Hamel P, Gray E, Lucchesi Mansur H, Piazzetta Valente T, Baladelli Ribeiro J. Infraestrutura Natural para Água no Sistema Guandu. Rio de Janeiro: WRI; 2018. https://wribrasil.org.br/sites/default/files/InfraestruturaNaturalGuanduRJ.pdf. [Accessed 20 November 2019].

Forest Trends. Alliances for green infrastructure: State of watershed investment 2016, https://www.forest-trends.org/wp-content/uploads/2017/03/2016SOWIReport121416.pdf; 2016. [Accessed 20 November 2019].

Georgetown Climate Center. Green infrastructure. Washington, DC https://www.georgetownclimate.org/adaptation/toolkits/green-infrastructure-toolkit/how-to-pay-for-green-infrastructure-funding-and-financing.html?chapter; 2015. [Accessed 12 November 2019].

Goldstein A, Turner WR, Gladstone J, et al. The private sector's climate change risk and adaptation blind spots. Nat Clim Chang 2019;9:18–25. https://doi.org/10.1038/s41558-018-0340-5.

Heineken. Water balancing and conservation agriculture in Mexico, https://www.theheinekencompany.com/our-sustainability-story/our-progress/case-studies/water-balancing-and-conservation-agriculture; February 15 2019. [Accessed 2 November 2020].

Kroeger T, Klemz C, Shemie D, Boucher T, Fisher JRB, Acosta E, Dennedy-Frank PJ, Cavassani AT, Garbossa L, Blainski E, Santos RC, Petry P, Giberti S, Dacol K. Assessing the return on investment in watershed conservation best practices approach and case study for the Rio Camboriú PWS program, Santa Catarina, Brazil. Arlington, VA: The Nature Conservancy; 2017. https://www.nature.org/content/dam/tnc/nature/en/documents/BrazilWaterROI_2.pdf. [Accessed 20 November 2019].

Leimona B, Negoro FS, Tanika L, Khususiyah N, Amaruzaman S, Lusiana B, Khasanah NM. Payment for ecosystem services for an urban water supply watershed: A business case of Rejoso Kita, Indonesia, http://www.worldagroforestry.org/publication/payment-ecosystem-services-urban-water-supply-watershed-business-case-rejoso-kita; 2018. [Accessed 9 July 2020].

Mayor T. Why ESG ratings vary so widely (and what you can do about it). Ideas made to matter. August 26, 2019 https://mitsloan.mit.edu/ideas-made-to-matter/why-esg-ratings-vary-so-widely-and-what-you-can-do-about-it; 2019. [Accessed 2 December 2019].

Money A. Plugging the water investment shortfall. Business Green; 2019. April 26, 2019 https://www.businessgreen.com/bg/opinion/3074648/plugging-the-water-investment-shortfall. [Accessed 2 December 2019].

Morgan Stanley Institute for Sustainable Investing. Sustainable signals: Individual investor interest driven by impact, conviction and choice, https://www.morganstanley.com/pub/content/dam/msdotcom/infographics/sustainable-investing/Sustainable_Signals_Individual_Investor_White_Paper_Final.pdf; 2019. [Accessed 20 November 2019].

Newborne P, Dalton J. Corporate water management and stewardship: signs of evolution towards sustainability. Brifing note. London, UK: ODI; 2019. https://cdn.odi.org/media/documents/12994.pdf.

Reinicke C. BANK OF AMERICA: These are the top 10 reasons investors and companies should care about ESG investing. Markets Insider; 2019. September 26, 2019 https://markets.businessinsider.com/news/stocks/10-reasons-to-care-about-esg-investing-bank-of-america-2019-9-1028557439#1-esg-can-generate-alpha1. [Accessed 20 November 2019].

Sarabi SE, Han QQ, Romme AG, Vries BD, Wendling LA. Key enablers of and barriers to the uptake and implementation of nature-based solutions in urban settings: a review. Resources 2019;8(3):121. https://doi.org/10.3390/resources8030121.

Sarni W, Share H. From corporate water risk to value creation. Global Water Intelligence; August 1 2019. https://www.globalwaterintel.com/news/2019/31/from-corporate-water-risk-to-value-creation. [Accessed 19 November 2019].

Scott O, Fatoric S, Merritt R, Fulton Lee V, Miley D. Engaging the private sector in green infrastructure development and financing: A pathway toward building urban climate resilience. USAID; 2018. https://www.chemonics.com/wp-content/uploads/2019/06/Engaging-the-Private-Sector-in-Green-Infrastructure.pdf. [Accessed 20 November 2019].

Soz Saiman A, Kryspin-Watson J, Stanton-Geddes Z. The role of green infrastructure solutions in urban flood risk management. World Bank; 2016. http://documents.worldbank.org/curated/en/841391474028584058/pdf/108291-WP-P156654-PUBLIC-ABSTRACT-SENT-WBUFCOPKnowledgeNoteGreenInfrastructureSolutions.pdf. [Accessed 20 November 2019].

Thygesen T. Everyone is talking about ESG: What is it and why should it matter to you? Forbes; November 8, 2019. https://www.forbes.com/sites/tinethygesen/2019/11/08/everyone-is-talking-about-esgwhat-is-it-and-why-should-it-matter-to-you/#111c949132e9. [Accessed 19 November 2019].

TNC (The Nature Conservancy). Upper Tana-Nairobi water fund business case. Version 2. Nairobi: The Nature Conservancy; 2015. https://www.nature.org/content/dam/tnc/nature/en/documents/Nairobi-Water-Fund-Business-Case_FINAL.pdf. [Accessed 20 November 2019].

TNC (The Nature Conservancy), Business Council Natural Infrastructure Working Group. Strategies for operationalizing nature-based solutions in the private sector, https://www.nature.org/content/dam/tnc/nature/en/documents/NBSWhitePaper.pdf; 2019a. [Accessed 25 October 2019].

TNC (The Nature Conservancy). Greater Cape Town Water Fund Business Case: assessing the return on investment for ecological infrastructure restoration; 2019b. https://www.nature.org/content/dam/tnc/nature/en/documents/GCTWF-Business-Case-April-2019.pdf [Accessed 26 April 2021].

UN Environment-DHI, UN Environment, IUCN. Nature-based solutions for water management: A primer, https://wedocs.unep.org/bitstream/handle/20.500.11822/32058/NBSW.pdf; 2018. [Accessed 2 December 2019].

UNGC CEOWM (UN Global Compact CEO Water Mandate). Water stewardship 101: The basics. n.d.-a https://ceowatermandate.org/university/101-the-basics/lessons/the-water-stewardship-journey/. [Accessed 26 October 2019].

UNGC CEOWM (UN Global Compact CEO Water Mandate). The six commitment areas. n.d.-b https://ceowatermandate.org/about/six-commitment-areas/. [Accessed 29 October 2019].

Universite De Montpellier, Universitas Gajah Mada and Danone Waters. Rejoso Project, East Java Indonesia: Output of hydrogeological research (2015–2019); 2019. presented on April 29, 2019.

US. PIRG Education Fund. Tax-increment financing: The need for increased transparency and account-ability in local economic development subsidies, https://uspirgedfund.org/sites/pirg/files/reports/Tax-Increment-Financing.pdf; 2011. [Accessed 12 November 2019].

Waage S. The benefit multiplier of investing in nature: Solving business problems and realizing multiple returns through working with ecological systems. Business Brief, San Francisco, CA: BSR; 2016. USA in collaboration with Restore the Earth Foundation, Ithaca, New York, USA.

WBCSD (World Business Council for Sustainable Development). The business case for natural infrastructure, https://www.naturalinfrastructureforbusiness.org/wp-content/uploads/2016/02/WBCSD_BusinessCase_jan2016.pdf; 2015. [Accessed 27 October 2019].

WWAP (United Nations World Water Assessment Programme). The United National world water development report 2018: Nature-based solutions for water. Paris: UNESCO; 2018.

CHAPTER 15

Funding and financing to scale nature-based solutions for water security

Sophie Trémolet[a], Brooke Atwell[a], Kathleen Dominique[b,*], Nathanial Matthews[c], Michael Becker[d], and Raul Muñoz[e]

[a]*The Nature Conservancy, Arlington, VA, United States,* [b]*OECD, Paris, France,* [c]*Global Resilience Partnership, Stockholm, Sweden,* [d]*Critical Ecosystem Partnership Fund, Arlington, VA, United States,* [e]*Inter-American Development Bank, Washington, DC, United States*

Introduction

If nature-based solutions (NBS) generate so many cobenefits, why are they not employed more often? There are plenty of good examples of NBS for water security around the world, but these experiences have yet to become mainstream despite growing evidence that they are cost-effective and can generate a host of cobenefits that help meet other objectives like the Sustainable Development Goals (SDGs) and nationally determined contributions. A more conducive enabling environment (policy framework and institutional arrangements) could support broader uptake of NBS (Coxon et al., 2021; Dominique et al., 2021).

This chapter starts by estimating the volume of funding to date for NBS for water security and examining where such funding has come from. We then identify the main challenges that have resulted in limited investment in NBS for water security. Some of these challenges are linked to the intrinsic nature of NBS, which can produce results that are more unpredictable than gray infrastructure, while others are directly influenced by the current status quo. Conventional gray infrastructure is the "norm" and the existing enabling environments (institutional arrangements and policies), and financing arrangements are conducive to their implementation. The status quo needs to be challenged so the sector considers NBS more systematically. We identify which funding sources can be mobilized and blended together to cover the costs of implementing and sustainably managing NBS over time, as part of a comprehensive strategy to improve water security.

The chapter ends with a series of recommendations on how to turn NBS for water security from a series of small-scale, pilot investments into a mainstream approach to delivering water

---

* The views expressed in the chapter are the author's own and do not necessarily reflect the views of OECD member countries.

Nature-Based Solutions and Water Security. https://doi.org/10.1016/B978-0-12-819871-1.00007-5

security. To that end, many stakeholders will need to change their minds and practices: policy makers will need to create a conducive enabling environment for NBS; project sponsors will need to define a sustainable funding strategy early in project development and do so with consideration for the implications of technical design and governance arrangement; and potential funders will need to actively engage with project developers to identify opportunities for NBS and support their development.

Where has funding for NBS for water security come from to date?

Few global reports capture the breadth of investment in NBS, and even fewer focus on estimating the value of investments in NBS for water security. Here we present the information that is currently available, though improving data to track the volume of investments in NBS for water security would be a good way to identify whether progress is being made in this area. However, reliable, comprehensive, and regular tracking does not currently exist.

Although comprehensive figures are largely unavailable, high-level data presented in the 2018 United Nations World Water Development Report suggest that, despite some countries increasing allocations toward NBS, current direct investments make up less than 1% of total water sector investments globally (WWAP/UN-Water, 2018).

A 2016 report by Forest Trends (Bennett and Ruef, 2016) sought to capture the size, scale, and scope of investment in NBS for water security. According to this report, investments in green infrastructure for watershed services[a] reached nearly USD 25B in 2015. Public funding (mostly via payments from national governments to landholders) was by far the largest source of funding, amounting to USD 23.7B.

Table 15.1, adapted from Forest Trends' 2016 report, highlights the two main sources of funding identified for investments in NBS for water security, how much was invested in 2015, and their area of impact in hectares. Area in 2015 is measured as "hectares under management." These estimates include land under agricultural or pastoral sustainable management; forest conservation; forest restoration/enhancement; forestry/agroforestry; grassland conservation; new habitat or green infrastructure creation; wetland restoration or enhancement; riverine/floodplain conservation; and riverine/floodplain restoration.

---

[a]   In their 2016 report, Forest Trends defines investment as "any transaction where financial value is exchanged for activities or outcomes associated with the maintenance, restoration, or enhancement of watershed services or natural areas considered important for watershed services." The activities that are included in this report under the term "green infrastructure" are equivalent to NBS; solutions that "restore, maintain, or mimic natural hydrological processes through natural and seminatural features and practices." Their definition of green infrastructure may cover activities outside of NBS as has been defined in this book. However, the trends remain a good representation of investment in NBS for water security.

Table 15.1: Funding sources and their impact (in hectares) in 2015.

Funding source	Definition	Value in 2015 (USD)	Area in 2015
Public funding for watershed protection	Funded by governments (occasionally with multilateral or donor support) to reward upstream land managers for enhancing or protecting ecosystem services for watershed protection. Usually acting on behalf of the public good. Operations tend to be at a large scale.	23.7B	426.7M ha
User-driven watershed investments	Payments from water users (companies or water utilities, acting as "buyers" of ecosystem services) to landholders or other parties (acting as "sellers") in exchange for conserving, restoring, managing, or creating green infrastructure. Buyers may contract directly with sellers or pay into a Collective Action Fund (for example, a Water Fund) that pools contributions for greater impact.	657M	11M ha

Modified from Bennett G., Ruef F. Alliances for green infrastructure: State of watershed investment 2016. Washington, DC: Forest Trends; 2016.

We know from the Forest Trends report and other investigations at the national level that most of the funding for NBS for water security has come, so far, in the form of rather fragmented public funding flows, with some limited user-driven watershed investments. Other funding sources, such as the sale of biodiversity or water quality credits were found to be more marginal on the global scale, even though they hold a lot of potential. They are discussed in a subsequent section of this chapter that outlines potential funding sources.

This chapter provides additional examples on these two main funding sources that have supported investment in NBS for water security so far.

Public funding for watershed protection

According to Forest Trends (Bennett and Ruef, 2016), an estimated USD 23.7B had been transferred in public subsidies for watershed management in 2015. Public funding was counted when such funding rewarded practices that support healthy watershed functions, which is markedly different than traditional subsidy programs that seek to influence commodity supply or pricing. Public funding coming from supranational, national, and state/provincial governments tend to target sustainable agricultural, pastoral, or forest management practices. Several governments provide targeted payments to the agriculture sector to produce environmental public goods such as protecting biodiversity or improving water quality. Switzerland, for example, provides a relatively high-level of support to its agricultural producers compared to other OECD countries. From 2014 to 2017, it undertook a major reform to adjust its direct payment scheme to better target the delivery of public goods, including environmental protection (OECD, 2017).

It is worth noting, however, that public funding for watershed protection remains meager compared to overall public support to farmers, globally (OECD, 2020). The OECD estimates that, in 2017–19, governments in 54 countries provided USD 629B in net transfers to the agriculture sector per year (OECD, 2020).[b] A recent joint report by The Nature Conservancy (TNC), the Paulson Institute and Cornell Atkinson Center for Sustainability estimated that each year, governments aid USD 421B in subsidies to the agricultural sector that are harmful for biodiversity (Deutz et al., 2020).

Most public subsidies for watershed protection were provided at the national scale. Europe, however, is a notable exception. Most subsidies in 2015 (over 50%) were budgeted for at the supranational scale by the European Union. Under the Common Agricultural Policy (CAP), subsidies are provided to farmers across the EU and matched at the national level. Some of these payments can support NBS with the potential to contribute to water security. For example, green payments were introduced in the CAP in 2013. These are direct payments to reward farmers for the public goods they provide when they adopt measures that contribute to soil health, water quality, and groundwater recharge, among others. The green payments account for EUR 12B a year, which corresponds to 30% of all direct payments under the CAP (Trémolet et al., 2019).

User-driven watershed investments

Water users—cities, companies, and water utilities—channeled USD 657M in 2015 through basin-scale programs to manage growing water risks (Bennett and Ruef, 2016). User-driven NBS programs, however, tend to be smaller in size than those funded via public budgets; in 2015, the median program size for these user-driven investments was USD 856K. Those programs often focus on small basins or subbasins that service high-demand users such as a city or a corporation. While NBS for water security programs are growing in popularity, there is little evidence to suggest that the size of transactions has grown. Small program sizes create challenges in terms of mobilizing funding and repayable financing, as detailed in the next section.

Most actors that have funded NBS—such as governments, water service providers, businesses, or other stakeholders—have done so because they see value in NBS that generate ecosystem services that, ultimately, provide value for them. Therefore, the spatial scale and location of the beneficiaries providing the funding would be highly relevant in terms of their incentive to contribute. A common example is that of a water utility who is interested in reducing the amount of sediment in a river so they can, in turn, reduce the amount of

---

[b] The analysis includes 37 OECD countries, the 5 non-OECD EU Member States, and 12 emerging economies.

money they spend on treatment. The utility may also be interested in generating associated biodiversity of carbon benefits. Removing sediment at the treatment plant might cost USD 200K per year, but paying local farmers to reduce sedimentation by planting trees or employing cover crops may cost USD 150K per year. Of 320 schemes surveyed by Forest Trends, 48% noted water risk as one of their top 3 motives for watershed investment, while reputational risk was a close second.

Again, the public sector—state and local governments—lead in spending in this area. In 2015, governments invested USD 194.3M, almost eight times more than the next highest investor: water utilities at USD 15.3M. Collective Action Funds[c]—also referred to as water funds (WFs)—are growing in popularity as a mechanism to channel investments toward NBS. An early example of such a fund is FONAG, a WF established in Quito in 2000 (Box 15.1) that paved the way for other WFs in at least 40 other locations (see TNC WFs Toolbox for more information, www.waterfundstoolbox.org).

In 2015, such funds accounted for 86% of user-driven transactions, though the primary contributors to these funds varied by region. In Europe, 45 cents of every dollar in the fund came from private contributions. In Latin America, NGOs and the private sector together nearly matched public contributions. It is worth noting, however, that NGO and donor dollars in Latin America outpaced private sector contributions by about 800% in 2015.

A good example in Latin America of innovative funding instruments that combine public, private, and donor resources is the Latin America Water Funds Partnership (LAWFP). Aware of the urgent need to create financial and governance mechanisms that allocate resources for water source conservation, TNC, FEMSA Foundation, the Global Environment Facility (GEF) and the Inter-American Development Bank, launched the LAWFP in 2011 to create and strengthen WFs across the region. In 2018, the International Climate Initiative (IKI) of the German government joined as a new partner. This mechanism offers opportunities to advance sustainable watershed management, increase urban water security, and strengthen community resilience to extreme climatic phenomena.

Since its creation, the alliance has committed approximately USD 41M to test, mature, and scale the WF model in 14 countries. Thirty-seven WFs have been developed, 18 of which are formally created and 19 of which are operating in Brazil, Mexico, Peru,

---

[c] Forest Trends defines collective action funds as "collective action partnerships that include a cooperative funding element in order to maintain, restore, or create green infrastructure in the watershed. Partners contribute financial and/or in-kind support to watershed protection activities, which are typically designed and implemented in consultation with the group. Collective action funds may establish an endowment or trust fund to manage contributions, as in the case of a number of 'water funds' in Latin America and the Caribbean; or they may use other systems for administering funds or operate on the principle of matching funds (where partners coordinate investments, but resources are never actually pooled)."

Box 15.1   FONAG, the first official "Water Fund," Quito, Ecuador.

A good example of a well-functioning WF is one of the oldest, namely Quito's Water Protection Fund, FONAG (*Fondo para la protección del Agua*) (http://www.fonag.org.ec/web/), created in 2000 with support from TNC and other partners. The city of Quito is home to approximately 2.6 million people, and its population—and that of the upper Guayllabamba basin—relies on a combination of surface water, groundwater, and interbasin transfers for their water supply. Over time the infrastructure tapping the aquifer deteriorated, and the quality of the water therein declined. As a result, the city changed its approach and increasingly relied on surface water as the main source for Quito. However, this merely shifted the risk. The water supply was now dependent on the health of high Andean wetlands and paramos (high-altitude tropical grasslands), which act as the source for much of the city's water supply. The ongoing deterioration of these ecosystems, which also harbored many threatened animal species, exposed the city of Quito to a new type of risk. As a result, the Quito WF became the driving mechanism for a multisectoral collaborative effort that focused on NBS.

FONAG is a financial and governance mechanism that oversees a coherent plan of work to protect the watershed, and that pools and safeguards the money required to support these activities. Most directly, FONAG runs programs and conservation projects that focus on ecological restoration. It also supports targeted environmental education, aiming to foster a new "culture of water" and garner support for integrated water resources management.

In 2007, the local water utility (EPMAPS) was allocating 1% of all sales income to FONAG to support its implementation and education efforts. That same year a municipal law (ordinance) was passed that required the allocation to increase by 0.25% over a 4-year period until it reached a 2% threshold, where it currently stands. This tariff now provides the bulk of FONAG's funding. Early in project development and operation, however, the contribution from EPMAPS was complemented by financial donations or investments from other partners, including an electricity company, a beverage firm, a bottled water company, a nongovernmental organization (NGO; TNC) and a bilateral international donor (SDC). As of December 2018, FONAG had an estate worth USD 18.7M in which the return plus proceeds from agreements generates an annual budget greater than USD 2M.

One of the Fund's challenges was quantifying the benefits of NBS interventions in order to attract more funding for water resources management. Estimating economic benefits and providing concrete evidence of positive returns have been requirements expressed by funders, who are also looking for documented financial performance (OECD, 2019b).

Ecuador, Guatemala, Colombia, Costa Rica, and the Dominican Republic. WFs have been able to leverage over USD 210M for conservation investments from a variety of public and private sources and convene over 300 local multisector partners. Nearly 119 million people will potentially benefit from watershed conservation projects implemented through these WFs. The total area to be conserved by these funds is more than 300,000 ha. This partnership demonstrates that solving water challenges requires joint work between different stakeholders and intelligent combinations between gray infrastructure and NBS, which is part of the principles of WFs.

Limited repayable financing has been mobilized to date

It has been possible to leverage repayable financing only in very few cases, mostly from development finance institutions and impact investors. There is a missed opportunity here: the pool of private capital looking for natural capital investment opportunities is potentially far greater than what is available from public funders.

In addition, private investors are actively looking for opportunities in this area. The report *Investing in Nature: Private Finance for Nature-Based Resilience* (https://www.environmental-finance.com/content/research/investing-in-nature-private-finance-for-nature-based-resilience.html) (Trémolet et al., 2019) showed existing and increasing interest of private investors for including environmental concerns in investment decisions, based on a global survey of institutional investors. The report documents a wide range of approaches to investing in natural capital and identified what has worked and what is still needed to carry momentum forward and bring these models to scale.

Reaching a far larger scale requires mobilizing more repayable financing (including from private capital) for NBS for water security. At present, however, there is no comprehensive estimate of the amount of repayable financing going toward NBS for water security. The recent decade has seen an influx of private investment in other types of NBS, primarily those that generate returns via real assets like forestry or agricultural commodities. Private capital investment in conservation grew from an average of USD 0.2B in 2004–08 to USD 2B in 2015, which amounted to a total investment of USD 8.2B over that decade. However, as shown in Fig. 15.1, investments in sustainable food and fiber production (USD 6.5B) dwarfed investments in habitat conservation (USD 1.3B) and water quality and quantity (USD 0.4B) during that time period (Bennett et al., 2017).

These sustainable food and fiber investments usually seek to improve farm productivity or produce premium products—for example, FSC certified wood or organic dairy—which can fetch a higher price on the market. From 2004 to 2015, sustainable forestry and agriculture were the largest investment subcategories, representing 44% and 32%, respectively, of the estimated USD 8.2B invested in conservation. In these programs, NBS such as cover crops, no till practices, and silvopastoral systems, are employed to ensure the sustainable management of these assets. This in turn increases the price consumers pay for the outputs while simultaneously generating cobenefits like carbon sequestration, improved soil health, and biodiversity habitat. While many of these NBS programs can also generate benefits for water security, such benefits are often neither the driver nor the focus of these investments.

What have been the key challenges in mobilizing funding for NBS?

The overall share of investments that are directed toward NBS for water security could be substantially increased. Mobilizing funding and financing for water-related investments is

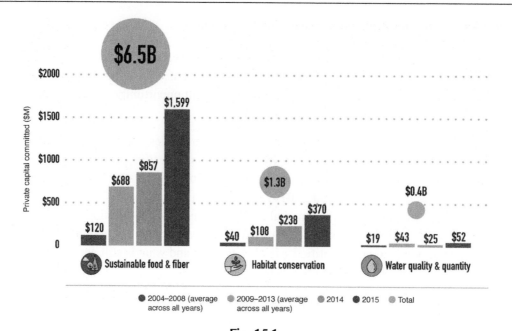

Fig. 15.1

Annual private capital committed by conservation category, 2004–15. *From Hamrick K. State of private investment in conservation 2016: A landscape assessment of an emerging market. Washington, DC: Forest Trends; 2016.*

generally challenging, as documented by numerous reports that have highlighted the financial challenges in the sector. Pervasive undervaluing of the resource and the benefits associated with investment constrains funding and financing opportunities. Water-related investments generate a mix of public and private benefits in terms of valued goods and services as well as reduced water-related risks. Many of the benefits from these investments cannot be easily monetized, undermining potential revenue flows that can support sustainable financing strategies (OECD, 2018a; Winpenny, 2015). For water supply and sanitation services, revenues typically do not recover the full cost of service provision (OECD, 2017).

Investments in water and sanitation services and water resources management have historically been paid for by the public sector, with concessional finance playing an important role in developing countries. Risk-return considerations and structural issues related to profitability of operating business models often limit opportunities to mobilize commercial investment (OECD, 2019a). Challenges with financing NBS can seem even more daunting, as these solutions to improving water security are still seen as new and innovative. None of them are insurmountable, however, and the last section of this chapter provides concrete ways forward.

Challenges to mobilizing funding for NBS broadly fall into two categories. On the one hand, NBS have intrinsic characteristics that raise specific difficulties (and opportunities) in terms of mobilizing funding. On the other hand, the existing status quo, which favors gray infrastructure investments, needs to evolve to accommodate greater innovation in an area that can generate multiple benefits over time and for future generations. We explore these challenges in further detail in the following subsections.

Challenges related to NBS characteristics

The characteristics of NBS and the extent to which they can help improve water security have been examined in detail in earlier chapters (Cassin and Matthews, 2021; Vigerstol et al., 2021). Due to their nature, NBS generate multiple benefits for multiple beneficiaries, as shown in Fig. 15.2. As a result, there is not always a "primary beneficiary" that would be willing to take the lead for designing and implementing an NBS for water security investment. Multiple revenue streams can be considered to monetize and capture the value generated by such investments, but understanding what potential revenues can be mobilized for a specific investment and coordinating them can be difficult. A later section in this chapter reviews in more detail what these alternative revenue streams consist of.

In some cases, there is a significant lag between the time when investments are made and the time when benefits can be reaped. In the case of tree planting, for example, benefits can take several decades to materialize. In addition, performance can be impacted by external factors that are difficult to control or foresee, such as climatic factors or natural disasters.

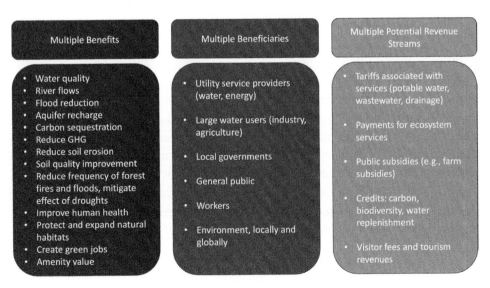

Fig. 15.2

Nature-based solutions are multifunctional, generating multiple benefits for multiple beneficiaries, resulting in multiple potential revenue streams. *From Authors.*

Some argue that there is a lack of scientific and technical evidence on what benefits NBS can deliver, with which degree of certainty, and at what cost. Many existing NBS compendia provide information on technical solutions, but evidence is not often documented in a way that makes NBS directly comparable to gray infrastructure solutions so as to drive investment decisions. There are several reasons for this. In some schemes, either because of the small size of the intervention or because there was no funding allocated at the planning stage, monitoring and evaluation protocols are not always in place. Therefore, data on actual costs and performance (how well the NBS can deliver the desired impact) in a range of conditions has not been systematically collected. This is partly due to the fact that NBS are very location-specific, which means that building an evidence base on their effectiveness can be fraught with difficulties. Contrary to carbon markets where transacted offsets used a third-party standard to guide project development and verify results, there is little standardization in watershed investments.

An effort to address this has been developed by TNC's WFs strategy, which employs NBS to address local water security issues. With over a decade of experience and 41 active WFs, they have systematically codified lessons learned and recommendations to address the monitoring and evaluation gap.

A Primer for Monitoring WFs (Higgins and Zimmerling, 2013) highlights the critical information needs common to WF projects and summarizes issues and steps to address when developing a monitoring program. To ensure that investments are having their anticipated impacts and to enable adaptive management, WFs must include robust monitoring programs to track the environmental, economic, and social impacts of their actions.

TNC later published *A Guide to Monitoring and Evaluating Water Funds* (Leisher et al., 2019), aimed at designers of new WFs and managers of existing ones. It is a blend of monitoring and evaluation guidance, new ideas and advice that can be applied to NBS for water security at large and beyond WFs.

In addition to expertise that can design scientifically robust monitoring programs, NBS for water security projects also need expertise in the various instruments, techniques, and software needed to properly collect, manage, and communicate results. Stakeholders may also have different reasons for investing and may, therefore, want to monitor additional benefits of the NBS program like jobs created, human health outcomes, or biodiversity benefits. Contracting or hiring the right capacity is important because adequate monitoring programs can look very different depending on the water security objective, NBS employed, and local ecological and hydrological factors.

Challenges for NBS adoption related to the status quo

The status quo tends to favor gray infrastructure investments. The many challenges to NBS adoption can, usefully, be grouped in thematic categories: inadequate policy and legal frameworks, fragmented governance arrangements (and associated fragmented funding streams) and inadequate understanding of what NBS can achieve by those making investment decisions or allocating funding (see Table 15.2).

Inadequate policy and legal frameworks

In some parts of the world, legal and policy frameworks may create barriers for NBS adoption or neglect to mention them explicitly, which makes funding their implementation difficult, if not impossible. This is particularly important for water utilities (or regulators) who are looking for cost-effective ways to supply water to customers. If regulation only permits gray infrastructure to secure a city's water supply, the set of potential solutions from which they can draw narrows considerably.

By contrast, an adequate legal and policy framework would specify broad water security outcomes and specifically refer to the role NBS can play in reaching those targets (Dominique et al., 2021). A conducive enabling environment (policy framework and institutional arrangements) would create a level playing field for conventional (gray) infrastructure,

Table 15.2: Summary of potential challenges for NBS adoption—By stakeholder types.

Stakeholders	Potential challenges
National governments and policymakers	• Limited awareness of available NBS options • Uncertainty about effectiveness and cost-effectiveness of NBS • Perception that NBS are riskier than conventional "gray" solutions • Concerns about compliance with environmental and water quality regulation
Local governments	• Procurement rules for service delivery contracts often focus on gray infrastructure or do not allow NBS • Limited knowledge on how to plan, adopt and monitor NBS • Local policy concerning multiple areas that can impact NBS (e.g., urban planning, biodiversity, water) may not be well-coordinated or are difficult to align due to silo approaches
Service providers (e.g., water and wastewater)	• Operating and procurement models are often focused on building gray infrastructure • Focus on regulatory compliance • Limited potential for experimentation or risk-taking
Regulators	• Risk-averse when it comes to granting permits • Favor short-term results and regulatory certainty over more sustainable long-term results
Private landowners	• Reluctant to sell land or water abstraction rights for conservation or to modify farming practices for fear of reductions in yield and income • Lack of land that can be set aside for conservation
Funders and financiers	• Lack of "at-scale" NBS investment projects • Lack of intermediaries and technical capacities to prepare investment projects

Modified from: Trémolet S, et al. Investing in nature for Europe water security. London: The Nature Conservancy, Ecologic Institute and ICLEI; 2019.

NBS, and/or their use in combination. Water security challenges are highly localized, and a framework that stipulates too many requirements would not provide the flexibility needed for local adoption and implementation. A recent review of European policy frameworks for water management found that they were broadly adequate to support NBS adoption (Trémolet et al., 2019). The Water Framework Directive, other related directives, the EU Biodiversity Strategy, and the Green Infrastructure Strategy all provide a conducive environment, coupled with substantial public funding.

Fragmented governance arrangements

There are often several government agencies or ministries in charge of managing the provisioning of water resources, and they may have different objectives and varying levels of influence. Programmatic approaches for jointly managing land and water are often lacking, which means that working at the level of a single project or initiative is often too late or too limited in terms of impact. As such, any one government agency may have a limited view on water management, and it can be challenging to incentivize cooperation and cross-collaboration. Government agency staff may lack technical capacity to assess and build NBS into their investment programs. They do not always understand that NBS, which generate many cobenefits, could be very attractive for governments looking to simultaneously meet development, climate, biodiversity, and human well-being goals.

Constraints on key stakeholders: Acquired behaviors and lack of adequate incentives

Stakeholders may not understand how investing in NBS can provide timely returns for water security—or they may be skeptical because they have not seen sufficient evidence (Garrick et al., 2020). They may not feel responsible for investing in—nor have the incentives to invest in—NBS, due to complex governance arrangements and the lack of clearly defined roles and responsibilities. Those who stand to benefit most from NBS investments (e.g., downstream water users in a watershed at risk of environmental degradation) may not be responsible for investing in NBS for water security.

Most water companies are better at building and managing gray infrastructure (e.g., a water treatment facility or pipeline) because they are more comfortable and familiar with the risk, returns, and benefits of these structures. In addition, water companies may not be allowed to invest in protecting or restoring the upstream basin despite a growing recognition that water service providers should act as "good stewards" of water resources, healthy ecosystems, and biodiversity (see also Coxon et al., 2021). They may be limited in their ability to invest in NBS due to: (1) asset capitalization accounting rules (e.g., they cannot include a planted or a protected forest in their asset base); (2) issues relative to land ownership (e.g., they do not own the land where they would need to deploy NBS); (3) limits on their ability to finance activities outside of their service area; (4) a perception of higher risks associated with NBS investments (see subsection on evidence base); or (5) difficulties engaging with multiple stakeholders via collective action or partnership structures, especially when some stakeholders may perceive them as competition for water resources. Their existing procurement rules are also frequently designed to purchase works and professional services, rather than purchase ecosystem services from a range of different actors. As risk-averse entities, they may be uncomfortable with engaging in what they perceive as complex and uncertain schemes that require them to collaborate with multiple actors, including farmers or environmental NGOs. Finally, the majority of water companies do not have staff who are trained in multistakeholder engagement, nor do they have capacity to draft and manage corresponding governance arrangements.

Local governments are, likewise, key investors in local services and often contract out when they cannot deliver the services themselves. However, if water supply or treatment services are contracted in a fairly rigid manner (e.g., requiring the delivery of specific outputs, such as a wastewater treatment plant, rather than specifying outcomes—such as tons of pollutants removed from waterways) or if local governments do not include NBS in their local development plans (see subsection on policy framework), they may hinder NBS adoption rather than encourage it.

Regulatory constraints

It is not uncommon for regulators to be unwilling to accept NBS projects because they view them as having a lower certainty of success, at least in the short term. In some cases, regulators may be excessively focused on regulatory compliance and may not give water providers the space to innovate and adopt NBS instead of—or as a complement to—gray infrastructure. In England and Wales, for example, water companies found that the Environment Agency (the regulator responsible for setting and enforcing environmental norms) sometimes restricted their ability to experiment with NBS to achieve the objectives of the Water Framework Directive. The economic regulator (Ofwat; responsible for approving investments so they can be funded by water charges or other revenue) is nevertheless supportive of NBS adoption, and current water companies' investment programs include ambitious goals for environmental investments by themselves and their partners.

Fragmented funding streams

The provisioning of funding often mirrors the governance framework, and if a local government is not coordinated across agencies, funds can respond to multiple—and possibly contradictory—objectives. Subsidies can go to farmers, NGOs or local governments in a way that is not coordinated to deliver well-defined environmental outcomes. Accessing funding, particularly from public grants programs or philanthropy can create significant transactional burdens for applicants. Conversely, there is frequently a lack of at-scale investment projects that could attract repayable financing and a lack of technical assistance providers with knowledge and understanding of NBS as a business to be implemented at a medium-large scale. This is one of the reasons why repayable financing has been extremely limited to support investment in NBS for water security so far.

In summary, whereas access to financial resources may often be perceived as a key challenge, it may often be symptomatic of deeper issues related to policy, governance, or incentives. For a project to be successful, these issues need to be addressed before project design even begins. If these issues can be addressed, significantly greater funding and financing from a variety of sources could be mobilized to accelerate investments in NBS for water security, as discussed in the next subsection.

What funding and financing sources exist, and how can they be blended?

Understanding the sources of funding available and how they can be combined is pivotal to the success of any NBS program. As shown in Fig. 15.2, these programs can generate multiple benefits for multiple beneficiaries. Not all such benefits can be "monetized," however, which means that only some of them may generate revenue streams.

To design an NBS investment program, it is therefore important to understand the benefits generated by the NBS project, which beneficiaries will receive these benefits, and how these benefits can be converted into revenue streams, for example, in the form of charges, tariffs, payments for ecosystem services, ecosystem compensation schemes, or the sale of circular economy products. From the outset, one should identify potential funding sources and how they can be combined by defining a sustainable funding strategy over the entire lifetime of the NBS project, also referred to as "the investment." The success of NBS interventions often depends on mobilizing adequate funding for project design (including governance and funding arrangements, which are key to long-term success), implementation, and ongoing maintenance of the initial investment. If a program includes multiple NBS, financial resources will need to be mobilized for the overall program rather than for each NBS in isolation, which adds an additional layer of complexity.

Identifying NBS benefits that can generate revenue streams

Prior to defining a funding strategy, the project team would need to conduct an economic and financial analysis to explicitly estimate the return that the project sponsor (and other beneficiaries) can expect to receive for their investment. The beginning of this chapter noted that water service providers (e.g., utilities) are prominent actors in user-driven investments. For example, if a utility is interested in reducing their treatment costs by investing in watershed protection to reduce the amount of sediment entering their treatment facilities, they would want to know the return on investment they can expect for every dollar they put into an NBS program that would prevent sediment build-up. Therefore, the economic analysis would need to focus on defining the value of that sediment reduction for the utility. For example, instead of spending USD 350K per year on treatment costs, the economic analysis might show that it makes better sense for their bottom line to invest USD 200K per year in an NBS program that would reduce sedimentation.

Often, an economic analysis will also define the cobenefits of an NBS program, including carbon sequestration, biodiversity, human health and well-being, increased agricultural production, and so forth. Only a subset of cobenefits, however, can be monetized. For example, of the cobenefits previously listed, carbon sequestration and increased agricultural production are likely to be the only monetizable cobenefits that an investor would care about. Carbon credits could be sold on the carbon market, and agricultural commodities can be processed and sold for profit. Stacking benefits can be one way to increase the level of private investment in NBS, as this can provide an additional opportunity to generate returns. While biodiversity and human health and well-being are highly desired cobenefits, it is often difficult to capture this value and generate a revenue stream for potential investors. Box 15.2 describes ways in which climate funding could be leveraged given the important role NBS for water security can play in climate adaptation.

Box 15.2   How NBS for water security benefits climate adaptation.

NBS for water security have a role to play in climate adaptation and have been highlighted in several recent reports and policy papers (OECD, 2020; Mason et al., 2019; Cooper and Matthews, 2020; Smith et al., 2019).

As stressed in a recent policy paper by OECD (2020), there is a significant and growing body of scientific evidence documenting that climate change will intensify the risks of water-related hazards. The paper focuses on the role of NBS in limiting and managing the current and future impacts of climate change, focusing on water-related risks, and specifically examining coastal flooding, riverine flooding, urban flooding, and drought.

As the pace of climate change quickens, source water protection is now also becoming a critical component to ensuring resilience (Matthews et al., 2019). Climate change presents a new range of threats, drivers, and uncertainties in how we interact with freshwater ecosystems, but natural approaches to cope with climate impacts are now widely seen as a necessary path forward.

Climate finance generally is growing in scope and many organizations have started using adaptation and climate risk as both a reason for implementing NBS and as a pathway for finance. Therefore, when we look at investments in NBS for water security, we also have to account for their role in adapting to climate change.

For example, multilateral development banks, including the World Bank and the Asian Development Bank (ADB), are working to generate client demand from countries and to mainstream this type of option within their work. The ADB is currently piloting a number of NBS for disaster risk reduction projects in Vietnamese cities, and, moreover, developing internal guidance for NBS investment as part of a more systematic and operational commitment to natural solutions. Internal guidance will align NBS as an additional set of options within the disaster risk and climate adaptation programs for water, urban and transport sector investments (Cooper and Matthews, 2020).

Mobilizing funding and repayable financing in a coordinated manner

The next step in program preparation entails projecting program costs over the lifetime of an NBS project. Most costs will typically materialize upfront—for example, to plant trees, build a constructed wetland, purchase land for permanent protection, and more—whereas revenues might take longer to come in, particularly for NBS that take years to mature.

One should then forecast potential funding sources—revenue streams that can cover the costs of the program—and assess whether the project currently has a gap in funding or is likely to have a funding gap in the future. Understanding a project's financial standing will help assess whether the project can deliver results at the expected scale and in a sustained manner. In the event of a gap, repayable financing may need to be mobilized to "bridge the financing gap" (Fig. 15.3).

Box 15.3 highlights how such a funding strategy was defined for the newly established Greater Cape Town WF, building on a business case (i.e., an economic analysis) that identified investing in nature (invasive plant removal) would be much more cost-effective than the next available "gray infrastructure" option, that is, building a desalination plant. The sustainable funding strategy mobilized regular contributions from the City of Cape Town, which provides water services in the region, to complement philanthropic and corporate donations.

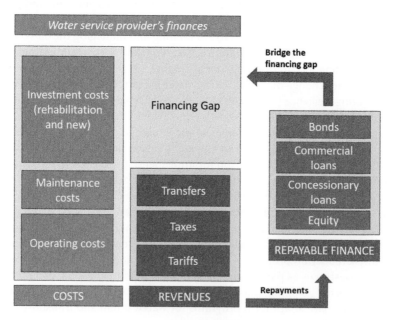

Fig. 15.3
Common sources of funding and repayable financing. *From Authors.*

Box 15.3   Defining a sustainable funding strategy for the Greater Cape Town WF.

Between 2015 and 2017, Cape Town experienced the worst drought in 100 years. In 2018, dam reservoirs dropped below 20%, and the city prepared for the day when the taps would run dry. The crisis was stayed by strong measures to limit water demand, groundwater exploration, water transfers from a neighboring storage scheme, and the arrival of the winter rains. The crisis would have affected four million people—40% of whom live below the poverty line. Still, it is estimated that Cape Town's water demand will outstrip available supply by 2021. To avert this threat, the city's water utility considered a range of water augmentation options, including desalinating seawater, exploring groundwater, and increasing storage capacity. Another cost-effective, sustainable option was to restore the city's watersheds by controlling the invasive plants that capture 55 million cubic meters of water each year—a massive long-term effort covering a vast land area. If no action is taken, however, water losses will double within just 20 years.

The NBS employed consists of removing invasive species from the Cape Town catchment area to augment water streams feeding into critical reservoirs for the city. These invasive trees (pine trees, acacias, and eucalyptus) have much deeper root systems than the native vegetation, fynbos, which is unique to the Greater Cape Town region; as a result, they suck up much more water. They also burn more easily augmenting fires in the region. A number of actors have conducted invasive species removal in the catchment for some time, although several of these programs were focused on job creation rather than seeking to maximize ecological benefits.

In 2018, TNC and partners released the Greater Cape Town business case (https:// waterfundstoolbox.org/regions/africa/cape-town-water-fund), which showed that removing invasive species in priority areas could save up to 2 months' worth of water at one-tenth of the cost of the next available option, a desalination plant. The business case provided a compelling basis for mobilizing funding from a range of sources, which has included a mix of philanthropy, corporate water replenishment funding, public funding, and a grant from the water company, with the possibility of levying a small charge as part of water tariffs in the future. A further grant from Water Unite (https://www.waterunite.org/) enabled the team to define more detailed governance arrangements—a basis for all participating stakeholders to make decisions—and to create management tools, such as a decision support system, which allows all participating stakeholders to track implementation progress and funding. Doing so required skills in multiple fields, including legal, financial, stakeholder engagement, and communication.

The following subsections discuss the types of funding, repayable financing sources, and financing vehicles that can mobilize diverse revenue streams to cover the full costs of an NBS program.

Mapping out potential funding sources

Funding for NBS may come from two primary sources: direct revenues generated from the project or external transfers, in the form of public grants and subsidies and private donations. Some examples of each type follows.

Direct revenues. Direct revenues are funds that the project or the project implementer can generate on their own. These revenue streams typically constitute the most reliable and sustainable form of funding because they are usually recurrent and relatively stable over time. Different types of revenue streams are included in this overall category. Recurrent revenue streams can come from tariffs and charges levied for water-related services, payments for ecosystem services, commercial revenues associated with the NBS investment, the sale of credits (including for carbon, biodiversity, water replenishment, or water quality), and visitor fees and revenues from eco-tourism. These are discussed in more detail later in this chapter.

Tariffs and charges. The NBS project implementer can generate revenue by selling a service provided by the NBS implemented in the program. The economic analysis will help identify which NBS services can be monetized and thereby could help fund the initiative. For example, this could consist of the sale of wastewater treatment services from a constructed wetland, which could attract a wastewater treatment charge.

Large-scale implementers of NBS projects such as water companies should also consider the extent to which investing in NBS can bring down their overall costs and therefore generate savings; savings could, in turn, help lower water tariffs. Several water companies that have included NBS in their investment programs have realized significant cost savings. For example, Eau du Grand Lyon, a Veolia subsidiary that provides and distributes water in the Grand Lyon metropolitan area under contract with the municipality, is actively protecting 375 ha around water fields in the heart of Lyon, France's third largest city. Based on ex-post calculation, they found that this "green infrastructure" (wellfields with source water protection) is more cost-effective than building a water filtration plant—and generates biodiversity benefits. Total annualized costs associated with a typical coagulation and filtration plant would cost EUR 52–74M per year, compared to the annualized costs of the existing green infrastructure (EUR32M per year). Significant savings are also achieved on operating costs: EUR 0.04 per m^3 for green infrastructure as opposed to EUR 0.15–0.25 per m^3 for a typical plant (Trémolet et al., 2019).

To fund the initial investments and ongoing costs, water companies typically charge water and wastewater service fees and can dedicate a portion of tariff revenues to cover the costs of NBS, particularly if they form an integral part of their investment program. Some WFs receive proceeds from a small charge for source water protection that is applied through the water tariffs (in the case of FONAG, Ecuador, discussed in Box 15.1 or nationwide in Peru).

Peru's legislation establishing a recurrent source of income for the protection of its water sources was the first of its kind in Latin America. The regulator of water and sanitation services, SUNASS, requests that all water service providers allocate 1% of their anticipated income to invest in natural infrastructure. In the case of Lima, an additional 3.5% (on average) is earmarked for risk mitigation and adaptation to climate change, which accounts for USD 25M and USD 92M, respectively, over the next 5 years. Nontariff public funds have also been successfully mobilized, such as a municipal tax in San Antonio, Texas (Abell et al., 2017). Box 15.4 provides an example of how the water company in Camboriú, Brazil, has raised funding for NBS through its water tariff.

Box 15.4   Camboriú, Brazil's water company adopts tariff to fund NBS program.

EMASA, the water company in Camboriú, Brazil, wanted to explore how NBS could be employed to reduce sediment concentrations at its municipal drinking water intake thereby reducing water treatment costs and water loss during the treatment process. The economic analysis found that reductions in sediment treatment costs and water losses could offset 80% of the public water company's investment in the NBS program over a 30-year time horizon. If expanded to 43 years or more, the company's return on investment could exceed one. This, in itself, would not have been sufficient to convince water company management to invest in the NBS program.

Fortunately, the cobenefits of the program—reduced flood risk and water supply shortages during the tourist high season—were of high concern to the two municipalities in the watershed, and cost-sharing with other stakeholders could lift the ROI of the program above one. Cost-sharing could be achieved by incorporating watershed conservation costs into water user fees or levying a conservation fee on visitors during the high season when water resources are strained. Studies in Brazil on the value of improved water supply security and flood control report an average household willingness to pay several orders of magnitude higher than what would be required to lift the ROI of the Camboriú program above one.

Therefore, even though the business case did not indicate a viable investment from EMASA's perspective, the finding that cobenefits accrue to stakeholders beyond EMASA alone supported the case for incorporating watershed conservation costs into calculating the water tariff.

Sunrise on Taquaras Beach in Balneário Camboriú, Santa Catarina, Brazil. *Photo Credit: © Fausto Da Rosa/TNC Photo Contest 2018.*

However, in locations where water tariffs are deliberately kept low (for affordability or political concerns), it may be difficult to increase tariffs to include a charge for water resource protection activities. Where tariffs or taxes can be increased, it typically takes time to ensure that tariff or local tax regulations are changed to cover water source protection costs and to ensure that a portion of tariff or local tax revenues is earmarked to that effect. In the Upper Tana WF, which serves the city of Nairobi, although a small charge has been applied as part of the water tariff in order to cover some of the WFs costs, proceeds from this charge account for a very small portion of the WF's total revenue, which is mostly covered to date by external transfers, both from public and private sources.

Payment for ecosystem services. Users (such as water companies) that benefit from an NBS investment may be willing to pay to ensure the associated ecosystem service is provided. They can either do so by providing in-kind assistance or by paying farmers directly (see Box 15.5). Other large users in the basin such as bottling plants, agri-food businesses, irrigation boards or hydropower plants, may also agree to contribute if the NBS increases their water security and/or helps fulfill water stewardship obligations. Such approaches reflect the "beneficiary pays principle," which can complement the "polluter-pays" principle, commonly accepted in OECD countries. According to this principle, the polluter should pay for the cost their pollution has on society, for example, through a tax. Decisions about how and when to apply these principles usually depend on what society deems acceptable in terms of incidence and levels of pollution in the context of productive activities (OECD, 2012).

Commercial revenues associated with NBS investment. Investing in NBS can also generate greater cash flows for other types of beneficiaries, such as farmers, than when agriculture is conventionally managed. For example, an NBS for water security program that aims to reduce sediment in nearby waterways, might work with farmers in the catchment to improve

---

Box 15.5   Eau de Paris—Paying farmers to adopt grass-fed cattle rearing and other sustainable practices.

Eau de Paris (France) is the public water service provider that serves three million consumers in Paris, France's capital. Since 2008, Eau de Paris has been supporting farmers with financial assistance programs to help them reduce fertilizer and pesticide use and adopt organic farming practices. Staff contribute to disseminating good agricultural practices. The company has helped develop market opportunities for farmers' products, including in Paris-managed school canteens. Eau de Paris has also purchased land where there is a specific risk of contamination, with a total of 574 ha acquired by 2018. Eau de Paris leases the land to farmers for one symbolic euro. In exchange, farmers engage in agricultural models that protect water quality, including organic farming and grass-fed cattle rearing. Since early 2020, Eau de Paris has secured the authorization from the European Commission to make direct "payments for ecological services" to the farmers that it works with. This was not previously allowed, as it was considered to be a public subsidy, which Eau de Paris is not allowed to provide.

soil retention on their land. Employing sustainable practices like cover crops, agroforestry, agroecology, and other conservation agriculture practices would improve the health of the farm's soil, thereby potentially increasing harvest yield (revenue) and decreasing the fertilizer they would need to apply (costs). If a farm has incorporated enough of the right sustainable practices, they could even pursue organic, FSC, or other sustainable certifications that fetch a premium price on the global market. This could lead to asset appreciation, that is, an increase in the value of the farm. An organic farm, for example, would typically sell for a higher price than a conventional farm of similar size.

Sale of credits for water replenishment, water quality, carbon, or biodiversity. Funding for the investment may be generated through the sale of carbon credits (e.g., if the NBS sequesters carbon or reduces emissions), water quality credits, biodiversity credits, or contributions from corporates looking to achieve their water replenishment goals. In the United Kingdom, for example, the forthcoming Environment Bill will mandate the adoption of a biodiversity net gain principle for all new housing developments. The principle will mandate that developers follow the different steps of the mitigation hierarchy (Fig. 15.4), including purchasing biodiversity credits to compensate for unavoidable losses on site and paying for additional gains to be achieved somewhere else. The principle aims to leave the environment in a better state than it would have been without the development.

Frameworks for offsetting biodiversity losses are in place in at least 33 countries around the world and have cumulatively restored and protected 8.3 million ha of land in 2016 (Bennett et al., 2017). About USD 4.8B in mitigation bank credits and financial compensation was transacted in 2016 across 99 regulatory programs. Nearly two-thirds of the costs for conservation programs were paid for by the private sector—primarily the energy,

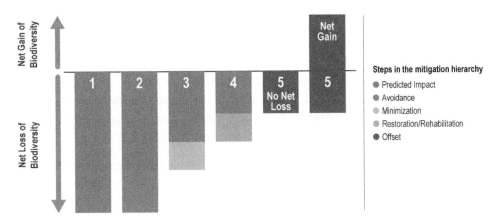

Fig. 15.4

The mitigation hierarchy: Avoidance, minimization, rehabilitation, offset, and net gain. *From Bennett G, Gallant M, Ten Kate K. State of biodiversity mitigation 2017 markets and compensation for global infrastructure development. Washington DC: Forest Trends; 2017.*

transportation, mining, and minerals sectors—and forest and wetlands projects dominated the market. The recent report from the Paulson Institute, TNC and the Cornell Atkinson Center for Sustainability (Deutz et al., 2020) presents a summary of global biodiversity conservation financing flows in 2019. During this year, the total funding flows for biodiversity were between USD 124B and USD143B. Of these, USD 6–9B came from biodiversity offsets.

The US Aquatic Resources Compensatory Mitigation program, which is focused on wetland and stream offset credits, is the largest banking market in the world. It transacted about USD 3.3B in bank credits in 2016, and the volume of transactions has grown by an average of 18% per annum since 2010. However, more than one-third of these programs, globally, did not accept third-party compensatory mitigation as a compliance option and placed responsibility on the shoulders of the permittees. Permittee-responsible offsets comprised 97% of overall global compensatory mitigation activity by area of habitat restored, protected, or created. In other words, the USD 4.8B in transactions to third-party mitigation providers documented in Forest Trends' report (which only covers circumstances in which there was a transaction), was only a fraction of actual global spending. In reality, the dollars flowing to biodiversity offsets and other forms of mitigation is far higher than what has been systematically recorded to date.

An example of a third-party involved in this market is Ecosystem Investment Partners (EIP), a private equity firm specialized in acquiring, entitling, restoring, and sustainably managing lands with potential to generate wetland, stream, and endangered species mitigation bank credits in the United States. Since 2007, EIP has raised nearly USD 1B from private investors in the United States and internationally for its four private equity funds, including the latest tranche of USD 454.5M in 2020 (Deutz et al., 2020).

EIP invests in large-scale ecosystem restoration and conservation with committed capital from institutional investors, such as pension funds and insurance companies. Restoration investments generate credits that can be sold to entities to offset their negative environmental impacts in order to comply with environmental legislation. To be effective, investments must provide permanent protection of land delivering improved ecosystems services that meet government-determined standards and are scientifically measurable. These developments illustrate how investments in large-scale ecological restoration can be financially viable for private investors and deliver environmental improvements (OECD, 2019b).

To deal with different forms of trading sources and contaminants, integrated models have been developed in the last decade to assess the economic and environmental benefits of implementing water quality trading programs (Corrales et al., 2014). By assessing the effect of different caps on the market potential, these programs have been used as a solution to reduce nutrient pollution and to minimize costs, potential cost savings, and credit supply and demand. A good example is found in Lake Okeechobee, Florida, where the cost-effectiveness of the water quality trading program over a command-and-control approach was assessed in two agricultural subwatersheds (Upper Kissimmee and Taylor Creek/Nubbin Slough). The water quality trading regime led to phosphorus reduction (46% and 32%, respectively) and net cost savings (USD 34.9M and USD 3.2M per year, respectively). This enabled them to reach water quality goals while providing additional benefits to the whole watershed (Corrales et al., 2017).

Visitor fees and revenues from eco-tourism. NBS that provide amenity value can sometimes generate visitor fees, for example, if they are located in a national park or turned into a tourist destination, like a wetland. In Africa, for example, the World Travel and Tourism Council estimates that 3.6 million people are employed in the nature-based tourism industry, worth USD 29B in 2018. Globally, natural protected areas receive eight billion visitors annually generating (up to) an estimated USD 600B. This includes wetlands and forests that are important sites for birdwatchers, game watching, and other forms of recreation. Revenues from eco-tourism could also be generated, but as most of these tend to accrue to private stakeholders, the latter either would need to invest directly or a tourism tax would need to be applied to generate revenues for investment in NBS with multiple benefits.

External transfers: public grants and subsidies. As discussed in the first section, the vast majority of funding for NBS for water and land management, to date, has come from public coffers, either in the country where the NBS program is implemented or via international transfers from entities such as the European Union or the GEF. Public funds come from different types of public funding programs, not all of which would specifically refer to NBS. For example, in the European Union, an estimated 99% of the funding for NBS for the conservation of watersheds comes from public funding sources (an average of EUR 5.5B per year between 2014 and 2020), via CAP subsidies (mentioned at the beginning of the chapter), dedicated EU grants, as well as grants from national and local governments or public agencies.

Public subsidies, research grants, and private donations can also play a key role in supporting and proving pilot projects when there is a perceived need to demonstrate the impacts of an innovative or novel technology. In the case of constructed wetlands, for example, a solution's ability to deal with various effluent qualities, different temperatures, and different technical designs may be tested for efficacy and robustness. The European Union has funded many such projects through its Horizon 2020 (research focused), LIFE, or InterReg programs. Government departments, local governments, or independent bodies (such as the river basin agencies in France) can provide sizable grants for NBS projects. However, such funding can be challenging to mobilize, with lengthy application processes.

External transfers: private donations. Common sources include funding from corporate or independent foundations, private companies' corporate social responsibility programs, or individual contributions, for example, membership fees, contributions to environmental NGOs or crowdfunding mechanisms. Donations may sometimes be provided—or pooled—as an endowment. If so, these funds can be invested, and returns from those investments can be used for ongoing operations and maintenance. The capital itself remains untouched to continue to generate returns over time.

Producers, cooperative, and commodity buyers can also be the driving force to implement and finance NBS along the agriculture supply chains. This is particularly important in countries like Brazil, where the maintenance of ecosystem services and agriculture development are strongly connected. The *Consorcio das Águas do Cerrado* in the coffee-producing region of the city of Patrocionio, Minas Gerais, Brazil, is one successful example of supply chain actors' involvement in reestablishing ecosystem services and promoting climate-smart solutions. A consortium was established between IUCN, Nespresso, Critical Ecosystem Partnership Fund, Lavazza, Nestlé, Expocaccer, Conservation International, and Cooxupé to implement climate-smart approaches addressing ecosystem risks at a property and landscape level. The Investment Program for the Conscious Producers was introduced to design and implement a long-term strategy to safeguard and enhance ecosystem services for all the producers whose properties are part of Feio river basin to make the regioń's landscape more climate resilient.

The program supported by the roasters and donations involved 67 producers, and a total of 43 Individual Property Projects were developed. These Property Projects were executed on 24 farms and formulated detailed recommendations on the resources necessary to implement climate-smart farming and restoration interventions. The producers have offered to cofinance 40% of the implementation costs of the restoration interventions and 88% of the implementation costs of the climate-smart agriculture interventions. In the end, the producers provided almost 50% cofinancing for total implementation costs (over USD 127K). The consortium provided 30,000 seedlings and over 14,000 seeds used in 12 different restoration strategies tailored for each property and implemented with the farms. The consortium defined a specific mix for restoration actions based on specialist knowledge of each area's particular vegetation characteristics, using up to 70 different native species. The goal is to reach 170 ha of rehabilitated area. The potential for this effort becomes clear when the whole Cerrado Mineiro coffee region is considered. A total of 4500 coffee producers share 234,000 ha of productive land and account for 12.7% of Brazil's coffee production. In this case, involving important actors in the supply chain, like cooperatives and international roasters, was crucial to financing the implementation of NBS.

Mobilizing repayable financing where appropriate

Mobilizing repayable financing can help bridge the financing gap and bring forward investments, which can then be repaid over time, usually with the additional cost of debt service or dividends to investors who provide the capital. Repayable finance can take several forms, including commercial or concessional loans, raising debt on capital markets with bonds, or equity investments that generate a return on the capital provided via asset appreciation and/or dividends. Accessing repayable finance requires an attractive risk-adjusted return, meaning the return must be commensurate with the risk of the investment. Financing must be repaid, typically via revenue streams generated from beneficiaries or public funding (OECD, 2018a). Mobilizing repayable financing can enable investments earlier in a project's lifetime and at a larger scale than may have otherwise been possible if simply relying on expected funding flows from public budgets or users. Indeed, if nature is not adequately taken care of, it can deteriorate to a point where restoration can become far costlier than protecting nature in the first place.

Although repayable financing for NBS for water security has increased in recent years, not all NBS programs will be good candidates to mobilize such financing. As previously noted, mobilizing repayable finance requires investments to generate financial returns to investors, which may be difficult, as not all benefits of NBS for water security programs can be monetized and used to generate revenue streams. Before seeking to access such finance, decision-makers should consider how the proposed NBS could generate revenues and/or lead to asset appreciation that can provide a financial return.

Investors are usually seeking market-rate returns, and they would like to see those returns as quickly as possible. One of the more common challenges with NBS projects is the time lag between when capital investments are made and when improvements that generate cash flows are realized. For example, when converting a property to organic production, a farmer must wait several years before they see price increases garnered by their organic products. The selection of financing mechanisms for such a project would need to include a strategy to address the lag between initial investment and when increased revenues materialize.

Many NBS for water security programs focus on land interventions that generate water security benefits over a prolonged period; typically, 15+ years. In this case, the benefits eligible for repayable financing are likely to be valuable cobenefits of the NBS program that can be realized in the short term. If a project has identified such benefits, potential ways to mobilize repayable financing include loans (either concessionary or commercial), bonds, or equity participations.

These are shown in Fig. 15.3 and discussed in more detail in this chapter. In addition to considering the attributes of the project itself and the related risk-adjusted return, the maturity of local capital markets merits attention as well. Notably, there is limited value exploring issuing a local bond to finance NBS for water security in a country where bond issuance for more standard investments is already limited.

Concessionary and commercial loans. The most common approach is to obtain a loan that would provide money upfront and can be repaid over time. Repayment would include the initial amount of capital provided, referred to as capital repayment or loan amortization plus interest. Loans are referred to as "concessionary" or "soft" when they are provided to the borrower on terms that are more advantageous than commercial loans. This may include an interest rate that is lower than the market rate, a longer maturity period (i.e., more time to repay) or a "grace" period (i.e., a period during which the borrower does not have to repay anything). A loan would be the easiest type of repayable financing for projects of any size to tap into, as long as they can demonstrate clear revenue streams to repay the loan from all of the funding sources previously mentioned.

Loans that incorporate an NBS or green infrastructure component are few and far between due to a number of factors, including lack of demand. However, more examples are now starting to emerge as exemplified in a recent European Investment Bank (EIB) loan to the city of Athens that financed green infrastructure projects aimed to increase the city's ability to adapt to climate change (Box 15.6).

Bonds. A bond is similar to a loan, except it is a market-based instrument. Instead of borrowing money directly from a financial institution, typically, from a bank, the borrower would issue a bond that gets sold to multiple investors, also referred to as "bond holders." Each investor, or "bond holder," would receive a regular "coupon" payment, which is essentially equivalent to an interest payment.

The green finance market has seen the emergence of "green bonds," for which specific rules apply. Beginning in 2007, bonds labeled "green bonds" or "climate bonds" were developed to alert potential investors that the investment had a positive environmental and/or climate

---

Box 15.6   EIB nature-based climate adaptation project in Athens.

In 2019, Athens began four innovative climate adaptation projects with EUR 5M in financing from the Natural Capital Finance Facility (the Facility), a program run by the EIB, in cooperation with the European Commission, which focuses on nature conservation, biodiversity, and adaptation to climate change through NBS. The Facility focuses on green and blue infrastructure measures that can support urban resilience by retaining water and slowing flows, lowering temperatures, cleaning the air, and promoting biodiversity.

The Facility aims to increase green space by at least 25% and to introduce eight different climate adaptation measures that will include, among many others, trees, bushes, and bird houses.

Athens is the first city to be financed under the Facility, and they intend to replicate the project in other locations. The Facility provides free technical assistance to the municipality to pay for an international consortium of consulting firms that will support the city in designing and developing their green infrastructure elements. The consulting team is managed and guided by both the municipality and the EIB.

The loan from the Facility is part of a EUR 55M loan from the EIB to support Athens' "2030 Resilience Strategy." The loan will primarily finance energy upgrades, earthquake fortification of municipal buildings, and development and waste management initiatives.

From Gkionaki M. Athens bets on green infrastructure and biodiversity. European Investment Bank Media Centre (stories); 2019. [Accessed October 2020]. https://www.eib.org/en/stories/athens-climate-adaptation.

adaptation or mitigation component. The borrower commits to using the money from the bond, also referred to as bond proceeds, specifically for "green" investments, which are defined in advance when the bond is issued.

The green and climate bonds market grew to an estimated USD 200B in 2018. As a result, several certification schemes have been developed to determine if the investments meet criteria for sustainability. In 2014, a consortium of NGOs including the Climate Bonds Initiative (CBI) (https://www.climatebonds.net/), the Alliance for Global Water Adaptation (AGWA) (https://www.alliance4water.org/), the World Resources Institute (WRI) (https://www.wri.org/), Ceres (https://www.ceres.org/), and CDP (formerly Carbon Disclosure Project) (https://www.cdp.net/en) worked to develop a set of Water Infrastructure Criteria. The criteria, which score the water resilience of both gray solutions and NBS, were created with consultation from about 150 experts and investors spanning five continents (CBI, 2018).

These criteria emphasize the use of basin-scale management, the importance of modeling ecological and hydrological qualities through the lens of robustness and flexibility, and the integration of adaptive governance and allocation systems (Gartner and Matthews, 2018; Matthews, 2018).

Since the Water Infrastructure Criteria were launched in 2016, many large-scale bonds have been issued by utilities, municipalities, and banks. To date, the CBI Water Infrastructure Criteria have been applied to approximately USD 12B of investments over six continents, including more than USD 5B for NBS (Mauroner et al., 2019). The first to issue a climate bond under these criteria was the San Francisco Public Utilities Commission, with subsequent climate bonds going toward financing new infrastructure assets or system improvements and upgrades for USD 240M (May 2016), USD 256M (December 2016), and USD 402M (July 2018) (Mauroner et al., 2019).

Bonds can be issued against more than one of CBI's sector criteria simultaneously. One example is the green bond issued by the City of Cape Town (USD 75M) in 2017, with proceeds designated toward water infrastructure and low-carbon transport.

Anglian Water, a water company in the United Kingdom, also finances most of its investment program through green bonds and specifically marks out its investment in NBS as the greenest. The company has acquired experience with NBS by engaging with local NGOs and is now looking to scale up by mobilizing financing, in part, via green bonds. Since launching its first green bond in 2017, Anglian Water has drawn down five additional green bonds, now in operation—the most recent having launched in 2020. It has funded 850 capital investment projects totaling GBP 811M, the bonds for which will mature between August 2025 and June 2039 (Anglian Water, 2019).

In May 2019, the Dutch State Treasury Agency issued a certified bond using the Water Infrastructure Criteria, which mobilized nearly USD 6.8B for low-carbon development and sustainable water management, including financing natural infrastructure solutions that would reduce flood risks in the country (Box 15.7). Also in the Netherlands, the NWB

Box 15.7 **The Netherlands issues one of the largest green bonds to date.**

In 2019, the Dutch government issued one of the largest green bonds to date, certified by the CBI Water Infrastructure Criteria for EUR 5.98B, equivalent to approximately USD 6.8B. The bond, issued for investments in low-carbon development and sustainable water management, finances natural infrastructure solutions that will reduce the country's flood risks in coastal and low-lying areas, in addition to other projects.

The green bond was received very positively by investors. Within 90 minutes of the bond's issuance, the government had received EUR 21.2B worth of orders for EUR 5.98B in certificates, oversubscribing the bond by three-and-a-half times the original amount. The prospect of a AAA-rated bond that credibly met investors' needs for certified and innovative green investments was a success. The Netherlands, for their part, gave preference to investors with registered sustainability credentials.

This response signals the current shift in conservation finance from a niche market to a mainstream, large-scale investment strategy.

This natural infrastructure approach is not new in the Netherlands, where a multigovernance platform was created to facilitate the scale-up of NBS. The multistakeholder Room for the River program implemented 34 projects across the Netherlands between 2007 and 2018. Projects were located in the catchments of the Netherlands' four main rivers and was entirely funded by the Dutch government with a total budget of EUR 2.3B (Trémolet et al., 2019).

From Anderson J, Gartner T, Mauroner A, Matthews J. Conservation finance takes off as the Netherlands issues one of the largest green bonds ever. Water Resources Institute (WRI); 2019. [Accessed October 2020]. https://www.wri.org/blog/2019/06/conservation-finance-takes-netherlands-issues-one-largest-green-bonds-ever.

(formerly known as Nederlandse Waterschapsbank), commonly referred to as the Dutch Water Bank, has issued EUR 3.6B in green bonds, which they refer to as "water bonds." Only a very small proportion of the proceeds, however, is used for investments in NBS (Trémolet et al., 2019).

Many other bond labels have emerged in recent years, such as sustainability, social, or sustainability-linked bonds. Bonds, however, can only be used by large creditworthy borrowers or for large projects. Organizing a bond issue generates significant transaction costs and are, therefore, not applicable to all circumstances, particularly where NBS programs are implemented by small actors with limited access to sustainable revenue streams.

Equity participations. Another way to mobilize repayable finance is by selling shares into a business. When purchasing shares, an investor provides capital and would expect a dividend in return. The capital would only be paid back if the company is wound down or when the investor sells their shares. Impact investors may be prepared to accept a slightly lower financial return on their investment, in exchange for social or environmental benefits, but most will still expect a near-market rate of return.

A "fund" is a common investment vehicle for equity participation that pools investor money—both public and private—into a centrally managed portfolio. The portfolio is managed by a fund manager who makes decisions with a Board and technical advisor about which projects to invest in. Funds—which each have their own investment criteria—do not only make equity investments. However, it is a common way for impact investors to amplify the results of their contribution: if five impact investors pool their funds, they can make a significant investment of USD 10M together rather than USD 2M separately (see Box 15.8).

Box 15.8   Moringa fund and technical assistance facility.

Managed by the Moringa Partnership, the Moringa Fund (https://www.moringapartnership.com/) is a EUR 84M investment fund that targets profitable, large-scale agroforestry projects with high environmental and social impacts in Latin America and sub-Saharan Africa. Launched in 2010, the Fund aims to increase land rehabilitation and reforestation while simultaneously creating sustainable jobs by incorporating small or medium farmers into the value chain. Their investments seek, wherever possible, to develop or create a nucleus plantation or processing facility that can improve production and provide greater access to the market; farmers can usually sell processed goods for a higher price than raw commodities. The Fund aims to create jobs, improve livelihood security for farmers, restore land, and generate 10%–12% return for investors.

The Fund is supported by public and private investors like FMO, the GEF, La Compagnie Benjamin de Rothschild (CBR), FISEA (FISEA is owned by *Agence Française de Développement* and managed by its subsidiary PROPARCO), the CAF (CAF is the development bank of Latin America with 18 member countries—16 in Latin America and the Caribbean, in addition to Spain and Portugal—and 14 private banks) and others. It makes equity and quasiequity investments of EUR 4–10M, alongside the Agroforestry Technical Assistance Facility (ATAF) (https://www.moringapartnership.com/agroforestry-technical-assistance-facility/), which provides grant-based technical assistance to investments of the Fund. The ATAF supports capacity building of the investees, outgrowers, smallholder farmers, and vulnerable communities located in the vicinity of the operation area, and it identifies potential new sources of revenue from products or byproducts of the operation. Its goal is to improve knowledge of agroforestry models among local producers in the area and hopefully diversify their revenues by encouraging a circular economy.

Thus far, the Fund has invested in 10 companies in 8 countries and supports more than 10,500 farmers through outgrowing programs; 2500 of these farmers are in underserved rural communities. They are implementing agricultural best management practices on more than 11,500 ha of land, 14% of which is preserved forest. Investments are also funneled through the ATAF, which has developed 15 projects supporting small producers and employees of their investee companies. One such investment is Tolaro Global, a leading cashew processing company in Benin. Investment from Moringa enabled Tolaro to implement best industrial practices and obtain organic, fair-trade, and BRC-certification in 2019. Tolaro Global collects cashews from a network of over 7000 farmers and comprises about 80% of national kernel exports.

From Moringa Partnership website: www.moringapartnership.com.

Blended finance: Combining funding and financing to ensure sustainability and increase scale

Funding and financing would typically need to be mobilized at different stages of program development and ideally in combination, using what is commonly referred to as "blended finance" approaches. "Blended finance" is defined as the strategic use of development finance for the mobilization of commercial finance toward sustainable development in developing countries (OECD, 2018b). It is a structuring approach that allows organizations with different objectives to invest alongside each other; it is not an investment instrument in itself. Some investors are risk-averse: their perceived risk of an investment is high, or they assess that returns are not commensurate with the level of risk. Such investors would be more comfortable investing alongside a public entity that may be willing to take on the risk of losing some of the invested funds. The approach applies to both developing and developed countries.

Many financial institutions are actively looking to deploy capital for good NBS projects, whereas NBS project implementers frequently identify a financing gap to get their programs off the ground. For these reasons, it is important to consider whether a blended finance approach can be deployed to bridge the financing gap.

Program developers should identify where catalytic support from public or philanthropic sources can help kick-start program preparation and which funding streams can be deployed to repay financing. These revenue streams should be as secure and reliable as possible, with commitments evidenced through contracts, agreements, pledges, and so on. The more complex an NBS project—considering technical, governance, and financing elements—the longer and costlier it might be to prepare. Project preparation would typically be funded through philanthropy or public subsidies, unless such costs are borne by a private entity that is willing to invest money upfront and recoup it via project revenues. When determining the full cost of a project—and therefore funding need—transaction and project preparation costs should be accounted for.

To provide secured revenue streams as repayments for program implementation, it would be preferable to identify a recurrent charge that can be applied to beneficiaries of the investment, such as a tariff or a recurrent public fund transfer.

At present, examples of blended finance approaches for NBS water and land management are few and far between. In 2018, a Forest Resilience Bond (FRB) (https://www.blueforest.org/forest-resilience-bond) was launched in California to mobilize financing to reduce fire risk by employing forest best management practices. Investors provide upfront capital to pay for restoration projects that protect forest health, while public and private beneficiaries agree to make payments based on the water, fire, and other benefits created by restoration activities. Water companies pay into this mechanism because they benefit from reduced sediment loads, which would otherwise increase after a forest fire. The FRB contracts with the beneficiaries to

---

Box 15.9   California, United States—Forest Resilience Bonds.

The FRB was first applied in the North Yuba River watershed. The aim of the FRB Yuba Project I LLC is to protect 15,000 acres of forest in California's Tahoe National Forest. The project is located 50 miles from the town of Paradise, the site of California's deadliest wildfire in history. A group of private investors—CSAA Insurance Group, Calvert Impact Capital, the Rockefeller Foundation, and the Gordon and Betty Moore Foundation—provided an initial investment of USD 4.6M. Both foundations accepted a below-market interest rate of 1%, whereas other financiers expect to earn a 4.5% interest rate. This financial mechanism was conceived by Blue Forest Conservation (BFC), a start-up launched by business school graduates of the University of California, Berkeley, with support from the WRI.

 The beneficiary group, which will repay investors and provide the user payments, includes the Yuba Water Agency (a water utility) and the California Climate Change Investment program (USD 2.6M in grant funding via a program called CalFire). The primary implementing partner is the National Forest Foundation, which organizes contracts with the US Forest Service (USFS) and the state. The FRB structure sped up implementation from an estimated 10 years to 4 years.

 The FRB Yuba Project I LLC is managed by a new nonprofit set up by BFC. Restoration is carried out through prescribed burning, ecologically based tree thinning, meadow restoration, and invasive species management, all specifically designed to reduce the risk of severe fire, improve watershed health, and protect water resources.

---

share in the costs of forest restoration while providing modest returns to investors. Additional detail on this structure is provided in Box 15.9.

Many of the blended finance structures that have been set up or are in the process of being set up share a common approach (Fig. 15.5), which look to put greater emphasis on results-based financing. Such investment vehicles allow project teams to mobilize financing upfront, which can be repaid based on evidence of outcomes. In this type of structure, private investors prefinance through a special purpose vehicle a series of activities that contribute to the delivery of a prespecified environmental outcome. If the agreed impact is achieved, an outcome payor (typically a public agency) reimburses the investor plus a preagreed return. These schemes always include external validators who assess the outcomes to determine repayment levels.

Recommendations: Accelerating investment in NBS for water security

This chapter has shown that investment in NBS for water security has mostly been funded by public subsidies to date, with limited investment from private actors. However, the cases of private investment we have seen have demonstrated the critical role of private finance and how NBS benefits can be quantified and monetized. Previous sections identified key challenges that currently limit a significant increase in investments in NBS for water security and distinguished between challenges that relate to the status quo and those that are linked to the intrinsic characteristics of NBS. By clearly identifying potential sources of funding early,

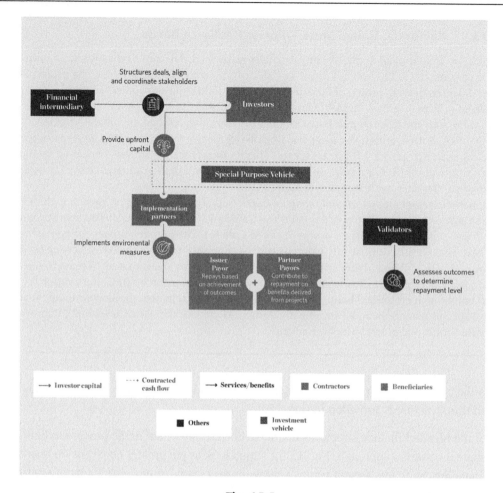

Fig. 15.5

Example of an outcome-based blended finance structure. *From Trémolet S, et al. Investing in nature for Europe water security. London: The Nature Conservancy, Ecologic Institute and ICLEI; 2019; originally adapted from Environmental Defense Fund (EDF). Financing resilient communities and coastlines: How environmental impact bonds can accelerate wetland restoration in Louisiana and beyond. Environmental Defense Fund Qualified Ventures; 2018, August.*

it is possible to mobilize a wide range of potential revenue streams to monetize both water-related benefits and other cobenefits, for example, carbon, biodiversity, public health, cultural values, or social cohesion. In cases where funding sources are insufficient to cover costs or where it is critical to front-load investments to deliver at scale, repayable financing could also be mobilized.

To accelerate investments in NBS for water security, action needs to be taken at multiple levels. Program developers need to consider the role of NBS in delivering water security outcomes, and they should prepare their NBS investment programs with consideration for how the program can be funded. In some cases, program developers should consider if costs of the investment program could be cut if the proposed NBS are part of a hybrid green-gray investment program. In doing so, they need to pay particular attention to how program preparation costs can be covered, particularly when the approach is innovative and can generate high transaction costs.

In this chapter we set out recommendations at two levels: first, relative to what program developers can do to formulate NBS investment programs; and second, identifying some systemic, sector-level changes that could create a more conducive environment for NBS investments.

Recommendations for NBS program design

Build a multiskilled team for NBS program preparation. Preparing a financially viable NBS program requires a well-trained, multiskilled team. Generally, projects advance through four key building blocks: (1) identifying potential technical solutions, (2) conducting an economic analysis of the program, (3) building a funding and financing strategy, and (4) defining the governance arrangements. Each stage benefits from different capacities—economist, hydrologist, financial assessment and structuring, communications, and so forth—therefore, many of the examples throughout this chapter were developed by multifaceted teams composed of NGOs, government agencies, financiers, utilities, corporations, and more.

Engage with stakeholders. Each building block will entail substantial stakeholder consultation and engagement, and successful projects are able to present the details of the proposal in a clear and transparent manner. The goal is to garner stakeholder support to collaboratively build a sound business case for the investment, with associated financing and governance arrangements. The latter needs to be established so as to give clarity on who is going to implement the investment, who is accountable for results, who is in charge of monitoring and evaluation, who is in charge of mobilizing and channeling funding, and who is responsible for repaying upfront financing if initially provided. This would usually best be done through establishing a temporary governance structure to manage the program preparation phase, with a clear definition of roles and responsibilities, distinguishing between partners that are central to the program (in their quality of program sponsors, implementers or financial contributors) and a broader set of beneficiaries, which need to be consulted but would not necessarily have the ability to make decisions about the investment itself.

Evaluate the need for establishing collective action platforms. While the multiple benefits of NBS projects is often touted as an advantage, this can sometimes create challenges for mobilizing investments if no single actor derives enough benefit to justify investing on their own. Setting up collective action platforms (such as WFs) allows all beneficiaries to work together toward a common NBS investment program, from which they all benefit. Where such structures are not already in place, beneficiaries should work together to establish governance and financing structures that enable joint planning, investment, management, and maintenance. Models for such platforms exist around the world, and they can often be a prerequisite to attract funding and financing from various sources. To the extent possible, these platforms should aim to develop sustainable funding strategies to assess how investments in NBS can be sustained and deliver results over time, as the example of Cape Town demonstrates (Box 15.3).

Identify potential funders and contributors. For any proposed NBS project, it is important to identify which stakeholders would benefit from the investments and which stakeholders could invest in the project; they may or may not overlap. During the stakeholder mapping process, program developers should discern whether users are incentivized—or able—to make investments themselves or in association with other stakeholders, like the local government. A local government could support the investment by allowing implementation within their jurisdiction, permitting NBS investment in their procurement rules or by providing in-kind or financial support.

Assess enabling legal framework. In assessing the feasibility of an NBS program, project developers should examine whether the proposed NBS are allowed or even recommended in the legal framework. In addition to the water management policy, it is recommended to look outside the water sector and consider policies governing urban planning, agriculture, biodiversity, climate adaptation, and/or green finance. If obstacles to implementation are identified, program implementers should consider forming a stakeholder coalition at the sector level to advocate for policy changes. In some cases, where a policy has been defined but without detail on implementation, an NBS program can provide a basis for "stress-testing" broad principles enshrined in the policy and clarifying implementation approaches.

Assess regulatory support. Before settling on the design of the project, developers should identify whether regulators would provide the necessary permits and enquire about the information they would need to authorize the investment. In most cases, they would likely need to provide evidence on effectiveness and cost-effectiveness of the proposed NBS or collection of NBS.

Provide evidence of efficacy. To garner financial, political, regulatory, and public support, program implementers should thoroughly document evidence on the effectiveness (and cost-effectiveness) of the proposed NBS, examine where it has been implemented in the past, and identify the outcomes and the cobenefits of the specific NBS or NBS program. If it was implemented as an alternative to gray infrastructure, they should obtain evidence on how the outcomes of the NBS program compared with that of the gray infrastructure and whether this comparison was based on monitoring or estimated via models. When the NBS is implemented as part of a hybrid package of green and gray investments, they should identify whether the NBS component delivered substantial cost reductions or generated additional cobenefits. This is not just about looking back; establishing strong monitoring and evaluation systems that capture both outcomes and costs could enable the team to derive cost-effectiveness estimates and build the evidence base for all future investments.

Identify available funds and preconditions for obtaining such funds. For an NBS program to be successful, managers must understand what funds are available and whether there are preconditions to accessing those funds. Whether seeking investment or looking to invest, it is important to understand the following about the NBS project or program:

- What benefits will be generated? Can value be assigned to any of these benefits?
- How much will the NBS project cost?
- Where will potential revenue streams for the investment come from?
- Who can pay and how much?
- Who will be responsible for delivering the project? Does mobilizing the potential funding create specific requirements in terms of governance arrangements?
- Is there a financing gap? How can such a gap be covered and from which financing sources?
- If repayable financing is mobilized, who will be responsible for repaying and how?

Recommendations for broader changes in sectoral approaches

In addition to what program developers can do at the level of specific programs, broader changes need to take place in order to create a more conducive environment for NBS investments. These changes, although potentially harder to achieve, could unlock much greater interest in NBS investments by a variety of actors.

Value natural capital on balance sheets. Natural capital should be accounted for in investment decisions so NBS for water security projects can be compared with gray infrastructure options on a more equal footing. A single NBS project can generate multiple benefits that should be reflected as assets. Water sector actors should, likewise, measure the impact of their investment decisions on natural capital and give priority to NBS for water security projects that increase natural capital values. For example, Anglian Water is in the process of adopting natural capital accounting, which reflects a greater focus on NBS investments as part of its overall investment program. In addition to valuing natural capital, projects should verify impacts and track costs so potential investors can better understand how NBS solutions perform when compared to alternative built infrastructure. Strong methods to accurately estimate cobenefits would need to be developed.

Build a pipeline of investable projects that could be eligible for repayable financing. Mobilizing repayable financing brings funding into projects earlier and can therefore prevent further land and water degradation. Private financing provides access to the substantial, liquid financing markets necessary for scale. Funders, currently, consider fragmented investment opportunities with limited coordination. Identifying where certain types of NBS can work on a landscape scale and building pipelines of NBS for water security projects can help mobilize a mix of private and public funding and financing. Importantly, funders are looking for large-scale transactions, and current NBS projects tend to be very small in scale. Shared pipelines of NBS programs should be developed, potentially with philanthropic or public funding and by running innovation prizes or setting up innovation labs.

Intermediaries have a key role to play. Given the need to build a pipeline of projects and to connect NBS project sponsors with funders and financiers, intermediaries—NGOs or consultancies—have a significant role to play to help structure such projects. They can help structure projects that match beneficiaries' demands with funders' interests and bridge the knowledge gap between water sector actors and financiers. Financial institutions with an interest in green finance are typically looking for "shovel-ready" projects and often lack funding and time for project preparation. They have their own internal metrics and priorities and would need to be convinced of a project's value compared to that of competing investments. Private financiers are not very familiar with water and land NBS and typically question the certainty of returns and revenue streams. Intermediaries are often the organizations who can best answer these questions and provide assurances and would need to be supported so as to develop a thriving "market" in which NBS program developers and funders can work together to accelerate and scale-up investments in NBS for water security.

References

Abell R, et al. Beyond the source: The environmental, economic and community benefits of source water protection. Arlington, VA: The Nature Conservancy; 2017.

Anglian Water. Green bond impact report 2019. Anglian Water Services Limited; 2019.

Bennett G, Gallant M, Ten Kate K. State of biodiversity mitigation 2017 markets and compensation for global infrastructure development. Washington DC: Forest Trends; 2017.

Bennett G, Ruef F. Alliances for green infrastructure: State of watershed investment 2016. Washington, DC: Forest Trends; 2016.

Cassin J, Matthews JH. Setting the scene: nature-based solutions and water security. In: Cassin J, Matthews JH, Lopez-Gunn E, editors. Nature-based solutions and water security: An action agenda for the 21st century. Elsevier; 2021.

Climate Bond Initiative (CBI), Alliance for Global Water Adaptation (AGWA), Carbon Disclosure Project (CDP), Ceres, World Resources Institute (WRI). Water infrastructure criteria under the climate bonds standard background paper: Background paper. London: Climate Bonds Initiative; 2018. https://www.wetlands. org/publications/mastering-disaster-changing-climate-reducing-disaster-risk-resilient-water-management/. Climate Bonds Initiative website https://www.climatebonds.net/standard/water.

Cooper R, Matthews J. Water finance and nature-based solutions. GSDRC, Alliance for Global Water Adaptation (AGWA); 2020. https://opendocs.ids.ac.uk/opendocs/bitstream/handle/20.500.12413/15592/857_Water_Finance_and_Nature-based_Solutions.pdf?sequence=1&isAllowed=y.

Corrales J, Naja G, Bhat M, Miralles-Wilhelm F. Modelling a phosphorus credit trading program in an agricultural watershed. J Environ Manag 2014;0301-4797. 143:162–72.

Corrales J, Naja G, Bhat M, Miralles-Wilhelm F. Water quality trading opportunities in two sub-watersheds in the northern Lake Okeechobee watershed. J Environ Manag 2017;0301-4797. 196:544–59.

Coxon C, Gammie G, Cassin J. Mobilizing funding for nature-based solutions: Peru's drinking water tariff. In: Cassin J, Matthews JH, Lopez-Gunn E, editors. Nature-based solutions and water security: An action agenda for the 21st century. Elsevier; 2021.

Deutz A, Heal GM, Niu R, Swanson E, Townshend T, Zhu L, Delmar A, Meghji A, Sethi SA, Tobinde la Puente J. Financing nature: closing the global biodiversity financing gap. The Paulson Institute, The Nature Conservancy, and the Cornell Atkinson Center for Sustainability; 2020.

Dominique K, Matthews N, Danielson L, Matthews JH. Why governments embrace nature-based solutions: the policy rationale. In: Cassin J, Matthews JH, Lopez-Gunn E, editors. Nature-based solutions and water security: An action agenda for the 21st century. Elsevier; 2021.

Garrick D, Iseman T, Gilson G, Brozovic N, O'Donnell E, Matthews N, Miralles-Wilhelm F, Wight C, Young W. Scalable solutions to freshwater scarcity: advancing theories of change to incentivize sustainable water use. Water Secur 2020;9:100055.

Gartner T, Matthews JH. Forests and wetlands are water infrastructure. New green bond helps finance their protection. World Resources Institute; 2018. May 22 https://www.wri.org/blog/2018/05/forests-and-wetlands-are-water-infrastructure-new-green-bond-helps-finance-their.

Higgins JV, Zimmerling A, editors. A primer for monitoring water funds. Arlington, VA: The Nature Conservancy; 2013.

Leisher C, et al. A guide to monitoring and evaluating water funds. Arlington, VA: The Nature Conservancy; 2019.

Mason N, et al. The untold story of water in climate adaptation part II: 15 countries speak synthesis report. Global Water Partnership (GWP) and Overseas Development Institute (ODI); 2019. https://www.gwp.org/globalassets/global/events/cop25/gwp_synthesisreport.pdf.

Matthews JH. Financing water resilience: the emergence of green and climate bonds for water. In: The United Nations world water development report 2018: Nature-based solutions. United Nations world water assessment programme. Paris: UNESCO; 2018.

Matthews J, Matthews N, Simmons E, Vigerstol K. Wellspring: Source water resilience and climate adaptation. Arlington, VA: The Nature Conservancy; 2019. https://www.nature.org/content/dam/tnc/nature/en/documents/Wellspring_FULL_Report_2019.pdf.

Mauroner A, Matthews J, Alliance for Global Water Adaptation (AGWA), USA, Gartner T, World Resources Institute (WRI), USA. Financing sustainable growth through climate bonds. Global Water Forum; 2019. https://globalwaterforum.org/2019/12/23/financing-sustainable-growth-through-climate-bonds/.

OECD. A framework for financing water resources management. OECD Studies on Water, Paris: OECD Publishing; 2012.

OECD. The political economy of biodiversity policy reform. Paris: OECD Publishing; 2017.

OECD. Financing water: Investing in sustainable growth. Policy Perspectives, OECD Environment Policy Paper No. 11 http://www.oecd.org/water/Policy-Paper-Financing-Water-Investing-in-Sustainable-Growth.pdf; 2018a.

OECD. Making blended finance work for the sustainable development goals. Paris: OECD Publishing; 2018b.

OECD. Making blended finance work for water and sanitation: Unlocking commercial finance for SDG 6. OECD Studies on Water, Paris: OECD Publishing; 2019a. https://doi.org/10.1787/5efc8950-en.

OECD. Summary of the 4th meeting of the roundtable on financing water, 26-27 June, 2020, http://www.oecd.org/water/Summary-RTmeeting-26and27June.pdf; 2019b.

OECD. Agricultural policy monitoring and evaluation 2020. Paris: OECD Publishing; 2020. https://doi.org/10.1787/928181a8-en.

Smith DM, Matthews JH, Bharati L, Borgomeo E, McCartney M, Mauroner A, Nicol A, Rodriguez D, Sadoff C, Suhardiman D, Timboe I, Amarnath G, Anisha N. Adaptation's thirst: Accelerating the convergence of water and climate action. Background Paper prepared for the 2019 report of the Global Commission on Adaptation, Rotterdam and Washington, DC. Available online at www.gca.org; 2019.

Trémolet S, et al. Investing in nature for Europe water security. London: The Nature Conservancy, Ecologic Institute and ICLEI; 2019.

Vigerstol K, Abell R, Brauman K, Buytaert W, Vogl A. Addressing water security through nature-based solutions. In: Cassin J, Matthews JH, Lopez-Gunn E, editors. Nature-based solutions and water security: An action agenda for the 21st century. Amsterdam: Elsevier; 2021.

Winpenny J. Water: Fit to finance? Catalyzing national growth through investment in water security. Marseille, France: World Water Council; 2015.

WWAP (United Nations World Water Assessment Programme)/UN-Water. The United Nations world water development report 2018: Nature-based solutions for water. Paris: UNESCO; 2018. https://unesdoc.unesco.org/ark:/48223/pf0000261424/PDF/261424eng.pdf.multi.

Further reading

Browder G, Ozment S, Rehberger Bescos I, Gartner T, Lange G. Integrating green and gray: Creating next generation infrastructure. Washington, DC: World Bank and World Resources Institute; 2019.

Coalition for Private Investment in Conservation. (CPIC). Conservation investment blueprint: Forest resilience bond. Blue Forest Conservation; 2019.

Cooper G, Trémolet S. Investing in nature: Private finance for nature-based resilience. London: The Nature Conservancy and Environmental Finance; 2019.

Dai L, Matthews J. China's green bonds finance climate resilience. New Security Beat. The Wilson Center; 2018. March 22 https://www.newsecuritybeat.org/2018/03/chinas-green-bonds-finance-climate-resilience/.

European Commission (EC). Common implementation strategy for the water framework directive (2000/60/EC) guidance document no 1 economics and the environment—The implementation challenge of the water framework directive produced by working group 2.6. Luxembourg: WATECO; 2003.

Forest Resilience Bond (FRB), Blue Forest. Introduction to the forest resilience bond (FRB) prepared for the climate finance alliance; 2019. Presentation.

Gray E, Ozment S, Carlos Altamirano J, Feltran-Barbieri R, Morales G. Green-gray assessment: How to assess the costs and benefits of green infrastructure for water supply systems. Working Paper, Washington, DC: World Resources Institute; 2019.

McCartney M. Is green the new grey? If not, why not? Water Science Policy; 2020. 22 May 2020 https://www.watersciencepolicy.com/2020/05/22/is-green-the-new-grey-if-not-why-n.

Moringa Partnership website. n.d. https://www.moringapartnership.com/moringa/.

Norfolk River Trust. River Ingol wetland creation. Brochure; 2018.

The International Water Association (IWA) and The Nature Conservancy (TNC). Nature-based solutions utility spotlight: Anglian water; 2018.

US Forest Service. How the forest resilience bond works; 2019. Video.

Moving to scale: What is needed

Mainstreaming nature-based solutions through insurance: The five "hats" of the insurance sector

Elena Lopez-Gunn[a], Monica A. Altamirano[b], Mia Ebeltoft[c], Nina Graveline[d], Roxane Marchal[e], David Moncoulon[e], Beatriz Mayor[a], Florentina Nanu[f], Nora van Cauwenbergh[g], Peter van der Keur[h], Josh Weinberg[i], Pedro Zorrilla Miras[a], and Jan Cassin[j]

[a]ICATALIST S.L., Madrid, Spain, [b]DELTARES, Delft, Netherlands, [c]Climate Risk Advisory, Oslo, Norway, [d]INRAE, Montpellier, France, [e]Caisse Centrale de Réassurance, Paris, France, [f]Business Development Group, Bucharest, Romania, [g]IHE Delft, Delft, Netherlands, [h]Geological Survey of Denmark and Greenland, Copenhagen, Denmark, [i]Stockholm International Water Institute, Stockholm, Sweden, [j]Forest Trends, Washington, DC, United States

Key messages

- The insurance sector is becoming, and will be in the future, an important global actor to help build societal disaster preparedness and reduce the costs of disasters, while helping to improve the ability of society to cope successfully with climate change and other shocks.
- Insurance can play five key roles in mainstreaming nature-based solutions (NBS): as data providers and risk modelers; as providers of insurance (risk transfer) products; as investors (funding and financing NBS implementation); as innovators in the new impact economy; and as partners in cross-sectoral, multipartner collective action for building resilience via NBS.
- Damage data and avoided costs are critical to understand the most vulnerable locations, sectors, and populations and exposed assets. Thus, data sharing agreements have tremendous potential to help mitigate individual, private, and collective/public risks.
- Insurance is quickly moving to provide valuable new insurance products and services like parametric insurance for drought, insuring green assets, other premium-based options, or knowledge-sharing opportunities to help reduce the protection gap through, for example, new insurance for natural assets.
- Globally, the insurance industry is a large institutional investor, together with pension funds, focusing on longer-term investments. Nature is a long-term investment and thus it is well-aligned with sector investment strategies, provided value from nature can be captured and that the enabling frameworks are well-designed to provide regulatory certainty.

Nature-Based Solutions and Water Security. https://doi.org/10.1016/B978-0-12-819871-1.00006-3

Introduction

Climate change and the growing vulnerability of modern societies to resulting perils is acknowledged by the insurance industry at the global level, increasingly mobilizing the sector as a call to action. Systemic risks from ecosystem degradation, climate change, and, in particular, water-related risks result in both challenges and opportunities that affect the insurance industry in two ways. First, natural disasters are increasingly impacting the viability of individual insurance companies and the industry as a whole. As catastrophes become more frequent and severe, costs of premiums go up, often to unsustainable levels that reduce the customer base for insurance companies (City of Copenhagen, 2015). Some insurance companies have gone out of business, as they are not able to cover the payouts required when large disasters strike. Second, these systemic risks are also creating new business opportunities and partnerships for the industry. Building on the sector's unique expertise and skills, there are opportunities new insurance products to address climate risks, new roles and products for forecasting and risk modeling, and innovative risk reduction strategies to reduce their own exposure. These are all areas where expanding the insurance industry's role in climate mitigation and adaptation can create new business prospects. The insurance industry is already responding to these challenges and opportunities as their traditional role of compensating municipalities, businesses, and individuals for losses due to natural hazards is changing. It is now accompanied by a dedicated focus on prevention and enhancing adaptation to climate change, including through ecosystem-based adaptation.

Nature-based solutions (NBS), or "solutions that are inspired by nature and use or mimic natural processes" to address societal challenges (Cassin and Matthews, 2021; Cohen-Shacham et al., 2016), are now widely understood to be critical elements of successful strategies for climate adaptation, disaster risk reduction, and building resilience in the face of future shocks (Kabisch et al., 2017; Renaud et al., 2016; Seddon et al., 2020; Smith et al., 2019). We argue that the insurance industry, including both public and private insurance and reinsurance schemes, can significantly contribute to mainstreaming NBS as part of developing and assuring resilience in the face of current and future shocks. This chapter will explore the roles insurance can play in mainstreaming NBS as elements in the portfolio of approaches to address disaster risk management and climate impacts.

We outline a conceptual framework to analyze how the insurance sector can and indeed should contribute to mainstreaming NBS, with specific reference to the context within the European Union (EU), but with relevance globally. The framework is based on work done as part of the NAIAD H2020 Project (http://www.naiad2020.eu/; Natural Insurance Value: Assessment and Demonstration), including a survey of 61 primary and (re)insurance companies in 10 European countries. We first introduce and distinguish the concepts of *assurance* and *insurance* values of nature to describe the different ways that NBS are relevant to building resilience in the context of climate change. We then explore the multiple roles (or hats) that the insurance industry can and should play (or wear) in mainstreaming NBS for disaster risk reduction. The five roles refer to the functions insurance entities perform for managing risk to ensure the resilience and insurability of the system (including natural systems): as data providers and risk modelers; as providers of insurance (risk transfer) products; as investors (funding and financing NBS implementation); as innovators in the new impact economy; and as partners in cross-sectoral, multipartner collective action for building resilience via NBS. We illustrate these roles through cases and discuss issues and challenges in mainstreaming NBS via the insurance industry.

The insurance sector: Uniquely positioned to mainstream nature-based solutions for climate resilience

The insurance industry (Box 16.1) already has significant potential to lead in mainstreaming NBS for resilience through three areas of sector expertise are particularly important: (1) a deep understanding of the insurance coverage gap and costs to the industry and society from natural disasters; (2) an in-depth understanding of risk and a unique expertise in risk assessment, risk modeling, and risk management; and (3) new tools, insurance products, and investments that are beginning to move the sector from a sole focus on compensating losses to prevention and risk reduction.

Understanding the scale of loss from natural disasters

Recent trends in damage losses illustrate the growing threat to insurance companies from increasingly frequent natural disasters. Worldwide the total reported economic losses for

---

Box 16.1   Insurance and (re)insurance.

Insurance is a tool for managing risk used to transfer risk from the owner of an asset (government, community, business, individual, or household) to the insurance provider at the cost of an insurance premium. The insurance buyer pays the premium in return for a guarantee that the insurer will cover the costs if the insured asset suffers damages, as specified in the insurance policy.

The insurance industry includes insurers that cover communities, noninsurance companies, and households, as well as *(re)insurance* companies that transfer risk from insurance companies to cover their losses in the event of large-scale or extreme disasters. In this way insurance companies faced with making payments that are potentially larger than their assets—for example, when a catastrophic earthquake or hurricane strikes—can insure their assets against a total loss.

Traditionally, insurance companies have used historical data to model the risks of catastrophic losses and set insurance premiums and compensation for damages. Payments from insurance companies are particularly important for postevent recovery following disasters, but insurers are increasingly involved in earlier stages of disaster recovery, including preparedness, early warning systems, and the potential for prevention to reduce the costs of damages.

natural disaster events in 2019 reached around $140 billion. It is estimated that $56 billion of the losses were insured (Swiss Re, 2019). In the case of Europe, out of the €19.5 billion total estimated economic losses due to natural catastrophes in 2018, €7.3 billion were covered by insurers (www.insuranceeurope.eu). This leaves an enormous coverage gap of uninsured losses that affects insurance companies and society at large (Box 16.2).

Risk assessment expertise

The insurance sector is well-positioned to help societies understand the emerging risk landscape with climate change, given their in-depth expertise in risk management and the extensive knowledge on how to assess and value risk. The industry is already contributing significantly to studying climate change impacts to understand the rapidly changing risk landscape (Surminski et al., 2016; Tanner et al., 2015; Warner et al., 2009). There are some examples of (re)insurance industry studies on the consequences of climate change insured damages (CCR et al., 2020; Gloor, 2019; Golnaraghi, 2018; Moncoulon et al., 2018). For example, in France, it is estimated that the insured damages could increase by 50% by 2050 (35% due to increased hazards and 15% due to increased concentration of assets within

---

Box 16.2   The case for action: The impact of climate change on the insurance industry.

Climate change is and will impact the insurance sector by increasing the consequences from natural disasters. More frequent and intense events will occur, such as locally intense rainfall events, prolonged droughts, damaging wildfires, and extreme storms. For the insurance industry, the first consequence of climate change are physical risks. The increased frequency and extreme events lead to a higher exposure for the industry. Climate change also has long-term consequences and uncertainties, which are critical issues that the insurance industry must address to better estimate their exposure and the exposure of their policyholders. The insurance industry itself rated climate change as a generator of physical, transition, and liability risks. Physical risks refer to the increase in uncertainty and the creation of future risks, which will be difficult to model. Transition risks relate to the move from the carbon economy to a lower-carbon economy. Liability risks encompass the impact of natural disasters, which may become uninsurable because of the lack of supply (i.e., availability of coverage) or demand (i.e., affordability of premiums).

   An in-depth understanding of these different exposures is required to guarantee the solvency and affordability for the insurance sector. Solvency refers to being capable of predicting the costs of natural disasters, whereas affordability refers to the capacity to be able to cover all people exposed. Climate change affects premium calculation due to more frequent, costly, and unpredictable risks: It impacts affordability of premiums for high-risk areas and the capacity to calculate risks to be solvent in case of disaster. Thus, the insurance industry is looking for solutions in terms of loss prevention, conceptualized as a "sustainable flywheel business for the (re)insurance future" (Costa et al., 2020).

high-risk areas). The (re)insurance industry in particular is beginning to develop new catastrophe models (CAT) and modeling approaches to better understand the impacts of climate change, improve knowledge around potential hazards, and account for uncertainties in modeling future risks (Marchal et al., 2019).

Innovations as the industry evolves to meet new challenges

Insurance has always had an element of prevention of damage to what is insured. The companies will: (1) inform customers/clients, especially larger companies/clients, on what to do to prevent a damage; (2) reduce the premium if clients actively install things or take actions that will lower the risk of damage and loss; and (3) have conditions in the insurance agreement that you do not get full compensations if you have not taken due care of the insurance object. Also, in many countries with partly private, partly public or fully public systems (e.g., France, Norway, Spain) close cooperation on prevention is strengthening. Traditionally, the insurance industry has focused more on vulnerability and exposure and less on the natural hazard reduction component. There is now an evolving paradigm from response-based measures toward ex-ante prevention actions before a disaster occurs.

The industry is actively innovating with new tools and approaches to insurance that are opening the door to more proactive prevention and risk reduction measures that incorporate NBS. As a key part of the European economy, the insurance sector is already investing in the transition to carbon neutrality, resource-efficiency, and a greater resilience to climate change in a shared effort with the other economic sectors (for example via the EU Green Deal). The EU has identified a yearly investment gap of about €260 billion that needs to be closed to achieve its 2030 climate and energy targets. Insurance Europe estimates the insurance industry is planning to allocate over €140 billion to sustainable investments. However, this significant financial potential is not yet matched with the availability of mature investment projects and new partnerships, and financial instruments will need to be developed to support scaled NBS investment strategies.

The industry is experimenting with regulations and incentives that go beyond the traditional role of compensating for losses. Regulations such as risk mapping zones, which forbid construction in floodplains or other high-risk areas, or incentives such as reductions in premiums for taking preventive measures can help avoid damages. Providing advice and collaboration (including data sharing) with policymakers is helping to raise risk awareness to encourage better management of climate-related risks. Insurance is poised to play a catalyzer role and drive the implementation of natural and hybrid infrastructure strategies. For example, by implementing risk-based premiums set by models that consider the resilience dividends of ecosystems, as institutional investors by requiring minimum resilience standards and consideration of climate and water risks from the projects they finance, and last but not least, offering new insurance schemes and products that allow for the monetization of the resilience dividends of ecosystems.

The assurance and insurance value of ecosystems

We argue that the insurance sector can engage in disaster risk resilience by adopting an assurance perspective: widening the scope of actions to include NBS and expanding the time horizon of interest, shifting from a short-term, insurer assets risk reduction only, to a long-term, system risk reduction. Conceptually it is important for understanding the role of the insurance sector and the "insurance value" of nature, to clarify the fundamental difference between assurance and insurance value. Theoretical aspects of the insurance value of ecosystems have been covered in the literature (Baumgärtner and Strunz, 2014; Pascual et al., 2010), but it has been unclear how the insurance sector can operationalize the concept (Denjean et al., 2017; López Gunn et al., 2020; Marchal et al., 2019).

The insurance value of nature has been defined as the capacity to maintain the output or flow of ecosystem services in the face of variability and disturbance (Pascual et al., 2010). The term "insurance value of ecosystems" as frequently used is therefore related to both the output or delivery of valued services themselves, and resilience, or the capacity of ecosystems to absorb and/or adapt in the face of shocks in a way that preserves basic functionality and the continued delivery of beneficial services (Walker et al., 2004). Both aspects are important in examining the roles that the insurance industry can play in mainstreaming NBS, but the two have different implications for these roles.

The resilience value of ecosystems, maintaining functionality and flows of services over the long-term, is aligned with no-regrets, precautionary, preventive, and risk reduction measures. This *assurance* value of ecosystems has been used to describe the capacity of ecosystems to reduce risk through maintaining system resilience (Marchal et al., 2019). Assurance is a guarantee, a promise of something (Cambridge Dictionary). In the EU NAIAD project, natural assurance schemes (López Gunn et al., 2020) are defined as ecosystem-based risk reduction measures that reduce the level of multiple risks in an area. The assurance value of an ecosystem is the role that nature plays in mitigating natural risks, such as the risk of biodiversity loss, while also providing long-term guarantees that the ecosystem itself will continue to deliver the flows of ecosystem services that enables social-ecological resilience. If these assurance values can be properly accounted and internalized, through for example the United Nations' System of Environmental-Economic Accounting, accrued resilience benefits could well be larger than short-term risk reduction benefits.

The risk transfer and compensation functions of insurance can be directly related to the output of individual ecosystem services—the *insurance* value of ecosystems. Loss of beneficial services involves costs to those who are benefiting from the delivery of those services. For example, a water utility experiences increased treatment costs when forests in source watersheds are removed and sediment and nutrient runoff increases, reducing the quality of raw water inputs to the utility. Maintaining forest cover (an NBS) is providing an insurance value to the water utility and this is a value that can be tied to traditional insurance products. For example, insuring the natural infrastructure of the forest and providing funds or payments to protect or restore the forest in the case of damages (see Coral Reef Insurance example in this chapter). Used in this sense, the insurance value of nature mimics the classical financial insurance instrument: an insurance service is a transfer mechanism between two parties—the insurance company and the subscriber that pays regular risk premiums in exchange for financial risk coverage—but there is no significant reduction in the global risk; it is just a reduction in the individual risk level (the damage will still occur).

We suggest a change in paradigm that encompasses both a narrower, insurance-only perspective and the more comprehensive assurance perspective (Table 16.1). The overall natural insurance value is thus wider than just the financial insurance value: the compensation of the timely loss of a particularly good or service is limited compared to the existence of an ecosystem and the continued provision of a suite of valued services. In the natural assurance scheme, nature insures some assets in real monetary terms against loss (e.g., coral reefs), and these schemes also assure, through protection or restoration, that the ecosystem is better able to withstand anthropocentric threats or natural disasters. There is a double "dividend" from the natural assurance scheme. In this perspective, the assurance value of NBS is the value provided to humans while maintaining critical levels of ecological functions and keeping ecosystems away from ecological thresholds. This approach emphasizes the self-insurance mechanism or resilience of the system per se (Gómez-Baggethun and Martín-López, 2015).

The roles of insurance in capturing the resilience value of nature

From the NAIAD H2020 research project based on 61 in-depth interviews from the insurance industry, we identified five roles that the sector can play in mainstreaming NBS (Marchal et al., 2019):

1. **As data owners:** Insurance has a good knowledge of risk, the value of their contracts, cost of damages or losses, the history of payouts, and how to anticipate their exposure to risk (i.e., the geography of their assets).

Table 16.1: Key differences between the assurance and insurance concepts in relation to nature.

Assurance concept characteristics	Insurance concept characteristics
Long-term Value pluralism Resilience enhancement perspective Generic protection (known and unknown goods or services)	Short term Money value Risk reduction perspective Specific protection (known/definite goods or services)

2. **As insurance providers**: The role of prevention in risk management matters for the insurance industry to decrease its risk exposure. It is particularly important for the industry to be solvent while keeping premiums affordable. We observe a paradigm change in the industry from response-based measures (ex-post action) toward more ex-ante prevention actions.

3. **As investors**: The insurance sector holds large (capital) assets derived from insurance premiums and other income, which must be available quite rapidly to provide compensation after extreme events. These large assets are often invested in property, which is quite stable, or in the stock market and bonds. A new investable asset could be natural capital and the NBS that sustain natural assets.

4. **As insurance innovators:** The industry continues to innovate with products, tools, and services, for example, as prevention-advisor to help limit the consequences of natural disasters, or as collaborators with governments to strengthen preventive measures in land-use planning and building codes.

5. **As insurance partners**: The industry can further its level of collaboration with multiple actors involved in all aspects of risk management. Public-private partnerships enabled by regulatory reforms are one example, as in the case in the French Natural Hazard Insurance scheme, linking compensation (insurance) to prevention (Barnier Fund).

Hat 1: Data owners and providers

The insurance industry holds extensive data on where and when losses from damages have occurred, where claims have been paid out, and in some cases the cause of the damage, and therefore where risks and exposure are most significant. The data are used internally to develop models and to understand consequences of a variety of risks, including more recently, climate change. Catastrophe risk modeling chains are one of the main tools of the (re)insurance industry. These models offer a comprehensive analysis of the entire risk loss chain: hazard, vulnerability, and damage. Beyond the compensation role of the (re)insurance industry, new roles in prevention are emerging, notably through risk modeling (Le Quesne et al., 2017; Quantin, 2018). These tools developed by the insurance industry for estimating and managing catastrophic insurance risks can be adapted to assess the impacts of extreme weather events predicted with climate change and evaluating the effectiveness of nature-based protective or preventive measures (CCR, 2018; Walker et al., 2016; DERRIS Project, 2018; Narayan et al., 2016).

Mainstreaming natural solutions into decision-making requires an expansion of high-quality demonstration projects provided through practical methodologies (indicators and criteria) and good documentations (Van Wesenbeeck et al., 2017). The integration of NBS in risk modeling remains a challenge, although there are now relevant examples on the use of risk modeling to assess the role of NBS (CCR, 2018; Beck and Lange, 2016; Beck et al., 2018; Cohn, 2017; Maynard et al., 2017; Reguero et al., 2018). Relying on the catastrophe loss risk structure, the (re)insurance industry could be able to identify hazard areas and the most exposed areas, quantify damages for different NBS scenarios, and compare solutions with methods such as cost–benefit analysis (CBA) and multicriteria analysis (MCA).

Finance Norway: Increasing urban resilience through better data

Losses and damages related to urban flooding and storms are likely to increase due to climate change. In the EU, economic losses from weather and climate-related extremes are already about €12 billion per year. On average, only 35% of the climate-related economic losses are insured, with proportions as low as 5% or less in Southern and Eastern Europe. Total insurance losses for weather-related events reached 0.1% of GDP in 2018, and this share is likely to increase in the future. Businesses and the financial sector need access to more granular climate physical risk data to build capacity to respond to the new demands linked to the implementation of the sustainable finance action plan and to better adapt their business models to climate risk (EU, 2020).

The insurance industry can potentially play a key role in addressing these losses and improving prevention by contributing to the understanding of risks associated with climate change. Adaptation is about understanding, planning, and acting to prevent the impacts in the first place, minimize their effects, and address their consequences. According to the new EU Adaptation Strategy (2021), it is vital to build on, expand upon, and share the wealth of experience accumulated thus far on climate risks. Increased awareness, better access to knowledge, and risk-specific and place-based data on individual and collective climate risks would help all levels of government and stakeholders to improve knowledge of climate impacts. By sharing data on the location of insurance claims associated with natural hazards, extreme rainfall, and other climate-related (insured) perils, the insurance industry can enable better-informed adaptation planning and risk management actions, including guidance on when and where NBS can contribute to risk prevention or management.

At the initiative of Norway's insurance association, Finance Norway, the "Insurance Loss Data Sharing Project for Climate-Resilient Municipalities" project was initiated to assess if municipalities' capacity to prevent and reduce climate- and weather-related losses could be strengthened if they had access to insurance company data on loss damages from small scale to extreme weather events. Based on a collaboration with Norway's 10 largest insurance companies, the Western Research Institute, the Norwegian University of Science and Technology (NTNU), and 10 pilot municipalities (Bærum, Grue, Kongsvinger, Løten, Nord-Odal, Ringsaker, Stavanger, Tromsø and Trondheim, and later Oslo), a dialogue was initiated that facilitated sharing asset-level loss data held by insurers with the cities' planning and infrastructure sectors.

Finance Norway collected and organized the disaster loss data from these 10 insurance companies, while Western Norway Research Institute and NTNU assisted cities with importing and analyzing the data. The pilot project helped the local decision makers better understand risks and prioritize management, renovation, and reinvestment in public infrastructure. The municipalities agreed that the dataset on damages provides useful additional information for flood risk management and/or adaptation to climate change. The combination of the damages dataset with digital terrain models provided valuable information about specific areas at risk and can help to identify future problems related to urban flooding, along with informing adaptation action in urban spatial planning to address risks such as cloud bursts.

The project also helped develop a method for using disaster loss data from insurers, clarify the methods and benefits of using the insurance industry's injury data, and outline the structure of a future system to use disaster loss insurance data. This helped municipalities and national authorities understand the costs involved in data sharing and strengthened the trust between and within municipalities, state agencies, and insurers on prevention and reduction of climate-related losses. Municipalities saw the dataset as useful for financial and spending plans (Hauge et al., 2018; Climate ADAPT, 2020). The cost overview of damages from the insured losses was a good basis for local politicians to improve budgeting for water management and to identify the need for new skill sets and work tools, such as software allowing working with this type of data in a spatial context. However, the insurance data will be more useful if geocoded.

As a side effect, the pilot project raised awareness on climate change and improved knowledge on how climate change affects society. The pilot triggered other research projects to investigate climate-related risk factors, risk awareness, risk management, and risk prevention. For example, the project led to a national collaboration between the Norwegian Directorate of Civil Protection (DSB), the National Flood Agency, the State Road directorate, and Finance Norway to establish a national loss data platform under DSB, "The Knowledge Bank," extending access to local, anonymized insurers' loss data to all the public at municipal level. The municipalities, the flood agency, and the regional governor, however, will have access to the data at house level. The goal is to achieve better overview and knowledge about undesirable events and disasters, and thus to strengthen work on societal safety, enhanced disaster prevention, and reduce losses.

Hat 2: New insurance products

The insurance industry continues to develop new insurance products to capture the insurance value of nature: insuring natural infrastructure based on the value of the ecosystem services it provides. For example, the Insurance Bureau of Canada is exploring ways to insure natural infrastructure, such as wetlands, based on the valued of avoided or reduced damages from flooding when wetland areas are restored (IBC, 2020). The Nature Conservancy has partnered with SwissRe, local governments, and businesses on a novel coral reef insurance product in coastal Mexico.

The example of Coral Reef insurance—Quintana Roo Coastal Zone Management Trust (Mexico)

Coral reefs help maintain the vitality of the tourism industry in the State of Quintana Roo, Mexico, by protecting the coastal area and beaches from storms, reducing beach erosion, producing the striking white sands, and attracting snorkelers and scuba divers. But the coral reefs are also at risk from damage from strong storms and hurricanes, which can cause a loss of up to 55% of coral cover (Berg et al., 2020). Damages from these storms are particularly likely given other anthropogenic impacts that are impacting the health of coral reefs worldwide—ocean acidification, nutrient and sediment pollution, and warming waters. Healthy reefs reduce wave energy by 97%, protecting beaches, shoreline property, and associated economic activities. Studies conducted by The Nature Conservancy estimated that "a loss of 1 meter of reef crest height would increase damages to built capital up to 300% in Puerto Morelos" (TNC, 2019, p. 1). The reef's natural infrastructure provides a more resilient solution than the typical coastal protection gray infrastructure works, such as breakwaters or seawalls.

In 2018, The Nature Conservancy, the state government of Quitana Roo, the local tourism industry and SwissRe pioneered a new type of insurance that provides innovative funding for conservation, insuring the natural infrastructure of 160 km of coral reef against damages from storms (TNC, 2019). As an example of parametric insurance, rather than indemnifying against an actual loss (as in traditional home insurance), the coral reef insurance provides preagreed payout amounts based on the occurrence and intensity of a hazard event, in this case tropical storms and hurricanes (Berg et al., 2020). The intensity of an event serves as a proxy for impacts and losses. In this case the intensity parameter is wind speed occurring in the area covered by the insurance policy. The insurance will pay 40% of maximum payout for damages between 100 and 130 knots. It will pay 80% of the maximum payout for damages occurring due to wind speed between 130 and 160 knots. And it will pay 100% of the maximum payout for wind speed higher than 160 knots. The maximum payout is USD 3.8 million over the 12 months policy (Beck and Lange, 2016; Beck et al., 2018).

To implement the scheme, payment of insurance premiums is paid into a trust fund, the Coastal Zone Management Trust, which is then managed by a bank and overseen by an advisory committee composed by local community members. The partners of the initiative include The Nature Conservancy, Puerto Morelos Hotel Owners' Association, the National Commission for Natural Protected Areas, and Mexican Universities. Hotels and other tourism businesses, shoreline property owners, and local municipalities pay into the trust, which then purchases the parametric catastrophe insurance on behalf of the state government and tourism sector. The trust fund contracts for reef restoration, maintenance, and resilience activities, and when a triggering event occurs, parametric insurance payouts are paid to the trust, which then uses the funds for immediate repair work on the reef (Berg et al., 2020). These parametric insurance schemes have succeeded in monetizing the protective service of reefs and beaches to the tourist and hotel sector in Cancún and Puerto Morelos. In other words, through this product they have managed to translate the insurance value of ecosystems into revenue streams that can pay for the maintenance activities of these ecosystems.

Reef damage after a storm is repaired by deploying work crews to remove debris from reefs, consolidate loose colonies and broken fragments, and grow broken fragments in coral nurseries for future transplanting. Restoring activities unfold for a period of 2 years following a storm event. The first payout of $800,000 was triggered by winds of 100 knots during Hurricane Delta in October of 2020, which allowed immediate deployment of brigades working in the Puerto Morelos Reef National Park. Repair work funded by the insurance payout helped stabilize about 1200 large coral colonies that were damaged and transplant about 9000 coral fragments (https://www.nature.org/en-us/newsroom/coral-reef-insurance-policy-triggered/). Lessons learned in this first implementation can inform future adjustments, for example, improving contract designs.

For protecting natural infrastructure, parametric insurance has several advantages over traditional (indemnity-based insurance) insurance. When natural infrastructure is degraded or damaged it can be challenging to determine the actual value of the loss (in monetary terms). By focusing on the hazard, itself, protection of assets that are public goods means that insurance benefits a broad range of stakeholders. Using measurable pre-defined event parameters that—if exceeded—trigger the immediate payout of predefined amounts, means that much lower transaction costs and time are required to process insurance claims compared to traditional insurance (no need for a third-party assessment to determine the damage and amount of loss). As seen in the payouts from Hurricane Delta, funds can be made available immediately to begin repairing damages and begin recovery. Finally, the insurance coverage is not limited to physical damage but can also cover the economic losses from the interruption in ecosystem services provided by natural infrastructure (Berg et al., 2020).

Hat 3: Insurance as investors

Insurance companies are large investors in financial markets due to their substantial financial assets from premium payments. As such, the sector has the potential to significantly influence the kinds of activities those investments support. An industry that has traditionally been involved in short-term investments is now moving toward longer-term investments due to climate change and unpredictable extreme weather events. Due to insurance companies' needs for liquidity to cover loss damages, however, there are constraints on long-term investments (e.g., in Europe the current Solvency II Directive). By investing in NBS, insurance companies can greatly increase the funding flows to NBS while diversifying their portfolios, reducing their risk exposure and the cost of claims through the protective, risk prevention measures that NBS can provide. While there is great potential for the insurance sector to invest in NBS, there are also significant barriers.

The bond markets are a good opportunity for the insurance industry to finance NBS projects. The Natural Capital Financing Facility of the European Investment Bank and the variety of green bonds and catastrophe bonds are financial instruments to promote the conservation of natural capital and adaptation to climate change. The catastrophe bonds (cat bonds) are a debt instrument to raise money in the event of natural disaster. The issuer (e.g., a city) issues a cat bond for earthquake coverage or cyclonic events. If the event occurs during the contract's lifetime (normally 3–5 years), the investors pay to repair the losses. If no event occurs during the defined period, the investors will have a return on investment (ROI). The green bond principles frame the categories of potential eligible "green projects." Insurance companies also may make NBS investments through several objectives related to their environmental, social, and governance commitments, such as reducing their carbon footprint through the task force on climate-related financial disclosures. For example, in 2019, the Netherlands issued a €5.98 billion green bond to fund NBS for coastal and river ecosystems to reduce the consequences of high flood risk exposure.

An insurance fund that could invest in loss prevention

In France, protection measures are funded by the Barnier Fund mechanism (FPRNM, Fonds de Prévention des Risques Naturels Majeurs). The mechanism was created by Law n°95-101 on February 2, 1995. It is funded by a 12% levy on the 12% premium linked to the compulsory extended natural catastrophe coverage on all property and car damage insurance contracts (Law 82-600 on July 13, 1982). The Barnier Fund is funded by taxpayers. Initially, it was created to pay for the expropriation cost of assets that were in high-risk areas (e.g., floodplains) and to finance other measures of risk reduction, primarily gray infrastructure. This fund is dedicated to reducing vulnerability of assets (local communities and homeowners) exposed to natural hazards. It is used for different risk reduction measures, such as structural protective measures, amiable land acquisition, and targeted communication to raise risk awareness. Implementation of physical measures (works) make up 50% of the funding.

The distribution of the funding allowed by the Barnier Fund are: 70% dedicated to dike rings, 20% toward engineering works aiming to slow down water flows, and 10% for adaptive measures. These allocations have been relatively stable over the past decade. It cannot, however, be used for maintenance or reconstruction, which would limit its use for NBS such as wetland or river restoration. Also, the fund is monospecific, which means that it will preferentially fund a prevention project that is dedicated to a single hazard risk reduction. It may be challenging to use the fund for NBS implementation due to the multifunctional nature of NBS to simultaneously address multiple hazards.

Barriers to greater NBS investments by the insurance sector

While there is great potential for insurance sector investments in NBS, there are several barriers. These include a lack of a robust evidence base for the efficacy of NBS as protective measures, a lack of "bankable" NBS projects to invest in, and "free-rider" risks to insurance companies that bear the cost of preventive measures that also benefit other insurance companies.

Building the evidence base for hazard reduction through NBS—Riverine flooding in France

The insurance industry already has experience financing protective measures, for example, fire or theft alarms for fire or theft prevention, and these can be models for NBS investments for risk reduction. When embracing these protective measures, insurance companies have a clear understanding that the costs of fire or theft prevention measures are more than balanced by the benefits of reduced claims. To invest in NBS, the industry will also need strong evidence of the benefits for disaster risk reduction from nature-based preventive measures. Collaboration between insurance, governments, and scientists on developing this evidence may offer a window of opportunity to support greater insurance sector investment in NBS.

In 2018, the French public reinsurer, Caisse Centrale de Réassurance (CCR) modeled damage and NBS prevention measures effectiveness during the flood event of January and February 2018 on the Seine and Marne Rivers (CCR, 2018). The estimated damages were around €225–350 million, affecting 560 local communities recognized as damaged by the flood event. In addition to that damage assessment, the effectiveness of the lake "reservoir," located upstream the Seine and Marne catchments, was performed. The storage of flood waters in four lakes significantly reduced the floodwater heights by 80 cm, which in turn helped protect downstream cities from flooding. This reduction in water height leads to a 30% reduction in damages, up to €90 million for the entire flood event. Within Paris, the lakes allowed a reduction in the flood peak by 60 cm, which reduced the direct impact on basements, some public buildings, electricity, and transportation infrastructure.

Bankable projects for NBS investment

The lack of funding/financing and bankable projects is seen as one of the major barriers to mainstreaming of NBS for climate adaptation and resilience (e.g., Droste et al., 2017; Nesshöver et al., 2015). NBS-based natural assurance schemes (NAS) are often not "investment ready," preparing NBS promoters to be aware of the precise needs of investors and to be able to reply through adequate information, and preparation is lacking (Fellnhofer, 2016; Van Cauwenbergh et al., 2020; Altamirano et al., 2021). Low levels of investment readiness have been linked to a mismatch between the concerns of investors at the supply side who want to see a clear business case and trustworthy ROI and the understanding of the investment rationale by proposers of innovations (such as NBS).

Investment readiness for NAS is challenging, as there is usually no predefined legal entity that investors are familiar with and there may be multiple owners of individual NBS "projects" (Van Cauwenbergh et al., 2020). To date, NBS projects and NAS are "owned" by ad-hoc multistakeholder platforms composed by different regional or municipal departments, private companies, community organizations, and nongovernmental organizations. This means that investment readiness is intrinsically and narrowly linked to governance and institutional arrangements, often with a need to have a legal entity receiving the investment on the demand side.

When zooming in on the role of the insurance sector as investors, a series of roundtables with representatives of the insurance sector, municipalities and regional government agencies revealed that investment in prevention is limited in the EU. Whereas the Barnier Fund in France presents a good example of best practices, other EU countries face several barriers to invest in prevention in general and NBS-based prevention in particular. In some countries, Barnier-like investment in green assets and infrastructure is also seen as an issue for agricultural insurance, whereas other countries mention a lack of sector consensus and political will. In general, the insurance sector would require further engagement and nudges to be more involved as a financial investor to influence resilience and risk reduction.

Even when insurance companies are willing to reduce the exposure of their assets to risk, there is one major reason that makes them reluctant to invest in NBS or in general in reducing the risk at an aggregated level (suburb or wider scale): they do not insure all the assets in the area. There is no geographical specialization of insurance companies, so that in one risk hotspot area, multiple insurance companies will share the market (and do not cover all the potential markets). One company alone will not make it worth the effort of reducing the risk for only a share in the return. A third party seems necessary in between insurance companies and the investment in risk protection measures at a territorial scale.

Hat 4: Innovators

The insurance industry has become increasingly and directly involved in developing, enabling, and supporting innovative approaches to risk prevention and adaptation. These efforts also involve new partnerships with municipalities and the private sector as the insurance sector takes on new roles that are outside their traditional function of compensation for losses.

Example: Copenhagens cloudburst response

On July 2, 2011, the City of Copenhagen was hit by a cloudburst event; more than 150 mm of rain fell in 2 hours—a 1000-year storm event (https://www.asla.org/2016awards/171784. html). Extensive flooding of houses, streets, and major roads caused damages that resulted in approximately €1 billion in insurance claims (City of Copenhagen, 2015). The large number of insurance claims—90,000—put a strain on insurance companies. Contractors working for the insurance companies were overburdened, and response times for pumping out water and drying out buildings resulted in an increase in mold damage, compounding the costs for insurers (GenRe, 2015). After the torrential rain on July 2, 2011, it was not possible to obtain the same insurance premium, and the cost of premiums rose substantially—following a similar cloudburst in the Municipality of Greve, insurance premiums rose by 20% but were lowered after implementation of cloudburst measures and anticipating reduced damages as a result.

In response to the July 2011 event and the recognition that cloudburst events were becoming more frequent and intense due to climate change, the city took several steps. The cloudburst event triggered the Cloudburst Management Plan as an extension of the Copenhagen Climate Adaptation Plan, which included adaptation initiatives for more than €500 million. Measures included a mix of traditional drainage pipes and retention basins with blue-green solutions such as parks, rain gardens, green streets, and retention boulevards that detain and infiltrate stormwater, reducing flood damages and preventing floodwaters from burdening the sewer system and overload its capacity. The city was divided into seven urban water catchments areas with detailed adaptation plans for each area and prioritized based on a risk assessment of where flooding causes the most damage, easiest to implement and synergistic effects with urban development. In total, 300 cloudburst management project proposals were prepared that are subject to approval for each municipal financial year.

The combined alternative cloudburst and stormwater management solution in Copenhagen costs a total of DKK 11 billion (€1.3 billion) (City of Copenhagen, 2015). Grey infrastructure for the management of water (hydraulic function: channels, cloudburst tunnels, and surface solutions) is financed through the water charges by the water utility company (HOFOR), while the funds for green solutions (urban space improvements) are financed continuously through the municipal yearly budget. The combined solution additionally necessitates private individuals investing in antiflood backflow valves and local removal of rainwater (City of Copenhagen, 2015). A solid business case was built with robust socio-economic figures of costs and gains showing that the green infrastructure-based plan was cheaper than traditional grey solutions (City of Copenhagen, 2015). It would lead to less flooding and therefore less damage, lower insurance costs while harvesting cobenefits, including a greener and more attractive city, less traffic, and a positive effect on real estate prices. It was concluded that adaptation was a good investment for the city, and that focusing on the interaction of adaptation with other urban development was positive overall.

The insurance industry has played a key role in developing innovative incentive programs that help fund the implementation of adaptation measures to reduce risks from future cloudburst events. The Danish insurance industry organization, Forsikring and Pension, formalized a public-private partnership with the city, private landowner associations, and HOFOR (Centre for Liveable Cities, 2016). Insurance companies help disseminate information on how to adapt houses to flooding, monitor flood risk and trends in insurance claims, implement requirements to hasten adaptation such as not insuring buildings that do not install backflow valves, and provide incentives for implementation of adaptations measures (such as installation of backflow valves) in the form of reduced insurance premiums. The implementation of the measures in the climate change adaptation and cloudburst management plans can dramatically reduce the risk of damages to buildings, and as a result, the insurance industry could also achieve a corresponding reduction in claim payouts from extreme rainfall events. The insurance companies, therefore, will study the trend in claims very closely in the future, and if risk declines, this is anticipated to affect the insurance premium prices and terms (City of Copenhagen, 2015).

Hat 5: Insurers as partners for sustainable finance

Overall, several barriers have been identified that hamper investment in prevention by the different EU countries. The following factors were mentioned as contributing to under investment in resilience/adaptation/risk reduction:

- Insurance assessments use a short time frame (often revised annually) and are constrained under the Solvency II Directive requirements on capital. Regulatory constraints on the insurance sector raised limits for solvency in the area of loss prevention making it more difficult to invest in start-up companies involved in research and development and long-term investments.
- Large infrastructure is often uninsured but guaranteed by government, so there would be potential, for example, for the insurance value of green assets and the assurance value under major sustainable long-term investment schemes based on avoided damages and cobenefits.
- Insurance alone will not engage; key additional partners are needed; a fund to work with all insurance companies in one basin; the state to control/regulate.
- There are challenges in the current investment model for paying back/earning revenues of investment (that provide risk reduction but not revenue flows).
- There are insufficient regulatory requirements for risk assessment/protection in building/planning codes or for cities and regions as a standardized procedure.
- There is a need for a risk culture change toward embracing the natural hazard component for insurers by partnering with cities and states to work on global risk reduction and capacity building; as well as to build new skills and profiles in the industry.

While the ROI is still a barrier, there are also examples of effective ROI on prevention coming from France, for example, as mentioned for building investment. When insurers will do a prior risk exposure analysis, or investments into specific funds for climate. Investments in green assets (under shared indicators) must be encouraged by a gain and/or a reduction of cost of capital. Some stakeholders highlighted PPPs as a potential solution, whereas others see that the insurance sector can stimulate investments in protection infrastructure via facultative insurance products by focusing on digitalization (digital interaction with consumers), the quality of the claims assessment and adding waivers or top up services to stimulate risk reduction measures. This can be further strengthened by making sure there are good loss data and loss curves and CBA to be able to look at risk reduction options with the cooperation of the whole sector.

Insurance companies could support the development of transformational investment pipelines by sharing their data on historic losses and damage with municipalities (as is happening in Norway) and their expertise. By leading the discussion and development of catastrophic models that consider the effect of ecosystems in systemic resilience, these could incentivize investors to look at their portfolio in a systemic way (Altamirano, 2019).

Increasing interest in loss prevention could also be fostered by participating in reducing the current protection gap in some countries. Indeed, a better assessment of risk exposure and assessment of loss prevention largely improves the knowledge and increases markets by incorporating green assets to capture the insurance value (and their assurance value). Nevertheless, there is a knowledge gap on modeling the effects of NBS on hazard reduction, monitored data on the effectiveness, and the integration of the hazard modeling within the damage estimate. During the interviews undertaken as part of the H2020 NAIAD project, we also observed a lack of knowledge on NBS, 58% of respondents declared not to have enough knowledge on the effects of ecosystems in risk reduction. Unlocking that knowledge will also unlock the investments. Cost-benefits analysis and monitoring aspects are also lacking or largely underestimated, thus greatly reducing the potential revenue flows.

In terms of data, it would be particularly important to share with the public statistics (not per asset but as granular as possible) all the losses, separating the various losses of ex-urban flooding and river flooding (pluvial and fluvial). This would give, in collaboration with, for example, municipalities, regions, river basin agencies, or the private sector itself, a strong basis for structured collaboration processes to build collective resilience and reduce overall anticipated costs from increased (and uncertain) risk exposure.

In addition (re)insurance companies can either "push" their clients and customers in the right direction to take preventive measures via better price (premium/the carrot) or "punish" them via conditions in the insurance agreement if they are careless, for example, with building standards or new green standards to be developed, poor maintenance, and so on.

Conclusions

Climate change will increase the frequency and intensity of storms, floods, and droughts. This will place at risk not only human lives and livelihoods but also enormous amounts of assets and properties (including our green infrastructure). It is natural for those outside the insurance industry to assume that insurance companies will consider the risks posed to assets they insure and have a material interest to reduce those risks and avoid costly payouts. Yet ensuring that people's homes, assets, and increasingly green infrastructure itself can be insured at all is a critical and overlooked issue. It is also the area where both the public good and the insurance sector at large have the most at stake—insurability reflects a basic security that can be provided to people and enables the business of insurance companies. NBS can be highly effective at reducing future persistent water risks that result from climate change, for example, from more frequent and intense flooding events, by providing, for example, natural solutions to drain and detain water.

While fundamental to the insurance sectors business, at the local scale, these risks will not necessarily be forecasted in advance. A decision not to cover flooding, for example, would follow the experience of the first cases of the event and a revision of the cost and risk profile of the asset. When looking at pathways to mainstream NBS, partnerships between public authorities and insurers should be placed front and center. This should involve evaluations of flood and water risks under potential climate changes, the projected impacts on the insurability of public and private assets under climate change scenarios, and last but not least, the capacity of NBS to reduce those risks to a level that can be securely insured. While insurers would not have the incentives to directly pay for those investments in risk reduction through NBS, they should have immediate interests to engage in a productive partnership to promote such investments. This will help enable their business to be viable under a changing climate. Insurance, through its different hats, has many roles to play in both assuring and insuring the value of nature to increase overall system resilience in the face of global and climate change.

References

Altamirano MA. Hybrid (green-grey) water security strategies: a blended finance approach for implementation at scale. In: Background paper session 3. Roundtable on financing water, regional meeting Asia Manila. OECD; 2019.

Altamirano MA, de Rijke H, Basco Carrera L, Arellano Jaimerena B. In: Handbook for the implementation of nature-based solutions for water security: Guidelines for designing an implementation and financing arrangement (DELIVERABLE 7.3). 1st. EU Horizon 2020 NAIAD Project, Grant Agreement No. 730497 Dissemination; 2021.

Baumgärtner S, Strunz S. The economic insurance value of ecosystem resilience. Ecol Econ 2014;101:21–32. https://doi.org/10.1016/j.ecolecon.2014.02.012.

Beck MW, Losada IJ, Pelayo M, Reguero BG, Díaz-Simal P, Felipe F. The global flood protection savings provided by coral reefs. Nat Commun 2018. https://doi.org/10.1038/s41467-018-04568-z.

Beck MW, Lange G-M. Managing coasts with natural solutions, guidelines for measuring and valuing the coastal protection services of mangroves and coral reefs. Waves technical paper 103340, Washington, DC: World Bank Group; 2016. http://documents.worldbank.org/curated/en/995341467995379786/Managing-coasts-with-natural-solutions-guidelines-for-measuring-and-valuing-the-coastal-protection-services-of-mangroves-and-coral-reefs.

Berg L, Bertolotti T, Bieri J, Bowman R, Braun J, Cardillo M, Chaudhury K, Falinski L, Geselbracht K, Hum C, Lustic E, Roberts SY, Way M. Insurance for natural infrastructure: Assessing the feasibility of insuring coral reefs in Florida and Hawai'i. Arlington, VA: The Nature Conservancy; 2020.

Cassin J, Matthews JH. Setting the scene: nature-based solutions and water security. In: Cassin J, Matthews JH, Lopez-Gunn E, editors. Nature-based solutions and water security: An action agenda for the 21st century. Elsevier; 2021.

CCR. Retour sur les inondations de janvier et février 2018, modélisation des dommages et évaluation des actions de prévention. Département Analyses et modélisation Cat. Caisse Centrale de Réassurance; 2018.

Centre for Liveable Cities. Cloudburst solutions in Copenhagen. Research Report https://www.clc.gov.sg/docs/default-source/reports/cloudburst-solutions.pdf; 2016.

Cohen-Shacham E, Walters G, Janzen C, Maginnis S, editors. Nature-based solutions to address global societal challenges. Gland: IUCN; 2016. https://doi.org/10.2305/IUCN.CH.2016.13.en.

Cohn C. Mangroves, coral reefs could cut flood insurance premiums: Lloyd's. Reuters; 2017. juin http://in.reuters.com/article/usinsuranceclimatechangeresearchidINKBN1932NF.

Costa MM, Marchal R, Moncoulon D, Martin EG. A sustainable flywheel: opportunities from insurance' business to support nature-based solutions for climate adaptation. Environ Res Lett 2020;15(11):111003. https://doi.org/10.1088/1748-9326/abc046.

Denjean B, Denjean B, Altamirano MA, Graveline N, Giordano R, Van der Keur P, Moncoulon D, Weinberg J, Máñez Costa M, Kozinc Z, Mulligan M, Pengal P, Matthews J, van Cauwenbergh N, López Gunn E, Bresch DN, Denjean B. Natural Assurance Scheme: a level playing field framework for Green-Grey infrastructure development. Environ Res 2017;159. https://doi.org/10.1016/j.envres.2017.07.006.

DERRIS Project. Guidelines for PA: Assessment and management of risks linked to climate change. Brussels: European Commission; 2018.

EU. Adaptation to climate change blueprint for a new, more ambitious EU strategy, https://ec.europa.eu/clima/sites/clima/files/consultations/docs/0037/blueprint_en.pdf; 2020.

Droste N, Schröter-Schlaack C, Hansjürgens B, Zimmermann H. Implementing nature based solutions in urban areas: Financing and governance aspects; 2017. https://doi.org/10.1007/978-3-319-56091-5_18. .

Fellnhofer K. Literature review: investment readiness level of small and medium sized companies. Int J Manag Financ Account 2016. https://doi.org/10.1504/IJMFA.2015.074904.

GenRe. Perspective: Cloudburst in Denmark—Just another tail event? October 22, 2015 https://www.genre.com/knowledge/blog/cloudburst-in-denmark-just-another-tail-event-en.html; 2015. [Accessed 19 January 2021].

Gloor M. Insurance in a world of climate extremes: what latest science tells us. In: Natural catastrophes & climate; 2019. Zurich https://www.swissre.com/institute/research/topics-and-risk-dialogues/natcat-and-climate/insurance-world-climate-extremes.html.

Golnaraghi M. Climate change and the insurance industry: Taking action risk managers and investors perspectives from C-level executives in the insurance industry. Zurich: The Geneva Association; 2018. https://www.genevaassociation.org/sites/default/files/research-topics-document-type/pdf_public/climate_change_and_the_insurance_industry_-_taking_action_as_risk_managers_and_investors.pdf.

Gómez-Baggethun E, Martín-López B. Ecological economics perspectives on ecosystem services valuation. In: Martínez-Alier J, Muradian R, editors. Handbook of ecological economics; 2015. https://doi.org/10.4337/9781783471416.00015.

IBC (Insurance Bureau of Canada). Nature-based solutions, http://www.ibc.ca/on/disaster/nature-based-solutions; 2020.

Kabisch N, Korn H, Stadler J, Bonn A. Integrating the grey, green, and blue in cities: Nature-based solutions for climate change adaptation and risk reduction; 2017. https://doi.org/10.1007/978-3-319-56091-5_6.

Le Quesne F, Tollmann J, Range M, Balogun K, Zissener M, Bohl D, Souvignet M, et al. The role of insurance in integrated disaster & climate risk management: evidence and lessons learned. In: UNU-EHS publication series 22. Bonn: MCII, United Nations University and GIZ; 2017. http://www.climate-insurance.org/fileadmin/user_upload/ACRI__2017_Role_of_Insurance_in_ICRM_online.pdf.

López Gunn E, Marcos C, Vay L, Burke S, Giordano R, Graveline N, Le Coent P, Mayor B, Marchal R, Moncoulon D, Mulligan M, Nanu F, Peña K. Natural assurance schemes: Moving earlier in the risk management cycle with nature-based solutions and strategies. Consorseguros Number 12; Spring 2020; 2020. p. 1–22.

Marchal R, Piton G, Lopez-Gunn E, Zorrilla-Miras P, der Keur PV, Dartée K, Pengal P, et al. The (re)insurance industry's roles in the integration of nature-based solutions for prevention in disaster risk reduction—insights from a European survey. Sustainability 2019;11(22):6212. https://doi.org/10.3390/su11226212.

Maynard T, Stanbrough L, Stratton-Short S, Hewitt B. Future cities: building infrastructure resilience. In: Emerging risk report. City infrastructure resilience—Designing the future. London: Lloyd's of London, Apur; 2017. https://www.lloyds.com/news-and-risk-insight/risk-reports/library/society-and-security/arup.

Moncoulon D, Desarthe J, Naulin J-P, Onfroy T, Tinard P, Wang Z, Hajji C, Veysseire M, Dequé M, Régimbeau F. Conséquences du changement climatique sur le coût des catastrophes naturelles en France à l'horizon 2050. Paris: Caisse Centrale de Réassurance & Météo France; 2018. https://www.ccr.fr/documents/35794/35836/Etude+Climatique+2018+version+complete.pdf/6a7b6120-7050-ff2e-4aa9-89e80c1e30f2?t=1536662736000.

Narayan S, Beck MW, Reguero B, GLosada IJ, van Wesenbeeck B, Pontee N, Sanchirico JN, Ingram JC, Lange GM, Burks-Copes K. The effectiveness, costs and coastal protection benefits of natural and nature-based defences. PLoS One 2016;11(5). https://doi.org/10.1371/journal.pone.0154735.

Nesshöver C, Prip C, Wittmer H. Biodiversity governance: a global perspective from the convention on biological diversity. In: Gaspartos A, Eillis KJ, editors. Biodiversity in the green economy. London: Routledge; 2015.

Pascual U, Muradian R, Brander L, Gómez-Baggethun E, Martín-López B, Verma M, et al. The economics of valuing ecosystem services and biodiversity coordinating lead authors. Available at http://africa.teebweb.org/wp-content/uploads/2013/04/D0-Chapter-5-The-economics-of-valuing-ecosystem-services-and-biodiversity.pdf; 2010.

Quantin A. Nouvelles compétences dans l'assurance. Préventique; 2018. décembre 2018.

Reguero BG, Beck MW, Bresch N, Calil D, J. and Meliane. I. Comparing the cost effectiveness of nature based and coastal adaptation: a case study from the Gulf Coast of the United States. PLoS One 2018. https://doi.org/10.17605/OSF.IO/D6R5U.

Renaud, et al. Ecosystem-based disaster risk reduction and adaptation in practice. Springer; 2016. https://doi.org/10.1007/978-3-319-43633-3_1.

Seddon N, Chausson A, Berry P, Gerardin CAJ, Smith A, Turner B. Understanding the value and limits of nature-based solutions to climate change and other global challenges. Philos Trans R Soc B 2020;375:20190120.

Smith DM, Matthews JH, Bharati L, Borgomeo E, McCartney M, Mauroner A, Nicol A, Rodriguez D, Sadoff C, Suhardiman D, Timboe I, Amarnath G, Anish N. Adaptation's thirst: Accelerating the convergence of water and climate action. Background Paper prepared for the 2019 report of the Global Commission on Adaptation. Rotterdam and Washington, DC www.gca.org; 2019.

Surminski S, Bouwer LM, Linnerooth-Bayer J. How insurance can support climate resilience. Nat Clim Chang 2016;6(4):333–4. https://doi.org/10.1038/nclimate2979.

Swiss Re. Global catastrophes caused USD 56 billion insured losses in 2019. Estimates Swiss Re Institute; 2019. 19 December 2019 https://www.swissre.com/media/news-releases/nr-20191219-global-catastrophes-estimate.html.

Tanner TM, Surminski S, Wilkinson E, Reid R, Rentschler J, Rajput S. The triple dividend of resilience: realising development goals through the multiple benefits of disaster risk management. Global Facility for Disaster Reduction and Recovery (GFDRR) at the World Bank and Overseas Development Institute (ODI); 2015. https://www.gfdrr.org/sites/default/files/publication/The_Triple_Dividend_of_Resilience.pdf.

TNC. Insurance nature to ensure a resilient future: Coastal Zone Management Trust. Retrieved from https://www.nature.org/content/dam/tnc/nature/en/documents/TNC-CoastalManagementTrust_Infographic_04.pdf; 2019.

Van Cauwenbergh N, Dourojeanni P, Mayor B, Altamirano M, Dartee K, Basco-Carrera L, Piton G, Tacnet JM, Manez M, Lopez-Gunn E. Guidelines for the definition of implementation and investment plans for adaptation with nature-based solutions; 2020. EUEU Horizon 2020 NAIAD Project, Grant Agreement N°730497.

Van Wesenbeeck BK, Ijff S, Jongman B, Breunig Balog-Way SA, Kaupa SM, Vuillemot Bosche L, Lange G-M, et al. Implementing nature based flood protection: Principles and implementation guidance. Working paper 120735, Washington, DC: World Bank Group; 2017. http://documents.worldbank.org/curated/en/739421509427698706/Implementing-nature-based-flood-protection-principles-and-implementation-guidance.

Walker B, Holling CS, Carpenter SR, Kinzig A. Resilience, adaptability, and transformability in social-ecological systems. Ecol Soc 2004;9(2):5.

Walker GR, Mason MS, Crompton RP, Musulin RT. Application of insurance modelling tools to climate change adaptation decision-making relating to the built environment. Struct Infrastruct Eng 2016;12(4):450–62. https://doi.org/10.1080/15732479.2015.1020498.

Warner K, Ranger N, Surminski S, Arnold M, Linnerooth-Bayer J, Michel-Kerjan E, Kovacs P, Herweijer C. Adaptation to climate change: Linking disaster risk reduction and insurance. Geneva: United Nations Office for Disaster Risk Reduction (UNISDR); 2009. https://www.undrr.org/publication/adaptation-climate-change-linking-disaster-risk-reduction-and-insurance.

CCR (Caisse Centrale de Réassurance), Météo-France, Risk Weather Tech. Évolution du risque cyclonique en Outre-Mer à horizon 2050. Paris: CCR; 2020.

Climate ADAPT; 2020. https://climate-adapt.eea.europa.eu/metadata/case-studies/use-of-insurance-loss-data-by-local-authorities-in-norway. [Accessed 6 October 2020].

City of Copenhagen. Climate change and investment statement (Part 1). Denmark: City of Copenhagen; 2015.

Hauge Å, Flyen C, Venås C, Aall C, Kokkonen A, Ebeltoft M. Attitudes in Norwegian insurance companies towards sharing loss data. Trondheim: Klima 2050 Report 11; 2018. ISBN 978-82-536-1590-5 (Attitudes in Norwegian insurance companies towards sharing loss data SINTEF Bokhandel).

CHAPTER 17

Operationalizing NBS in low- and middle-income countries: Redefining and "greening" project development

Alex Mauroner[a],[*], Nureen F. Anisha[b],[*], Ernesto Dela Cruz[c],[†], Eugenio Barrios[d],[†], and Sujith Sourab Guntoju[a],[†]

[a]*Alliance for Global Water Adaptation, Corvallis, OR, United States,* [b]*Oregon State University, Corvallis, OR, United States,* [c]*Asian Development Bank, Manila, Philippines,* [d]*Mexican National Water Commission (CONAGUA), Mexico City, Mexico*

Introduction

Nature-based solutions (NBS) are means by which all countries—regardless of their socioeconomic status—can develop toward a greener and more water secure future while mitigating harm to ecosystems. NBS are already being deployed at large scales in certain instances, yet there is a need for more widespread operationalization of these types of approaches to meet the scale of water management challenges being faced today.

In many instances, developed countries are increasingly considering NBS over traditional gray infrastructure, after having learned from decades of trial and error with rigid, single-purpose infrastructure assets. Low- and middle-income countries (LMICs) have the advantage of adopting NBS earlier in their development pathway as they aspire for long-term growth, thereby potentially avoiding making the same mistakes as developed nations. This chapter will examine the rationale for mainstreaming NBS within economic development pathways, identifying the processes and actions by which LMICs can more systematically bring NBS to scale as a complement to traditional or "gray" infrastructure.

---

[*] Lead authors.
[†] Contributing authors.

Nature-Based Solutions and Water Security. https://doi.org/10.1016/B978-0-12-819871-1.00009-9

Understanding the development context

To identify the potential role of NBS in LMICs, it is important to first establish a general understanding of the development context for these countries—where they are now and how they arrived. All LMICs are aspiring toward economic development, with poverty alleviation and increasing standards of living as top priorities. A path toward alleviating poverty requires clean water, affordable energy, health care, and strong economies.

The environmental problems faced by LMICs are much different from those in industrialized economies. Communities in LMICs, especially rural ones, are directly tied to the use of local natural resources for their livelihoods (Barbier, 2007). Healthy ecosystems and ecosystem services are linked to requirements of a good quality of life, having implications for food production (e.g., agriculture, fisheries) and livelihoods, especially for the rural poor (Kapos et al., 2019). At the same time, LMICs face rapid land use change and loss of ecosystems (e.g., deforestation, draining of wetlands) as part of their export economies. Population growth further exacerbates the strain on natural resources. Yet, countries such as Brazil, Indonesia, and the Democratic Republic of the Congo are still home to some of the largest tracts of forest on the planet. How, then, can ecosystems be properly valued and put to use while minimizing harm?

In the conventional route of economic development used by high-income countries over the last century, economic growth and social achievements were widely recognized as complementary. However, economic development and environmental sustainability were often perceived as antithetical. The development model of more industrialized countries has been exported to LMICs; under this paradigm, economic development looks like concrete infrastructure. This model was largely reinforced by the financial flows of development funds and the policies of international governance systems until relatively recently.

As populations grow and cities expand in LMICs, there is a gap between infrastructure needs and existing capacity to deliver necessary services. This "infrastructure deficit" means that infrastructure is being developed and constructed at a rapid rate and at high volume. As government agencies and ministries plan, design, and construct new assets to solve water management needs, there is an opportunity for mainstreaming "green" or "greener gray" solutions.

Opportunities for a shift to NBS
Changing the development paradigm

Many of the development schemes used by high-income countries have come at the cost of environmental degradation, greater pollution, loss of biodiversity, risks to human health, overexploitation of natural resources, and unsustainable solutions to problems (Gill et al.,

2018). In fact, this approach had been reinforced by the Environmental Kuznets Curve (EKC) theory that is based on the idea that environmental quality first deteriorates with growth and then improves (Kuznets, 1955)—that economic growth is both the cause and solution to environmental harm.

The EKC hypothesis exerted a significant impact on the economic policies of the developing and developed countries. Webber and Allen highlighted that "EKC hypothesis has important inferences that developing countries should peruse for fast economic growth instead of implementing pro-environment policies. Because economic growth eventually leads to attaining both environmental and economic goals, whereas pro-environment policies just slow down the economic growth" (Webber and Allen, 2004). But the assumption of the EKC theory that environmental losses of economic growth can be recovered at the later stages of economic development was also criticized by many researchers in the last two decades (Gill et al., 2018). Although the EKC theory has been popularly adopted in many developed countries, empirical evidence supporting its validity is neither consistently supportive nor statistically strong (Karsch, 2019). There is no evidence that eventual environmental improvement occurs as a result of all affluent nations investing in eco-friendly and cleaner technologies. Further, there is no proof that every wealthy society begins to demand environmental quality once their basic needs have been met (Raymond, 2004).

The EKC theory fails to consider environmental limitations. One of its major flaws is the implication that economies can grow infinitely without acknowledging that economies are mainly based on finite natural resources. Several critics have also outlined how it is falsely believed that all forms of environmental damage can be recovered; environmental systems are incapable of returning to their original conditions once certain environmental thresholds have been crossed (He, 2007). The loss of some natural capital can be substituted with other types of capital in the short term but not in the long term. For example, increasing the use of fertilizer to compensate for soil degradation—a short-term solution that is not sustainable over the long term (World Bank, 2012).

Several developed countries have acted upon these lessons and gradually shifted toward working with nature rather than working against nature. The increasing recognition of nature-based and "green" solutions, particularly regarding infrastructure, has led a number of countries (for example, Netherlands, the United Kingdom, Germany, and the United States) and international institutions to the development of policies and guidelines over the last two decades to operationalize NBS (Schoonees et al., 2019). The UK Shoreline Management Plan has developed strategies to reduce the threat of flooding and erosion that are beneficial to the environment, society, and the economy (DEFRA, 2006). The US Army Corps of Engineers (USACE) adopted the Engineering with Nature program beginning in 2010 with the expressed purpose of promoting "the intentional alignment of natural and engineering processes to efficiently and sustainably deliver economic, environmental and social benefits through collaborative processes." (Bridges et al., 2016). The "Building with Nature" concept in the Netherlands aims to utilize the potentials of natural processes to develop multifunctional solutions that are aligned with the interests of nature and project stakeholders (De Vriend and Van Koningsveld, 2012). China has also shifted from Mao Zedong's concept of overcoming and changing nature (i.e., to obtain freedom from nature) to Xi Jinping's thought of positioning China as an ecological civilization and realizing the sustainable development of economic construction and environmental protection (Ho, 2003; Xiang-Chao, 2018).

Curbing infrastructure deficit is an important aspect of economic growth for LMICs (Rozenberg and Fay, 2019). NBS approaches enable a paradigm shift from the status quo, introducing greener, more cost-effective, and more efficient infrastructure. While developed countries are increasingly considering NBS over traditional gray solutions—a point they reached through decades of trial and error with gray solutions—LMICs have the advantage to adopt NBS as they aspire for long-term growth without having made the same mistakes (Nesshöver et al., 2017; Karsch, 2019). Webber and Allen's "Green Solow Model" of economic growth states that environmental policies can increase economic output directly by improving environmental conditions and utilizing natural capital (Webber and Allen, 2004; World Bank, 2012). The multibenefit and low-cost features of NBS make them greatly suitable for LMICs who are aspiring for fast economic growth through poverty alleviation, increases in per capita income, environmental preservation, and curbing infrastructure deficit without having to greatly deplete existing natural capital.

NBS as a tool to address uncertainty

It is important to understand what lies in the future in order to develop long-term solutions (Haasnoot et al., 2018). But the combined impacts of climate change and economic, social, and technical developments impose higher uncertainties that make prediction of the future difficult and decisions about the future more complicated. A solution can be considered sustainable when it can not only achieve economic, environmental, and social objectives but can also be robust and is able to be adapted over time to changing and uncertain future conditions (Walker et al., 2013). For example, a robust wastewater treatment system will maintain functionality for a few decades under conditions of changing water demand and discharge, changing wastewater composition, higher effluent water quality standards, and wear and tear. Adaptivity is the property of a design to deliver functionality by changing itself. Flexibility is another concept that entails changes in structure, scale, functionality, and operating objectives after the solution has been implemented. Robustness is an essential part of flexible systems and solutions (Spiller et al., 2015).

Natural ecosystems are usually well adapted to their natural disturbance regimes such as episodes of flooding, drought, storms, or wildfires (Seddon et al., 2020). Natural systems often demonstrate evidence of a balance between stability (resistance to change) and ability to adapt. While natural systems are able to cope within a range of conditions without changing structure or functionality (robustness), they also have the ability to adapt by altering their structure and operating conditions when an environmental condition exceeds a threshold (Folke, 2006). This ability to be robust while also being able to change is an essential property of a sustainable solution (Holling and Gunderson, 2002). Hence, natural systems are more appropriate to address uncertainty compared to rigid gray infrastructure.

NBS can combine green and gray solutions. Gray infrastructure can be rigid, though these assets are designed to address predicted changes (for example, dikes for coastal flood management). Integration of flexible natural elements (for example, planting of mangroves for coastal flood management), which can change as necessary in response to future uncertainties, can provide a cost-effective way of sustaining the possibilities for addressing uncertain climate-induced pressures (Pathirana et al., 2018). Several studies have outlined how NBS can perform as well or better compared to traditional gray solutions regarding flood mitigation, agricultural technology, disaster risk reduction, water pollution control, climate change adaptation, and urban water systems (Van den Hoek et al., 2012; Ourloglou et al., 2020; Gusev, 2020; Onuma and Tsuge, 2018; Liquete et al., 2016; Seddon et al., 2020; Ashley et al., 2016).

Connecting systems with multibenefit solutions

NBS are inherently multifunctional and multipurpose and provide cobenefits, such as improvements to livelihoods, recreation, and the growth of green jobs (Raymond et al., 2017). The multipurpose nature of NBS creates an opportunity for coordination and collaboration between sectors and ministries, including the pooling of funds and expertise. For example, China's Sponge City Construction Project tackled issues of flood management through close coordination between upstream farmers and land managers alongside city planners and municipal government agencies. In addition, this type of coordination is necessary due to the fact that green and hybrid infrastructure may be operating across regional, national, and sectoral boundaries. The challenge and opportunity lies in the large scale of projects. Use of NBS demonstrates the interconnections between hydrological systems within cities and across a catchment area. Boxes 17.1 and 17.2 provide examples of coordinated planning between sectors and jurisdictions.

Climate change adaptation and disaster risk reduction

While traditional infrastructure is effective at reducing impacts of specific hazards in the short term, NBS provide a wider range of ecosystem services and can therefore offer protection from a number of hazards occurring either simultaneously or through more gradual, slow-onset progression (Matthews and Endo, 2020; Kapos et al., 2019). Since many

Box 17.1   Sponge cities in People's Republic of China.

Many cities today have large areas of land covered by hard and impervious surfaces, leading to increased amounts of stormwater runoff during monsoons. The traditional response to managing stormwater runoff has been to convey the stormwater away from urban areas using gray infrastructure—pushing the water downstream. During instances of "urban waterlogging," this strategy fails as runoff exceeds the design capacity of urban drainage networks.

In December 2013, People's Republic of China President Xi Jinping spoke about "building Sponge Cities where stormwater can be naturally conserved, infiltrated, and purified." Sponge Cities recognize the interconnected hydrological systems within and outside of cities. This catchment-level thinking was embedded within a comprehensive solution called the Sponge City Construction Project (SCCP), designed to deal with the three main urban water problems: storm floods, water pollution, and water shortages across the whole landscape.

A Sponge City captures stormwater and uses it to supplement existing water supply. Sponge Cities are able to reduce runoff and pollution and restore downstream ecology through the combination of several low-impact development methods with gray infrastructure, large-scale flood control projects, and rehabilitation (e.g., flood control engineering plus afforestation projects). The approach is part of a broader cultural and programmatic shift that emphasizes more broad and cross-cutting problem definition—moving away from single-purpose infrastructure toward more holistic solutions for water management issues.

Interventions are designed as part of an interconnected system, inclusive of stakeholders both upstream and downstream (e.g., farmers, land managers, city planners). The resulting Sponge Cities feature NBS and hybrid solutions to mitigate flood risk; they can also be used for other benefits, such as recreation, fishing, livelihoods, and watershed protection.

Natural systems perform a variety of functions across the Sponge Cities program. The integrated river rehabilitation and flood risk management component (river component) has various features and water management functions, including green embankments, publicly accessible river greenways, floodplain protection, wetlands rehabilitation, and wetland parks for stormwater retention. Other NBS components include building permeable pavements, green roofs, sunken greenbelts, vegetated swales, bio-retention ponds, detention tanks, and other green "sponge bodies" to slow the release of stormwater into drainage systems and to promote groundwater recharge.

China's Sponge Cities were implemented rapidly on a large scale in 2015 through initiation of the SCCP in 16 pilot cities, including Pingxiang. In 2016, 14 cities were selected as the second group of demonstration sites for the project. Overall, the project plans to transform 20% of China's built-up area into sponge city regions by 2020, with the ultimate goal of reaching 80% of China's built-up area by 2030. The local Pingxiang Municipal Government served as the executing agency, while an outside consultant firm created the sponge city design proposition. County and district governments are heavily involved, acting as the implementing agencies of the project.

Box 17.2   Urban NBS for flood risk reduction in Udon Thani, Thailand.

As one of the fastest growing cities in Southeast Asia, Udon Thani in Thailand has developed ambitious plans to become an economic hub and gateway to the Mekong region. City managers aim to double Udon Thani's size and population in the coming decades. The projected growth in population has led to increasing concerns about stressing the water supply, particularly in the dry season, and increased flood impacts during the rainy season. In their efforts to make the city an economic hub and a livable space for its citizens— without losing the green spaces its residents value—the city is opting for a nature-based infrastructure strategy to enhance its robustness and maintain adaptive capacity for future resilience.

The city is already equipped with sophisticated flood risk reduction infrastructure operated and managed by the Public Works Department and constructed by the Department of Transportation. With the aim of future development, however, new infrastructure solutions would be necessary to enhance the capacity for temporary storage and peak reduction of stormwater due to the rapid development in the peri-urban areas. These new developments, which are spanning across different municipal authorities, would mean additional water routed downstream—a major concern for the communities further downstream. Residents and the mayor wanted to avoid "overdeveloping" through more concrete infrastructure if at all possible.

An urban planning firm introduced a number of NBS and hybrid green-gray infrastructure strategies to reduce current and future flood risks for Udon Thani in 2017. Potential NBS interventions included the connection of wetlands, development of urban recreation detention basins, and widening of canals. A number of parties took part in a collaborative process to identify risks, define suitable options, and maximize outcomes for multiple benefits. Active participation came from Udon Thani's mayor, Public Works Department, business owners, and local stakeholders. All stakeholder groups are engaged in an open and iterative communication framework, with the city staff and leadership continually updated on the concerns of financial, social, and political requirements of the NBS solutions.

The city is currently in the design and construction (USD $25 million) of the first phase of development of green infrastructure solutions that integrate urban storm water storage and diversion with recreation in the downtown area.

---

sectors, such as energy, transport, and irrigation, are vulnerable to the same climate hazards, NBS interventions represent opportunities for cross-cutting solutions for mutual benefit (Box 17.3).

Timing is important, too. NBS can only help mitigate the effects of disasters and extreme weather events, such as hurricanes and typhoons, if the measures are in place. In many cases, extreme weather events will instead serve as an impetus to shift toward green and hybrid solutions as climate adaptation or disaster risk reduction (DRR) strategies. For example, following 2012's Hurricane Sandy in the United States, the USACE "began to take clues from communities and ecosystems that have adapted over time to changing

---

Box 17.3   Nationwide implementation of NBS with Mexico's water reserve program.

Mexico has proven to be another positive example of operationalizing NBS for water and climate challenges at a large scale. For over 10 years, the Mexico national water agency (CONAGUA), supported by WWF Mexico, the Inter-American Development Bank, and more than 50 specialists from academic institutions and nongovernmental organizations (NGOs), had been working to build a strong Mexican water management framework. The framework enshrines the use of "water reserves," a form of NBS that sets aside a volume of water in a watershed or river to set the foundation for healthy, functioning rivers and ecosystems.

The former Mexican Government's National Water Reserves Program (NWRP) represents a new approach to the management of rivers. The NWRP created a framework to systematically and legally define and protect the inputs needed for functioning ecosystems. The program exemplifies the use of NBS as a comprehensive, cross-cutting solution to water management challenges behind the principle that hydrology, biodiversity conservation, ecosystem services, and human wellbeing are all integrated and support each other. Functioning and well-established water reserves protect natural habitats (e.g., wetlands, riparian corridors) and protect vital ecosystem services (e.g., flood protection, mass and energy transportation, connectivity, etc.). The NBS approach enhances human livelihoods, benefiting fisheries and floodplain agriculture, while serving as an effective climate adaptation tool.

In addition to protecting the river ecosystems and their services, the water reserve concept has become an important management instrument in the water allocation process, considering both present and future needs. The result has been the protection of water for human consumption; particularly for indigenous and rural communities. The program has already resulted in sustaining water security for more than 45 million people, as well as for future generations.

CONAGUA and implementing partners saw success in operationalizing the NWRP for a number of reasons. From the start, the program required effective collaboration between government, NGOs, and academic institutions working toward a common goal—in this case, to "save water for the environment now to avoid problems in the future." The participatory process was key in building stakeholder and institutional support, developing an understanding of hydro-ecology, water management procedures, and instruments, and in creating policies such as a national environmental flow (e-flow) standard.

This innovative approach and the experiences from the scientific, technical, and policy fields to develop a coordinated water management strategy for biodiversity conservation has now entered into full implementation at a wider scale involving even more basins. Already having been applied in 295 of Mexico's 760 basins, a new national policy for 2020–24 includes the National Water Program with an aim to extend the NBS approach to cover the whole country.

---

conditions" by expanding their risk reduction approaches to include natural and nature-based systems (USACE, 2015). While natural disasters and extreme weather events are an unfortunate reality, they also represent an opportunity to mainstream NBS. As programs dedicate funds toward recovery efforts, natural systems can be prioritized as a means of building resilience into the system and mitigating the impacts of future climatic events.

NBS and a postpandemic recovery

As countries around the world battle the COVID-19 crisis, there is another significant opportunity to operationalize NBS in order to "build back better." Nations have assembled (and continue to assemble) relief packages for social and economic recovery—in some instances in the range of trillions of US dollars. A forward-looking approach to investments would prioritize ecosystem restoration, sustainable agriculture, and future-proofed natural or hybrid infrastructure (Conservation International, 2020). These approaches lead to both environmental and social benefits. NBS approaches also represent opportunities for job creation and growth across multiple sectors, with restoration projects supporting as many as 33 jobs per $1 million invested (Edwards et al., 2013).

Disruption of ecosystems can exacerbate infectious diseases. It can harm the very biodiversity and natural resources used to develop new advances in pharmaceuticals. Climate change is also compounding the issue. Increasing temperatures, disruptions to safe drinking water, and changes in distribution of disease vectors (e.g., mosquitoes) are creating their own set of underlying health issues. The protection and restoration of natural systems will not only build resilience into infrastructure systems and aid in economic recovery from COVID-19 but also help lessen the likelihood of future pandemic outbreaks (Edwards et al., 2020).

Processes and actions—Prerequisites for success

Laying the groundwork

Redefining successful infrastructure

A concerted effort to operationalize NBS in LMICs requires significant changes in terms of both perception and practice. Culturally, ecosystems are thought of as largely separate from people and the places they inhabit—relegated to parks or managed lands. This mindset is apparent in the purposes for which built infrastructure have been designed and operated. The mainstreaming of NBS will require a change in this paradigm.

Over the past century or more, traditional gray infrastructure has served to make ecosystem services more reliable (e.g., channelization of rivers, damming of reservoirs). Now, scientists and engineers have a better understanding of how ecosystems function, thanks to technological advances and increasing data availability. Ecohydrological landscapes (i.e., natural systems) can be treated as assets that provide distinct and defined functions, comparable and complementary to the role of traditional gray infrastructure over the past century (see also Cassin and Matthews, 2021).

The infrastructure development process will require revisions, too, if ministries and other agencies are to be more inclusive of NBS. One identified barrier to uptake and implement NBS is the "path dependency" of organizational decision-making, which confines decision makers by their active memory based on past experiences and often leads to resistance

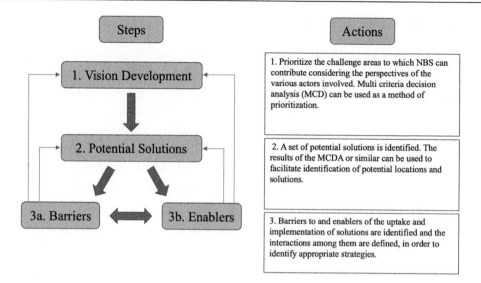

Fig. 17.1

NBS strategic planning framework. *From Ershasd Sarabi S, Han QL, Romme AG, de Vries B, Wendling L. Key enablers of and barriers to the uptake and implementation of nature-based solutions in urban settings: a review. Resources 2019;8(3):121.*

to change (Ershad Sarabi et al., 2019). Instead of a conventional prescriptive, top-down approach, explorations of NBS options utilize a more iterative problem-solving and planning process. A strategic planning framework, such as the one in Fig. 17.1, illustrates the feedback mechanisms involved and the role of actors across different institutional levels in planning for NBS.

Mainstreaming will require an understanding and normalization of NBS within existing decision-making processes at a number of levels so that green and hybrid approaches are viewed as comparable and/or competitive alternatives to traditional gray solutions (Matthews and Endo, 2020). Systems thinking—as opposed to narrowly defining problems—will lead to a more comprehensive understanding of the environmental, social, and economic factors of sustainable development (Brown and Mijic, 2019). A cross-sectoral, system-based approach also identifies linkages between sectors and the associated cross-cutting benefits of NBS interventions (see Box 17.2, Udon Thani Case). A simple mapping framework, such as the one proposed by Brown and Mijic (2019), provides an analytical tool for assessing NBS benefits across sectors.

Understanding the role of natural capital

In NBS, the potential of natural processes is utilized to address a particular issue or set of issues. Before developing any NBS intervention, a thorough understanding of the natural capital and the natural processes is required. A preliminary step is developing an inventory of natural capital, see, for example, Cassin (2021).

Knowing the capacity of the existing natural resources and their respective vulnerabilities is essential to designing a sustainable solution. Equally important is determining whether multiple NBS projects within a boundary could have competing or conflicting functions. For example, Song et al. (2019) suggest that the use of NBS for contaminated land remediation or redevelopment of brownfield sites may depend on the choices made at other brownfield sites in the same city or metropolitan area. If a given brownfield site is converted into a greenspace for public access, it may discourage the same choice for other sites. Similarly, conversion of brownfield sites into industrial heritage parks may depend on whether the metropolitan area has other industrial heritage parks and whether the nation has other facilities of similar nature (if originating from the same industrial sector and maintained with similar historical features). They also outline that native or imported plant species, or even genetically modified species, can be used to address contaminated land issues through a process known as phytoremediation (Song et al., 2019). In order to proceed with an effective phytoremediation, it is essential to have an inventory of the potential plants.

NBS interventions cover a wide spectrum in terms of form and function; however, they are not a necessary or appropriate solution to every problem. The appropriateness, design, and cobenefits of green solutions are largely context-specific (Kapos et al., 2019); their functions are interwoven into a web of overlapping systems. Larger scale assessments, such as landscape or catchment-level, are necessary to understand the complex dynamics of (and potential for) green solutions.

Identifying the right financial infrastructure

The 2018 UN World Water Development Report states, "NBS do not necessarily require additional financial resources but usually involve redirecting and making more effective use of existing financing" (WWAP and UN Water, 2018). Governing institutions and project planners may need to explore alternatives to traditional financial instruments, but a diverse set of financial instruments are already available to support NBS, with new approaches continually emerging. A more comprehensive assessment of financial tools for implementing NBS is available in Tremolet et al. (2021).

Reforming policies and decision-making

Whenever possible, LMICs should work to ensure that NBS are systematically incorporated into decision-making, especially through national-level priority setting. National-level commitments such as nationally determined contributions (NDCs) and national adaptation plans (NAP) can serve as the connective tissue or mandate for collaboration between departments and ministries. Collaboration is a prerequisite for cross-cutting NBS interventions.

NBS present an opportunity to achieve objectives of multiple parallel or overlapping processes such as climate change adaptation or DRR. This ability to achieve multiple policy goals with one intervention make NBS particularly effective and appealing public investments (OECD, 2020). National and international policy implications for NBS are further explored in Dominique et al. (2021) and Timboe et al. (2021).

Creating a coordinated upstream-downstream strategy

Developing high-level strategies

The processes outlined earlier in this section identify some of the ways in which barriers to mainstreaming can be removed. Beyond the removal of barriers, there must also be a concerted effort to internalize greener approaches within institutions' and ministries' core strategies. Minimally, this would mean the inclusion of "do no harm" principles, maintaining existing environmental regulations, and ensuring transparency within infrastructure programs (Conservation International, 2020). A more proactive approach includes writing in mandates for exploration of nature-based or greener alternatives to traditional gray infrastructure, wherein programmatic strategies are revised to facilitate the consideration of NBS as competitive alternatives to traditional infrastructure solutions in LMICs (see, for example, the case of Peru's drinking water sector, Coxon et al., 2021). This includes aligning NBS with higher-level priorities and agendas that guide the types of work being done at the national level all the way down to cities and rural communities. In this way, sustainability and environmental commitments are demonstrated and codified formally.

The framing and scope of problems being addressed directly shapes the types of solutions proposed by engineers. As demonstrated in the case studies earlier, NBS are a means to address multiple issues (e.g., drought, water security, flooding) with one coordinated solution (e.g., managed aquifer recharge). The large spatial scales required of NBS (due to the interwoven nature of ecosystems mentioned earlier) means that these approaches cross jurisdictions. The issues of scale are tractable with enough planning and coordination through a number of actions, including identifying the appropriate agencies to promote NBS or redesigning funding mechanisms to require cooperation (European Commission, 2017).

At the city level, new frameworks for urban planning, such as those proposed by Bush and Doyon (2019) articulate the need to better integrate new relationships and partnerships (with, for example, urban ecologists and horticulturalists) into the urban planning and spatial planning processes. Opportunities to operationalize NBS can be found in regional and national strategy-settings. These approaches can manifest in the policymaking process through mechanisms such as NDCs, NAPs, and transboundary water agreements. LMICs are already making strategic commitments to use NBS as part of their adaptation plans: all low-income countries refer to NBS in the adaptation components of their NDCs (Kapos et al., 2019).

Developing successful NBS projects

As mentioned in the previous section, green and hybrid solutions can be integrated into system-level thinking within decision-making agencies. With support in place at the higher level, the next step is to create an enabling environment for inclusion of NBS as common and viable options within the "project pipeline" of those responsible for soliciting and ultimately approving infrastructure projects. In practice, it comes down to setting up the processes that will allow for engineers to evaluate green options alongside traditional gray solutions. Opportunities to promote greener infrastructure projects will be present across the project cycle (regardless of the sector): project solicitation, preparation/design, approval, implementation, and evaluation.

Mechanisms need to allow for direct comparisons and tradeoffs in terms of costs, efficacy, resilience, cobenefits, and other variables of interest when evaluating a suite of green to gray project options (Matthews and Endo, 2020). If this is done properly, green and hybrid solutions can be explored in the project design stage along with traditional gray solutions. In this way, the suite of possible solutions is expanded.

The regulatory environment also plays a direct role in whether or not NBS can be carried out, especially during the project planning and development stages. Some of the policy elements that will impact the feasibility of NBS include: land-use regulation and zoning, permitting, safety and performance codes and standards, procurement policies, land rights, and environmental protection regulation (OECD, 2020). These regulations can create bias toward traditional gray infrastructure. Government ministries and regulatory agencies would need to assess and potentially modify legal frameworks in order to remove burdens or restrictions limiting NBS uptake (WWAP and UN Water, 2018). For example, design guidelines and incentives can encourage or even require engineers to consider green alternatives or supplements to traditional infrastructure.

System-level recommendations

The following section outlines many of the practical and procedural changes that will help facilitate the mainstreaming of NBS within infrastructure development programs. The recommendations will not result in the selection of NBS as a solution for every problem; rather, they are meant to remove barriers to their consideration alongside traditional infrastructure solutions. A summarized version is illustrated in Table 17.1.

Consider a multipronged portfolio of NBS projects

Deemphasize isolated pilot projects and instead focus on systemic solutions

To date, most water security investments in NBS have been carried out as pilot projects at relatively small scales (Trémolet et al., 2019). This ad hoc approach, while encouraging in terms of generating success stories, faces limitations of its own. Often, the ecosystem services of NBS depend on processes taking place at a larger scale and beyond the confines of a project (World Bank, 2017). Mainstreaming NBS will require a more coordinated effort, focusing more on systemic challenges and objectives and less on ad hoc projects.

Table 17.1: Recommendations for facilitating the inclusion of NBS in infrastructure development programs.

Recommendation	Examples
Consider a multipronged portfolio of NBS projects	• Deemphasize isolated pilot projects and instead focus on systemic solutions • Use multiple NBS solutions in tandem or as a complement to gray infrastructure
Coordinate planning for multibenefit solutions	• Address multiple problems through coordination between ministries and sectors, possibly involving outside actors (public and private)
Shape project constraints for engineers to promote consideration of NBS	• Avoid narrowly defined problems • Emphasize long-term success and flexibility
Engage with stakeholders early and often	• Use stakeholder input to identify problems and consideration of solutions/consequences
Broaden the scope of environmental assessments and feasibility studies	• Consider the health and function of existing ecosystems, plus the unintended consequences of NBS
Use and support innovative finance mechanisms	• Explore alternatives to traditional project finance
Allow for alternative economic evaluation mechanisms	• Find other ways to evaluate success aside from cost-benefit analyses
Prioritize NBS in national, regional, and city-level policy	• Use existing frameworks and planning tools to promote NBS

For example, the Asian Development Bank's (ADB) NBS initiative in the towns and cities of Greater Mekong Subregion demonstrated a holistic approach to address the aforementioned concerns (ADB, 2016). This NBS approach is based upon a "green plan" that essentially forms connected networks across the towns' areas and is implemented across different sectors. In this case, the natural infrastructure systems are developed as a strong green fabric or network which binds a town together and requires close collaboration between developers, local planners, stakeholders, and affected communities.

Use multiple NBS solutions in tandem or as a complement to gray infrastructure

As long as they are properly integrated into planning and operations, NBS can support and complement gray infrastructure, helping to increase their efficiency, reliability, or operational lifetime while lowering the cost of services (Browder et al., 2019; Matthews and Endo, 2020). For example, wetlands and healthy watersheds can support hydropower facilities through their water filtration abilities. Introducing NBS later into project development or operations is more challenging; early planning and coordination are key.

Coordinate planning for multibenefit solutions

Address multiple problems through coordination between ministries and sectors, possibly involving outside actors (public and private)

Projects involving NBS will potentially span the jurisdictions or responsibilities of several ministries or sectors (e.g., energy, irrigation, urban planning). Vertical and horizontal coordination mechanisms are integral to develop more comprehensive and cross-cutting solutions (OECD, 2020; WWAP and UN Water, 2018). Collaborative governance approaches are encouraged as means to connect with citizens, businesses, civil society (including NGOs), and community-led organizations in order to capitalize on the broader interest in sustainability and create partnerships for action (Kabisch et al., 2016; Kapos et al., 2019). Responsibilities for planning and management are thus shared across ministries, sectors, and outside actors.

These types of partnerships are also opportunities for innovation and the sharing of risk while achieving mutually beneficial outcomes.

Shape project constraints for engineers to promote consideration of NBS

Avoid narrowly defined problems

The scope of issues to be addressed within any infrastructure project will directly shape the suite of possible solutions. Traditional infrastructure is optimized to perform well for a narrow set of processes, yet a narrow definition of problems will lead to solutions that create their own new and unintended consequences (Matthews and Endo, 2020).

Engineers have the technical expertise and knowledge to implement NBS. Many utility groups and engineering societies have already developed NBS design guidance. The barrier to more widespread operationalization of NBS from an engineering perspective relates to the types of solutions requested by project funding and management institutions. To provide a more conducive environment for the consideration of green and hybrid approaches, funding agencies need to broaden the way problems are defined and seek out multipurpose projects.

Emphasize long-term success and flexibility

NBS or hybrid approaches can be more complex or follow a longer trajectory than traditional project development cycles (Matthews and Endo, 2020). Their benefits may take longer to accrue and may need to be supplemented with hybrid or engineered solutions (Kapos et al., 2019). Project leads and engineers need to be assured by ministries and funding agencies that these factors are understood, anticipated, and acceptable.

Engage with stakeholders early and often

Use stakeholder input to identify problems and consideration of solutions/consequences

The active support of local stakeholders and landowners is often required for the long-term success and viability of NBS interventions (Browder et al., 2019). Engaging with relevant stakeholders early in the process yields multiple benefits in the project design, selection, and operations phases. Stakeholder input will help identify common priorities, ensure early buy-in and continued support over the long run, all while mitigating any potential negative social impacts of the intervention. Local individuals and coalitions can also be involved in the implementation and operations of a green project, for example, through vegetative management or structural maintenance. To enable stakeholder engagement and collaborative governance, policies to incentivize NBS and remove administrative barriers should be enacted (Kabisch et al., 2016). This will remove some of the barriers inhibiting business, local citizens, and civil society from actively participating in NBS mainstreaming.

Broaden the scope of environmental assessments and feasibility studies

Consider the health and function of existing ecosystems, plus the unintended consequences of NBS

An ecosystem is a web of interrelations among biological, physical, and chemical components. Abrupt and incoherent natural solutions can be detrimental for ecosystems and human wellbeing. Eggermont et al. (2015) suggest that NBS can utilize ecosystem services (ES) in three ways: (1) consisting of no or minimal intervention in ecosystems, with the objectives of maintaining or improving the delivery of a range of ES both inside and outside of preserved ecosystems (for example, mangrove protection in coastal area); (2) implementation of management approaches that develop sustainable and multifunctional ecosystems and landscapes (extensively or intensively managed), which improves the delivery of selected ES compared to what would be obtained with a more conventional intervention (for example, innovative planning of agricultural landscapes to increase their multifunctionality); and (3) managing ecosystems in very intrusive ways or even creating new ecosystems (for example, artificial ecosystems with new assemblages of organisms for green roofs and walls to mitigate city warming and clean polluted air).

It is recommended that any proposed NBS or hybrid intervention involves careful consideration of ecological complexity and connections with surrounding ecosystems to provide biodiversity benefits (Eggermont et al., 2015). Solutions must also be grounded in robust understanding of the geographical distribution of the biomes of the world, the value of their biodiversity, and their ecological resilience (Seddon et al., 2020).

Consider changes to the ecosystem over an operational lifetime

Although natural systems can be considered robust and adaptive, the element of time is important. The increasing frequency and intensity of disturbances, such as drought, flooding, storms, or wildfires resulting from climate change, combined with human activities such as land use change, are causing disturbances to recur before natural systems have a chance to recover. This affects the adaptive capacity of the ecosystem and can alter the ecosystem to the point that it may lose its functionality for which it is being utilized as the NBS (Folke, 2006; Seddon et al., 2020). Planning and design of green or hybrid systems must attempt to incorporate projected climate impacts over the lifetime of these systems and build in the necessary robustness or flexibility to adapt and maintain their functionality (Kapos et al., 2019).

Use and support innovative finance mechanisms

Explore alternatives to traditional project finance

Traditional finance mechanisms may not always be an appropriate solution for supporting green or hybrid approaches. Alternative financial concepts (e.g., payment for watershed services) or markets (e.g., green or climate bonds) offer vast potential across project scales and types (WWAP and UN Water, 2018). Private capital can also be leveraged through innovative options. Public Private Partnerships and other cofunding options are available for some interventions and can be implemented through coordination with relevant industry or business groups (see also Tremolet et al., 2021).

Allow for alternative economic evaluation mechanisms

Find other ways to evaluate success aside from cost-benefit analyses

Current standard economic evaluation mechanisms such as cost-benefit analyses (CBAs) tend to favor traditional gray infrastructure since these solutions are optimized to perform well based upon narrower purposes and criteria. Conversely, NBS perform better across a spectrum of distinct purposes and when quantitative performance measures may be difficult to define with high confidence (Matthews and Endo, 2020). While cobenefits and other metrics of success (e.g., avoided costs) are of great interest to the public, the government, and affected communities, they are not adequately reflected in traditional assessments (OECD, 2020). Alternative economic evaluation mechanisms such as multicriteria analysis or true cost accounting should be explored and promoted as complements to CBAs (see also Tremolet et al., 2021).

Prioritize NBS in national, regional, and city-level policy

Use existing frameworks and planning tools to promote NBS

The coordination across ministries and agencies required of successful NBS projects is made possible through supportive policy and governance arrangements (OECD, 2020). Individual ministries and agencies are bound by their own vision, resources, and procedures. Policy mechanisms at the national through city level can help in the alignment of priorities between sectors and even jurisdictions. Examples of potential policy tools include NDCs to the Paris Climate Agreement, NAPs, or even city master plans. Dominique et al. (2021) and Timboe et al. (2021) provide deeper analyses of the role of policy on NBS.

Synthesis and closing

Nature-based and hybrid infrastructure represent a viable alternative or complement to traditional gray infrastructure for addressing water management challenges. Developed and some middle-income countries have begun to pilot and even mainstream NBS within governance and project development processes. This change comes after a century or more of reliance upon single-purpose infrastructure optimized to solve narrowly defined problems. Complex global challenges such as climate change and biodiversity loss, combined with population growth and other drivers, require more cross-cutting and multipurpose solutions. NBS represent the potential to achieve sustainable development and resilient societies without compromising the ecosystems upon which countries depend.

The de facto approach to development for LMICs is still dominated by traditional gray infrastructure. Operationalizing NBS or "greener gray" solutions in LMICs will require paradigm shifts—both culturally and in practice. The guidelines referenced in this chapter represent a starting point for elevating NBS to a level on par with standard engineering solutions. Through more coordination across sectors and broader system-level thinking, LMICs have the opportunity to achieve economic development and increased standards of living while avoiding the development mistakes of more industrialized nations.

References

Ashley MR, Digman CJ, Horton B. Demonstrating and monetizing the multiple benefits from using sustainable drainage. In: World water congress. Brisbane: IWA; 2016.

Asian Development Bank (ADB). Nature-based solutions for building resilience in towns and cities: Case studies from the Greater Mekong Subregion. Mandaluyong City: Asian Development Bank; 2016.

Barbier EB. Natural resources and economic development. Cambridge: Cambridge University Press; 2007.

Bridges TS, Banks CJ, Chasten MA. Engineering with nature: advancing system resilience and sustainable development. Mil Eng 2016;108(699):52–4.

Browder G, Ozment S, Bescos I, Gartner T, Lange G-M. Integrating green and gray: Creating next generation infrastructure. Washington, DC: World Bank and World Resources Institute; 2019.

Brown K, Mijic A. Integrating green and blue spaces into our cities: Making it happen. Grantham Institute Briefing paper, London: Imperial College London; 2019.

Bush J, Doyon A. Building urban resilience with nature-based solutions: how can urban planning contribute? Cities 2019;95:102483.

Cassin J. Nature-based solutions for source water protection in North America. In: Cassin J, Matthews JH, Lopez-Gunn E, editors. Nature-based solutions and water security: An action agenda for the 21st century. Amsterdam: Elsevier; 2021.

Cassin J, Matthews J. Setting the scene: nature-based solutions and water security. In: Cassin J, Matthews J, Lopez Gunn E, editors. Nature-based solutions and water security: An action agenda for the 21st century. Elsevier; 2021.

Conservation International. Leveraging nature as the foundation of resilient societies: Recommendations for post-COVID 19 recovery plans. Conservation International; 2020. https://www.conservation.org/docs/default-source/publication-pdfs/recovery_recommendations_final.pdf.

Coxon C, Gammie G, Cassin J. Mobilizing funding for nature-based solutions: Peru's drinking water tariff. In: Cassin J, Matthews J, Lopez Gunn E, editors. Nature-based solutions and water security: An action agenda for the 21st century. Elsevier; 2021.

De Vriend HJ, Van Koningsveld M. Building with nature: Thinking, acting and interacting differently. Dordrecht: EcoShape; 2012.

DEFRA. Shoreline management plan guidance. Aims and requirements, vol. 1. London: Department for Environment, Food, and Rural Affairs; 2006.

Dominique K, Matthews N, Danielson L, Matthews JH. Why governments embrace nature-based solutions: the policy rationale. In: Cassin J, Matthews J, Lopez Gunn E, editors. Nature-based solutions and water security: An action agenda for the 21st century. Elsevier; 2021.

Edwards PET, Sutton-Grier AE, Coyle GE. Investing in nature: restoring coastal habitat blue infrastructure and green job creation. Mar Policy 2013;38:65–71.

Edwards G, Garcia Salinas A, Watkins G. What is the link between COVID-19 and the ecological and climate emergencies? IDB Blogs; March 25, 2020. https://blogs.iadb.org/sostenibilidad/en/what-is-the-link-between-covid-19-and-the-ecological-and-climate-emergencies/.

Eggermont H, Balian E, Azevedo JM, Beumer V, Brodin T, Claudet J, et al. Nature-based solutions: new influence for environmental management and research in Europe. GAIA Ecol Perspect Sci Soc 2015;24:243–8.

Ershad Sarabi S, Han QL, Romme AG, de Vries B, Wendling L. Key enablers of and barriers to the uptake and implementation of nature-based solutions in urban settings: a review. Resources 2019;8(3):121.

European Commission. Assessing adaptation knowledge in Europe: Ecosystem-based adaptation. Final Report for DG CLIMATE https://ec.europa.eu/clima/sites/clima/files/adaptation/what/docs/ecosystem_based_adaptation_en.pdf; 2017.

Folke C. Resilience: the emergence of a perspective for social–ecological systems analyses. Glob Environ Chang 2006;16:253–67.

Gill AR, Viswanathan KK, Hassan S. The Environmental Kuznets Curve (EKC) and the environmental problem of the day. Renew Sustain Energy Rev 2018;81:1636–42.

Gusev EM. Evolution of agricultural technologies: from "gray" to "green". Arid Ecosyst 2020;10(1):1–9.

Haasnoot M, van't Klooster S, Van Alphen J. Designing a monitoring system to detect signals to adapt to uncertain climate change. Glob Environ Chang 2018;52:273–85.

He J. Is the Environmental Kuznets Curve hypothesis valid for developing countries? A survey. Cahiers de Recherche (No. 07-03), 3; 2007. p. 1–44.

Ho P. Mao's war against nature? The environmental impact of the grain-first campaign in China. China J 2003;50:37–59.

Holling CS, Gunderson LH. Resilience and adaptive capacity. In: Gunderson LH, Holling CS, editors. Panarchy: Understanding transformations in human and natural systems. Washington, DC: Island Press; 2002. p. 25–62.

Kabisch N, Frantzeskaki N, Pauleit S, Naumann S, Davis M, Artmann M, et al. Nature-based solutions to climate change mitigation and adaptation in urban areas: perspectives on indicators, knowledge gaps, barriers, and opportunities for action. Ecol Soc 2016;21(2):39–44.

Kapos V, Wicander S, Salvaterra T, Dawkins K, Hicks C. The role of the natural environment in adaptation. Background Paper for the Global Commission on Adaptation, Rotterdam and Washington, DC: Global Commission on Adaptation; 2019.

Karsch NM. Examining the validity of the environmental Kuznets curve. Consilience 2019;21:32–50.

Kuznets S. Economic growth and income inequality. Am Econ Rev 1955;45(1):1–28.

Liquete C, Udias A, Conte G, Grizzetti B, Masi F. Integrated valuation of a nature-based solution for water pollution control. Highlighting hidden benefits. Ecosyst Serv 2016;22:392–401.

Matthews J, Endo I. A practitioner's guide to nature-based solutions. ADB Working Paper, Manila: Asian Development Bank; 2020.

Nesshöver C, Assmuth T, Irvine KN, Rusch GM, Waylen KA, Delbaere B, et al. The science, policy and practice of nature-based solutions: an interdisciplinary perspective. Sci Total Environ 2017;579:1215–27.

OECD. Nature-based solutions for adapting to water-related climate risks. OECD Environment Policy Paper No. 21, Paris: OECD Publishing; 2020. https://doi.org/10.1787/2257873d-en.

Onuma A, Tsuge T. Comparing green infrastructure as ecosystem-based disaster risk reduction with gray infrastructure in terms of costs and benefits under uncertainty: a theoretical approach. Int J Disaster Risk Reduct 2018;3:22–8.

Ourloglou O, Stefanidis K, Dimitriou E. Assessing nature-based and classical engineering solutions for flood-risk reduction in urban streams. J Ecol Eng 2020;21(2):46–56.

Pathirana A, Radhakrishnan M, Ashley R, Quan NH, Zevenbergen C. Managing urban water systems with significant adaptation deficits—unified framework for secondary cities: part II—the practice. Clim Change 2018;149(1):57–74.

Raymond L. Economic growth as environmental policy? Reconsidering the Environmental Kuznets Curve. J Publ Policy 2004;24(3):327–48.

Raymond CM, Frantzeskaki N, Kabisch N, Berry P, Breil M, Nita MR, et al. A framework for assessing and implementing the co-benefits of nature-based solutions in urban areas. Environ Sci Policy 2017;77:15–24. https://doi.org/10.1016/j.envsci.2017.07.008.

Rozenberg J, Fay M, editors. Beyond the gap: How countries can afford the infrastructure they need while protecting the planet. Washington, DC: World Bank; 2019.

Schoonees T, Mancheño AG, Scheres B, Bouma TJ, Silva R, Schlurmann T, Schüttrumpf H. Hard structures for coastal protection, towards greener designs. Estuar Coasts 2019;42(7):1709–29.

Seddon N, Chausson A, Berry P, Giradin CA, Smith A, Turner B. Understanding the value and limits of nature-based solutions to climate change and other global challenges. Philos Trans R Soc B 2020;375(1794). https://doi.org/10.1098/rstb.2019.0120.

Song Y, Kirkwood N, Maksimović Č, Zheng X. Nature based solutions for contaminated land remediation and brownfield redevelopment in cities: a review. Sci Total Environ 2019;663:568–79.

Spiller M, Vreeburg JH, Leusbrock I, Zeeman G. Flexible design in water and wastewater engineering–definitions, literature and decision guide. J Environ Manage 2015;149:271–81.

Timboe I, Phar K, Freeman S. Nature-based solutions in international policy instruments. In: Cassin J, Matthews J, Lopez Gunn E, editors. Nature-based solutions and water security: An action agenda for the 21st century. Elsevier; 2021.

Trémolet S, et al. Investing in nature for Europe water security. London: The Nature Conservancy, Ecologic Institute, and ICLEI; 2019.

Tremolet S, Atwell B, Dominique K, Matthews N, Becker M, Munoz R. Funding and financing to scale nature-based solutions for water security. In: Cassin J, Matthews J, Lopez Gunn E, editors. Nature-based solutions and water security: An action agenda for the 21st century. Elsevier; 2021.

US Army Corps of Engineers (USACE). North Atlantic coast comprehensive study: Resilient adaptation to increasing risk. Washington, DC: USACE; 2015.

Van den Hoek RE, Brugnach M, Hoekstra A. Shifting to ecological engineering in flood management: introducing new uncertainties in the development of a building with nature pilot project. Environ Sci Policy 2012;22:85–99.

Walker WE, Haasnoot M, Kwakkel JH. Adapt or perish: a review of planning approaches for adaptation under deep uncertainty. Sustainability 2013;5(3):955–79.

Webber DJ, Allen DO. Environmental Kuznets Curves: Mess or meaning? Working Papers 0406, Bristol: Department of Accounting, Economics and Finance, Bristol Business School, University of the West of England; 2004.

World Bank. Inclusive green growth: The pathway to sustainable development. Washington, DC: World Bank; 2012.

World Bank. Implementing nature-based flood protection: Principles and implementation guidance. Washington, DC: World Bank; 2017.

WWAP (United Nations World Water Assessment Programme), UN-Water. The United Nations world water development report 2018: Nature-based solutions for water. Paris: UNESCO; 2018.

Xiang-Chao P. Research on Xi Jinping's thought of ecological civilization and environment sustainable development. Earth Environ Sci 2018;153(6):062067.

CHAPTER 18

Nature-based solutions: Action for the 21st century

Jan Cassin[a], John H. Matthews[b], Elena Lopez-Gunn[c], Leah L. Bremer[d], Cheyenne Coxon[a], Kathleen Dominique[e], Marta Echavarria[f], Gena Gammie[a], Roxane Marchal[g], Alex Mauroner[b], Naabia Ofosu-Amaah[h], Denielle M. Perry[i], Eleanor Sterling[j], Ingrid Timboe[b], Paula Vandergert[k], Kari Vigerstol[h], and Sophie Trémolet[h]

[a]Forest Trends, Washington, DC, United States, [b]Alliance for Global Water Adaptation, Corvallis, OR, United States, [c]ICATALIST S.L., Madrid, Spain, [d]University of Hawai'i at Mānoa, Honolulu, HI, United States, [e]OECD, Paris, France, [f]Ecodecision, Quito, Ecuador, [g]Caisse Centrale de Réassurance, Paris, France, [h]The Nature Conservancy, Arlington, VA, United States, [i]School of Earth and Sustainability, Northern Arizona University, Flagstaff, AZ, United States, [j]Center for Biodiversity and Conservation, American Museum of Natural History, New York, NY, United States, [k]Sustainability Research Institute, University of East London, London, United Kingdom

The opportunities and challenges of nature-based solutions

The contributors to this volume have approached nature-based solutions (NBS) from many different angles. Our collective experience with NBS includes applied research but primarily comprises practical experience supporting, developing, and implementing NBS policies and guidance for governments and businesses; advising local to international entities on NBS including on developing enabling regulatory frameworks; helping assess and develop new funding and financing instruments; developing science-based tools for planning and designing NBS interventions; and collaborating with communities and local nongovernmental organizations (NGOs) in planning and implementing on-the-ground projects. From this diversity of perspectives and experience, we have reflected on "what is next" for NBS, the potential pitfalls to avoid, and the opportunities that can lead to mainstreaming NBS for water security.

What is the role of nature-based solutions in a post—2020 world?

Water security is not a new concept, but new priorities for water security are emerging now. Even at the time of the formulation of the Dublin Principles in 1992, the linkages between

Nature-Based Solutions and Water Security. https://doi.org/10.1016/B978-0-12-819871-1.00010-5

water, equity, economic integration, and ecosystems had long been valued, at least as aspirations. Concepts such as Integrated Water Resource Management (IWRM)—a precursor or early cognate for water security—have long treated "ecosystems" and "water" as separate entities. For decades, we have acted as if water could be managed independently from the landscapes and climate that produce water and as if hydrology was separate from ecology; is water simply an assemblage of hydrogen and oxygen, or is it a medium that can carry our values and hopes? Reducing the impacts of our water management decisions on ecosystems or negotiating tradeoffs between infrastructure and ecosystems is fundamentally different than integrating dynamic eco-hydrological landscapes within governance, economic, and infrastructure frameworks and designs. If IWRM and water security are early attempts to create a grand vision of water-centric sustainability and economic growth, more recent terms such as "water resilience" and "nature-based solutions" suggest attempts to renegotiate our core values through the medium of water itself.

A proliferation of nature-based initiatives

Interest in NBS has been growing for decades, with a new surge seeming to appear in the late 2010s. The reasons for the heightened focus are not clear, but the signs are widespread. Major initiatives gaining widespread attention include a €5 billion sovereign green bond issued by the Dutch government in 2019 for a massive "room for the river" project (Anderson et al., 2019), publications aligning NBS with water sector priorities among development banks (Browder et al., 2019), and the announcement by the UK government, host of UNFCCC COP 26, that adaptation and NBS will be major priorities for negotiators. Many other examples are documented in this volume. NBS are being embraced wholeheartedly by supranational institutions such as the EU, permeating academic discourse and conservation group priorities, and being incorporated into international initiatives around climate, biodiversity, sustainable development, and disaster risk management. A dramatic illustration of the reach of NBS is the principles for a water resilience portfolio in California:

This water resilience portfolio (established by California Governor Gavin Newsom on 29 April 2019) shall embody these principles:

a. Prioritize multi-benefit approaches that meet multiple needs at once.
b. Utilize natural infrastructure such as forests and floodplains.
c. Embrace innovation and new technologies.
d. Encourage regional approaches among water users sharing watersheds.
e. Incorporate successful approaches from other parts of the world.
f. Integrate investments, policies and programs across state government.
g. Strengthen partnerships with local, federal, and tribal governments, water agencies, and irrigation districts, and other stakeholders.

California Executive Order NO-10-19

Possible reasons for the recent explosion of interest in NBS may stem from their potential as systemic solutions to the many issues facing society, along with the growing sense of urgency—made more evident by the events of 2020—that we are not acting rapidly enough.

The urgency of emerging fragilities

During 2020, dramatic shifts occurred in societal assumptions about the robustness of our social-ecological systems driven by local to global impacts of the COVID-19 pandemic. This pandemic exposed widespread fragilities in our economic, health, social, and political systems. While people around the world have been consumed by the struggle against a deadly pandemic, deforestation and biodiversity loss have increased sharply and greenhouse gas emissions, while initially dropping in early 2020, have begun to rise again, bringing us ever closer to an existential climate emergency. The world needs a major reset to transform how we operate: in the way we live, how we interact with each other and the planet, the way we consume, and the way our economies operate. Working with nature can serve as a path for achieving such a reset, although it is only part of the change we need. As the world builds back from the current pandemic, putting nature at the center of resilient water management will help address many of the world's linked, systemic challenges—environmental degradation, water insecurity, food insecurity, inequities in health and well-being, and climate change.

Water and nature: A broader conception of health and resilience

The global COVID-19 pandemic gave the world a renewed and deeper appreciation of the links between nature and our vulnerability to emerging pathogens (Schmeller et al., 2020). But this pandemic has also highlighted nature's positive role in human well-being. COVID-19 put a spotlight on the critical role that water security—especially equitable access to clean water and sanitation—plays in health outcomes. NBS for water may be the thread that can connect resilient water management and water security with a much broader conception of human well-being—building resilience in the face of shocks such as pandemics and climate change. There is emerging recognition and evidence of the dynamic connections and feedbacks between nature, water, and mental and physical health and well-being. The link between green spaces and health is well-established, whether in green urban areas or forested natural areas (Van den Bosch and Sang, 2017). More recently, the role of water itself has become better understood, and access to healthy aquatic ecosystems—rivers, oceans, lakes, springs, and wetlands—brings a heightened sense of well-being that can be directly attributed to being in blue spaces (Sandifer et al., 2015).

The core concept behind NBS—maintaining and capitalizing on the beneficial functions of ecosystems—can help us link nature and people more holistically, balancing well-being, livelihoods, and economic development while building resilience. Maintaining the ability of ecosystems to continue to deliver ecosystem services is at the heart of social-ecological resilience— the ability of individuals and communities to cope with, adjust to, and/or reorganize in the face of change in ways that allow them to continue to thrive. The positive outcomes associated with effective NBS can affect individuals (greater food security, sense of spiritual healing/resilience), communities (water security), societal systems, and economies (resilient infrastructures, supply chains). Similarly, a focus on NBS can broaden conventional understandings of well-being to include the principle that individual well-being is inseparable from community and environmental well-being.

Reflections on what is next for nature-based solutions

NBS are being actively embraced at an increasing rate, perhaps most aggressively by the water sector, as the major initiatives presented in this volume show. These initiatives come from around the world, from the European Union, Latin America, Asia, small island states, Africa, and North America. Uptake by businesses, local and national governments, and multilateral and bilateral development institutions show the broad appeal of NBS. While a few years ago, NBS proponents focused on building awareness and advocacy, what is needed now is advancement to the next level of mainstreaming NBS across and within multiple sectors. Attaining this step will include implementation of practices at scale, moving beyond the core groups and sectors in water and conservation that have already embraced NBS, and enacting the enabling policies, institutional transformation, and funding models that are discussed in this volume.

For example, over the next few years, countries around the world will need to spend significant public funds to "build back better" from the health and economic impacts of COVID-19, all while addressing how to curb and adapt to climate change and address social justice challenges more broadly. What role should NBS and the NBS community play in ensuring that: (1) this money is spent effectively and equitably; and (2) mistakes of the past are avoided (e.g., sinking large sums of money into inflexible, long-lasting infrastructure that has significant environmental and health impacts)? What should different constituencies and sectors do, and what are the avenues through which NBS approaches can help ensure we really will "build back better," even as we try to answer carefully, for whom are we building back?

The current acceleration of interest in NBS presents both risks and opportunities. For instance, large international organizations may come to dominate how NBS are framed and defined and where resources go, potentially sidelining local actions and knowledge and the rich diversity of place-based approaches that arise from contextual traditions. An explicit emphasis on equity, justice, and a broad community of decision-making can ensure that we can encourage a rich, complex vision of the role of NBS at all scales. There are exciting examples of how diverse interests and sectors are bridging local to global levels of action in this volume, but many questions remain about how to do this effectively in more places.

There is a need to connect local actions and innovations to the international narrative around NBS, as international legitimacy and recognition can help support and value local efforts. But only if these international efforts are focused on enabling and building local, bottom-up efforts rather than controlling or directing them. How can financing be ramped up from the corporate, national, and international entities that represent the largest sources of investments in NBS while supporting strong, community-led, local efforts? Can we effectively learn from and replicate the kinds of tools, capacity-building models, and policies highlighted in this volume that can help streamline implementation and scaling-up while prioritizing the flexibility and support that works locally?

As the contributors to this volume have reflected together on the path ahead for NBS, we offer the following observations for realizing the promise of NBS and avoiding some of the potential pitfalls that can occur with rapid scaling, recognizing that there is not a single path to scale but some common elements and principles.

Importance of local knowledge and action
Learning from and respecting the rights and knowledge of indigenous peoples

"Nature-based solutions" may be a new term and a useful concept (if rigorously defined and applied), but it is critical to recognize that the ideas and approaches that comprise NBS are not new. Indigenous and local communities have long engaged in NBS created from their place-based knowledge-practice-values systems, based on traditional ecological knowledge. Ancestral or Indigenous water technologies, regenerative agriculture, and ecosystem and agroecosystem management (e.g., fire or wetland management) have proven to be successful and enduring solutions for water security (see Cassin and Ochoa-Tocachi, 2021).

Importance of grassroots, local action

As cases in this volume show, local actions, local knowledge, and local leadership are key to moving NBS into mainstream decision-making. Municipal water utilities, Indigenous peoples, rural and urban community groups, and businesses have galvanized local initiatives that created the critical mix of innovation, demonstration actions, enabling policies, and cross-sectoral partnerships that have inspired adaptation and replication at regional, national, and international levels. Locally led, community initiatives that use NBS to protect a community's resources through collective action mechanisms and provide multiple benefits across the wider community have proven successful and are likely to be the most equitable and durable actions (Ochoa-Tocachi et al., 2019).

Avoiding the "sustainability" and "greenwashing" traps

NBS have definitely become a trend over the past couple of years. There is a risk that the concept will be applied vaguely, without thoughtful planning. If actions are labeled as NBS without rigorous definitions and criteria, then there is a risk of undermining the credibility of the concept and of the NBS that have already been successfully implemented. Clearly defining what is credible and effective on the ground is essential for governments, NGOs, businesses, and other potential actors to be able to engage with NBS. The NBS community needs to ensure that claims about NBS are grounded in specifics and supported by sound science and local and expert knowledge. Clear definitions of what can be considered NBS and why, rigorous identification and definition of problems and how NBS can effectively address them, and effective monitoring and evaluation of NBS outcomes are ways the NBS community can build a larger evidence base and greater confidence in these approaches over time.

The NBS community is a growing and diverse group. A connected community of practice can help provide more clarity and consistency around NBS as a means to avoid greenwashing traps. Some key questions remain largely undefined: What is the most appropriate role for NBS in water governance? What is the collective opinion emerging on what "good" or "best practice" NBS looks like and how it performs? How can we make sure we refine best practices over time as we gain experience with NBS, and, in particular, allow for place-based, local definitions of "good" NBS without diluting the overall meaning and relevance? A strong NBS community of practice can protect the relevance and meaning of NBS through robust dialogue and consensus-building, while demonstrating what strong and just NBS for water and climate security look like.

The NBS community is making progress in this area. The IUCN Global NBS Standard (https://www.iucn.org/theme/nature-based-solutions/resources/iucn-global-standard-nbs), Conservation International's *Practical Guide to Implementing Green-Gray Infrastructure (https://www.conservation.org/docs/default-source/publication-pdfs/a-practical-guide-to-implementing-green-gray-infrastructure_aug2019.pdf?Status=Master&sfvrsn=7c25b40b_2)*, and more locally driven efforts, such as Peru's *Guidance on Effective, Equitable, and Sustainable Natural Infrastructure* (https://www.forest-trends.org/wp-content/uploads/2020/12/Guia-EES-1.pdf; in Spanish), are three examples. A number of typologies of NBS are also helping more explicitly define what is or is not an NBS (e.g., The Nature Conservancy https://www.nature.org/content/dam/tnc/nature/en/documents/TNC_ResilientEuropeanCities_NBSWater.pdf), and new platforms and initiatives connecting practitioners, compiling evidence, and providing resources on NBS continue to emerge at a rapid pace (e.g., the NBS Initiative, https://www.naturebasedsolutionsinitiative.org/; the EU's thinknature platform, https://www.think-nature.eu/).

Finally, NBS themselves are vulnerable to changing conditions and climate shocks. Development and land use changes in the larger landscape in which NBS are embedded, changes in the hydrological cycle due to climate change, and biodiversity loss can affect the functionality and viability of individual NBS. To avoid risks to the viability of NBS approaches, NBS proponents and practitioners must consider and be sensitive to these potential changes in biodiversity, water, and climate. Just as hydrologists have recognized that the past is no longer a sound guide to how freshwater systems will perform in the future, NBS practitioners can no longer use past or current ecological conditions as solid guides for future NBS performance. We must plan for flexibility, recognizing that as climate change deepens, ecosystems and the planned performance of NBS may change in novel ways. For NBS to be flexible and durable, their planning and design must consider and incorporate the characteristics that enhance ecosystem resilience, such as connectivity, heterogeneity, and diversity (Grantham et al., 2019; Matthews et al., 2019; Poff, 2018).

A real opportunity for recovery through effective integration with gray solutions

The world needs gray infrastructure and NBS for water security. However, gray infrastructure does not come without social and environmental costs. Moreover, access to and the quality of gray water infrastructure remains highly unequal throughout the world as a result of both current and historical legacies of racial and environmental injustice. Despite this, gray infrastructure, while having many disadvantages that are described in this volume, has proven essential in improving health and livelihoods around the world. While NBS are powerful tools for achieving water security, there are limitations to what NBS can do. The real potential lies in effective protection and management of existing natural infrastructure, along with the integration of NBS into mixed portfolios of "green and gray" infrastructure. For example, the city of Paris relies on gray infrastructure (dikes) within the city for critical protection from Seine River floods, but they work in combination with the natural infrastructure of lakes and wetlands upstream that help mitigate the extent of flooding that occurs in the city.

Identifying the appropriate role for both NBS and gray infrastructure in a specific context will require a commitment to collaboration and cross-learning between the engineering and NBS communities as well as local stakeholders and policymakers. Well-designed NBS with effective participatory processes have the potential to address current and legacy inequities in existing water infrastructure systems. By combining NBS with improved gray infrastructure, poor access to water and sanitation in underserved rural and urban communities around the world can be addressed while providing additional cobenefits. At a minimum, NBS proponents must be aware of and address existing inequities in access to water and sanitation as NBS are embraced to build back better.

Practitioners and advocates of NBS need to be able to clarify exactly when and how NBS adds value to infrastructure planning and resource management. Cost-effectiveness, resilience, flexibility, adaptability, energy efficiency, greenhouse gas offsetting, multifunctionality, cobenefits, and the precautionary nature of NBS all should be considered when planning water infrastructure. An often overlooked but important function of NBS is its role in protecting gray infrastructure—not only water infrastructure but also transportation and energy infrastructure. Integrating NBS with gray infrastructure requires systems thinking on the part of advocates for both gray infrastructure and NBS and the kinds of steps for operationalizing NBS in development that are outlined in this volume (see Mauroner et al., 2021).

Can NBS for water be the force for eroding and breaking down silos that prevent effective action?

Many of the global and local issues we face are exacerbated by siloed governance systems, artificially separating interconnected components of human and natural systems. These silos make effective systems thinking and action difficult. We manage natural resources separately within separate institutions, scientific disciplines, and professions (e.g., forests, water, endangered species). Even within organizations working on NBS, there can be a lack of integration or connection, for example, in work on biodiversity and water, or water and climate, which inhibits finding synergies and systemic solutions.

One of the great strengths of NBS for water is their multifunctionality, which means these solutions can simultaneously address several issues, providing different benefits to a range of stakeholder groups. As such, NBS are a natural way to encourage cross-sectoral, cross-stakeholder collaboration and collective action. The multiple benefits of NBS are fundamentally different from the indirect or associated benefits that gray infrastructure provides. While there are a few examples of gray infrastructure designed for more than one purpose (e.g., dams that provide water for irrigation while also serving to store flood water, or below-level garages that can also store floodwaters in cities) these are the exception, and importantly, they often provide different functions in sequence, at different times, rather than simultaneously.

Gray infrastructure is more often designed to provide a single targeted benefit, such as drinking water of an acceptable standard at the tap. This results in many important indirect benefits that spin off from access to clean water, such as lowered disease burdens and increased health, or time and labor saved that can be applied to other activities, such as education or participating in economic enterprises. The indirect benefits accrue primarily to the individual with access to clean drinking water. With NBS, such as protecting a forested watershed to provide drinking water of an acceptable standard, the same indirect benefits are delivered, and at the same time, a host of other direct benefits are provided. Sequestering and storing carbon, providing recreational spaces, supplying nontimber forest products (which can be essential food or medicines for local and Indigenous communities), and protecting biodiversity are direct benefits from the single NBS. These additional direct benefits accrue to the individuals with access to clean drinking water but also to many other local individuals and the global community. This is not to say that there are no tradeoffs among various benefits from NBS, and when trade-offs occur they need to be accounted for and managed.

The multifunctional aspects of NBS make it a powerful tool for breaking down the silos and social barriers that limit systems thinking and more effective action. Water security has been the most common entry point for NBS to date, and it is now well-accepted that water (too much or too little) is the primary medium through which the effects of climate change are being and will be felt (WMO, 2020). Investing in resilient water management via NBS, therefore, can be a way to contribute to a green and resilient post–2020 recovery, generating benefits for a multitude of other sectors. Water security is inseparable from other types of global development challenges, but it also impacts all of them—climate adaptation and resilience, biodiversity conservation, disaster risk management, food security, and economic development.

The three chains to scale nature-based solutions

Finally, we propose three "chains" that suggest a framework that can help guide paths to scaling NBS, reflecting both opportunities and needs: (1) the chain of resilience; (2) the chain of integration; and (3) the chain of alignment.

The chain of resilience represents the role of NBS in building resilience across individual, community, and societal levels. It can also help build resilience across sectors and the multiple domains that are needed now—pandemic resilience, natural capital resilience, climate resilience, and social resilience. How is resilience built within and across these domains, and what are the common attributes of resilience that can be strengthened through NBS?

The chain of integration reflects the need to bridge multiple scales of knowledge and action, from local to national to international scales, and from individual sectors to institutions and governance. Doing so will help bring NBS within the mainstream of legal, governance and economic systems.

The chain of alignment represents the importance of consensus on what the NBS community of practice considers effective, beneficial, and equitable NBS. There is a need for more rapid alignment around what "good," resilient NBS looks like, while acknowledging that the vision for this must account for a diversity of perspectives and local contexts.

References

Anderson J, Gartner T, Mauroner A, Matthews JH. Conservation finance takes off as the Netherlands issues one of the largest green bonds ever. World Resources Institute Blog; 2019. June 21, 2019 https://www.wri.org/blog/2019/06/conservation-finance-takes-netherlands-issues-one-largest-green-bonds-ever.

Browder G, Ozment S, Rehberger Bescos I, Gartner T, Lange G. Integrating green and gray: Creating next generation infrastructure. Washington, DC: World Bank and World Resources Institute; 2019.

Cassin J, Ochoa-Tocachi BF. Learning from indigenous and local knowledge: the deep history of nature-based solutions. In: Cassin J, Matthews JH, Lopez-Gunn E, editors. Nature-based solutions and water security: An action agenda for the 21st century. Elsevier; 2021.

Grantham TE, Matthews JH, Bledsoe BP. Managing freshwater ecosystems for ecological resilience in a changing climate. Water Secur 2019;8:100049.

Matthews JH, Matthews N, Simmons E, Vigerstol K. Wellspring: Sources water resilience and climate adaptation. Arlington, VA: The Nature Conservancy; 2019. https://www.nature.org/content/dam/tnc/nature/en/documents/Wellspring_FULL_Report_2019.pdf.

Mauroner A, Anisha NF, de la Cruz E, Barrios E, Guntoju SS. Operationalizing NBS in low- and middle-income countries: redefining and 'greening' project development. In: Cassin J, Matthews JH, Lopez-Gunn E, editors. Nature-based solutions and water security: An action agenda for the 21st century. Elsevier; 2021.

Ochoa-Tocachi BF, Bardales JD, Antiporta J, Perez K, Acosta L, Mao F, Zulkafli Z, Gil-Rios J, Angulo O, Grainger S, Gammie G, DeBievre B, Buytaert W. Potential contributions from pre-Inca infiltration infrastructure to Andean water security. Nat Sustain 2019;2:584–93.

Poff NL. Beyond the natural flow regime? Broadening the hydro-ecological foundation to meet environmental flows challenges in a non-stationary world. Freshw Biol 2018;63(8):1011–21.

Sandifer PA, Sutton-Grier AE, Ward BP. Exploring connections among nature, biodiversity, ecosystem services, and human health and well-being: opportunities to enhance health and biodiversity conservation. Ecosyst Serv 2015;12:1–15.

Schmeller DS, Courchamp F, Killeen G. Biodiversity loss, emerging pathogens and human health risks. Biodivers Conserv 2020;29:3095–102.

Van den Bosch M, Sang AO. Urban natural environments as nature-based solutions for improved public health—a systematic review of reviews. Environ Res 2017;158:373–84.

WMO (World Meteorological Organization). United in science 2020, https://public.wmo.int/en/resources/united_in_science; 2020.

Index

Note: Page numbers followed by *f* indicate figures, *t* indicate tables, *b* indicate boxes, and *np* indicate footnotes.

CPI Antony Rowe
Eastbourne, UK
July 31, 2021